CONSUL-TE
831
Palos |
Ph. 708-923-10

STATES OF MATTER

DAVID L. GOODSTEIN

Professor of Physics and Applied Physics
California Institute of Technology

Dover Publications, Inc.
New York

Copyright © 1975, 1985 by David L. Goodstein.
All rights reserved under Pan American and International Copyright Conventions.

Published in Canada by General Publishing Company, Ltd., 30 Lesmill Road, Don Mills, Toronto, Ontario.
Published in the United Kingdom by Constable and Company, Ltd.

This Dover edition, first published in 1985, is an unabridged and corrected republication of the work first published by Prentice-Hall, Inc., Englewood Cliffs, New Jersey, in 1975.

Manufactured in the United States of America
Dover Publications, Inc., 31 East 2nd Street, Mineola, N.Y. 11501

Library of Congress Cataloging in Publication Data

Goodstein, David L., 1939–
   States of matter.

   Previously published: Englewood Cliffs, N.J. : Prentice-Hall, 1975.
   Includes bibliographies and index.
   1. Matter—Properties. I. Title.
QC173.3.G66   1985   530.4   85-6801
ISBN 0-486-64927-X (pbk.)

Nel mezzo del cammin di nostra vita....

TO JUDY

CONTENTS

PREFACE xi

1 THERMODYNAMICS AND STATISTICAL MECHANICS 1

1.1 Introduction: Thermodynamics and Statistical Mechanics of the Perfect Gas *1*

1.2 Thermodynamics *10*
    a. The Laws of Thermodynamics *10*
    b. Thermodynamic Quantities *13*
    c. Magnetic Variables in Thermodynamics *19*
    d. Variational Principles in Thermodynamics *23*
    e. Examples of the Use of Variational Principles *25*
    f. Thermodynamic Derivatives *29*
    g. Some Applications to Matter *34*

1.3 Statistical Mechanics *41*
    a. Reformulating the Problem *43*
    b. Some Comments *49*
    c. Some Properties of Z and $\mathscr{L}$ *53*
    d. Distinguishable States: Application to the Perfect Gas *55*
    e. Doing Sums over States as Integrals *61*
    f. Fluctuations *71*

1.4 Remarks on the Foundations of Statistical Mechanics   83
    a. Quantum and Thermodynamic Uncertainties   83
    b. Status of the Equal Probabilities Postulate   87
    c. Ensembles   89
Appendix A. Thermodynamic Mnemonic   90

2 PERFECT GASES   98

2.1 Introduction   98
2.2 The Ideal Gas   99
2.3 Bose-Einstein and Fermi-Dirac Statistics   105
2.4 Slightly Degenerate Perfect Gases   110
2.5 The Very Degenerate Fermi Gas: Electrons in Metals   114
2.6 The Bose Condensation: A First Order Phase Transition   127

3 SOLIDS   142

3.1 Introduction   142
3.2 The Heat Capacity Dilemma   144
    a. The Einstein Model   147
    b. Long Wavelength Compressional Modes   152
    c. The Debye Model   154
3.3 Normal Modes   160
3.4 Crystal Structures   175
3.5 Crystal Space:
Phonons, Photons, and the Reciprocal Lattice   183
    a. Diffraction of X rays   185
    b. The Reciprocal Lattice and Brillouin Zones   186
    c. Umklapp Processes   191
    d. Some Orders of Magnitude   194
3.6 Electrons in Crystals   195
    a. A Semiqualitative Discussion   195
    b. The Kronig-Penney Model   202
    c. The Fermi Surface: Metal or Nonmetal?   212

4 LIQUIDS AND INTERACTING GASES   227

4.1 Introduction   227
4.2 The Structure of a Fluid   231
    a. Measuring the Structure   232

      b. The Statistical Mechanics of Structure   *237*
      c. Applicability of This Approach   *246*
4.3 The Potential Energy   *248*
      a. The Pair Potential   *250*
      b. The Approximation of Pairwise Additivity   *252*
4.4 Interacting Gases   *260*
      a. Cluster Expansions and the Dilute Gas   *260*
      b. Behavior and Structure of a Dilute Gas   *263*
      c. The Liquefaction of Gases and the Joule-Thomson Process   *266*
      d. Higher Densities   *271*
4.5 Liquids   *283*
      a. The Yvon-Born-Green Equation   *284*
      b. The Hypernetted Chain and Percus-Yevick Equations   *288*
      c. Comments and Comparisons   *302*

5 SOME SPECIAL STATES   320

5.1 Introduction   *320*
5.2 Superfluidity   *321*
      a. Properties of Liquid Helium   *321*
      b. The Bose Gas as a Model for Superfluidity   *327*
      c. Two-Fluid Hydrodynamics   *335*
      d. Phonons and Rotons   *342*
      e. The Feynman Theory   *348*
      f. A Few Loose Ends   *360*
5.3 Superconductivity   *371*
      a. Introduction   *371*
      b. Some Thermodynamic Arguments   *375*
      c. Electron Pairs   *386*
      d. The BCS Ground State   *392*
      e. The Superconducting State   *404*
5.4 Magnetism   *411*
      a. Introduction   *411*
      b. Magnetic Moments   *412*
      c. Paramagnetism   *419*
      d. Ferromagnetism   *421*

6 CRITICAL PHENOMENA AND PHASE TRANSITIONS   436

6.1 Introduction   *436*
6.2 Weiss Molecular Field Theory   *438*

6.3　The van der Waals Equation of State　*443*
6.4　Analogies Between Phase Transitions　*453*
6.5　The Generalized Theory　*457*
　　a. Equilibrium Behavior　*459*
　　b. Fluctuations　*463*
6.6　Critical Point Exponents　*473*
6.7　Scaling Laws: Physics Discovers Dimensional Analysis　*478*
6.8　Evaluation and Summary: Testing the Scaling Laws　*484*

INDEX　494

PREFACE

This book is based on lectures prepared for a year-long course of the same title, designed as a central part of a new curriculum in applied physics. It was offered for the first time in the academic year 1971-1972 and has been repeated annually since.

There is a good deal of doubt about precisely what applied physics is, but a reasonably clear picture has emerged of how an applied physics student ought to be educated, at least here at Caltech. There should be a rigorous education in basic physics and related sciences, but one centered around the macroscopic world, from the atom up rather than from the nucleus down: that is, physics, with emphasis on those areas where the fruits of research are likely to be applicable elsewhere. The course from which this book arose was designed to be consistent with this concept.

The course level was designed for first-year graduate students in applied physics, but in practice it has turned out to have a much wider appeal. The classroom is shared by undergraduates (seniors and an occasional junior) in physics and applied physics, plus graduate students in applied physics, chemistry, geology, engineering, and applied mathematics. All are assumed to have a reasonable undergraduate background in mathematics, a course including electricity and magnetism, and at least a little quantum mechanics.

The basic outline of the book is simple. After a chapter designed to start everyone off at the same level in thermodynamics and statistical mechanics, we have the basic states—gases, solids, and liquids—a few special cases, and, finally, phase transitions. What we seek in each case is a feeling for the essential nature of the stuff, and how one goes about studying it. In general, the book should help give the student an idea of the language and ideas that are reasonably current in fields other than the one in which he or she will specialize. In short, this is an unusual beast: an advanced survey course.

All the problems that appear at the ends of the chapters were used as either homework or examination problems during the first three years in which the course was taught. Some are exercises in applying the material covered in the text, but many are designed to uncover or illuminate various points that arise, and are actually an integral part of the course. Such exercises are usually referred to at appropriate places in the text.

There is an annotated bibliography at the end of each chapter. The bibliographies are by no means meant to be comprehensive surveys even of the textbooks, much less of the research literature of each field. Instead they are meant to guide the student a bit deeper if he wishes to go on, and they also serve to list all the material consulted in preparing the lectures and this book. There are no footnotes to references in the text.

The history of science is used in a number of places in this book, usually to put certain ideas in perspective in one way or another. However, it serves another purpose, too: the study of physics is essentially a humanistic enterprise. Much of its fascination lies in the fact that these mighty feats of the intellect were performed by human beings, just like you and me. I see no reason why we should ever try to forget that, even in an advanced physics course. Physics, I think, should never be taught from a historical point of view—the result can only be confusion or bad history—but neither should we ignore our history. Let me hasten to acknowledge the source of the history found in these pages: Dr. Judith Goodstein, the fruits of whose doctoral thesis and other research have insinuated themselves into many places in the text.

Parts of the manuscript in various stages of preparation have been read and criticized by some of my colleagues and students, to whom I am deeply grateful. Among these I would like especially to thank Jeffrey Greif, Professors T. C. McGill, C. N. Pings and H. E. Stanley, David Palmer, John Dick, Run-Han Wang, and finally Deepak Dhar, a student who contributed the steps from Eq. (4.5.20) to Eq. (4.5.24) in response to a homework assignment. I am indebted also to Professor Donald Langenberg, who helped to teach the course the first time it was offered. The manuscript was typed with great skill and patience, principally by Mae Ramirez and Ann Freeman. Needless to say, all errors are the responsibility of the author alone.

Pasadena, California DAVID L. GOODSTEIN
April 5, 1974

ONE

THERMODYNAMICS AND STATISTICAL MECHANICS

1.1 INTRODUCTION: THERMODYNAMICS AND STATISTICAL MECHANICS OF THE PERFECT GAS

Ludwig Boltzmann, who spent much of his life studying statistical mechanics, died in 1906, by his own hand. Paul Ehrenfest, carrying on the work, died similarly in 1933. Now it is our turn to study statistical mechanics.

Perhaps it will be wise to approach the subject cautiously. We will begin by considering the simplest meaningful example, the perfect gas, in order to get the central concepts sorted out. In Chap. 2 we will return to complete the solution of that problem, and the results will provide the foundation of much of the rest of the book.

The quantum mechanical solution for the energy levels of a particle in a box (with periodic boundary conditions) is

$$\varepsilon_\mathbf{q} = \frac{\hbar^2 q^2}{2m} \qquad (1.1.1)$$

where m is the mass of the particle, $h = 2\pi\hbar$ is Planck's constant, and $\mathbf{q}$ (which we shall call the wave vector) has three components, x, y, and z, given by

$$q_x = \left(\frac{2\pi}{L}\right)\ell_x, \text{ etc.} \qquad (1.1.2)$$

where

$$\ell_x = 0, \pm 1, \pm 2, \text{ etc.} \qquad (1.1.3)$$

and
$$q^2 = q_x^2 + q_y^2 + q_z^2 \qquad (1.1.4)$$

L is the dimension of the box, whose volume is L^3. The state of the particle is specified if we give three integers, the quantum numbers ℓ_x, ℓ_y, and ℓ_z. Notice that the energy of a particle is fixed if we give the set of three integers (or even just the sum of their squares) without saying which is ℓ_x, for example, whereas ℓ_x, ℓ_y, and ℓ_z are each required to specify the state, so that there are a number of states for each energy of the single particle.

The perfect gas is a large number of particles in the same box, each of them independently obeying Eqs. (1.1.1) to (1.1.4). The particles occupy no volume, have no internal motions, such as vibration or rotation, and, for the time being, no spin. What makes the gas perfect is that the states and energies of each particle are unaffected by the presence of the other particles, so that there are no potential energies in Eq. (1.1.1). In other words, the particles are noninteracting. However, the perfect gas, as we shall use it, really requires us to make an additional, contradictory assumption: we shall assume that the particles can exchange energy with one another, even though they do not interact. We can, if we wish, imagine that the walls somehow help to mediate this exchange, but the mechanism actually does not matter much as long as the questions we ask concern the possible states of the many-particle system, not how the system contrives to get from one state to another.

From the point of view of quantum mechanics, there are no mysteries left in the system under consideration; the problem of the possible states of the system is completely solved (although some details are left to add on later). Yet we are not prepared to answer the kind of questions that one wishes to ask about a gas, such as: If it is held at a certain temperature, what will its pressure be? The relationship between these quantities is called the *equation of state*. To answer such a question—in fact, to understand the relation between temperature and pressure on the one hand and our quantum mechanical solution on the other—we must bring to bear the whole apparatus of statistical mechanics and thermodynamics.

This we shall do and, in the course of so doing, try to develop some understanding of entropy, irreversibility, and equilibrium. Let us outline the general ideas briefly in this section, then return for a more detailed treatment.

Suppose that we take our box and put into it a particular number of perfect gas particles, say 10^{23} of them. We can also specify the total energy of all the particles or at least imagine that the box is physically isolated, so that there is some definite energy; and if the energy is caused to change, we can keep track of the changes that occur. Now, there are many ways for the particles to divide up the available energy among themselves—that is, many possible choices of ℓ_x, ℓ_y, and ℓ_z for each particle such that the total energy comes out right. We have already seen that even a single particle generally has a number of possible states of the same energy; with 10^{23} particles, the number of possible quantum states of the set of particles that add up to the same energy can become astronomical. How does the system decide which of these states to choose? The answer depends, in general, on details that we have not yet specified: What is the past history that is, how was the energy injected into the box? And how does the system change from one state to another—that is, what is the nature of the interactions? Without knowing these details, there is no way, even in principle, to answer the question.

At this point we make two suppositions that form the basis of statistical mechanics.

1. If we wait long enough, the initial conditions become irrelevant. This means that whatever the mechanism for changing state, however the particles are able to redistribute energy and momentum among themselves, all memory of how the system started out must eventually get washed away by the multiplicity of possible events. When a system reaches this condition, it is said to be in equilibrium.

2. For a system in equilibrium, all possible quantum states are equally likely. This second statement sounds like the absence of an assumption—we do not assume that any particular kind of state is in any way preferred. It means, however, that a state in which all the particles have roughly the same energy has exactly the same probability as one in which most of the particles are nearly dead, and one particle goes buzzing madly about with most of the energy of the whole system. Would we not be better off assuming some more reasonable kind of behavior?

The fact is that our assumptions do lead to sensible behavior. The reason is that although the individual states are equally likely, the number of states with energy more or less fairly shared out among the particles is enormous compared to the number in which a single particle takes nearly all the energy. The probability of finding approximately a given situation in the box is proportional to the number of states that approximate that situation.

The two assumptions we have made should seem sensible; in fact, we have apparently assumed as little as we possibly can. Yet they will allow us to bridge the gap between the quantum mechanical solutions that give the physically possible microscopic states of the system and the thermodynamic questions we wish to ask about it. We shall have to learn some new language, and especially learn how to distinguish and count quantum states of many-particle systems, but no further fundamental assumptions will be necessary.

Let us defer for the moment the difficult problem of how to count possible states and pretend instead that we have already done so. We have N particles in a box of volume $V = L^3$, with total energy E, and find that there are Γ possible states of the system. The entropy of the system, S, is then defined by

$$S = k \log \Gamma \qquad (1.1.5)$$

where k is Boltzmann's constant

$$k = 1.38 \times 10^{-16} \text{ erg per degree Kelvin}$$

Thus, if we know Γ, we know S, and Γ is known in principle if we know N, V, and E, and know in addition that the system is in equilibrium. It follows that, in equilibrium, S may be thought of as a definite function of E, N, and V,

$$S = S(E, N, V)$$

Furthermore, since S is just a way of expressing the number of choices the system has, it should be evident that S will always increase if we increase E, keeping N and V constant; given more energy, the system will always have more ways to divide it. Being thus monotonic, the function can be inverted

$$E = E(S, N, V)$$

or if changes occur,

$$dE = \left(\frac{\partial E}{\partial S}\right)_{N,V} dS + \left(\frac{\partial E}{\partial V}\right)_{S,N} dV + \left(\frac{\partial E}{\partial N}\right)_{S,V} dN \qquad (1.1.6)$$

The coefficients of dS, dV, and dN in Eq. (1.1.6) play special roles in thermodynamics. They are, respectively, the temperature

$$T = \left(\frac{\partial E}{\partial S}\right)_{N,V} \qquad (1.1.7)$$

the negative of the pressure

$$-P = \left(\frac{\partial E}{\partial V}\right)_{S,N} \qquad (1.1.8)$$

and the chemical potential

$$\mu = \left(\frac{\partial E}{\partial N}\right)_{S,V} \qquad (1.1.9)$$

These are merely formal definitions. What we have now to argue is that, for example, the quantity T in Eq. (1.1.7) behaves the way a temperature ought to behave.

How do we expect a temperature to behave? There are two requirements. One is merely a question of units, and we have already taken care of that by giving the constant k a numerical value; T will come out in degrees Kelvin. The other, more fundamental point is its role in determining whether two systems are in equilibrium with each other. In order to predict whether anything will happen if we put two systems in contact (barring deformation, chemical reactions, etc.), we need only know their temperatures. If their temperatures are equal, contact is superfluous; nothing will happen. If we

1.1 Introduction: Thermodynamics and Statistical Mechanics of the Perfect Gas

separate them again, we will find that each has the same energy it started with.

Let us see if T defined in Eq. (1.1.7) performs in this way. We start with two systems of perfect gas, each with some E, N, V, each internally in equilibrium, so that it has an S and a T; use subscripts 1 and 2 for the two boxes. We establish thermal contact between the two in such a way that the N's and V's remain fixed, but energy is free to flow between the boxes. The question we ask is: When contact is broken, will we find that each box has the same energy it started with?

During the time that the two systems are in contact, the combined system fluctuates about among all the states that are allowed by the physical circumstances. We might imagine that at the instant in which contact is broken, the combined system is in some particular quantum state that involves some definite energy in box 1 and the rest in box 2; when we investigate later, these are the energies we will find. The job, then, is to predict the quantum state of the combined system at the instant contact is broken, but that, of course, is impossible. Our fundamental postulate is simply that all states are equally likely at any instant, so that we have no basis at all for predicting the state.

The precise quantum state is obviously more than we need to know in any case—it is the distribution of energy between the two boxes that we are interested in. That factor is also impossible to predict exactly, but we can make progress if we become a bit less particular and ask instead: About how much energy is each box likely to have? Obviously, the larger the number of states of the combined system that leave approximately a certain energy in each box, the more likely it is that we will catch the boxes with those energies.

When the boxes are separate, either before or after contact, the total number of available choices of the combined system is

$$\Gamma_t = \Gamma_1 \Gamma_2 \qquad (1.1.10)$$

It follows from Eq. (1.1.5) that the total entropy of the system is

$$S_t = S_1 + S_2 \qquad (1.1.11)$$

Now suppose that, while contact exists, energy flows from box 1 to box 2. This flow has the effect of decreasing Γ_1 and increasing Γ_2. By our argument, we are likely to find that it has occurred if the net result is to have increased the total number of available states $\Gamma_1 \Gamma_2$, or, equivalently, the sum $S_1 + S_2$. Obviously, the condition that no net energy flow be the most likely circumstance is just that the energy had already been distributed in such a way that $\Gamma_1 \Gamma_2$, or $S_1 + S_2$, was a maximum. In this case, we are more likely to find the energy in each box approximately unchanged than to find that energy flowed in either direction.

The differences between conditions in the boxes before contact is established and after it is broken are given by

$$\delta E_1 = T_1 \, \delta S_1 \qquad (1.1.12)$$

$$\delta E_2 = T_2 \, \delta S_2 \qquad (1.1.13)$$

from Eqs. (1.1.6) to (1.1.9), with the N's and V's fixed, and

$$\delta(E_1 + E_2) = 0 \qquad (1.1.14)$$

since energy is conserved overall. Thus,

$$T_1 \, \delta S_1 + T_2 \, \delta S_2 = 0 \qquad (1.1.15)$$

If $S_1 + S_2$ was already a maximum, then, for whatever small changes do take place, the total will be stationary,

$$\delta S_1 + \delta S_2 = 0 \qquad (1.1.16)$$

so that Eq. (1.1.15) reduces to

$$T_1 = T_2 \qquad (1.1.17)$$

which is the desired result.

It is easy to show by analogous arguments that if the individual volumes are free to change, the pressures must be equal in equilibrium, and that if the boxes can be exchange particles, the chemical potentials must be equal. We shall, however, defer formal proof of these statements to Secs. 1.2f and 1.2g, respectively.

We are now in a position to sketch a possible procedure for answering the prototype question suggested earlier: At a given temperature, what will the pressure be? Given the quantities E, N, V for a box of perfect gas, we count the possible states to compute S. Knowing $E(S, V, N)$, we can then find

$$T(S, V, N) = \left(\frac{\partial E}{\partial S}\right)_{V,N} \qquad (1.1.18)$$

$$-P(S, V, N) = \left(\frac{\partial E}{\partial V}\right)_{S,N} \qquad (1.1.19)$$

and finally arrive at $P(T, V, N)$ by eliminating S between Eqs. (1.1.18) and (1.1.19). That is not the procedure we shall actually follow—there will be more convenient ways of doing the problem—but the very argument that that procedure could, in principle, be followed itself plays an important role. It is really the logical underpinning of everything we shall do in this chapter. For example, Eqs. (1.1.6) to (1.1.9) may be written together:

$$dE = T \, dS - P \, dV + \mu \, dN \qquad (1.1.20)$$

1.1 Introduction: Thermodynamics and Statistical Mechanics of the Perfect Gas 7

Our arguments have told us that not only is this equation valid, and the meanings of the quantities in it, but also that it is integrable; that is, there exists a function $E(S, V, N)$ for a system in equilibrium. All of equilibrium thermodynamics is an elaboration of the consequences of those statements.

In the course of this discussion we have ignored a number of fundamental questions. For example, let us return to the arguments that led to Eq. (1.1.7). As we can see from the argument, even if the temperatures were equal, contact was not at all superfluous. It had an important effect: we lost track of the exact amount of energy in each box. This realization raises two important problems for us. The first is the question: How badly have we lost track of the energy? In other words, how much uncertainty has been introduced? The second is that whatever previous operations put the original amounts of energy into the two boxes, they must have been subject to the same kinds of uncertainties: we never actually knew exactly how much energy was in the boxes to begin with. How does that affect our earlier arguments? Stated differently: Can we reapply our arguments to the box now that we have lost track of its exact energy?

The answer to the first question is basically that the uncertainties introduced into the energies are negligibly, even absurdly, small. This is a quantitative effect, which arises from the large numbers of particles found in macroscopic systems, and is generally true only if the system is macroscopic. We cannot yet prove this fact, since we have not yet learned how to count states, but we shall return to this point later and compute how big the uncertainties (in the energy and other thermodynamic quantities as well) actually are when we discuss thermodynamic fluctuations in Sec. 1.3f. It turns out, however, that in our example, if the temperatures in the two boxes were equal to start with, the number of possible states with energies very close to the original distribution is not only larger than any other possibility, it is also vastly greater than all other possibilities combined. Consequently, the probability of catching the combined system in any other kind of state is very nearly zero. It is due to this remarkable fact that statistical mechanics works.

The simple answer to the second question is that we did not need to know the exact amount of energy in the box, only that it could be isolated and its energy fixed, so that, in principle, it has a definite number of available states. Actually, there is a deeper reason why we cannot speak of an exact energy and an exact number of available states. At the outset we assumed that our system was free, in some way, to fluctuate among quantum states of the same energy. The existence of these fluctuations, or transitions, means that the individual states have finite lifetimes, and it follows that the energy of each state has a quantum mechanical uncertainty, $\delta E \geq \hbar/\tau$, where τ is the lifetime. τ may typically be estimated, say, by the time between molecular collisions, or whatever process leads to changes of quantum state. We cannot imagine our box of gas to have had an energy any more definite than δE. δE is generally small compared to macroscopic energies due to the smallness of $\hbar$. The fact remains, however, that we must always expect to find a quantum uncertainty in the number of states available to an isolated system.

Having said all this, the point is not as important as it seems; there is actually less than meets the eye. It is true that there are both quantum and thermodynamic uncertainties in the energy of any system, and it is also true that the number of states available to the system is not as exact a concept as it first appeared to be. However, that number of states, for a macroscopic system, turns out to be such a large number that we can make very substantial mistakes in counting without introducing very much error into its logarithm, which is all we are interested in. For example, suppose that we had $\Gamma \sim 10^{100}$. Then even a mistake of a factor of ten in counting Γ introduces an error of only 1% in log Γ, the entropy, which is what we are after, since $S = k \log (10^{100}$ or $10^{101}) = (100$ or $101)$ log 10. In real systems Γ is more typically of order 10^N, where N is the number of particles, so even an error of a factor of N in Γ—that is, if we make a mistake and get an answer 10^{23} times too big (nobody is perfect)—the result is something like $10^{23} \times 10^{10^{23}} = 10^{(10^{23}+23)}$ and the error in the logarithm is immeasurably small.

The concept that an isolated system has a definite energy and a corresponding definite number of states is thus still a useful and tenable one, but we must understand "definite" to mean something a bit less than "exact." Unfortunately, the indeterminacy we are speaking of makes it even more difficult to formulate a way to count the number of states available to an isolated system. However, there is an alternative description that we shall find very useful, and it is closely connected to manipulations of the kind we have been discussing. Instead of imagining a system isolated with fixed energy, we can think of our sample as being held at a constant temperature—for example, by repeatedly connecting it to a second box that is so much larger that its temperature is unaffected by our little sample. In this case, the energy of our sample is not fixed but instead fluctuates about in some narrow range. In order to handle this situation, instead of knowing the number of states at any fixed energy, it will be more convenient to know the number of states per unit range of energies—what we shall call the *density of states*. Then all we need to know is the density of states as a function of energy, a quantity that is not subject to quantum indeterminacy. This realization suggests the procedure we shall actually adopt. We shall imagine a large, isolated system, of which our sample is a small part (or subsystem). The system will have some energy, and we can make use of the fact that it has, as a result, some number of states, but we will never calculate what that number is. The sample, on the other hand, has a fixed temperature rather than a fixed energy when it is

1.1 Introduction: Thermodynamics and Statistical Mechanics of the Perfect Gas 9

in equilibrium, and we will wind up describing its properties in terms of its density of states. This does not mean that the sample cannot be thought of as having definite energies in definite states—we certainly shall assume that it does—but rather that we shall evade the problem of ever having to count how many states an isolated system can have.

A few more words about formalities may be in order. We have assumed, without stating it, that T and S are nonzero; our arguments would fall through otherwise. Arguments of the type we have used always assume that the system actually have some disposable energy, and hence more than one possible choice of state. This minor but necessary point will later be enshrined within the Third Law of Thermodynamics.

Furthermore, we have manipulated the concept of equilibrium in a way that needs to be pointed out and underlined. As we first introduced it, equilibrium was a condition that was necessary before we could even begin to discuss such ideas as entropy and temperature; there was no way, for example, that the temperature could even be defined until the system had been allowed to forget its previous history. Later on, however, we found ourselves asking a different kind of question: What is the requirement on the temperature that a system be in equilibrium? This question necessarily implies that the temperature be meaningful and defined when the system is not in equilibrium. We accomplished this step by considering a restricted kind of disequilibrium. We imagined the system (our combined system) to be composed of subsystems (the individual boxes) that were themselves internally in equilibrium. For each subsystem, then, the temperature, entropy, and so on are well defined, and the specific question we ask is: What are the conditions that the subsystems be in equilibrium *with each other*? When we speak of a system not in equilibrium, we shall usually mean it in this sense; we think of it as composed of various subsystems, each internally in equilibrium but not necessarily in equilibrium with each other. For systems so defined, it follows by a natural extension of Eqs. (1.1.10) and (1.1.11) that the entropy of the system, whether in equilibrium or not, is the sum of the entropies of its subsystems, and by an extension of the succeeding arguments that a general condition that the system be in equilibrium is that the temperature be uniform everywhere.

We have also seen that a system or subsystem in equilibrium is not in any definite state in the quantum mechanical sense. Instead, it is free to be in any of a very large number of states, and the requirement of equilibrium is really only that it be equally free to be in any of those states. The system thus fluctuates about among its various states. It is important to remember that these fluctuations are not fluctuations out of equilibrium but rather that the equilibrium is the averaged consequence of these fluctuations.

We now wish to carry out our program, which means that we must learn how to count the number of states it is possible for a system to have or, more precisely, how to avoid having to count that number. This is a formidable task, and we shall need some powerful tools. Accordingly, we shall devote the next section to discussing the quantities that appear in thermodynamics, and the methods of manipulating such quantities.

1.2 THERMODYNAMICS

a. The Laws of Thermodynamics

Thermodynamics is basically a formal system of logic deriving from a set of four axioms, known as the *Laws of Thermodynamics*, all four of which we arrived at, or at least flirted with, in our preliminary discussion of the previous section. We shall not be concerned here with formalities, but let us, without rigor and just for the record, indicate the sense of the four laws. Being a logical system, the four laws are called, naturally, the Zeroth, First, Second, and Third. From the point of view of thermodynamics, these laws are not to be arrived at, as we have done, but rather are assumptions to be justified (and are amply justified) by their empirical success.

The *Zeroth Law* says that the concept of temperature makes sense. A single number, a scalar, assigned to each subsystem, suffices to predict whether the subsystems will be found to be in thermal equilibrium should they be brought into contact. Equivalently, we can say that if bodies A and B are each separately found to be in thermal equilibrium with body C (body C, if it is small, may be called a thermometer), then they will be in equilibrium with each other.

The *First Law* is the thermodynamic statement of the principle of conservation of energy. It is usually stated in such a way as to distinguish between two kinds of energy—heat and work—the changes of energy in a system being given by

$$dE = dQ + dR \tag{1.2.1}$$

where Q is heat and R is work.

We can easily relate Eq. (1.2.1) to Eq. (1.1.6). Suppose that our box of perfect gas were actually a cylinder with a movable piston of cross-sectional area A as in Fig. 1.2.1. It requires work to push the piston. If we apply a force $\mathscr{F}$ and displace the piston an amount dx, we are doing an amount of mechanical work on the gas inside given by

$$dR = \mathscr{F}\, dx$$

This can just as well be written in terms of the pressure, $P = \mathscr{F}/A$, and

1.2 Thermodynamics

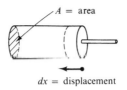

Fig. 1.2.1

volume, $dV = -A\,dx$ (the sign tells us that the volume decreases when we do positive work on the gas),

$$dR = \frac{\mathscr{F}}{A}(A\,dx)$$

$$= -P\,dV \qquad (1.2.2)$$

If the process we have described was done in isolation, so that energy was unable to leak in or out in any other form, Eqs. (1.2.1) and (1.2.2) together tell us that

$$\left(\frac{\partial E}{\partial V}\right)_Q = -P \qquad (1.2.3)$$

Comparing this result to Eqs. (1.1.6) and (1.1.8) of the previous section, we see that the pressure we are using here, which is just the force per unit area, is the same as the pressure as defined there, provided that, for our fixed number of particles, holding Q constant means that S has been held constant. We can show that such is the case. As we push the piston in, the quantitative values of the energies of the single-particle states, given by some relation like Eqs. (1.1.1) and (1.1.2), will change because the dimensions of the box (the value of L in the x direction) are changing. However, the enumeration of the single-particle states, the number of them and their separate identities, does not change. Consider a particular state of the system—a particular distribution of the particles among their various single-particle states, using up all the available energy—before the displacement. When the displacement occurs, each single-particle state shifts its energy a bit, but we can still identify one single-particle state of the new system with the state it came from in the old. If we induce the displacement slowly, each particle will stay in the state it is in, and so the work done just goes into changing the energies of all the occupied single-particle levels. The same statement is true of each of the possible states of the system, and so although the energy of the system changes in the process, the number of possible states, and hence the entropy, does not.

Now let us suppose that (applying the appropriate force) we hold the piston in a fixed position, so that the volume does not change, and add a bit of energy to the gas by other means (we can imagine causing it to absorb some light from an external source). No work (of the $\mathscr{F}\,dx$ type) has been done, and, furthermore, the single-particle states are not affected. But the amount of energy available to be divided among the particles has increased, the number of ways of dividing it has increased as well, and thus the entropy has increased. From Eqs. (1.2.1) of this section with $dR = 0$, and (1.1.6) and (1.1.7) of the previous section, we see that

$$dE)_R = dQ = dE)_{N,V} = T\,dS \qquad (1.2.4)$$

for changes that take place in equilibrium, so that (1.1.6) is applicable. Thus, for a fixed number of particles, we can write the first law for equilibrium changes in the form

$$dE = T\,dS - P\,dV \qquad (1.2.5)$$

Although changes in heat are always equal to $T\,dS$, there are kinds of work other than $P\,dV$: magnetic work, electric work, and so on. However, it will be convenient for us to develop the consequences of thermodynamics for this kind of work—that is, for mechanical work alone—and return to generalize our results later in this chapter.

According to the celebrated *Second Law* of thermodynamics, the entropy of a system out of equilibrium will tend to increase. This statement means that when a system is not in equilibrium, its number of choices is restricted—some states not forbidden by the design of the system are nevertheless unavailable due to its past history. As time goes on, more and more of these states gradually become available, and once a state becomes available to the random fluctuations of a system, it never again gets cut off; it always remains a part of the system's repertory. Systems thus tend to evolve in certain directions, never returning to earlier conditions. This tendency of events to progress in an irreversible way was pointed out by an eleventh-century Persian mathematician, Omar Khayyám (translated by Edward Fitzgerald):

> The Moving Finger writes; and, having writ,
> Moves on: nor all thy Piety nor Wit
> Shall lure it back to cancel half a line,
> Nor all thy Tears Wash out a Word of it.

There have been many other statements of the Second Law, all of them less elegant.

If disequilibrium means that the entropy tends to increase, equilibrium must correspond to a maximum in the entropy—the point at which it no longer can increase. We saw this in the example we considered in the previous section. If the two boxes, initially separate, had been out of equilibrium, one of them (the hotter one) would have started with more than its fair share of the energy of the combined system. Owing to this accident of history, the

1.2 Thermodynamics 13

number of states available to the combined system would have been smaller than if the cooler box had a fairer portion of the total energy to share out among its particles. The imbalance is redressed irreversibly when contact is made. In principle, a fluctuation could occur in which the initially cooler box had even less energy than it started with, but such a fluctuation is so utterly unlikely that there is no need to incorporate the possibility of it into our description of how things work in the real world. Quite the opposite, in fact: we have dramatic success in describing the real world if we assume that such fluctuations are impossible. That is just what the entropy principle does.

The *Third* and final *Law* states that at the absolute zero of temperature, the entropy of any body is zero. In this form it is often called *Nernst's theorem*. An alternative formulation is that a body cannot be brought to absolute zero temperature by any series of operations. In this form the law basically means that all bodies have the same entropy at zero degrees. According to the earlier statement, if a body had no disposable energy, so that its temperature were zero, it would have only one possible state: $\Gamma = 1$ and thus $S = k \log \Gamma = 0$. This amounts to asserting that the quantum ground state of any system is nondegenerate. Although there are no unambiguous counterexamples to this assertion in quantum mechanics, there is no formal proof either. In any case, there may be philosophical differences, but there are no practical differences between the two ways of stating the law. Among other things, the Third Law informs us that thermodynamic arguments should always be restricted to nonzero temperatures.

b. Thermodynamic Quantities

As we see in Eq. (1.2.5), the energy of a system at equilibrium with a fixed number of particles may be developed in terms of four variables, which come in pairs, T and S, P and V. Pairs that go together to form energy terms are said to be thermodynamically conjugate to each other. Of the four, two, S and V, depend directly on the size of the system and are said to be extensive variables. The others, T and P, are quite independent of the size (if a body of water is said to have a temperature of 300°K, that tells us nothing about whether it is a teaspoonful or an ocean) and these are called intensive variables.

We argued in Sec. 1.1 that if we know the energy as a function of S and V, then we can deduce everything there is to know, thermodynamically speaking, about the body in question: $P = -(\partial E/\partial V)_S$, $T = (\partial E/\partial S)_V$, and $T(V, S)$, together with $P(V, S)$, gives us $T(P, V)$; if we want the entropy, $T(V, S)$ can be inverted to give $S(T, V)$ and so on. If, on the other hand, we know the energy as a function, say of T and V, we generally do not have all the necessary information. There may, for example, be no way to find the entropy. For this reason, we shall call S and V the proper independent variables of the energy: given in terms of these variables the energy tells the whole story.

Unfortunately, problems seldom arise in such a way as to make it easy for us to find the functional form, $E(S, V)$. We usually have at hand those quantities that are easy to measure, T, P, V, rather than S and E. It will, therefore, turn out to be convenient to have at our disposal energylike functions whose proper independent variables are more convenient. In fact, there are four possible sets of two variables each, one being either S or T and the other either P or V, and we shall define energy functions for each possible set. The special advantages of each one in particular kinds of problems will show up as we go along.

We define F, the free energy (or Helmholz free energy), by

$$F = E - TS \qquad (1.2.6)$$

Together with (1.2.5), this gives

$$dF = -S\,dT - P\,dV \qquad (1.2.7)$$

so that

$$S = -\left(\frac{\partial F}{\partial T}\right)_V \qquad (1.2.8)$$

and

$$P = -\left(\frac{\partial F}{\partial V}\right)_T \qquad (1.2.9)$$

In other words, $F = F(T, V)$ in terms of its proper variables, and T and V are obviously a convenient set to use. F also has the nice property that, for any changes that take place at constant temperature,

$$\delta F)_T = -P\,\delta V)_T = \delta R \qquad (1.2.10)$$

so that changes in the free energy are just equal to the work done on the system. Recall for comparison that

$$\delta E)_S = -P\,\delta V)_S = \delta R \qquad (1.2.11)$$

the work done is equal to the change in energy only if the entropy is held constant or, as we argued in Sec. 1.2a, if the work is done slowly with the system isolated. It is usually easier to do work on a system well connected to a large temperature bath, so that Eq. (1.2.10) applies, than to do work on a perfectly isolated system, as required by Eq. (1.2.11).

The function $F(T, V)$ is so useful, in fact, that we seem to have gotten something for almost nothing by means of the simple transformation, Eq. (1.2.6). However, we have a fairly clear conceptual picture of what is meant by $E(S, V)$ in terms of the counting of states, whereas the connection between the enumeration of the states of a system and $F(T, V)$ is much less obvious. Our job in Sec. 1.3 will be to develop ways of computing $F(T, V)$ from the

1.2 Thermodynamics

states of a system, so that we can take advantage of this very convenient function.

The Gibbs potential (or Gibbs free energy), Φ, is defined by

$$\Phi = F + PV = E - TS + PV \tag{1.2.12}$$

which, combined with Eq. (1.2.7) or (1.2.5), gives

$$d\Phi = -S\,dT + V\,dP \tag{1.2.13}$$

It follows that

$$S = -\left(\frac{\partial \Phi}{\partial T}\right)_P \tag{1.2.14}$$

$$V = \left(\frac{\partial \Phi}{\partial P}\right)_T \tag{1.2.15}$$

and, consequently, $\Phi = \Phi(T, P)$ in terms of its proper variables. Φ will be the most convenient function for problems in which the size of the system is of no importance. For example, the conditions under which two different phases of the same material, say liquid and gas, can coexist in equilibrium will depend not on how much of the material is present but rather on the intensive variables P and T.

A function of the last remaining pair of variables, S and P, may be constructed by

$$W = \Phi + TS = E + PV \tag{1.2.16}$$

so that

$$dW = T\,dS + V\,dP \tag{1.2.17}$$

$$T = \left(\frac{\partial W}{\partial S}\right)_P \tag{1.2.18}$$

$$V = \left(\frac{\partial W}{\partial P}\right)_S \tag{1.2.19}$$

W is called the enthalpy or heat function. Its principal utility arises from the fact that, for a process that takes place at constant pressure,

$$dW)_P = T\,dS)_P = dQ \tag{1.2.20}$$

which is why it is called the heat function.

A fundamental physical fact underlies all these formal manipulations. The fact is that for any system of a fixed number of particles, everything is determined in principle if only we know any two variables that are not conjugate to each other and also know that the system is in equilibrium. In other words, such a system in equilibrium really has only two independent variables—at a given entropy and pressure, for example, it has no choice at all about what volume and temperature to have. So far we have expressed this fundamental point by defining a series of new energy functions, but the point can be made without any reference to the energy functions themselves by considering their cross-derivatives. For example,

$$\left(\frac{\partial P}{\partial T}\right)_V = \frac{\partial}{\partial T}\left(-\left(\frac{\partial F}{\partial V}\right)_T\right)_V = \frac{\partial}{\partial V}\left(-\left(\frac{\partial F}{\partial T}\right)_V\right)_T = \left(\frac{\partial S}{\partial V}\right)_T \quad (1.2.21)$$

Suppose that we have our perfect gas in a cylinder and piston, as in Fig. 1.2.1, but we immerse our system in a temperature bath, so that changes take place at constant T. If we now change the volume by means of the piston, the entropy will change owing to the changes in the energies of all the single-particle states. Since energy is allowed to leak into or out of the temperature bath at the same time, it would seem difficult to figure out just how much the entropy changes. Equation (1.2.21) tells us how to find out: we need only make the relatively simple measurement of the change in pressure when the temperature is changed at constant volume. It is important to realize that this result is true not only for the perfect gas but also for any system whatsoever; it is perfectly general. Conversely, the fact that this relation holds for some given system tells us nothing at all about the inner workings of that system.

Three more relations analogous to (1.2.21) are easily generated by taking cross-derivatives of the other energy functions. They are

$$\left(\frac{\partial T}{\partial V}\right)_S = -\left(\frac{\partial P}{\partial S}\right)_V \quad (1.2.22)$$

$$\left(\frac{\partial V}{\partial S}\right)_P = \left(\frac{\partial T}{\partial P}\right)_S \quad (1.2.23)$$

$$\left(\frac{\partial S}{\partial P}\right)_T = -\left(\frac{\partial V}{\partial T}\right)_P \quad (1.2.24)$$

Together these four equations are called the *Maxwell relations*.

A mnemonic device for the definitions and equations we have treated thus far is given in Appendix A of this chapter.

The machinery developed on these last few pages prepares us to deal with systems of fixed numbers of particles, upon which work may be done only by changing the volume—more precisely, for systems whose energy is a function of entropy and volume only. Let us now see how to generalize this picture.

If the number of particles of the system is free to vary, then, in addition to work and heat, the energy will depend on that number as well. We have already seen, in Eqs. (1.1.6) and (1.1.9), that in this case,

$$dE = T\,dS - P\,dV + \mu\,dN \quad (1.2.25)$$

1.2 Thermodynamics

μ is the chemical potential, which we may take to be defined by Eq. (1.2.25), so that

$$\mu = \left(\frac{\partial E}{\partial N}\right)_{S,V} \qquad (1.2.26)$$

If we now retain all the transformations among the energy functions as we have defined them, then

$$dF = d(E - TS) = -S\,dT - P\,dV + \mu\,dN \qquad (1.2.27)$$

Similarly,
$$d\Phi = -S\,dT + V\,dP + \mu\,dN \qquad (1.2.28)$$

$$dW = T\,dS + V\,dP + \mu\,dN \qquad (1.2.29)$$

Thus,
$$\mu = \left(\frac{\partial F}{\partial N}\right)_{T,V} = \left(\frac{\partial \Phi}{\partial N}\right)_{P,T} = \left(\frac{\partial W}{\partial N}\right)_{P,S} \qquad (1.2.30)$$

In other words, the effect of adding particles is to change any one of the energy functions by $\mu\,dN$ if the particles were added holding its own proper independent variables fixed:

$$\mu\,\delta N = (\delta E)_{S,V} = (\delta F)_{T,V} = (\delta \Phi)_{T,P} = (\delta W)_{S,P} \qquad (1.2.31)$$

Notice that the first member of Eq. (1.2.31) is not necessarily relevant. If we manage to change the energy at constant S and V (so that E depends on something besides S and V) and we retain the transformations, Eqs. (1.2.6), (1.2.12), and (1.2.16), then the other functions will change according to

$$(\delta E)_{S,V} = (\delta F)_{T,V} = (\delta \Phi)_{T,P} = (\delta W)_{S,P} \qquad (1.2.32)$$

We shall make use of this result later on (to have an application to think about now, you might consider changing the masses of the particles of a perfect gas; see Prob. 1.2).

The introduction of μ and changes in N thus serve as an example of how to generalize to cases where the energy can depend on something besides S and V, but we wish to retain the transformations between E, F, Φ, and W. However, the chemical potential is a particularly important function in thermodynamics, and its properties deserve further elaboration.

Like P and T, μ is an intensive variable, independent of the size of the system (this point can be seen from the result of Prob. 1.1a, where we see that two subsystems in equilibrium must have the same μ regardless of their sizes), and it is conjugate to an extensive variable, N. Suppose that we think of Eq. (1.2.25) as applying to a piece of matter, and we rewrite it to apply to another piece that differs from the first only in that it is λ times bigger. It could have been made by putting together λ of the small pieces. Then all the extensive quantities, E, S, V, and N, will simply be λ times bigger:

$$d(\lambda E) = T\, d(\lambda S) - P\, d(\lambda V) + \mu\, d(\lambda N)$$

or

$$\lambda\, dE + E\, d\lambda = \lambda(T\, dS - P\, dV + \mu\, dN) + (TS - PV + \mu N)\, d\lambda \quad (1.2.33)$$

Since λ is arbitrary, it follows that

$$E = TS - PV + \mu N \quad (1.2.34)$$

If we differentiate this result and subtract Eq. (1.2.25), we get

$$d\mu = -\frac{S}{N} dT + \frac{V}{N} dP \quad (1.2.35)$$

so that μ, like Φ, is a proper function of T and P. In fact, comparing Eq. (1.2.34) to (1.2.12) (which, remember, is still valid), we have

$$\Phi(P, T) = \mu N \quad (1.2.36)$$

As we have seen, the general conditions for a system to be in equilibrium will be that T, P, and μ be uniform everywhere.

Equation (1.2.31) tells us that μ is the change in each of the energy functions when one particle is added in equilibrium. That observation might make it seem plausible that μ would be a positive quantity, but it turns out instead that μ is often negative. Why this is so can be seen by studying Eq. (1.2.26). To find μ, we ask: How much energy must we add to a system if we are to add one particle while keeping the entropy and volume constant? Suppose that we add a particle with no energy to the system, holding the volume fixed, and wait for it to come to equilibrium. The system now has the same energy it had before, but one extra particle, giving it more ways in which to divide that energy. The entropy has thus increased. In order to bring the entropy back to its previous value, we must extract energy. The chemical potential—that is, the change in energy at constant S and V—is therefore negative. This argument breaks down if it is impossible to add a particle at zero energy, owing to interactions between the particles; the chemical potential will be positive when the average interaction is sufficiently repulsive, and energy is required to add a particle. It will be negative, for example, for low-density gases and for any solid or liquid in equilibrium with (at the same chemical potential as) its own low density vapor.

Now that we have a new set of conjugate variables, it is possible to define new transformations to new energy functions. For example, the quantity $E - \mu N$ would be a proper function of the variables S, V, and μ, and $W - \mu N$ of S, P, and μ. We do not get a new function from $\Phi - \mu N$, which is just equal to zero: P, T, and μ are not independent variables. Of the possible definitions, one will turn out to be most useful. We define the Landau potential, Ω,

$$\Omega = F - \mu N \tag{1.2.37}$$

so that

$$d\Omega = -S\,dT - P\,dV - N\,d\mu \tag{1.2.38}$$

that is, the proper variables of Ω are T, V, and μ. Notice that it follows from Eqs. (1.2.37) and (1.2.34) that

$$\Omega = -PV \tag{1.2.39}$$

Furthermore, since we have not altered the relations between the other energy functions in defining Ω, arguments just like those leading to Eq. (1.2.32) will yield

$$(\delta\Omega)_{T,V,\mu} = (\delta E)_{S,V,N} = (\delta F)_{T,V,N} = (\delta\Phi)_{T,P,N} = (\delta W)_{S,P,N} \tag{1.2.40}$$

c. Magnetic Variables in Thermodynamics

The total energy content of a magnetic field is

$$E_m = \frac{1}{8\pi}\int B^2\,dV \tag{1.2.41}$$

where B is the fundamental magnetic field, produced by all currents, and the integral extends over all space. In order to introduce magnetic work into our equilibrium thermodynamic functions, we need to know how much work is done on a sample when magnetic fields change—that is, the magnetic analog of $(-P\,dV)$. Equation (1.2.41) fails to tell us this for two reasons. First, it does not sort out contributions from the sample and from external sources; B is the total field. Second, there is no way to tell from Eq. (1.2.41) how much work was done in producing B; the work may have been much more than E_m with the excess dissipated in heat. Only if all changes took place in equilibrium will E_m be equal to the work done in creating the field. In other words, E_m is the minimum work (by sample and external sources together) needed to produce the distribution of fields in space.

Even assuming that all changes take place in equilibrium, we must have a way of distinguishing work done on the sample from other contributions— that is, work done on an external source during the same changes. To do so, we decompose B into two contributions

$$\mathbf{B} = \mathbf{H} + 4\pi\mathbf{M} \tag{1.2.42}$$

where $\mathbf{H}$ is the field due to currents running in external sources and $\mathbf{M}$ is the response of the sample, called the *magnetization*.

To keep things simple, let us set up a definite geometry, which we shall always try to return to when analyzing magnetic problems. We shall imagine $\mathbf{H}$ to be a uniform field arising from current in a long solenoid and our sample to be a long cylinder along the axis of the solenoid, as sketched in Fig. 1.2.2. There is no magnetic field outside the solenoid (we can, if necessary, imagine it closed upon itself in a torus), so that all contributions to Eq. (1.2.41) come from inside the windings. We shall call the total volume inside V_t. In V_t the H field is given by

$$H = \frac{4\pi}{c} \frac{N_0 I}{L} \tag{1.2.43}$$

where c is the speed of light, N_0/L the number of turns per unit length of the solenoid, and I the current in the coil. H is always parallel to the axis, and so we need not write it as a vector.

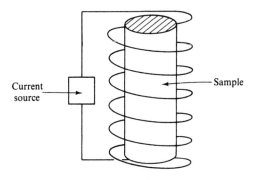

Fig. 1.2.2

The magnetization of the sample does not produce any fields outside the sample (just as current in the solenoid does not produce any field outside the solenoid). Furthermore, H is the same inside the sample and out regardless of M; it depends only on I through Eq. (1.2.43). All of this gives us a reasonably clean division of the fields. For example, if the sample is a superconducting material under certain circumstances, it will turn out that currents will flow in the surface of the sample that exactly cancel any applied field. An observer inside such a superconductor never detects any magnetic field at all. Then the situation is this: H is uniform everywhere inside the solenoid, sample included. Between the windings and the sample, $M = 0$, $B = H$. Inside the sample, $M = -(1/4\pi)H$, and $B = 0$. The total magnetic energy in this situation is given by

$$E_m = \frac{1}{8\pi} H^2 (V_t - V_s) \quad \text{(superconductor)} \tag{1.2.44}$$

where V_s is the volume of the superconducting sample. Our problem in this case is, assuming the sample stayed in equilibrium as the field was applied,

1.2 Thermodynamics

so that E_m is the work done, how much of this work was done by (or on) the current source in Fig. 1.2.2, and how much on (or by) the sample? In Sec. 1.2a we found how much work was done if a thermally isolated sample changed its volume under a constant applied force. The analogy here is to see how much work is done if a thermally isolated sample changes its magnetic state under a constant applied H field. Suppose that with constant current, I, in the solenoid, there is a small uniform change, δM, in the magnetization of the sample. The result is a change in the flux, φ_0, linked by the solenoid, inducing a voltage V_0 across it. Power, IV_0, is extracted from the current source or injected into it. If R_0 is the work done on the current source,

$$\delta R_0 = \int IV_0 \, dt = I \int V_0 \, dt \qquad (1.2.45)$$

since I is constant. The voltage is given by Faraday's law of electromagnetic induction:

$$V_0 = -\frac{1}{c} \frac{\partial \varphi_0}{\partial t} \qquad (1.2.46)$$

Then

$$\delta R_0 = -\frac{I}{c} \int \frac{\partial \varphi_0}{\partial t} \, dt$$

$$= -\frac{I}{c} \int d\varphi_0 = -\frac{I}{c} \delta \varphi_0 \qquad (1.2.47)$$

The sign convention is: if flux is expelled, work is done on the source; that is, if $\delta \varphi_0$ is negative, δR_0 is positive. The flux is given by

$$\varphi_0 = N_0 \int B \, dA \qquad (1.2.48)$$

where the integral is taken over a cross-sectional area of the solenoid. The change in M produces a change in B only inside the sample, so that

$$\delta \varphi_0 = N_0 A_s (\delta B)_s = 4\pi N_0 A_s \, \delta M \qquad (1.2.49)$$

where A_s is the cross-sectional area of the sample. Then

$$\delta R_0 = -\frac{4\pi I N_0 A_s}{c} \delta M \qquad (1.2.50)$$

Substituting $4\pi N_0 I/c = HL$ from Eq. (1.2.43) gives

$$\delta R_0 = -H(LA_s) \, \delta M \qquad (1.2.51)$$

But LA_s is just V_s, the volume of the sample, so

$$\delta R_0 = -V_s H \, \delta M \qquad (1.2.52)$$

Magnetization in the same direction as H extracts work from the source, lowering R_0; M opposing H (as in the superconductor) does work on the source, increasing R_0.

Now, since the sample is isolated thermally, the only kind of energy ever exchanged is work, and conservation of energy then requires

$$\delta R + \delta R_0 = 0 \text{ in equilibrium} \quad (1.2.53)$$

where δR is the work done on the sample. Equations (1.2.52) and (1.2.53) then give

$$\delta R = V_s H \, \delta M \quad (1.2.54)$$

If the volume V_s had changed at constant M, that, too, would have changed the linked flux, thereby doing work on the source. It is easy to see that a more general way of writing the work is

$$\delta R = H \, \delta \int M \, dV \quad (1.2.55)$$

and the First Law of Thermodynamics may be written

$$dE = T \, dS + H \, d \int M \, dV \quad (1.2.56)$$

For the simplest case, a uniformly magnetized sample,

$$dE = T \, dS + H \, dMV \quad (1.2.57)$$

$$E = E(S, MV) \quad (1.2.58)$$

With this as a starting point, E as a unique function of S, and MV in equilibrium, we can define magnetic analogs of all the energy functions; we write the following:

$$F = E - TS \quad (1.2.59)$$

so

$$dF = -S \, dT + H \, dMV \quad (1.2.60)$$

In this way F retains its central property: for changes at constant T, F is the work done. Its proper variables are

$$F = F(T, MV) \quad (1.2.61)$$

We now redefine Φ:

$$\Phi = F - HMV \quad (1.2.62)$$

$$d\Phi = -S \, dT - MV \, dH \quad (1.2.63)$$

$$\Phi = \Phi(T, H) \quad (1.2.64)$$

1.2 Thermodynamics

The volume integrals may easily be replaced in transforms like Eq. (1.2.62) and, of course, in

$$W = \Phi + TS = E - HMV \quad (1.2.65)$$

$$dW = T\,dS - MV\,dH \quad (1.2.66)$$

$$W = W(S, H) \quad (1.2.67)$$

In all of this, H behaves the way that the pressure did before; both may usually be thought of as uniform external "forces" to which the system must respond; its response is to change either its volume or its magnetization. There is a difference in sign due to a geometrical accident: the work done on a body is proportional to $-dV$ (you squeeze it to do work on it) but to $+dMV$. The reader can easily write down the Maxwell relations and revise the mnemonic (Appendix A) for this kind of work.

d. Variational Principles in Thermodynamics

Fundamentally there is only one variational principle in thermodynamics. According to the Second Law, an isolated body in equilibrium has the maximum entropy that physical circumstances will allow. However, given in this form, it is often inconvenient to use. We can find the equilibrium conditions for an isolated body by maximizing its entropy, but often we are more interested in knowing the equilibrium condition of a body immersed in a temperature bath. What do we do then? We can answer the question by taking the body we are interested in, together with the temperature bath and an external work source, to be a closed system. The external source does work to change the macroscopic state of the body. If more work is done than necessary—that is, if the body is out of equilibrium—the excess work can be dumped into the temperature bath as heat, thus increasing its entropy. We ensure that this step happens by requiring that the entropy of the combined system increase as much as it can. What is left over tells us the equilibrium conditions on the body itself.

Let us make the argument in detail for magnetic variables, using the formalism and geometry we have already set up (the analogous argument for P-V variables is found in Landau and Lifshitz, *Statistical Physics*, pp. 57–59; see bibliography at end of chapter). We now imagine that the region between the solenoid and the sample is filled with some nonmagnetic material ($M = 0$ always), which, however, is capable of absorbing heat and is large enough to do so at constant temperature. This is the temperature bath; we shall also refer to it as the medium. It is always to be thought of as being in internal equilibrium. That means that its energy E' and its entropy S' are connected by

$$dE' = T\,dS' \quad (1.2.68)$$

There are no other contributions, since no magnetic work can be done on it.

We imagine the work source to be thermally isolated from the sample and medium, so that no heat can get in or out of it, and its changes in energy are simply δR_0. By the arguments that led to Eq. (1.2.52),

$$\delta R_0 + H\,\delta(MV_s) = 0 \qquad (1.2.69)$$

if M is uniform in V_s. Equation (1.2.69) does not imply that the sample is in equilibrium, for it can still exchange heat with the medium. The combined system—sample, medium, and work source—is isolated, and thus its total energy is conserved:

$$\delta(E + E' + R_0) = 0 \qquad (1.2.70)$$

where E is the energy of the sample. Notice that E cannot be written as a function of S, the entropy of the sample, and M, since it is a unique function of those variables only in equilibrium. However, even with the sample not in equilibrium, we can put Eqs. (1.2.68) and (1.2.69) into (1.2.70) to get

$$\delta E + T\,\delta S' - H\,\delta(MV_s) = 0 \qquad (1.2.71)$$

We now apply the entropy principle to the system as a whole: in whatever changes take place,

$$\delta S + \delta S' \geq 0 \qquad (1.2.72)$$

Substituting into (1.2.71) and rearranging, we obtain

$$\delta E \leq T\,\delta S + H\,\delta(MV_s) \qquad (1.2.73)$$

If the sample is in equilibrium, the equality holds and (1.2.73) reduces to (1.2.57). What we have accomplished in (1.2.73) is an expression, in the form of an inequality, for changes in the sample variables, valid even when the sample is out of equilibrium and connected to a bath.

Suppose that changes take place at fixed T and fixed MV_s. The constant T may be taken inside the δ sign, and $\delta(MV_s) = 0$. We have then

$$\delta(E - TS) \leq 0 \qquad (\text{const. } T,\ MV_s) \qquad (1.2.74)$$

$E - TS$ is the free energy. As random fluctuations occur, F will decrease or remain constant with time,

$$\frac{\partial F}{\partial t} \leq 0 \qquad (1.2.75)$$

When F reaches the lowest value it can possibly have, no further changes in it can take place; under the given conditions, equilibrium has been reached. The equilibrium condition is

$$F = \text{minimum}$$

The constraint we have applied—that MV_s be constant—really means, through Eq. (1.2.69), that R_0 is constant; we are considering changes in which the

1.2 Thermodynamics

sample does no external work. The fact that we considered magnetic work here is clearly irrelevant; we could have written δR everywhere for $H \, \delta M V_s$. The general principle (including $-P \, \delta V$ and other kinds of work) is as follows: With respect to variations at constant temperature, and which do no work, a body is in equilibrium when its free energy, $F = E - TS$, is a minimum.

We can, if we wish, keep H constant and allow MV_s to change. The constant H then comes into the δ sign, and the quantity to be minimized is

$$E - TS - HMV_s = \Phi \quad (T, H \text{ constant}) \quad (1.2.76)$$

The analogous result for P-V work is, when T and P are held constant, to minimize

$$E - TS + PV = \Phi \quad (1.2.77)$$

Notice that the transformation from F to Φ simply subtracts the work done on the external source, $\delta(HMV_s)$. If no work is done, we minimize F; if work is done, we subtract it off and minimize what is left; it amounts to the same thing.

Without further ado, we can write the thermodynamic variational principle in a very general way: for changes that take place in any body,

$$\delta E \leq T \, \delta S + H \, \delta \int M \, dV - P \, \delta V + \mu \, \delta N + \cdots \quad (1.2.78)$$

where we have left room at the end for any other work terms that may be involved in a given problem. In Eq. (1.2.78) the intensive variables, T, P, H, and μ, are those of the medium, while the extensive variables, E, V, $\int M \, dV$, and N, are those of the subsystem of interest. The appropriate variational principle for any situation can easily be generated by applying the kind of arguments we have just given to Eq. (1.2.78). For example, in a nonmagnetic problem ($H = M = 0$), if V is fixed but N varies at constant T and μ,

$$\Omega = F - \mu N = \text{minimum}$$

e. Examples of the Use of Variational Principles

We can now apply our variational principles to a few examples to see how they work (needless to say, these are not just examples; the results we get will be useful later on).

First, consider the magnetic behavior of a perfect conductor. A perfect conductor is a metal whose electrical resistance is zero. If a magnetic field is applied to it, currents are induced that, by Lenz's law, oppose the applied field and prevent it from penetrating. Since there is no resistance, these currents persist, and we always have, inside, $B = 0$, or $M = -(1/4\pi)H$, just as in a superconductor. However, unlike the superconductor, $M =$

$-(1/4\pi)H$ is, in this case, a dynamical result, not necessarily the thermodynamic equilibrium state. In order to apply thermodynamics, we must make an assumption very similar to the one that was needed to apply thermodynamics to the perfect gas: even though there is no resistance, we assume that the current, and hence the magnetization, of the sample fluctuates about, seeking its most favorable level.

At a given T and H we seek the equilibrium magnetization, which we may as well take to be uniform over the fixed volume V_s of the sample. From Eq. (1.2.78) the quantity to be minimized is

$$\Phi = E - TS - HMV_s \qquad (1.2.79)$$

or, in other words,

$$\left(\frac{\partial \Phi}{\partial M}\right)_{T,H} = 0 \qquad (1.2.80)$$

In equilibrium, Φ does not depend on M, only on T and H. To solve the problem, we must construct the quantity $\Phi = E - TS - HMV_s$ when the body is not restricted to equilibrium and then use Eq. (1.2.80) to find the equilibrium dependence of M on T and H. To do so, we take the energy of the sample to be

$$E = E_0 + E_m \qquad (1.2.81)$$

where E_0 is whatever internal energy the sample has that is not associated with magnetic fields:

$$dE_0 = T\,dS - P\,dV + \cdots \qquad (1.2.82)$$

Since it is irrelevant to our problem, we have taken the nonmagnetic part of the sample to be in equilibrium. Putting Eqs. (1.2.81) and (1.2.82) into (1.2.79), we find

$$\Phi = E_0 - TS + E_m - HMV_s \qquad (1.2.83)$$

There is no heat associated with E_m, Eq. (1.2.41), so the sample entropy, S, is all in E_0. From Eqs. (1.2.41) and (1.2.42),

$$E_m = \frac{V_s}{8\pi}(H + 4\pi M)^2$$

$$= V_s\left(\frac{H^2}{8\pi} + HM + 2\pi M^2\right) \qquad (1.2.84)$$

Then

$$\Phi = E_0 - TS + V_s\left(\frac{H^2}{8\pi} + 2\pi M^2\right)$$

$$= F_0 + V_s\left(\frac{H^2}{8\pi} + 2\pi M^2\right) \qquad (1.2.85)$$

1.2 Thermodynamics

Since we are only considering magnetic work, F_0, the nonmagnetic part of the free energy depends only on T. Therefore,

$$\left(\frac{\partial \Phi}{\partial M}\right)_{H,T} = 4\pi V_s M = 0$$

$$M = 0 \qquad (1.2.86)$$

(Notice that $\partial^2 \Phi / \partial M^2 = 4\pi V_s > 0$, so this is really a minimum.) The result is that, at equilibrium, there is no magnetization. $B = H$ inside; the applied field penetrates completely. No shielding currents flow in equilibrium.

We could have seen by means of a different kind of argument that $M = 0$ is necessarily the right result. We know that in a real conductor, with electrical resistance, eddy currents produced by applying a magnetic field quickly die out, dissipated by the resistance, and the field penetrates unopposed. A dissipative effect, such as electrical resistance, turns work into heat (called Joule heating in this case), which is just what is needed to drive a system toward thermodynamic equilibrium, but it plays no role in the equilibrium state. Since they differ only in the mechanism for changing states, there can be no difference between the equilibrium of a perfect conductor and a real conductor. For a similar reason, the equilibrium of a real gas and a perfect gas will be the same, provided that the real gas interactions have little effect on the possible energy levels of each particle.

As we have already said, the equilibrium magnetization of a superconductor is

$$M = -\frac{1}{4\pi} H \qquad (1.2.87)$$

which we now know to be an unfavorable state of affairs; certainly, as we have just seen, if Eq. (1.2.87) holds in equilibrium, it is not a consequence of the fact that superconductors have no electrical resistance. As our second example of an application of the variational principles, let us examine this situation. If you do not yet know what a superconductor is, relax. Part of the beauty of thermodynamics is that you do not have to know much about what is going on inside.

Equation (1.2.87) is the equation of state of what we shall later call a type I superconductor. For internal reasons of its own (which we shall investigate much later), it chooses to go into this unfavorable magnetic condition, in which surface currents maintain $B = 0$ inside. The larger the applied H field, the more unfavorable the situation becomes until, at some value of H, the effort is no longer worth it and the material ceases to be superconducting, turns normal, and lets the field in. The applied field at which this situation occurs, H_c, is a definite function of temperature, $H_c = H_c(T)$, and is shown in Fig. 1.2.3. The process that occurs at $H_c(T)$ is a phase transition.

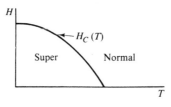

Fig. 1.2.3

At $H = H_c(T)$, the superconducting phase, whose equation of state is Eq. (1.2.87), and the normal phase, whose equation of state is $M = 0$, can coexist in equilibrium.

Suppose that we think of our sample as a piece of superconducting material (type I) held at constant T and constant $H = H_c(T)$. The two phases can coexist, but, in equilibrium, how much of each is present? Can we apply our variational principle to find out?

Since T and H are held fixed, it is once again Φ we wish to minimize. In particular,

$$\frac{\partial \Phi}{\partial V_{sc}} = 0 \qquad (1.2.88)$$

where V_{sc} is the volume of the superconducting part of the material, and the variations are subject to

$$dV_{sc} + dV_n = dV_s = 0 \qquad (1.2.89)$$

where V_n is the normal portion's volume. We can construct a generalized Φ of the sample that depends on V_{sc}, using the magnetic energy, E_m, and then put in the equations of state of the two parts. Proceeding as before, we have

$$\Phi = F_0 + \frac{1}{8\pi} \int_{\text{sample}} dV(H + 4\pi M)^2 - \int HM\, dV \qquad (1.2.90)$$

However, we cannot assume that F_0 is the same per unit volume of superconductor and normal conductor; after all, we know that something special is going on inside the superconductor. Let us write the nonmagnetic parts of the free energy densities of the phases as $f_{0,\text{sc}}$ and $f_{0,n}$. Then

$$\Phi = f_{0,\text{sc}} V_{sc} + f_{0,n} V_n + \frac{1}{8\pi} \int dV(H + 4\pi M)^2 - \int HM\, dV$$

$$= f_{0,\text{sc}} V_{sc} + f_{0,n} V_n + \frac{H^2}{8\pi} V_s + 2\pi M^2 V_{sc} \qquad (1.2.91)$$

1.2 Thermodynamics

In the last step we used the fact that $M = 0$ in the volume V_n. We now take $(\partial \Phi/\partial V_{sc})_{H,T}$, remembering that $\partial V_n/\partial V_{sc} = -1$, $\partial V_s/\partial V_{sc} = 0$, and $f_{0,sc}$ and $f_{0,n}$ depend only on T,

$$\frac{\partial \Phi}{\partial V_{sc}} = f_{0,sc} - f_{0,n} + 2\pi M^2 = 0 \tag{1.2.92}$$

Since $M = -(1/4\pi)H_c$, we have

$$f_{0,sc} = f_{0,n} - \frac{H_c^2}{8\pi} \tag{1.2.93}$$

Curiously enough, we did not find out how big V_{sc} is—that remains undetermined. But we have found out that the nonmagnetic (or zero field) part of the free energy density of a superconductor is lower than that of a normal conductor, which is why materials like to be superconductors, and the difference is related to H_c at the same temperature. We know nothing (yet) about the nature of the difference between the superconducting and normal states, but we do know how to measure the difference in free energy between the two.

f. Thermodynamic Derivatives

It is convenient to give names to certain derivatives of the thermodynamic quantities. The temperature, pressure, and volume are relatively easy quantities to measure directly. The entropy is much more difficult and cannot, in general, be measured directly. However, changes in the entropy may be measured by putting heat in under specified conditions. In doing so, we are said to be measuring the heat capacity. The two most commonly encountered heat capacities are those at constant volume and at constant pressure, defined, respectively, by

$$C_V = T\left(\frac{\partial S}{\partial T}\right)_V \tag{1.2.94}$$

and

$$C_P = T\left(\frac{\partial S}{\partial T}\right)_P \tag{1.2.95}$$

Since many microscopic theories of matter are designed to allow us to count states and thus arrive at the entropy, as outlined in Sec. 1.1, we often find that the resulting prediction to be compared to experiment will be the heat capacity. From the definitions, we may see that

$$C_V = \left(\frac{\partial E}{\partial T}\right)_V \tag{1.2.96}$$

That is, it is the total change in energy if no work is done. It is also given by

$$C_V = -T\left(\frac{\partial^2 F}{\partial T^2}\right)_V \tag{1.2.97}$$

and C_P is given by

$$C_P = -T\left(\frac{\partial^2 \Phi}{\partial T^2}\right)_P \tag{1.2.98}$$

We have previously considered only the cross second derivatives of the energy functions, which we did when deriving the Maxwell relations. Aside from the heat capacities, there are two other direct second derivatives; let us define the isothermal compressibility,

$$K_T = -\frac{1}{V}\left(\frac{\partial V}{\partial P}\right)_T \tag{1.2.99}$$

and the adiabatic compressibility,

$$K_S = -\frac{1}{V}\left(\frac{\partial V}{\partial P}\right)_S \tag{1.2.100}$$

The first is given by

$$K_T = -\frac{1}{V}\left(\frac{\partial^2 \Phi}{\partial P^2}\right)_T \tag{1.2.101}$$

and the second by

$$K_S = \left[V\left(\frac{\partial^2 E}{\partial V^2}\right)_S\right]^{-1} \tag{1.2.102}$$

Of course, there are a number of other ways of constructing these quantities from second derivatives of the energy functions.

There are definite relations between these quantities. For example, since a body's thermodynamic variables are fixed if P and T are fixed (for a given N), we can write S as a function of P and T: $S = S(P, T)$, or

$$dS = \left(\frac{\partial S}{\partial P}\right)_T dP + \left(\frac{\partial S}{\partial T}\right)_P dT \tag{1.2.103}$$

so that

$$\frac{C_V}{T} = \left(\frac{\partial S}{\partial T}\right)_V = \left(\frac{\partial S}{\partial P}\right)_T \left(\frac{\partial P}{\partial T}\right)_V + \frac{C_P}{T} \tag{1.2.104}$$

The quantity $(\partial S/\partial P)_T$ is inaccessible to measurement, so we eliminate it by means of the Maxwell relation, Eq. (1.2.24),

$$\left(\frac{\partial S}{\partial P}\right)_T = -\left(\frac{\partial V}{\partial T}\right)_P$$

1.2 Thermodynamics

This gives

$$\frac{C_P - C_V}{T} = \left(\frac{\partial V}{\partial T}\right)_P \left(\frac{\partial P}{\partial T}\right)_V \qquad (1.2.105)$$

We can put the difference between C_P and C_V in terms of K_T by use of the chain rule,

$$\left(\frac{\partial V}{\partial T}\right)_P \left(\frac{\partial T}{\partial P}\right)_V \left(\frac{\partial P}{\partial V}\right)_T = -1 \qquad (1.2.106)$$

This gives us a choice of two relations; eliminating $(\partial V/\partial T)_P$ or $(\partial P/\partial T)_V$, we get

$$C_P - C_V = TVK_T\left[\left(\frac{\partial P}{\partial T}\right)_V\right]^2 \qquad (1.2.107)$$

or

$$C_P - C_V = \frac{T}{VK_T}\left[\left(\frac{\partial V}{\partial T}\right)_P\right]^2 \qquad (1.2.108)$$

For magnetic systems, where P and V are replaced by H and M, we have by obvious analogy the quantities C_H, C_M, and the isothermal and adiabatic susceptibilities,

$$X_T = \left(\frac{\partial M}{\partial H}\right)_T \qquad (1.2.109)$$

$$X_S = \left(\frac{\partial M}{\partial H}\right)_S \qquad (1.2.110)$$

You can easily show, by reapplying the same arguments with these variables, that

$$C_H - C_M = \frac{TV}{X_T}\left[\left(\frac{\partial M}{\partial T}\right)_H\right]^2 \qquad (1.2.111)$$

The heat capacities, compressibilities, and susceptibilities are all derivatives of the thermodynamic quantities with respect to their own conjugate variables and are sometimes collectively called response functions. The heat capacities and compressibilities are always positive. We can see this by applying our extremum principle, Eq. (1.2.78). Consider, for example, a nonmagnetic system of fixed N, and take T and V to be constant, so that the equilibrium condition is F = minimum. We imagine that the system is all made of the same stuff, but let it be divided into two parts by a movable partition, as in Fig. 1.2.4. The partition is free to find its own equilibrium position, which it does by minimizing F with respect to its position, x. However, since $dx \propto dV_1$, we may as well minimize F with respect to V_1, subject to the condition

$$dV_1 + dV_2 = dV = 0 \qquad (1.2.112)$$

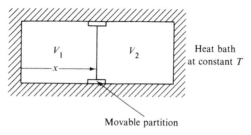

Fig. 1.2.4

We have

$$\frac{\partial F}{\partial V_1} = \frac{\partial}{\partial V_1}(F_1 + F_2) = \frac{\partial F_1}{\partial V_1} + \frac{\partial F_2}{\partial V_1} = \frac{\partial F_1}{\partial V_1} - \frac{\partial F_2}{\partial V_2} = 0 \quad (1.2.113)$$

Since these derivatives are taken at constant T and $(\partial F/\partial V)_T = -P$, we have just found out that $P_1 = P_2$, a result promised in Sec. 1.1. However, we know something else: F is not only stationary under variations in V_1 [used in Eq. (1.2.113)], but it is also a minimum. Its second derivative must be positive:

$$\frac{\partial^2 F}{\partial V_1^2} = \frac{\partial}{\partial V_1}(P_1 - P_2) = -\frac{\partial P_1}{\partial V_1} - \frac{\partial P_2}{\partial V_2} > 0$$

In other words,

$$\frac{1}{V_1 K_{T_1}} + \frac{1}{V_2 K_{T_2}} > 0 \quad (1.2.114)$$

But V_1 and V_2 contain the same substance at the same pressure and temperature. It follows that $K_{T_1} = K_{T_2}$, so that

$$K_T \geq 0 \quad (1.2.115)$$

This result is perfectly general: we never did mention what was in the box.

We can show that $C_V \geq 0$ by a closely analogous argument. Consider an isolated system, so that E and V (instead of T and V) are constant. Then S will be a maximum. Divide it into two parts so that $E_1 + E_2 = E =$ constant. Maximize S with respect to E_1:

$$\frac{\partial S}{\partial E_1} = \frac{\partial S_1}{\partial E_1} - \frac{\partial S_2}{\partial E_2} = \frac{1}{T_1} - \frac{1}{T_2} = 0 \quad (1.2.116)$$

Notice that we used quite precisely the same argument with slightly different

1.2 Thermodynamics

words in Sec. 1.1. Now get the second derivative, which must, in this case, be negative.

$$\frac{\partial^2 S}{\partial E_1^2} = \frac{\partial}{\partial E_1}\left(\frac{1}{T_1}\right)_V + \frac{\partial}{\partial E_2}\left(\frac{1}{T_2}\right)_V$$

$$= -\frac{1}{T_1^2}\frac{\partial T_1}{\partial E_1} - \frac{1}{T_2^2}\frac{\partial T_2}{\partial E_2}$$

$$= -\frac{1}{T_1^2 C_{V_1}} - \frac{1}{T_1^2 C_{V_2}} \leq 0 \qquad (1.2.117)$$

If we choose the two volumes to be equal, we see immediately that

$$C_V \geq 0 \qquad (1.2.118)$$

The case for C_P can be made by an additional argument of the same type, but it is easier to notice that since the right-hand side of Eq. (1.2.108) cannot be negative,

$$C_P \geq C_V \qquad (1.2.119)$$

It follows that C_P is positive.

We cannot make the same kind of argument for the magnetic susceptibilities, because we cannot apply to the magnetization a condition analogous to Eq. (1.2.112). In fact, there are indeed cases of materials with both negative and positive susceptibilities.

Finally, we should note that the two conditions we have just derived, $K_T \geq 0$ and $C_V \geq 0$, are both eminently reasonable—in fact, really quite obvious on intuitive physical grounds. First take $K_T > 0$, which means

$$\left(\frac{\partial P}{\partial V}\right)_T < 0 \qquad (1.2.120)$$

Imagine a substance in a cylinder and piston with $\partial P/\partial V > 0$ instead. Increasing the volume by pulling the piston out causes the pressure to increase, which pushes the piston out farther, causing additional increase in the pressure, and so on. Such a system would be explosively unstable, a condition incompatible with thermodynamic equilibrium. The other condition, $C_V > 0$, is simply that

$$\left(\frac{\partial E}{\partial T}\right)_V > 0 \qquad (1.2.121)$$

It merely says, in other words, that warming a body always requires increasing its energy.

g. Some Applications to Matter

Up to now the perfect gas and perfect conductor have been our principal examples of the workings of thermodynamics. Let us begin here to discuss some general ideas about real matter in thermodynamic terms. Figure 1.2.5 is a schematic pressure-temperature phase diagram of a typical simple substance. The fact that there can be only two independent variables for any system in equilibrium makes it possible to draw such a diagram.

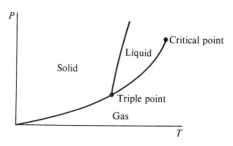

Fig. 1.2.5

In the general region of high pressures and low temperatures, the substance is a solid; at higher temperatures and lower pressures it is a gas, and there is a liquid region at intermediate values of P and T. The lines that separate these phases are called *coexistence curves*. In particular, that between the liquid and gas is the vapor pressure curve, between the solid and gas the sublimation curve, and between the solid and liquid the melting curve. On these coexistence curves, the phases can exist in equilibrium with each other, but for any value of P and T that does not fall on one of these curves, the substance, if it is in equilibrium, must be in a single, specific, homogeneous phase. Given a fixed number of particles (say, one mole) it is not necessary for us to specify the volume (or, equivalently, the density) in order to determine what phase the substance will be in; it is quite enough to specify P and T. Furthermore, if we know that two phases are coexisting in equilibrium, then either P or T suffices; at any pressure, the temperature is determined and vice versa. We say, for example, that the vapor pressure is a function of temperature only. That is why it may be represented as a single curve in a P-T plane. Usually, but not always, the coexistence curves have positive slope in the P-T plane. We shall see the thermodynamic significance of this point later on. If we cross a coexistence curve from one region to another, the substance is said to undergo a phase transition.

Two points on the plot are of special interest: the triple point and the critical point. At the triple point, the intersection of the coexistence curves, the three phases may all exist in mutual equilibrium. There is only one pressure and one temperature at which this situation can occur. The critical point is the endpoint of the vapor pressure curve, and it also occurs at a particular pressure P_c and at a particular temperature T_c. At any temperature higher than T_c, or any pressure higher than P_c, there is no distinction between liquid and gas; there is only a homogeneous fluid phase and (usually at very much higher pressure) a solid phase. The existence of a critical point, in fact, means that there is no way, in principle, to distinguish unambiguously between a gas and a liquid unless they happen to be coexisting in equilibrium. If they are coexisting, there will be a clear interface between regions of different but uniform density. But a homogeneous liquid can always be transformed into a homogeneous gas without passing through a phase transition, by following a path around the critical point. There is no known example of a critical point on a melting curve, so that, as far as we know, it is always possible to distinguish between the fluid phases on the one hand and the solid on the other. Apparently the kind of long-range order that characterizes the solid state is like pregnancy: either you have it or you do not; you cannot have a little bit of it or arrive at it gradually. The properties of matter at the gas-liquid critical point and other similar phenomena will be discussed at length in Chap. 6.

The same phases are shown in different perspective in the P-V diagram, Fig. 1.2.6. Here we find the solid at high pressures and low volumes (recall

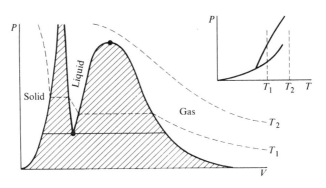

Fig. 1.2.6

that the number of particles is fixed), the gas at low pressure and large volume, and, again, the liquid in between. In the cross-hatched regions, more than one phase can coexist. The kind of path followed if we change V at constant temperature is shown by the dashed curves, for two temperatures T_1 and T_2, also indicated on the P-T diagram in the inset. Curves of constant temperature are called *isotherms*. The isotherm at T_1, which is below T_c, passes through all three phases. It must cross the coexistence curve at a fixed pressure, which means in the $P\text{-}V$ plot that V changes from its value in one phase to its value in the other at constant pressure; that is, the isotherm becomes horizontal. No horizontal portion is shown on the supercritical isotherm, at $T_2 > T_c$, although it presumably has one where it crosses the melting curve at much higher pressure.

The triple point of the $P\text{-}T$ diagram corresponds, on the $P\text{-}V$ diagram, to the single horizontal isotherm that touches all three phases. The critical point, however, is a point on this diagram as well, occurring at the maximum of the gas-liquid coexistence curve. There is thus a critical volume, V_c, as well as a critical temperature and pressure. At the critical point, the volumes of a specific amount of liquid and gas become equal to each other (or, more simply stated, the liquid and gas densities become equal). At the triple point, these volumes (or densities) may differ widely. The specific volumes of coexisting liquids and solids never become equal to each other, but they never differ by large factors either.

As mentioned above, coexistence curves in the $P\text{-}T$ plane are occasionally found to have negative slopes, even in very common materials. For example, the melting curve of water slopes backward, as sketched in Fig. 1.2.7. A

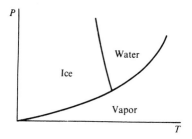

Fig. 1.2.7

coexistence curve with a negative slope is always associated with some sort of anomalous—that is, unusual—behavior in at least one of the phases. In the case of water, it happens because the liquid is denser than the solid in equilibrium, although the opposite is true for most substances. We shall return later to the thermodynamic connection between the densities and the melting curve.

There is one simple substance, helium, whose phase diagram is strikingly different from that of anything else (Fig. 1.2.8). Helium never freezes under its own vapor pressure; it is unique among all substances in that respect. The phases marked He II and He I that separate the solid from the gas are two different kinds of liquid, these, in turn, being separated by a further phase transition (notice there are two triple points on the diagram). The phase

1.2 Thermodynamics

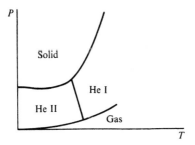

Fig. 1.2.8

diagram would be almost normal topologically if He II were a solid, but it is not. It is a very special liquid, called the *superfluid*, to which we shall return in Chap. 5. The melting curve starts out with zero slope at zero temperature, turns down, and goes through a minimum before finally turning up as shown. Liquid He I and the gas share a more or less normal critical point. Like the liquid and gas at the critical point, the specific volumes of liquid He I and II are equal to each other at the phase transition separating them.

Some progress in understanding these diagrams can be made by using the thermodynamic machinery we have developed. Suppose that we have a sealed container in which two phases, say liquid and gas, of the same material are coexisting, as in Fig. 1.2.9. Gravity has a weak influence on this problem,

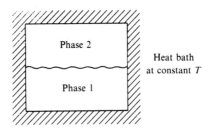

Fig. 1.2.9

which we shall neglect except insofar as it makes the denser of the two phases sink to the bottom of the container. We can think of this as a system at fixed T, divided into two subsystems, phase 1 and phase 2, which share the same total volume and can exchange particles. We have already worked out the conditions for equilibrium for this situation: for example, it differs from the situation in Fig. 1.2.4 only in that the movable partition is the interface between the two phases, so we know immediately that the temperatures and pressures of the two phases must be equal. There is an additional condition, since particles can be exchanged; we obtained it in Prob. 1.1a but let us do it now by our variational principle. For the two-phase system, Eq. (1.2.78) tells us that

$$\delta F = 0$$

will be the equilibrium condition. The total number of particles, $N_1 + N_2$, is fixed, but N_1 (or N_2) can vary:

$$0 = \frac{\partial F}{\partial N_1} = \frac{\partial}{\partial N_1}(F_1 + F_2) = \frac{\partial F_1}{\partial N_1} - \frac{\partial F_2}{\partial N_2} = \mu_1 - \mu_2 \quad (1.2.122)$$

The third condition is, as anticipated in Sec. 1.1, that the chemical potentials be equal.

It is clear from Eq. (1.2.35) or (1.2.36) that $\mu = \mu(P, T)$ quite generally. We may rewrite Eq. (1.2.122) as

$$\mu_1(P, T) = \mu_2(P, T) \quad (1.2.123)$$

This is a relationship between the temperatures and pressures at which coexistence can occur: one equation in two unknowns, which may be solved, say, for the pressure as a function of temperature, $P_0(T)$. Equation (1.2.123) is, in other words, the equation of one of the curves in Fig. 1.2.5. If we seek the conditions for three phases in simultaneous equilibrium, the pressures and temperatures must, as usual, be the same in all phases, and

$$\mu_1(P, T) = \mu_2(P, T) = \mu_3(P, T) \quad (1.2.124)$$

two equations in two unknowns that may be solved for P_3, T_3, the unique pressure and temperature of the triple point.

Chapters 2, 3, and 4 will concern themselves, respectively, with the gas, solid, and liquid phases of matter. A suitable program for us would be to work out from microscopic theory the general relation $\mu(P, T)$ for each of the phases in its own chapter and then at the end set them equal pair by pair and produce Fig. 1.2.5. Unfortunately, we cannot at present (nor perhaps ever) succeed in such a program. We shall, actually, succeed quite well in finding $\mu(P, T)$ for the gas, less well, under restricted conditions, for the solid, and poorly indeed for the liquid.

However, the fact that we are not clever enough to write down explicit relations of the form Eq. (1.2.123) does not diminish the importance of the fact that they exist in principle. Because they do, we can learn something useful about the behavior of matter. Suppose that we change the temperature of the system in Fig. 1.2.9 by a small amount, dT, and let it come into equilibrium. The pressure of each phase must change by the same amount, dP, and each chemical potential will change from its previous (unknown) value to a new (unknown) value. Since μ_1 and μ_2 must be equal to each

1.2 Thermodynamics

other both before and after the change, they must both suffer the same change:
$$d\mu_1 = d\mu_2 \tag{1.2.125}$$
or, using Eq. (1.2.35),
$$-s_1 \, dT + v_1 \, dP = -s_2 \, dT + v_2 \, dP \tag{1.2.126}$$
where we have written $s = S/N$ and $v = V/N$; we can call these quantities the molar entropy and molar volume. We solve for the slope of the coexistence curve,
$$\frac{dP_0}{dT} = \frac{s_2 - s_1}{v_2 - v_1} \tag{1.2.127}$$

Equation (1.2.127) is known as the Clausius-Clapeyron equation: the slope of the coexistence curve is given by the ratio of the change in molar entropy to the change in molar volume as we go from one phase to the other. The quantity $L = T(s_2 - s_1) = T \Delta s$ has the form of a heat and is called the latent heat of the transition. The Clausius-Clapeyron equation is sometimes written
$$\frac{dP_0}{dT} = \frac{L}{T(v_2 - v_1)} \tag{1.2.128}$$

Now that we know that the change in s and the change in v are important quantities, let us return to Fig. 1.2.5 again and imagine what we might generally expect to happen as we move across the diagram from lower right (rarified gas) toward the upper left, through the liquid phase and on into the compressed solid. Crudely speaking, we are lowering T and so should generally be moving into regions of lower s, and we are raising P and so should be causing the density to increase; thus v, like s, should be decreasing. To repeat: as we cross from gas to liquid to solid, we should, generally speaking, expect to be crossing into phases of successively lower entropy and lower volume. Since the change in entropy and the change in volume will usually have the same sign, we see from Eq. (1.2.127) that (dP_0/dT) should generally be positive, which accounts for our earlier remark that such is the case.

The exceptions, when the slope of the coexistence curve is not positive, mean that either s or v is not changing in the expected direction. The most familiar example is the melting curve of ice (solid H_2O), which has a negative slope as we have already seen in Fig. 1.2.8. It occurs because, along the melting curve, ice is slightly less dense than water (recall that ice cubes float— just barely—in water). Among the consequences is the fact that it is possible to ice-skate or ski; ice may be melted at constant temperature by compression, so that skis or ice-skate blades are self-lubricating. If the melting curve departed from the triple point with positive slope, it would be no more possible to skate on ice than on glass.

Another interesting example has already been mentioned. There is a minimum in the melting curve of helium (Fig. 1.2.8); starting at zero temperature, the slope is zero, negative, zero, then positive again. In this case, it is the entropy, rather than the density, that behaves anomalously. The zero slope at zero degrees means that the two phases have the same entropy, as they must if the Third Law is to be obeyed. The liquid, therefore, is able to become just as ordered as the solid, a unique circumstance that will bear further study later. Below the temperature of the minimum, in fact, the liquid actually has lower entropy than the solid at the same temperature and pressure, giving rise to the negative slope. At the minimum in the melting curve, which occurs at about 0.7°K, both phases have the same specific entropy, and at higher T the solid is the more ordered phase.

In Sec. 1.2e we investigated the conditions for equilibrium between the normal and superconducting states of a metal. Although the techniques used seem different from the ones we have applied to the gas-liquid-solid phase equilibria, the argument could have been carried out in a way that is strictly analogous to the way we have done it here (see Prob. 1.4). Equation (1.2.93) relating the super and normal free energy densities plays much the same role as Eq. (1.2.23) for the chemical potentials, and we may use it to derive something analogous to the Clausius-Clapeyron equation. We have, according to Eq. (1.2.93).

$$f_{0,\text{sc}} = f_{0,n} - \frac{H_c^2}{8\pi}$$

where $f_{0,\text{sc}}, f_{0,n}$, and H_c are all functions of T only. It follows that

$$\frac{df_{0,\text{sc}}}{dT} = \frac{df_{0,n}}{dT} - \frac{H_c}{4\pi}\frac{dH_c}{dT} \qquad (1.2.129)$$

But by definition, for either of the f_0's,

$$\frac{df_0}{dT} = \frac{1}{V}\left(\frac{\partial F}{\partial T}\right)_{H=0} \qquad (1.2.130)$$

so that Eq. (1.2.129) may be written

$$s_n - s_{\text{sc}} = -\frac{H_c}{4\pi}\frac{dH_c}{dT} \qquad (1.2.131)$$

where $s_n = S_n/V_n$ and $s_{\text{sc}} = S_{\text{sc}}/V_{\text{sc}}$. The latent heat of the transitions is

$$L = T(s_n - s_{\text{sc}}) \qquad (1.2.132)$$

so the superconducting equivalent of the Clausius-Clapeyron equation is

$$L = -\frac{TH_c}{4\pi}\frac{dH_c}{dT} \qquad (1.2.133)$$

1.3 Statistical Mechanics

A transition that has a latent heat is said to be a first-order phase transition. We may see from Fig. 1.2.3 that dH_c/dT is always negative, so L is always positive. However, it is also found that dH_c/dT is always finite, even in the limit as $H_c \to 0$, so that in that limit, $L \to 0$ as well. The transition is not first order if it occurs in zero field. To find out what happens in that case, we take the next derivative of Eq. (1.2.131):

$$\left(\frac{\partial s_n}{\partial T}\right)_{H=0} - \left(\frac{\partial s_{sc}}{\partial T}\right)_{H=0} = -\frac{1}{4\pi}\left[\left(\frac{dH_c}{dT}\right)^2 + H_c \frac{d^2H_c}{dT^2}\right]_{H=0}$$

$$= -\frac{1}{4\pi}\left(\frac{dH_c}{dT}\right)^2 \quad (1.2.134)$$

Although the entropies are equal, their slopes are not; there is a discontinuity in the zero-field specific heat:

$$c_H = T\left(\frac{\partial s}{\partial T}\right)_H \quad (1.2.135)$$

where $c_H = C_H/V$ and

$$c_{Hsc} - c_{Hn} = \frac{T_c}{4\pi}\left[\frac{dH_c}{dT}\right]^2_{T_c} \quad (1.2.136)$$

where T_c is the transition temperature in zero field. Equation (1.2.136) has been well verified experimentally. Figure 1.2.10 is a sketch of the heat capacity of a typical superconductor in zero field, showing the discontinuity at the transition.

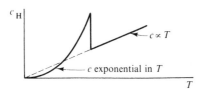

Fig. 1.2.10

1.3 STATISTICAL MECHANICS

The job of statistical mechanics is basically to count—or, more realistically, to find ways to avoid counting—the number of equally probable ways in which a system may divide up its energy. In principle, one might imagine writing down all the quantum solutions to the problem, then counting how many there are. For the perfect gas, for example, the procedure would be to choose a set of single-particle states, each with a definite energy, each to be occupied by some number of particles. A list would be drawn up of all the ways in which that could be done, using up all the particles and a fixed total energy. It is the number of such solutions that we are seeking. If the number of particles, and the energy, are macroscopic quantities, then the number we seek is so incredibly large that the job we have outlined is utterly unthinkable. The real job, then, is not literally to count but rather to interpret numbers. We shall have to find ways of breaking the problem down into factors that are, in some sense, countable.

The business of interpreting the meaning of numbers is a tricky one. Even the experts are sometimes fooled. It nevertheless goes on constantly, all around us, in all phases of life. An interesting example is the game of baseball.

There is little action in a baseball game. Those who criticize baseball on that account, however, do not understand the real nature of the enterprise. Most of the time is spent with all the men on the field going through a carefully prescribed ritual, with absolute, deeply satisfying predictability. On the other hand, every event occurring that is not perfectly predictable, every ball, every strike, every hit, every out, is recorded with meticulous and loving care for the benefit and enlightenment of posterity. The men who follow baseball professionally become true experts at interpreting the great mass of statistical data that emerges, masters at drawing the last atom of meaning out of each fluctuation.

Toward the end of the 1971 season, the San Francisco Giants were playing the Los Angeles Dodgers in a game televised in Los Angeles. In the first inning, Willie Mays, approaching the end of his illustrious carrer, hit a home run. Home runs in the first inning, one expects intuitively, are somewhat unusual events. The pitcher is fresh and strong, the batter's reflexes and muscular responses perhaps not yet entirely warmed up, there may be some mutual feeling out to be done between hitter and pitcher, and, besides, these explosive heroics are somehow more fitting late in the game. In any case, Willie hit a home run and that triggered a typical baseball statistician's response. The datum was produced that, of the 646 home runs Mays had hit, 122 of them had been hit in the first inning: 19 percent! In that most unlikely one-ninth of the innings, he had hit nearly one-fifth of his many home runs.

That item, for some reason, captured the fancy of the local baseball reporting community. It was discussed repeatedly during the remainder of the telecast and received much subsequent publicity in the newspapers. It meant, in the words of the Giants' publicity director, that "... Willie was always surprising pitchers in the first inning by going for the long ball before they were really warmed up." One sees the power of analysis of statistics to draw out hidden truths about the actions of men.

The datum, undoubtedly, was correct, but the interpretation was quite wrong. Throughout Mays' career, he had almost always batted third in the Giants' lineup (occasionally, he batted fourth). That meant that he almost always batted in the first inning. He averaged about four at bats per game,

1.3 Statistical Mechanics

meaning that approximately one-quarter of his at bats came in the first inning, an inning, you will recall, in which he could only manage to hit one-fifth of his home runs.

We do not wish to enter here into the reasons for Mays' special difficulties in trying to keep up a respectable production of first-inning home runs. More interesting is the universal misinterpretation of the meaning of the datum, by men who spend their professional lives interpreting just that kind of information. Of the millions of people in Los Angeles and San Francisco who must have heard or read the item, not one came to the aid of the befuddled sportswriters to point out their failure. It was simple, really. In the language we shall find ourselves using later in this section, they had the levels right but got the density of states all wrong.

Let us, now, return to physics. We shall begin our study of the statistical problem by reformulating it in a way that has two important advantages. First, it will allow us to deal with systems held at a given temperature rather than isolated with a given energy. Thus, starting from an enumeration of the quantum solutions, we shall have ways of writing the free energy in terms of its proper variables. Second, by turning our attention away from the isolated system, we shall eventually be able to avoid the difficulties of quantum uncertainties in the exact energies and numbers of states of systems, brought up in Sec. 1.1. The results of this reformulation of the problem constitute the essence of equilibrium statistical mechanics.

a. Reformulating the Problem

We began this chapter by imagining a body, such as a box of perfect gas, whose physical description involved specifying its volume V and number of particles N. Then, for each total energy E, there would be some number of possible quantum states the body could be in, Γ. In equilibrium, all these states would be equally likely, and the system would have entropy $S = k \log \Gamma$. Thus, in equilibrium, $S = S(E, V, N)$ or, inverting this, $E = E(S, V, N)$. Knowing E as a function of these variables, all other thermodynamic quantities of interest could be deduced. All that was left to do in order to make the connection between thermodynamics and quantum mechanics was to learn how to count the number of possible states at a given total energy.

We have already made a great deal of progress using this formulation; all of our thermodynamic arguments basically followed from just knowing that it was possible, in principle, to do the counting, without actually doing it. The counting of the possible states is the job of statistical mechanics.

Given a body to study with a certain V and N, we will generally want to know its thermodynamic properties not as a function of E but rather as a function of T, the temperature. What we want, in other words, instead of $E(S, V, N)$ is $F(T, V, N)$ or $\Omega(T, V, \mu)$. Of course, if we have $E(S, V, N)$, we can compute the other quantities, but it will make our work easier if we have a formulation in which F or Ω is given directly in terms of the quantum mechanical solutions for the body in question. We shall now work out such a formulation.

For this purpose, in order to establish a temperature and, if necessary, a chemical potential, we must imagine, as we have before, that the body or object of interest to us is embedded in a very large medium, whose T and μ will not be appreciably affected by what the object we are interested in is doing. We thus imagine the system as a whole, which is isolated, to consist of a subsystem—the object of our interest—and the medium, which is everything left over.

The system as a whole is subject to the formulation we started with. It has some volume V_0, number of particles N_0, and, being isolated, a total energy E_0. In equilibrium, it has Γ_0 possible states, all equally likely, and an entropy $S_0 = k \log \Gamma_0$. All the variables with subscript zero are properties of the system as a whole, and all are to be thought of strictly as constants, independent of whatever happens in the subsystem. If the subsystem is not in equilibrium with the medium, the entropy of the combined system will be less than S_0; we will call it S_t, so that $S_t \leq S_0$. Similarly, the number of possible states will be $\Gamma_t \leq \Gamma_0$.

The medium is to be thought of as being always internally in equilibrium. It has variables N', V', and E', giving Γ' choices and entropy S'. These quantities are connected by equilibrium equations; that is,

$$dE' = T\,dS' - P\,dV' + \mu\,dN' \tag{1.3.1}$$

but we should keep in mind that E', S', V', and N' all depend on the state of the subsystem, since the total E_0, V_0, and N_0 are fixed and, of course, S' depends on E', V', and N'. In other words, if the subsystem is in a state in which it has much more than its fair share of the energy, the medium, having less energy than it ought to, has fewer choices of what to do with it and, consequently, less entropy than it would otherwise have.

For the subsystem (which we have set out to study), we shall use unadorned variables, E, N, V, Γ, and S. We can take our subsystem to have a fixed volume within the medium (the subsystem cannot be defined unless we fix something). Energy and particles can enter and leave the subsystem as it fluctuates among its possible states. However, we must put in the restriction that the subsystem is small so that T and μ (of the medium) remain fixed during these fluctuations. We will use the approximation that, for each possible quantum state of the subsystem, α,

$$\begin{aligned} E_\alpha &\ll E_0 \\ N_\alpha &\ll N_0 \end{aligned} \tag{1.3.2}$$

Here α is the index of a single quantum state, in which the subsystem has

1.3 Statistical Mechanics

N_α particles and energy E_α. Formally, states of the system violating Eq. (1.3.2) should be allowed, but, for a small subsystem, there are so few of them that little error is introduced by ignoring them. Since we are making these restrictions, even though S', E', and N' depend on E_α and N_α, they will change very little, and we are safe in assuming that their derivatives, $\partial E'/\partial S' = T$ and $\partial E'/\partial N' = \mu$, are constant.

Under any specified set of conditions, in which the subsystem has Γ choices and the medium Γ', the total system will have

$$\Gamma_t = \Gamma\Gamma' \qquad (1.3.3)$$

and

$$S_t = S + S' \qquad (1.3.4)$$

In equilibrium, $\Gamma_t = \Gamma_0$, $S_t = S_0$, and the probability of the system as a whole being in any particular quantum state is

$$w_{eq} = \frac{1}{\Gamma_0} \qquad (1.3.5)$$

Every possible state has this same probability. Suppose, however, that instead of imagining the system to be in equilibrium, we specify that the subsystem is in one particular quantum state, α. Then the number of choices of the medium depends on E_α and N_α; let us call Γ'_α the number of choices of the medium when the subsystem is in the state α. Since we have specified the quantum state of the subsystem, it has only one possibility, so that, for the system as a whole,

$$\Gamma_{t\alpha} = 1 \cdot \Gamma'_\alpha \qquad (1.3.6)$$

In other words, of the Γ_0 possible states of the system as a whole, there are $\Gamma_{t\alpha} = \Gamma'_\alpha$ that find the subsystem in the particular state α. Since, in equilibrium, all Γ_0 states are equally likely, the random probability of the subsystem finding itself in state α is just the number of ways that can happen, Γ'_α, divided by the number of ways that anything can happen, Γ_0:

$$w_\alpha = \frac{\Gamma'_\alpha}{\Gamma_0} \qquad (1.3.7)$$

where w_α is the probability of finding the subsystem in state α under equilibrium fluctuations. Notice that, unlike w_{eq} of Eq. (1.3.5), w_α is most certainly not the same for all α. For example, as mentioned above, the more energy E_α the subsystem has in state α, the less energy is left for the medium, the fewer choices Γ'_α it has, and so the less likely is the state α.

The quantity w_α is of central importance in statistical mechanics. For example, if some quantity f (f might be the energy of the subsystem, say, or the number of particles) has some definite value in each state α, f_α, then the thermodynamic, or average, value of f will be

$$\bar{f} = \sum_\alpha w_\alpha f_\alpha \qquad (1.3.8)$$

subject to the condition that

$$\sum_\alpha w_\alpha = 1 \qquad (1.3.9)$$

Equation (1.3.8) simply says that the average value of f is obtained by weighing its value in each state by the probability of the state, whereas Eq. (1.3.9) requires that the subsystem always be in some state.

If the subsystem is in the state α, the medium has entropy S'_α,

$$S'_\alpha = k \log \Gamma'_\alpha \qquad (1.3.10)$$

where the notation S'_α simply means

$$S'_\alpha = S'(E_0 - E_\alpha, N_0 - N_\alpha) \qquad (1.3.11)$$

This follows from Eq. (1.3.1), which tells us that, for a subsystem of constant volume, $S' = S'(E', N')$. Since $S_0 = k \log \Gamma_0$, we can use Eq. (1.3.10) to write

$$S_0 - S'_\alpha = -k \log \frac{\Gamma'_\alpha}{\Gamma_0} = -k \log w_\alpha \qquad (1.3.12)$$

where we have used Eq. (1.3.7) in the last step. The quantity $S_0 - S'_\alpha$ is not the entropy of the subsystem, for we know the subsystem to be in the state α, so that it has no choices—its entropy is zero. However, if we average S'_α over all states of the subsystem, we do get the equilibrium entropy of the medium, and the subsystem equilibrium entropy S,

$$S = \overline{S_0 - S'_\alpha} = S_0 - \overline{S'_\alpha} = \overline{-k \log w_\alpha}$$

$$= \sum_\alpha w_\alpha(-k \log w_\alpha) \qquad (1.3.13)$$

Here we have used Eq. (1.3.8) in the last step, regarding the quantity $S_0 - S'_\alpha$ to be a property of each state of the subsystem.

From Eq. (1.3.12), before an average is taken, we have

$$w_\alpha = \exp\left(-\frac{S_0 - S'_\alpha}{k}\right) = A \exp\left(\frac{S'_\alpha}{k}\right) \qquad (1.3.14)$$

where A is a constant, independent of α but determined by the normalization condition, Eq. (1.3.9). Using Eq. (1.3.11), we can write S'_α as

$$S'_\alpha = S'(E_0 - E_\alpha, N_0 - N_\alpha)$$

$$= S'(E_0, N_0) - \left(\frac{\partial S'}{\partial E'}\right)_{V',N'} E_\alpha - \left(\frac{\partial S'}{\partial N'}\right)_{V',E'} N_\alpha \qquad (1.3.15)$$

1.3 Statistical Mechanics

We have made use of the approximations, Eq. (1.3.2), in writing Eq. (1.3.15). Evaluating $\partial S'/\partial E'$ and $\partial S'/\partial N'$ from Eq. (1.3.1), we get

$$S'_\alpha = \text{constant} - \frac{E_\alpha}{T} + \frac{\mu N_\alpha}{T} \quad (1.3.16)$$

where constant = $S'(E_0, N_0)$ is the entropy that the medium would have if it had all the energy and particles, and does not depend on α. Substituting Eq. (1.3.16) into Eq. (1.3.14) gives

$$w_\alpha = B \exp\left(-\frac{E_\alpha - \mu N_\alpha}{kT}\right) \quad (1.3.17)$$

where B is another constant, again subject to the normalization condition, Eq. (1.3.9):

$$B = \frac{1}{\sum_\alpha \exp\left[-(E_\alpha - \mu N_\alpha)/kT\right]} \quad (1.3.18)$$

Equations (1.3.17) and (1.3.18) allow us to find thermodynamic average quantities of the subsystem if we have enumerated the quantum states α, with E_α and N_α for each one. For example, the average number of particles is

$$N = \frac{\sum_\alpha N_\alpha \exp\left[-(E_\alpha - \mu N_\alpha)/kT\right]}{\sum_\alpha \exp\left[-(E_\alpha - \mu N_\alpha)/kT\right]} \quad (1.3.19)$$

and the average energy is

$$E = \frac{\sum_\alpha E_\alpha \exp\left[-(E_\alpha - \mu N_\alpha)/kT\right]}{\sum_\alpha \exp\left[-(E_\alpha - \mu N_\alpha)/kT\right]} \quad (1.3.20)$$

However, Eq. (1.3.19) formally gives us $N(T, V, \mu)$ (where V is implicit in the enumeration of the E_α's) and Eq. (1.3.20) gives $E(T, V, \mu)$. These are not the proper variables; the proper variables of E are S, V, and N. We can do better if we make use of Eqs. (1.3.13) and (1.3.17).

$$\begin{aligned} S &= -k \sum_\alpha w_\alpha \log w_\alpha \\ &= -k \log B \sum_\alpha w_\alpha + \frac{1}{T} \sum_\alpha w_\alpha E_\alpha - \frac{\mu}{T} \sum_\alpha w_\alpha N_\alpha \\ &= -k \log B + \frac{E - \mu N}{T} \end{aligned} \quad (1.3.21)$$

or
$$kT \log B = E - TS - \mu N = F - \mu N = \Omega \quad (1.3.22)$$

Substituting Eq. (1.3.18) into Eq. (1.3.22), we have

$$\Omega = -kT \log \sum_\alpha \exp\left(-\frac{E_\alpha - \mu N_\alpha}{kT}\right) \quad (1.3.23)$$

Equation (1.3.23) gives us $\Omega(T, V, \mu)$ for the subsystem, which is just what we need. If the quantum mechanical solutions, (E_α, N_α), are known, we can find all the thermodynamic functions.

In the formulation we have just worked out, V, T, and μ are held fixed, while E, N, and S are to be found by averaging over the possible states of the subsystem in equilibrium [the pressure P is fixed, since, in the medium, $P = P(T, \mu)$]. If we wish instead to deal with a body in which N as well as T and V are fixed, it is not necessary to start over again. Instead, we simply specify that

$$N_\alpha = N \quad \text{for all } \alpha \qquad (1.3.24)$$

Then Eq. (1.3.23) becomes

$$\Omega = -kT \log \exp\left(\frac{\mu N}{kT}\right) \sum_\alpha \exp\left(-\frac{E_\alpha}{kT}\right)$$

$$= -\mu N - kT \log \sum_\alpha \exp\left(-\frac{E_\alpha}{kT}\right) \qquad (1.3.25)$$

or since $F = \Omega + \mu N$,

$$F = -kT \log \sum_\alpha \exp\left(-\frac{E_\alpha}{kT}\right) \qquad (1.3.26)$$

Equation (1.3.26) gives us $F(T, V)$ for bodies of a given N. Instead of Eqs. (1.3.17), (1.3.18), and (1.3.20) we have, respectively,

$$w_\alpha = C \exp\left(-\frac{E_\alpha}{kT}\right) \qquad (1.3.27)$$

$$C = \frac{1}{\sum_\alpha \exp(-E_\alpha/kT)} \qquad (1.3.28)$$

$$E = \frac{\sum_\alpha E_\alpha \exp(-E_\alpha/kT)}{\sum_\alpha \exp(-E_\alpha/kT)} \qquad (1.3.29)$$

Depending on whether the number of particles is taken as variable or not, Eqs. (1.3.17) and (1.3.18) or Eqs. (1.3.27) and (1.3.28) are known as the Gibbs distribution.

Let us define the quantities

$$Z = \sum_\alpha \exp\left(-\frac{E_\alpha}{kT}\right) \qquad (1.3.30)$$

and

$$\mathcal{L} = \sum_\alpha \exp\left(-\frac{E_\alpha - \mu N_\alpha}{kT}\right) \qquad (1.3.31)$$

1.3 Statistical Mechanics

Z is called the *partition function*, and $\mathscr{L}$ is sometimes called the *grand partition function*. In terms of the quantities we have used up to now,

$$F = -kT \log Z \tag{1.3.32}$$

and
$$\Omega = -kT \log \mathscr{L} \tag{1.3.33}$$

or
$$Z = e^{-F/kT} = \frac{1}{C} \tag{1.3.34}$$

and
$$\mathscr{L} = e^{-\Omega/kT} = \frac{1}{B} \tag{1.3.35}$$

We shall make extensive use of the properties of Z and $\mathscr{L}$.

The arguments we have just given form the core of statistical mechanics. The basic approach is simple and worth keeping in mind. In equilibrium, all states of the system are equally likely. However, the states of the subsystem are not equally likely—the more energy (and particles) the subsystem has, the fewer are left for the medium, and hence the less likely the state is. This is the crucial part of the argument; we found the probability of a given quantum state of the subsystem as a function of temperature not by watching what the subsystem was doing but by watching what the medium was doing. We got the message from the medium.

b. Some Comments

In order to simplify the discussion we wish to make here, let us consider only bodies of fixed N; the entire derivation of the previous subsection is easily reduced to this case.

The results we obtained all depend on finding w_α, the probability of finding the subsystem in the state α. In the state α, the subsystem has energy E_α, but α is not, in general, the only state of the subsystem with energy E_α; there will usually be many states with the same energy. In other words, the subsystem, which can be a macroscopic body, will usually have a high degree of quantum degeneracy—that is, many states at the same energy. In fact, finding the number of states with the same energy in a macroscopic body was just the counting problem we started out with. Nevertheless, it must be stressed that w_α is the probability that the body is in one single state; it is not the probability that the body have energy E_α but rather the probability that it be in a particular one of the states that have energy E_α.

Yet we found the value of w_α by studying the number of choices left to the medium Γ'_α and S'_α, and these quantities depend only on E_α, not on α itself. How can the medium tell us the probability for a single state when it does not know what state the body is in? Perhaps we can help clarify this issue by starting out again but this time seeking $w(E_\alpha)$, which we define to be the probability that the subsystem have energy E_α rather than the probability that it be in any particular state with this energy. Let us suppose that there are ρ_{E_α} states of the subsystem that have energy E_α. By our fundamental postulate, all these states are equally likely when the subsystem has energy E_α; that is, the probability of any of the others is the same as the probability of α, or w_α. It follows that

$$w(E_\alpha) = \rho_{E_\alpha} w_\alpha \qquad (1.3.36)$$

Now, when the subsystem has energy E_α, the medium is left with Γ'_{E_α} choices, but as we have already argued,

$$\Gamma'_{E_\alpha} = \Gamma'_\alpha \qquad (1.3.37)$$

That is, the entropy of the medium is the same whether we specify that the subsystem is in state α or merely that it have energy E_α. On the other hand, if we specify only the energy of the subsystem, the subsystem itself now has ρ_{E_α} choices, so for the combined system,

$$\Gamma_{tE_\alpha} = \rho_{E_\alpha} \Gamma'_\alpha \qquad (1.3.38)$$

Thus, the probability that the subsystem have energy E_α under equilibrium fluctuations is

$$w(E_\alpha) = \frac{\Gamma_{tE_\alpha}}{\Gamma_0} = \frac{\rho_{E_\alpha} \Gamma'_\alpha}{\Gamma_0} \qquad (1.3.39)$$

Comparison of Eqs. (1.3.36) and (1.3.39) confirms that

$$w_\alpha = \frac{\Gamma'_\alpha}{\Gamma_0}$$

is actually the probability that the subsystem be in one particular state.

The rest of the derivation of the last subsection now goes through as before. However, we can use ρ_{E_α} to help give us some insight into how the averaging over states is actually being done. For example, according to Eq. (1.3.27), w_α is an exponentially decreasing function of E_α; at any finite temperature, the probability of any state falls monotonically as the energy of the state goes up. Yet we know perfectly well that the most likely energy of a body is not its lowest possible energy—if it were, we would constantly be finding macroscopic bodies in their ground states. Although the probability of any particular state, α, declines with energy, the number of states with a particular energy, ρ_{E_α}, rises rapidly with energy, so that the probability of the body having a given energy

$$w(E_\alpha) = \frac{1}{Z} \rho_{E_\alpha} \exp\left(-\frac{E_\alpha}{kT}\right) \qquad (1.3.40)$$

will usually have a very sharp maximum value at $E_\alpha = E$, the average energy; just how sharp that maximum is we shall compute a bit later.

1.3 Statistical Mechanics

In forming the partition function, Eq. (1.3.30), we sum over all states of the body. At each energy E_α, there will be ρ_{E_α} equal terms, $e^{-E_\alpha/kT}$. Thus, instead of summing over states α, we could sum over energies E_α,

$$Z = \sum_\alpha \exp\left(-\frac{E_\alpha}{kT}\right) = \sum_{E_\alpha} \rho_{E_\alpha} \exp\left(-\frac{E_\alpha}{kT}\right) \qquad (1.3.41)$$

Most of the contribution to the partition function will thus come from that same very small range of energies where $w(E_\alpha)$ is pcaked.

For a macroscopic body at finite temperature, the energy levels that are important in forming $w(E_\alpha)$ and Z will be so closely spaced that very little error is introduced by thinking of E_α as a continuous variable. We will use the notation $\mathscr{E}$ for this purpose (reserving E for the equilibrium average value of the energy). Then the proper language to use (in accordance with the remarks at the end of Sec. 1.1) is as follows.

The probability of the body having energy between $\mathscr{E}$ and $\mathscr{E} + d\mathscr{E}$ is $w(\mathscr{E}) \, d\mathscr{E}$, normalized by

$$\int_0^\infty w(\mathscr{E}) \, d\mathscr{E} = 1 \qquad (1.3.42)$$

The number of states in the same range is $\rho(\mathscr{E}) \, d\mathscr{E}$, where $\rho(\mathscr{E})$ is called the density of states. $w(\mathscr{E})$ is given in terms of $\rho(\mathscr{E})$ by

$$w(\mathscr{E}) = \frac{1}{Z} \rho(\mathscr{E}) e^{-\mathscr{E}/kT} \qquad (1.3.43)$$

and

$$Z = \int_0^\infty \rho(\mathscr{E}) e^{-\mathscr{E}/kT} \, d\mathscr{E} \qquad (1.3.44)$$

The average value of a quantity $f(\mathscr{E})$ is

$$\bar{f} = \int_0^\infty f(\mathscr{E}) w(\mathscr{E}) \, d\mathscr{E} = \frac{1}{Z} \int_0^\infty f(\mathscr{E}) \rho(\mathscr{E}) e^{-\mathscr{E}/kT} \, d\mathscr{E} \qquad (1.3.45)$$

A comment in passing: by means of the approximations, Eq. (1.3.2), the various sums and integrals are limited to values of $\mathscr{E}$ or E_α that are small compared to E_0. However, little error is introduced by taking them up to infinite values of the energy because the exponential, $e^{-\mathscr{E}/kT}$, will always cut them off; in fact, as we have seen, most of the contributions will come from a very narrow range about a finite value of the energy.

With the energy written as a continuous variable, the free energy becomes

$$F = -kT \log \int_0^\infty \rho(\mathscr{E}) e^{-\mathscr{E}/kT} \, d\mathscr{E} \qquad (1.3.46)$$

and so the entire problem of statistical mechanics has been reduced to finding the function $\rho(\mathscr{E})$. (Notice that $\mathscr{E}$ itself is now just a dummy variable of integration.) That might seem to be a significant advance, but actually it is not. In fact, it is identical to the problem as formulated in Sec. 1.1. We could then do our thermodynamics if we could count the number of states of a body at any given energy. $\rho(\mathscr{E})$ is simply the number of states of the body as a function of energy—we are left with almost the same counting problem that we had in the beginning.

Before finishing these comments, we would like to pull one more tidbit out of the arguments of the last subsection. Since, as we have stressed, all the states of the subsystem that play an important role are concentrated in a narrow range of the parameter $\mathscr{E}$ or E_α, it should be possible to take advantage of that fact and develop a formulation that concentrates on these states. This we shall now do.

Under any given set of circumstances, the entropy of the system as a whole is given by

$$S_t = k \log \Gamma_t \qquad (1.3.47)$$

According to Eq. (1.3.38), if the subsystem has energy E_α, the system entropy will be

$$S_t(E_\alpha) = k \log \rho_{E_\alpha} \Gamma'_\alpha \qquad (1.3.48)$$

so that

$$S_0 - S_t(E_\alpha) = -k \log \frac{\rho_{E_\alpha} \Gamma'_\alpha}{\Gamma_0} = -k \log w(E_\alpha) \qquad (1.3.49)$$

In other words,

$$w(E_\alpha) = A \exp\left[\frac{S_t(E_\alpha)}{k}\right] \qquad (1.3.50)$$

where $A = e^{-S_0/k}$ is the same constant used in Eq. (1.3.14). We may thus write the probability of the subsystem having E_α in terms of the entropy of the total system when we specify that the subsystem has E_α. Equation (1.3.50) is true not only for the energy but also for any quantity whose value in the subsystem affects the medium. Thus, if some quantity x is a property of the subsystem, and we can write

$$S'_x = S'_{\bar{x}} - \frac{\partial S'}{\partial x}(\bar{x} - x) \qquad (1.3.51)$$

where $\bar{x}$ is the average value of x, then the arguments we have given for E_α follow for x as well, and we can write

$$w(x) = A \exp\left[\frac{S_t(x)}{k}\right] \qquad (1.3.52)$$

Now, according to the arguments we have given, $S_t(x)$ will have a sharp maximum at $x = \bar{x}$:

$$\left(\frac{\partial S_t}{\partial x}\right)_{x=\bar{x}} = 0 \qquad \left(\frac{\partial^2 S_t}{\partial x^2}\right)_{x=\bar{x}} < 0 \qquad (1.3.53)$$

1.3 Statistical Mechanics

We can therefore expand S_t about $\bar{x}$,

$$S_t(x) = S_t(\bar{x}) + \left(\frac{\partial S_t}{\partial x}\right)_{x=\bar{x}} (x - \bar{x}) + \frac{1}{2}\left(\frac{\partial^2 S_t}{\partial x^2}\right)_{x=\bar{x}} (x - \bar{x})^2$$

$$= S_t(\bar{x}) - \tfrac{1}{2}\beta(x - \bar{x})^2 \qquad (1.3.54)$$

where we have used (1.3.53) in the last step; the coefficient β is positive. Substituting Eq. (1.3.54) into (1.3.52), we have

$$w(x) = D \exp\left[-\frac{\beta(x - \bar{x})^2}{2k}\right] \qquad (1.3.55)$$

subject to

$$\int w(x)\, dx = 1 = D\sqrt{\frac{2\pi k}{\beta}} \qquad (1.3.56)$$

which fixes the constant D. This is a Gaussian form, with most of the contribution to $w(x)$ coming from $x \approx \bar{x}$ as expected. We shall return to it later because it will be useful in calculating the mean probabilities of fluctuations of various quantities from their equilibrium values. This formulation, basically Eqs. (1.3.55) and (1.3.56), was the first publication, in 1904, of a 25-year-old patent clerk named Albert Einstein.

c. Some Properties of Z and $\mathscr{L}$

The quantities Z and $\mathscr{L}$, once they have been formed from the microscopic states, contain within them all the thermodynamic information about the systems to which they refer. From Eqs. (1.3.32) and (1.3.33), $F(T, V, N)$ and $\Omega(T, V, \mu)$ may be written directly in terms of $Z(T, V, N)$ and $\mathscr{L}(T, V, \mu)$, respectively. Given these, it is easy to write formulas for any of the other thermodynamic quantities. For example, to form the energy E from Z, we write

$$E = F + TS$$

$$S = -\frac{\partial F}{\partial T} = k \log Z + \frac{kT}{Z}\frac{\partial Z}{\partial T} \qquad (1.3.57)$$

or

$$E = -kT \log Z + kT \log Z + \frac{kT^2}{Z}\frac{\partial Z}{\partial T}$$

$$E = \frac{kT^2}{Z}\left(\frac{\partial Z}{\partial T}\right)_{V,N} \qquad (1.3.58)$$

Equation (1.3.58) gives us $E(T, V, N)$, which are not the proper variables, but Eq. (1.3.57) gives us $S(T, V, N)$, so, in principle, T may be eliminated between these two equations to give $E(S, V, N)$. As another example, we can find N from $\mathscr{L}(T, V, \mu)$:

$$N = -\frac{\partial \Omega}{\partial \mu} = \frac{kT}{\mathscr{L}}\left(\frac{\partial \mathscr{L}}{\partial \mu}\right)_{T,V} \quad (1.3.59)$$

where N is the average number of particles; recall that N is variable in the Ω formulation.

Equations (1.3.58) and (1.3.59) do not give us new results in terms of the microscopic states; they merely confirm the self-consistency of our formulas. Taking the temperature derivative of Z from Eq. (1.3.30), we obtain

$$\frac{\partial Z}{\partial T} = \frac{1}{kT^2} \sum_\alpha E_\alpha \exp\left(-\frac{E_\alpha}{kT}\right) \quad (1.3.60)$$

which may be substituted into Eq. (1.3.58) to give

$$E = \frac{1}{Z} \sum_\alpha E_\alpha \exp\left(-\frac{E_\alpha}{kT}\right) \quad (1.3.61)$$

This last equation is identical to Eq. (1.3.29). Similarly, taking the μ derivative of Eq. (1.3.31) gives

$$\frac{\partial \mathscr{L}}{\partial \mu} = \frac{1}{kT} \sum_\alpha N_\alpha \exp\left(-\frac{E_\alpha - \mu N_\alpha}{kT}\right) \quad (1.3.62)$$

which, upon substitution into Eq. (1.3.59), gives

$$N = \frac{1}{\mathscr{L}} \sum_\alpha N_\alpha \exp\left(-\frac{E_\alpha - \mu N_\alpha}{kT}\right) \quad (1.3.63)$$

and this tells us, reasonably enough, that N is the average value of N_α. In other cases, we may get less obvious results. For example,

$$P = -\frac{\partial F}{\partial V} = \frac{kT}{Z}\frac{\partial Z}{\partial V} = -\frac{1}{Z}\sum_\alpha \frac{dE_\alpha}{dV}\exp\left(-\frac{E_\alpha}{kT}\right) \quad (1.3.64)$$

so that the pressure turns out to be the average value of the quantity dE_α/dV (you should have found this out for the perfect gas in Prob. 1.1b). In performing these manipulations, we have made use of the fact that the E_α and N_α never depend on quantities like T and μ. They are, instead, definite, non-statistical properties of individual quantum states.

The partition function Z is called that because of an interesting and useful property it has: it partitions under certain circumstances. If the energies E_α are made up of independent contributions, say,

$$E_\alpha = H_i + G_j \quad (1.3.65)$$

1.3 Statistical Mechanics

so that α is a notation for two sets of quantum numbers, i and j, needed to specify the state, then Eq. (1.3.30) becomes

$$\begin{aligned} Z &= \sum_{i,j} \exp\left(-\frac{H_i + G_j}{kT}\right) \\ &= \sum_i \sum_j \exp\left(-\frac{H_i}{kT}\right) \exp\left(-\frac{G_j}{kT}\right) \\ &= \sum_i \exp\left(-\frac{H_i}{kT}\right) \sum_j \exp\left(-\frac{G_j}{kT}\right) \\ &= Z_H Z_G \end{aligned} \qquad (1.3.66)$$

The partition function has become a product of separate partition functions for the H and G contributions to the energy. H and G could be, for example, kinetic and rotational energies of the body as a whole, or they could be the energies of two independent parts of which the body is composed. This property of the partition function is rather analogous to the behavior of the wave function in quantum mechanics. If the Hamiltonian has independent contributions in quantum mechanics, the wave function is a product of corresponding independent factors. The partition function partitions into independent factors if the energy, which is the expectation value of the Hamiltonian, can be written as a sum of independent parts.

d. Distinguishable States: Application to the Perfect Gas

In order to apply our equations to the perfect gas, we must somehow organize the problem by deciding how we are going to sort out and describe all the possible individual quantum states of the many-particle system. There are two principal techniques, each of which has certain drawbacks. We can concentrate either on the particles themselves, giving the quantum numbers of each particle to describe a state, or we can concentrate on the single-particle states (i.e., the possible sets of quantum numbers), giving the number of particles, or populations, of each single-particle state to describe the state of the many-particle system.

In the first method we are basically considering the particles to be independent subsystems of the gas, with the state of the gas given by the state of each of its subsystems. Unfortunately, this procedure leads us to count each possible state of the gas far too many times, for we wind up counting separately states that differ only in that some individual particles have exchanged quantum numbers with each other. Since the particles are identical, there is, from the point of view of the many-particle system, no distinction between these states, and they should not be counted separately.

The second method, in which we say only how many particles have each set of quantum numbers, is not subject to that difficulty. The difference is that in the first method we decide which particles are in each single-particle state, thus distinguishing between the particles, whereas in the second method we decide only how many are in each state, not distinguishing between them. The second method leads unambiguously to a correct enumeration of the possible quantum states of the many-particle system. The drawback to this procedure is that it cannot be generalized for application to any problem other than the perfect gas. The procedure depends on having single-particle states that are quite independent of what all the particles are doing. In any interacting system the possible states of each particle depend on the state of all the other particles (in particular, where they are spatially located) and thus this kind of enumeration is not possible.

The first method, which is more easily generalized, can be patched up and made to work under certain circumstances. Suppose that the system has enough energy per particle so that the number of single-particle states in the range of energies that will make a contribution to the statistical properties is very much larger than the number of particles available to occupy them. We shall investigate the quantitative criteria for this condition in the gas shortly, but it will generally be the case when the temperature is reasonably high and the density reasonably low. Under these conditions, each single-particle state will usually be unoccupied, and its probability of being occupied by more than one particle at a time is negligibly small. If there are N particles in the gas, they occupy N different single-particle states. In constructing a many-body state from the particles, we have N choices of which single-particle state to put the first particle in, $N - 1$ choices of where to put the second particle, and so on; there are $N!$ ways of constructing that kind of many-particle state from the N particles. Thus, if we consider all possible states of N particles individually, we wind up, under these conditions, counting each distinct many-body state $N!$ times. Knowing how many times we have overcounted the many-particle states, we can easily correct the error. This argument does not depend on each single-particle state being independent of what the other particles are doing. It works equally well for interacting systems. So long as we can ignore the possibility of two or more particles doing exactly the same thing, $N!$ is still the difference between how many particles and which particles are doing each thing.

To summarize, then, the situation is this: for the perfect gas at high temperature and low density, or for most interacting systems of atoms (or molecules), we can count up all the possible states of the individual particles separately and then divide by $N!$ to enumerate the many-body states correctly. We cannot do the same for the perfect gas when the temperature is low or the density high. However, the perfect gas, under any circumstance, can be handled by simply specifying the populations of all single-particle states as a means of describing each distinct many-body state. Broadly speaking, the $N!$ approach is applicable to matter when it behaves classically, and some

1.3 Statistical Mechanics

other method is necessary when quantum effects become important. This problem of the indistinguishability of identical particles does not arise in the case of solids, where the particles may not be distinguished but the lattice sites to which they are attached may be. The most important example in nature of a system that cannot be handled by either the $N!$ or the populations-of-states method is liquid helium, which is a quantum fluid. We shall deal with that problem separately in Chap. 5.

Let us consider the perfect gas in the approximation in which either of the two methods ought to work. In order to have clear names for our separate models and approximations, we shall call this special case of the perfect gas, when the probability of multiple occupation of any single-particle state is negligible, the *ideal gas*. We shall later study cases where the gas is perfect but nonideal.

If we think of one particle of the ideal gas as a subsystem, we have for the single-particle partition function

$$Z_1 = \sum_q \exp\left(-\frac{\varepsilon_q}{kT}\right) \quad (1.3.67)$$

where ε_q is given by Eqs. (1.1.1) to (1.1.4) and $\sum_q$ denotes a sum over all possible values of the set of three quantum numbers, ℓ_x, ℓ_y, and ℓ_z. Making use of Eq. (1.3.66), the many-body partition function will have in it N factors identical to Z_1, but in forming it this way, we will have counted each distinct state with no multiple occupation $N!$ times. Other distinct states (i.e., the ground state of the whole system) are counted fewer times, but we expect them to make little contribution to the thermodynamic properties under the conditions we wish to investigate [the many-body density of states, ρ_{E_α}, in Eq. (1.3.41) will be so small for these states that we ignore them]. We therefore approximate the many-body partition function by

$$Z = \frac{1}{N!}(Z_1)^N \quad (1.3.68)$$

so that $\quad F = -NkT \log Z_1 + kT \log N! \quad (1.3.69)$

The term containing $N!$ will contribute (as it should) to the entropy, $S = -\partial F/\partial T$, but not to the equation of state,

$$P(T, V) = -\left(\frac{\partial F}{\partial V}\right)_T = \frac{NkT}{Z_1}\frac{\partial Z_1}{\partial V} \quad (1.3.70)$$

Using Eqs. (1.3.67) and (1.1.1) to (1.1.4), we get

$$\frac{\partial Z_1}{\partial V} = -\frac{1}{kT}\sum_q \frac{\partial \varepsilon_q}{\partial V}\exp\left(-\frac{\varepsilon_q}{kT}\right) = \frac{2/3}{VkT}\sum_q \varepsilon_q \exp\left(-\frac{\varepsilon_q}{kT}\right) \quad (1.3.71)$$

where we have used $L = V^{1/3}$ in Eq. (1.1.2). Putting Eq. (1.3.71) into (1.3.70) gives

$$P = \frac{2}{3}\frac{N}{V}\bar{\varepsilon} = \frac{2}{3}\frac{E}{V} \tag{1.3.72}$$

where
$$\bar{\varepsilon} = \frac{1}{Z_1}\sum_q \varepsilon_q \exp\left(-\frac{\varepsilon_q}{kT}\right) \tag{1.3.73}$$

Equation (1.3.72), $PV = \frac{2}{3}E$, could be shown earlier from much simpler (and more general) considerations (see Prob. 1.1b). In order to obtain the equation of state, $P = P(T, V)$, we must perform the sums in Eq. (1.3.73) to get $\bar{\varepsilon}(T, V)$. In the notation of continuous variables, Eq. (1.3.73) becomes

$$\bar{\varepsilon} = \frac{\int \varepsilon \rho(\varepsilon) e^{-\varepsilon/kT}\, d\varepsilon}{\int \rho(\varepsilon) e^{-\varepsilon/kT}\, d\varepsilon} \tag{1.3.74}$$

where ε is the continuous version of the single-particle energy and $\rho(\varepsilon)$ is the single-particle density of states. We must still perform a sum or integral over states, but the problem has been greatly simplified, for now we need only do it for single-particle rather than many-particle states.

In the alternative method we focus our attention on the single-particle states, rather than on the particles themselves, by choosing a single-particle state as a subsystem. The number of particles in the state n_q must be thought of as a variable, and the energy of the subsystem is $n_q \varepsilon_q$, where ε_q is the energy of a single particle in that state and is independent of n_q. Using the variable-number formalism, we write for the Ω of the state q,

$$\Omega_q = -kT \log \sum_{\text{states}} \exp\left(-\frac{n_q \varepsilon_q - n_q \mu}{kT}\right) \tag{1.3.75}$$

The $\sum_{\text{states}}$ is a sum over the states of the subsystem, which is itself a single-particle state. The only thing that can change is n_q itself, so n_q is the summation index:

$$\Omega_q = -kT \log \sum_{n_q} \exp\left[-\frac{n_q(\varepsilon_q - \mu)}{kT}\right] \tag{1.3.76}$$

where $n_q = 0, 1, 2, \ldots$. This choice of subsystem is no more peculiar than was the choice of one particle as a subsystem earlier. In assigning thermodynamic average properties to one of these essentially microscopic entities, we can imagine, for example, that we are averaging over a very long time. Once we compute Ω_q by means of Eq. (1.3.76), we can find Ω for the system as a whole from

$$\Omega = \sum_q \Omega_q \tag{1.3.77}$$

1.3 Statistical Mechanics

where $\sum_\mathbf{q}$ has the same meaning it has in Eq. (1.3.73); it is a sum over single-particle states. This formulation avoids the problem of multiple counting of the same many-particle state. Furthermore, we have not yet made the ideal gas approximation; Eqs. (1.3.76) and (1.3.77) are generally applicable to the perfect gas.

Let us now make the ideal gas approximation. For each $\mathbf{q}$, the probability that $n_\mathbf{q} = 0$ is nearly one; the probability that $n_\mathbf{q} = 1$ is nearly zero, that $n_\mathbf{q} = 2$ is much smaller still, and so on. The general expression for the probability of the subsystem being in a particular state comes from Eq. (1.3.17), together with Eq. (1.3.22):

$$w_\alpha = \exp\left[\frac{\Omega - (E_\alpha - \mu N_\alpha)}{kT}\right] \quad (1.3.78)$$

For the probability of $n_\mathbf{q}$ in a single-particle state $\mathbf{q}$, this expression reduces to

$$w_{n_\mathbf{q}} = \exp\left[\frac{\Omega_\mathbf{q} - n_\mathbf{q}(\varepsilon_\mathbf{q} - \mu)}{kT}\right] \quad (1.3.79)$$

In particular,

$$w_0 = \exp\left(\frac{\Omega_\mathbf{q}}{kT}\right) \simeq 1 \quad (1.3.80)$$

Then

$$w_1 = \exp\left(\frac{\Omega_\mathbf{q}}{kT}\right)\exp\left(-\frac{\varepsilon_\mathbf{q} - \mu}{kT}\right)$$

$$\simeq \exp\left(-\frac{\varepsilon_\mathbf{q} - \mu}{kT}\right) \ll 1 \quad (1.3.81)$$

$$w_2 = \exp\left[-\frac{2(\varepsilon_\mathbf{q} - \mu)}{kT}\right] = (w_1)^2 \ll w_1 \quad (1.3.82)$$

and so on. From here on we shall retain only leading-order terms in the small quantity $w_1 = \exp[-(\varepsilon_\mathbf{q} - \mu)/kT]$.

Before going on, let us see under what conditions our approximation is satisfied. It is clear from Eqs. (1.3.81) and (1.3.82) that the larger $\varepsilon_\mathbf{q}$ is, the smaller w_1, w_2, and so on will be, so the approximation will be satisfied for all the subsystems if it is satisfied for the one with the smallest value of $\varepsilon_\mathbf{q}$—namely, the single-particle ground state, $\varepsilon_\mathbf{q} = 0$. For this state, the condition $w_1 \ll 1$ is just

$$e^{\mu/kT} \ll 1 \quad (1.3.83)$$

or, in other words, μ/kT must be large and negative. An alternative statement of the ideal gas approximation, then, is that it is a perfect gas with a large, negative chemical potential. Recall that we argued earlier that the chemical potential of a perfect gas is usually negative.

The average value of n_q is obtained, using Eq. (1.3.63), by

$$\overline{n_q} = \frac{\sum_{n_q} n_q \exp\left[-n_q(\varepsilon_q - \mu)/kT\right]}{\sum_{n_q} \exp\left[-n_q(\varepsilon_q - \mu)/kT\right]} = \exp\left(-\frac{\varepsilon_q - \mu}{kT}\right) \quad (1.3.84)$$

where we have kept leading-order terms top and bottom. Doing the same in Eq. (1.3.76) we have

$$\Omega_q = -kT \log\left[1 + \exp\left(-\frac{\varepsilon_q - \mu}{kT}\right)\right]$$

$$= -kT \log(1 + \overline{n_q})$$

$$= -kT\overline{n_q} \quad (1.3.85)$$

since $\log(1 + x) = x$ when $x \ll 1$. For the gas as a whole,

$$\Omega = -PV = -kT \sum_q \overline{n_q} \quad (1.3.86)$$

$$\Omega = -kTN \quad (1.3.87)$$

where we have used Eqs. (1.2.39) and (1.3.77), and

$$N = \sum_q \overline{n_q} \quad (1.3.88)$$

which is basically a normalization condition. We have thus discovered that

$$PV = NkT$$

which is the equation of state of the ideal gas.

In discussing the $N!$ formulation, starting with Eq. (1.3.70) we attempted to find the equation of state but failed to do so, lacking as yet the means for doing the sum over single-particle states, Eq. (1.3.73). Here we set out to find $\Omega(T, V, \mu)$ and once again failed to do so for the same reason, but we did stumble across the equation of state without really seeking it. To get $\Omega(T, V, \mu)$, we must perform the sum over states in Eq. (1.3.88), which will give us $N = N(T, V, \mu)$, to be substituted into Eq. (1.3.87).

Notice how little we have actually had to assume in order to arrive at $PV = NkT$. The particles must be noninteracting (so that the ε_q's are independent of how the particles are distributed) and all the n_q's must be small. We never assumed, for example, that the energies of the particles are related to their momenta by $\varepsilon = p^2/2m$. Even particles obeying quite different dynamics would follow the ideal gas equation of state, provided only that the temperature is high and the density low (conditions that ensure the validity of both assumptions we have made).

In the latter half of the nineteenth century it was well established that matter was composed of atoms, and attention turned to the question of the composition of atoms. A suggestion arose that seems peculiar to us today;

it was proposed that atoms were vortex rings in the aether. (The aether, of course, was the medium that transmitted electromagnetic waves, such as light.)

Vortex rings are very strange beasts. We shall have occasion to return to them in Chap. 5, since they play a role in the physics of superfluid helium. The oddity about them is that instead of having their energies and momenta related by the usual $\varepsilon \propto p^2$, they have $\varepsilon \propto p^{1/2}$ (we are speaking loosely; see Chap. 5). That means their velocities are given by

$$v = \frac{\partial \varepsilon}{\partial p}$$

$$\propto p^{-1/2}$$

$$\propto \frac{1}{\varepsilon}$$

In other words, the more energy a vortex ring has, the slower it goes. How could such intelligent men—Maxwell, Kelvin, and many others were among the believers—how could they have supposed that matter was made up of vortex rings? One of the most convincing elements of that belief was the known fact that a rarified gas of vortex rings would obey $PV = NkT$.

The proof given then (in an essay by J. J. Thomson, written in 1879) was by means of difficult dynamical arguments. We have, in effect, proved the same thing, since, in deriving our result, we did not assume that atoms are not vortex rings. But we have also shown the result to be, in effect, trivial. Any rarified gas will behave that way, no matter how queer the dynamics of its particles.

e. Doing Sums over States as Integrals

Whatever formulation we use, our problem eventually reduces itself to a sum over all possible quantum states, either of the sample (giving $\mathscr{L}$ or Z directly) or of some astutely chosen subsystem. It is usually possible to do the sum as an integral, provided that we know the appropriate density of states. In fact, as pointed out in Sec. 1.3b, finding the density of states was just the problem we started out with, arising directly from our assumption that all states are equally likely in equilibrium. In classical statistics the assumption of equal probability of all quantum states makes no sense, but the equivalent statement is that, expressed as a function of momenta and coordinates rather than energies, the density of states is simply a constant. Using our quantum statistics, that constant is easily evaluated, and our counting problem—at least for the perfect gas—is then solved. Let us therefore take this opportunity to discuss some of the ideas that lay at the basis of classical statistics and compare them with their quantum counterparts.

In classical mechanics the microscopic state of a system of many particles is given by describing the system—that is, the nature and interactions of the particles, constraints such as boundaries or external force fields—and by giving all the positions and momenta of the particles at some instant of time. Given the necessary description, the energy of the system, $\mathscr{E}_0$, a continuous variable, is some function of all the positions r_i and momenta p_i,

$$\mathscr{E}_0 = \mathscr{E}_0(\{p_i\}, \{r_i\}) \tag{1.3.89}$$

where we use the notation $\{p_i\}$ to mean the set of all values of p_i. If there are N_0 particles in three dimensions, the index i runs from 1 to $3N_0$, three values of each of p and r being needed for each particle. In general, we shall say that

$$i = 1, 2, \ldots, f_0$$

where f_0 (equal to $3N_0$ for N_0 particles in three dimensions) is called the number of degrees of freedom of the system. In quantum mechanics f_0 corresponds to the number of quantum numbers that must be specified in order to determine a particular quantum state (it is also $3N_0$ for N_0 particles in three dimensions). Clearly, in the classical case, there are many possible sets of the $\{p_i\}$ and $\{r_i\}$ that will give the same $\mathscr{E}_0$, corresponding to the many equally likely-states in quantum mechanics. The problem, classically, is that there is no way, even in principle, of counting the number of energetically degenerate sets of $\{p_i\}$ and $\{r_i\}$, since neither the p_i nor the r_i is a discrete variable; the degree of degeneracy cannot be expressed as a number. We need an artificial construction in order to quantify the degeneracy.

Imagine a $6N_0$-dimensional space, one dimension for each of the r_i and one for each of the p_i needed to specify a state of the system. In such a space the instantaneous state of the N-particle system is represented by a single point (with $6N_0$ coordinates needed to specify where it is). This generalized space is known as *phase space*. A point in phase space, representing the instantaneous state of a system, follows a trajectory with time whose future course is completely determined—in classical mechanics—by its position at any instant. The detailed path followed by the point is governed by the laws of classical mechanics applied to each of the N_0 particles, but there is a broader restriction on where the point can go that is of interest to us. The conservation of energy, expressed by Eq. (1.3.89), gives one equation connecting the $6N_0$ variables, restricting its possible paths to a surface of $6N_0 - 1$ dimensions. The classical statement analogous to our equal probability of allowed quantum states is that the phase point of the system is equally likely to be found anywhere on that surface.

Given that assumption, it is clear that the entropy of the system must be given by the logarithm of something proportional to the area of that surface,

1.3 Statistical Mechanics

but to quantify it, we will need to divide that continuous surface into a number of distinct states. This step can be done by considering any one of the r_i, say r_k, and its conjugate momentum, p_k. We arbitrarily divide the entire length of r_k and p_k into equal segments of length Δr_k and Δp_k and call the product τ.

$$\tau = \Delta r_k \, \Delta p_k \tag{1.3.90}$$

All the coordinate-momentum pairs are to be divided into segments with the same product. By doing so, we have divided phase space into cells of equal volume. If we tell in which segment of each of the r_i and p_i a phase point is to be found, then we have localized the state of the system to a cell in phase space whose volume is τ^{f_0}. Let us choose τ small enough so that only one state fits in each cell. Doing so is inconsistent with classical concepts, so if we expect classical mechanics to work (we do not, of course), we should anticipate letting τ shrink to zero at the end of our argument. Now we have a way of counting the number of possible states of the system. The number, what we called Γ_0 in Sec. 1.3a, is given by the number of cells through which the surface, Eq. (1.3.89), passes.

$$\Gamma_0 = \frac{1}{\tau^{f_0}} \int \cdots \int_{f_0}' d\{p_i\} \, d\{r_i\} \tag{1.3.91}$$

where the prime on the integral means that we restrict the range of integration to values of r_i and p_i that lie within the same cells as values that satisfy Eq. (1.3.89). By introducing some uncertainty into the locality of each phase point, we have inevitably introduced some uncertainty into the energy, $\mathscr{E}_0$, of the system, and Eq. (1.3.91) gives Γ_0 proportional to the volume of a thin shell, rather than the area of a surface in phase space. Presumably, these are formal difficulties that will vanish when we let $\tau \to 0$. (Actually, they will not vanish, of course, but it will turn out to make no difference. We will return to discuss why more fully at the end of this chapter.)

We can now repeat the formal manipulations of Sec. 1.3a. Among the results will be

$$S_0 = k \log \int \cdots \int_{f_0}' d\{p_i\} \, d\{r_i\} - k \log \tau^{f_0} \tag{1.3.92a}$$

and

$$S_\alpha' = k \log \int \cdots \int_{f'}'' d\{p_i\} \, d\{r_i\} - k \log \tau^{f'} \tag{1.3.92b}$$

where f' is the number of degrees of freedom of the medium and the double prime on the integral restricts its range to the variables of the medium (i.e., to volume V') and to energy

$$\mathscr{E}' = \mathscr{E}_0 - \mathscr{E}_\alpha \tag{1.3.93}$$

where $\mathscr{E}_\alpha$ is the energy of the subsystem when we restrict it to a single cell of its phase space. We can imagine the subsystem to have a fixed V and N. Then the average entropy of the subsystem will be

$$S = S_0 - \bar{S}'_\alpha$$
$$= k \log \int \cdots \int_{f_0}^{f'} d\{p_i\} d\{r_i\} - k \log \int \cdots \int_{f'}^{f''} d\{p_i\} d\{r_i\}$$
$$- k \log \tau^{(f_0 - f')} \tag{1.3.94}$$

Thus the entropy of any subsystem, regardless of composition, has in it an additive term of the form

$$-k \log \tau^f = -fk \log \tau \tag{1.3.95}$$

where $f = f_0 - f'$ is the number of degrees of freedom of the subsystem. If we let τ shrink to zero, all entropies become infinite, an intolerable situation. On the other hand, the same quantity, τ, appears in the entropy of every system or sample (gas, liquid, solid, argon, water), implying that it is actually a universal constant. Thus, we are in a position to guess that classical physics was missing a universal constant—namely, the fundamental volume element of phase space, whose dimensions, according to Eq. (1.3.90), are those of action (momentum times distance or energy times time). It is not hard to guess that τ will be related to Planck's constant.

We can now perform the sum over states of any system that is not too close to its ground state as an integral over volume in phase space, provided that our division of phase space into cells of one state each is consistent with quantum mechanics. The transformation is given by

$$\sum_{\text{states}} \to \int \rho(\mathscr{E}) d\mathscr{E} = \frac{1}{\tau^f} \int \cdots \int_f d\{p_i\} d\{r_i\} \tag{1.3.96}$$

where the integrals on both sides of the equality are restricted to the same range of phase space. Restricted integrals are difficult to do, but we have already developed the formalism that generally relieves us of having to do them. Using equations like (1.3.30) or (1.3.31), we shall almost always end up finding what we want to know by integrating over all of phase space, which is a much easier task. This step also gets us around the dilemma of the uncertainty we have had to introduce into the energy. What we usually need to know is $\rho(\mathscr{E})$, the number of states per unit range of energy, a well-defined quantity, rather than the number of states at a fixed energy, which is not well defined.

We must still evaluate τ. Since it is a universal constant, we can obtain it by finding the correct quantum mechanical equivalent of Eq. (1.3.96) for

1.3 Statistical Mechanics

the simplest possible case. We shall do so for one perfect gas particle in one dimension. The energy of the particle is

$$\varepsilon_q = \frac{\hbar^2 q^2}{2m}$$

where
$$\mathbf{q} = \frac{2\pi}{L}\ell \quad (\ell = 0, \pm 1, \pm 2, \ldots)$$

L is the length of the line to which it is restricted, but in every state it has equal probability of being anywhere on the line (the wave functions are of the form $\psi \sim e^{-iqx}$, so that $|\psi|^2$ does not depend on its coordinate x). Its number of states is simply counted by the index ℓ; that is, the number of states between ℓ and $\ell + \Delta\ell$ is simply $\Delta\ell$ (we should formally imagine ℓ to be very large, since we wish to take the classical limit). Thus the number of states between p and $p + \Delta p$ is

$$\Delta\ell = \frac{L}{2\pi}\Delta q = \frac{L}{2\pi\hbar}\Delta p \tag{1.3.97}$$

The same number is now to be computed classically from Eq. (1.3.96). For this case, $f = 1$, and we restrict our integral over phase space to x anywhere in L, and p in Δp:

$$\frac{1}{\tau}\int_L \int_{\Delta p} dp\, dr = \frac{L\,\Delta p}{\tau} \tag{1.3.98}$$

Comparing Eq. (1.3.97) to (1.3.98), the universal constant τ is given by

$$\tau = 2\pi\hbar = h \tag{1.3.99}$$

τ is just equal to Planck's constant, or, alternatively, we have found a new interpretation of Planck's constant: it is the volume of the fundamental cell in phase space.

When we perform a sum over states by doing an integral over phase space instead, we are basically performing a classical operation on our quantum system. Before doing so, however, we have decided that we will not specify the state of the system anymore closely than to assign it to a cell of finite volume in phase space. This means that we are leaving some uncertainty in each component of each coordinate and momentum of the system. According to Eqs. (1.3.90) and (1.3.99), that uncertainty is

$$\Delta p_k\, \Delta r_k = 2\pi\hbar \tag{1.3.100}$$

which is just the lower limit allowed by the uncertainty principle. The formulation is thus consistent with quantum mechanics; states are packed into phase space to the maximum density allowed. As a matter of fact, we can now see that the uncertainty we have had to introduce into the energy in writing Eq. (1.3.91) is simply the same quantum uncertainty discussed in Sec. 1.1.

Let us now make some use of our new prescription for doing sums over states. In Sec. 1.3d we saw that the ideal gas problem could, in one way or another [i.e., Eq. (1.3.73) or (1.3.88)], be reduced to a sum over a special set of states, the states of a single particle in a box. We shall do so as a first example.

For a single particle, the number of states in a region $d^3p\, d^3r$ of its phase space (the notation here is $d^3p = dp_x\, dp_y\, dp_z$) is

$$\frac{d^3p\, d^3r}{(2\pi\hbar)^3} \tag{1.3.101}$$

For the perfect gas, nothing depends on r, so the coordinate integration can be extracted as V, the volume. Furthermore, the problem is isotropic, so that

$$d^3p = 4\pi p^2\, dp \tag{1.3.102}$$

Thus, the number of states between p and $p + dp$ is

$$\frac{4\pi V}{(2\pi\hbar)^3}\, p^2\, dp \tag{1.3.103}$$

The range of momentum dp corresponds to the range of energies $d\varepsilon = [d\varepsilon(p)/dp]\, dp$; therefore, the number of states per unit range of energy, which we have called $\rho(\varepsilon)$, is given by

$$\rho(\varepsilon)\, d\varepsilon = \frac{4\pi V}{(2\pi\hbar)^3}\, p^2\, \frac{dp}{d\varepsilon}\, d\varepsilon \tag{1.3.104}$$

Substituting in the relation between ε and p,

$$\varepsilon = \frac{p^2}{2m} \tag{1.3.105}$$

we get

$$\rho(\varepsilon)\, d\varepsilon = \frac{4\pi\sqrt{2}\, Vm^{3/2}}{(2\pi\hbar)^3}\, \varepsilon^{1/2}\, d\varepsilon \tag{1.3.106}$$

As an example of the use of this formula, let us calculate the average energy, $\bar{\varepsilon}$, of a single particle in a box at fixed temperature, according to Eq. (1.3.74).

$$\bar{\varepsilon} = \frac{\int \varepsilon\rho(\varepsilon)e^{-\varepsilon/kT}\, d\varepsilon}{\int \rho(\varepsilon)e^{-\varepsilon/kT}\, d\varepsilon} = \frac{\int \varepsilon^{3/2}e^{-\varepsilon/kT}\, d\varepsilon}{\int \varepsilon^{1/2}e^{-\varepsilon/kT}\, d\varepsilon} \tag{1.3.107}$$

Here we change variables to $x = \varepsilon/kT$.

$$\bar{\varepsilon} = kT\, \frac{\int_0^\infty x^{3/2}e^{-x}\, dx}{\int_0^\infty x^{1/2}e^{-x}\, dx} \tag{1.3.108}$$

1.3 Statistical Mechanics

The integrals are Γ functions

$$\Gamma(n+1) = \int_0^\infty x^n e^{-x}\, dx \qquad (1.3.109)$$

which are generalized factorials, having the property that

$$\frac{\Gamma(n+1)}{\Gamma(n)} = n \qquad (1.3.110)$$

The result is

$$\bar{\varepsilon} = \tfrac{3}{2}kT$$

which, substituted into Eq. (1.3.72), gives

$$PV = NkT \qquad (1.3.111)$$

a result we had already obtained by other arguments.

Making use of Eqs. (1.3.76) and (1.3.77), we can solve not only the ideal gas problem but, more generally, the problem of any perfect gas by reducing the computation to a sum over the states of a single particle, which we have, in turn, reduced to an integral over the density of states. The only requirement of the perfect gas is that the particles be noninteracting, so that the single-particle states be independent of their occupation numbers. It is not necessary that the energy and momentum be connected by $\varepsilon = p^2/2m$. For example, we can imagine a system of extremely relativistic particles, with

$$\varepsilon \gg m_0 c^2 \qquad (1.3.112)$$

where m_0 is the rest mass. Then Eq. (1.3.105) is replaced by

$$\varepsilon = cp \qquad (1.3.113)$$

where c is the speed of light. Substitution of this into Eq. (1.3.104) yields a density of states:

$$\rho(\varepsilon)\, d\varepsilon = \frac{4\pi V}{(2\pi\hbar c)^3} \varepsilon^2 \, d\varepsilon \qquad (1.3.114)$$

In particular, a gas of photons (e.g., the radiation in a cavity) could be described in this way.

There are other possibilities as well. In a number of instances, it is possible to take a system of interacting particles and describe its quantum states as being various numbers of (perhaps imaginary) particles of a new kind, which, however, each have some energy, some momentum, and which to a good approximation can be said not to interact with each other. Such "particles," which are actually collective states of a system, are called *quasiparticles*. One example, in fact, is to describe the excited states of an electromagnetic field as consisting of photons. Once we have made such a description, and have found the appropriate relation between ε and p for the quasi-particles, we can, by means of Eq. (1.3.104), easily calculate the thermodynamic properties of the original system. In other words, many problems in studying the states of matter can be reduced to the properties of a perfect gas of quasiparticles. Examples of quasiparticle gases that we shall study include phonons in crystalline solids, electrons in metals, phonons and rotons in superfluid helium, electron pairs in superconductors, and spin waves in ferromagnetic materials.

Unfortunately, not all the thermal behavior of interacting matter can be reduced to one kind or another of noninteracting quasiparticles. In particular, no way has been found to describe classical liquids, and interacting fluids in general, in that way, which is one reason why fluids are the least well-understood state of matter. We, nevertheless, can formulate the general problem of the statistical mechanics of interacting fluids and shall do so now to finish out this section.

It is adequate to describe the fluids we will study in classical terms. To start with, no two particles can ever have the same positions and momenta, so we can compute the distinguishable volume of phase space by taking the total volume of all the single-particle phase spaces and dividing by $N!$:

$$\rho(\mathscr{E})\,d\mathscr{E} = \frac{1}{N!} \frac{d\{p_i\}\,d\{r_i\}}{(2\pi\hbar)^f} \qquad (1.3.115)$$

where $\mathscr{E}$, as usual, is a system, or many-body, energy, and i runs from 1 to $f = 3N$. The partition function is then given by

$$Z = \int \rho(\mathscr{E}) e^{-\mathscr{E}/kT}\,d\mathscr{E} = \frac{1}{(2\pi\hbar)^{3N} N!} \int \cdots \int e^{-\mathscr{E}/kT}\,d\{p_i\}\,d\{r_i\} \qquad (1.3.116)$$

where

$$\mathscr{E}(\{p_i\}, \{r_i\}) = \sum_{i=1}^{3N} \frac{p_i^2}{2m} + U(\{r_i\}) \qquad (1.3.117)$$

Here $U(\{r_i\})$ is the potential energy of the system, which depends, in general, on the coordinates of all the particles. The first term on the right is, of course, just the kinetic energy. Since the kinetic energy depends only on the momenta, and the potential energy only on the coordinates, the multiple integral in Eq. (1.3.116) breaks into a product of integrals over momenta and integrals over coordinates:

$$Z = \frac{1}{(2\pi\hbar)^{3N} N!} \int \cdots \int \exp\left(-\sum \frac{p_i^2}{2mkT}\right) \exp\left(-\frac{U}{kT}\right) d\{p_i\}\,d\{r_i\}$$

$$= \frac{1}{(2\pi\hbar)^{3N} N!} \int \cdots \int \exp\left(-\sum \frac{p_i^2}{2mkT}\right) d\{p_i\} \int \cdots \int \exp\left(-\frac{U}{kT}\right) d\{r_i\}$$

$$(1.3.118)$$

1.3 Statistical Mechanics

The momentum integrals similarly consist of N identical factors of the form

$$\int e^{-\varepsilon/kT} d^3p \quad (1.3.119)$$

where

$$\varepsilon = \frac{p^2}{2m} = \frac{1}{2m}(p_x^2 + p_y^2 + p_z^2) \quad (1.3.120)$$

We have already solved this problem. In fact, the entire factor

$$\frac{1}{N!(2\pi\hbar)^{3N}} \int \cdots \int \exp\left(-\sum \frac{p_i^2}{2mkT}\right) d\{p_i\}$$

$$= \frac{1}{N!(2\pi\hbar)^{3N}} \left[\int \exp\left(-\frac{p^2}{2m}\right) d^3p\right]^N$$

$$= \frac{1}{V^N N!} \left[\iint e^{-\varepsilon/kT} \frac{d^3p\, d^3r}{(2\pi\hbar)^3}\right]^N$$

$$= \frac{Z_{IG}}{V^N} \quad (1.3.121)$$

where Z_{IG} is the partition function of an ideal gas. Let us define the quantity

$$Q = \int \cdots \int \exp\left[-\frac{U(\{r_i\})}{kT}\right] d\{r_i\} \quad (1.3.122)$$

which is called the configurational integral. We can then write the partition function as

$$Z = \frac{Z_{IG}}{V^N} Q \quad (1.3.123)$$

Notice that the fluid reduces to the ideal gas if $U(\{r_i\}) = 0$.

We have never actually evaluated Z_{IG} but that is easily done. In Eq. (1.3.121) we have

$$\int \exp\left(-\frac{p^2}{2mkT}\right) d^3p = 4\pi \int e^{-\varepsilon/kT} p^2\, dp$$

$$= 4\pi\sqrt{2}\, (mkT)^{3/2} \int e^{-x} x^{1/2}\, dx$$

$$= 4\pi\sqrt{2}\, \Gamma\left(\frac{3}{2}\right)(mkT)^{3/2} \quad (1.3.124)$$

With this result and the fact that $\Gamma(3/2) = \sqrt{\pi}/2$, Eq. (1.3.121) may be written

$$Z_{IG} = \frac{1}{N!}\left(\frac{V}{\Lambda^3}\right)^N \quad (1.3.125)$$

where
$$\Lambda = \frac{2\pi\hbar}{\sqrt{2\pi mkT}} \quad (1.3.126)$$

is a characteristic length, called the thermal de Broglie wavelength of the particles. The partition function for the fluid may be written

$$Z = \frac{Z_{IG}}{V^N} Q = \frac{Q}{N!\,\Lambda^{3N}} \quad (1.3.127)$$

Then the free energy may be written

$$F = F_{IG} + NkT \log V - kT \log Q \quad (1.3.128)$$

and the problem of fluids has been reduced to evaluating Q.

It will be useful for us to have in hand also the grand partition function for a fluid. In this case, when we sum over states, one of the set of quantum numbers over which we must sum is the number of particles.

$$\mathscr{L} = \sum_{\alpha} \exp\left(-\frac{E_\alpha - \mu N_\alpha}{kT}\right) = \sum_{N_\alpha} \exp\left(\frac{\mu N_\alpha}{kT}\right) \sum_{\alpha'} \exp\left(-\frac{E_\alpha}{kT}\right) \quad (1.3.129)$$

where we are using α' to denote the remainder of the quantum numbers in α. The first member of Eq. (1.3.129) is Eq. (1.3.31). For each value of N_α, the quantity we must consider is

$$z^{N_\alpha} \sum_{\alpha'} \exp\left(-\frac{E_\alpha}{kT}\right) = \frac{z^{N_\alpha}}{N_\alpha!} \int \exp\left(-\frac{\mathscr{E}}{kT}\right) d\{p_i\}_{3N_\alpha} d\{r_i\}_{3N_\alpha} \quad (1.3.130)$$

where we have defined

$$z = e^{\mu/kT} \quad (1.3.131)$$

a quantity called the fugacity, and the sets $\{p_i\}$ and $\{r_i\}$ in Eq. (1.3.130) range over i from 1 to $3N_\alpha$. The result is

$$\frac{z^{N_\alpha} Q_{N_\alpha}}{N_\alpha!\,\Lambda^{3N_\alpha}}$$

where Q_{N_α} is the configurational integral for N_α particles. Substituting back into Eq. (1.3.129), we obtain

$$\mathscr{L} = \sum_N \frac{z^N Q_N}{N!\,\Lambda^{3N}} \quad (1.3.132)$$

1.3 Statistical Mechanics

where we have dropped the subscript α from what is now a dummy variable, N. We will make use of these formulas in Chap. 4 when we study the liquid state.

f. Fluctuations

The usefulness of the apparatus we have developed so far is based on the assertion, as yet undemonstrated, that quantities such as the temperature of an isolated system or the energy of a body in a temperature bath, although purely statistical results of random processes, will turn out to be quite accurately predictable. In this section we shall investigate the uncertainties in that kind of prediction.

In order to get some feeling for the situation we are considering, let us start by returning to the example of the classical ideal gas, which we imagine to be immersed in a medium at temperature T. Rather than make use of the formulas we have worked out in terms of single-particle states, let us consider the N-particle gas as a whole. The energy of the gas is given by

$$\mathscr{E} = \frac{1}{2m} \sum_{i=1}^{3N} p_i^2 \tag{1.3.133}$$

This is an example of Eq. (1.3.89) for the surface in phase space to which a system of fixed energy is restricted. It is, in this case, the surface of a $3N$-dimensional sphere in momentum space, of radius P^*, given by

$$P^{*2} = 2m\mathscr{E} = \sum_i p_i^2 \tag{1.3.134}$$

(Any difficulty in visualizing the surface of a $3N$-dimensional sphere is simply a failure of imagination. The surface of a one-dimensional sphere is two points on a straight line; of a two-dimensional sphere, a circle; a three-dimensional sphere is a sphere through which any two-dimensional cross section is a circle or two-dimensional sphere; a four-dimensional sphere is a figure through which any three-dimensional cross section is a three-dimensional sphere; and so on.)

For the gas in a temperature bath, the probability of having energy $\mathscr{E}$ is [see Eq. (1.3.43)]

$$w(\mathscr{E}) = \frac{1}{Z} \rho(\mathscr{E}) e^{-\mathscr{E}/kT} \tag{1.3.135}$$

Here $\rho(\mathscr{E})$ is the many-body density of states, proportional to the volume of a shell in phase space at energy $\mathscr{E}$,

$$\rho(\mathscr{E}) \, \delta\mathscr{E} \propto \delta V^*(\mathscr{E}) \tag{1.3.136}$$

where V^* is the volume of the sphere of Eq. (1.3.134)

$$V^* \propto P^{*3N} \tag{1.3.137}$$

The distinguishable volume in phase space is really $V^N V^*/N!$, where V is the volume of the gas in real space, since V^* is constructed out of the momentum spaces of N indistinguishable particles, but we shall not have to worry about that here. Using Eqs. (1.3.134) and (1.3.137) in (1.3.135), we get

$$\rho(\mathscr{E}) \, \delta\mathscr{E} \propto \frac{\partial V^*}{\partial \mathscr{E}} \, \delta\mathscr{E}$$

$$\propto P^{*3N-1} \frac{\partial P^*}{\partial \mathscr{E}} \, \delta\mathscr{E}$$

$$\propto (\mathscr{E}^{1/2})^{3N-1} \mathscr{E}^{-1/2} \, \delta\mathscr{E}$$

$$\propto \mathscr{E}^{(3N/2)-1} \, \delta\mathscr{E} \qquad (1.3.138)$$

This result gives us the energy dependence of the density of states (there is a $V^N/N!$ and other factors in the equation). Notice that for $N = 1$, Eq. (1.3.138) reduces correctly to

$$\rho(\varepsilon) \propto \varepsilon^{1/2} \qquad (1.3.139)$$

as in Eq. (1.3.106).

After we lump together $1/Z$ and the constants in $\rho(\mathscr{E})$, Eq. (1.3.135) becomes

$$w(\mathscr{E}) = C_N \mathscr{E}^{(3N/2)-1} e^{-\mathscr{E}/kT} \qquad (1.3.140)$$

subject to

$$1 = \int w(\mathscr{E}) \, d\mathscr{E} = C_N \int \mathscr{E}^{(3N/2)-1} e^{-\mathscr{E}/kT} \, d\mathscr{E} \qquad (1.3.141)$$

or

$$\frac{1}{C_N} = (kT)^{3N/2} \int_0^\infty x^{(3N/2)-1} e^{-x} \, dx$$

$$= \Gamma\left(\frac{3N}{2}\right)(kT)^{3N/2} \qquad (1.3.142)$$

so that

$$w(\mathscr{E}) = \frac{1}{\Gamma(3N/2)} \left(\frac{\mathscr{E}}{kT}\right)^{(3N/2)-1} \frac{1}{kT} e^{-\mathscr{E}/kT} \qquad (1.3.143)$$

For a single particle, this reduces to

$$w(\varepsilon) = \frac{1}{\Gamma(\frac{3}{2})(kT)^{3/2}} \varepsilon^{1/2} e^{-\varepsilon/kT} \qquad (1.3.144)$$

The probability that the gas will have a given energy is the product of two factors. One is the probability for the gas to be in some particular microscopic state, proportional to $e^{-\mathscr{E}/kT}$, an exponentially declining function of $\mathscr{E}$. The other is the number of microscopic states at energy $\mathscr{E}$, proportional to $\mathscr{E}^{(3N/2)-1}$, a function that starts at zero (the formula is classical; quantum

1.3 Statistical Mechanics

mechanically there is one state at energy zero) and rises with $\mathscr{E}$. The product of the two factors starts at zero (or one) at zero energy and rises at first but must eventually fall, since $e^{-\mathscr{E}/kT} \to 0$ faster than any power of $\mathscr{E}$ rises, for sufficiently large $\mathscr{E}$. There is thus a maximum in $w(\mathscr{E})$, which represents the most likely energy of the gas. The problem we are investigating is, how sharp is this maximum? In other words, what is the relative probability of finding the gas with an energy significantly different from its most probable energy?

One point should be obvious enough: if the maximum of $w(\mathscr{E})$ is extremely sharp, the most probable energy will also be the average energy, since we will almost never observe any other value. These quantities are easily calculated from Eq. (1.3.143). The most probable energy, $\mathscr{E}_m$, is the solution of

$$0 = \frac{\partial w(\mathscr{E})}{\partial \mathscr{E}} \propto \frac{\partial}{\partial \mathscr{E}}[\mathscr{E}^{(3N/2)-1}e^{-\mathscr{E}/kT}] \qquad (1.3.145)$$

or

$$\mathscr{E}_m = \left(\frac{3N}{2} - 1\right)kT \qquad (1.3.146)$$

The average energy is given by

$$\bar{\mathscr{E}} = \int \mathscr{E} w(\mathscr{E})\, d\mathscr{E}$$

$$= C_N \int \mathscr{E}^{3N/2} e^{-\mathscr{E}/kT}\, d\mathscr{E}$$

$$= C_N(kT)^{(3N/2)+1} \int x^{3N/2} e^{-x}\, dx$$

$$= C_N(kT)^{(3N/2)+1} \Gamma(\tfrac{3}{2}N + 1)$$

$$= \frac{3N}{2} kT \qquad (1.3.147)$$

Comparing Eqs. (1.3.146) and (1.3.147), we see that $\mathscr{E}_m$ and $\bar{\mathscr{E}}$ are equal only if N is large—that is, for a macroscopic system. For a single particle,

$$\bar{\varepsilon} = \tfrac{3}{2}kT \qquad (1.3.148)$$

(a result we had earlier), and

$$\varepsilon_m = \tfrac{1}{2}kT \qquad (1.3.149)$$

which immediately tells us that the distribution $w(\varepsilon)$ is not very sharply peaked. The reason is that the maximum occurs at $\varepsilon \sim kT$ when $e^{-\varepsilon/kT}$ is not changing very fast. The product $\varepsilon^{1/2} e^{-\varepsilon/kT}$ is sketched in Fig. 1.3.1. For many particles, the maximum occurs at $\mathscr{E} \sim NkT$, where $e^{-\mathscr{E}/kT}$ is falling very rapidly, and, moreover, $\rho(\mathscr{E}) \sim \mathscr{E}^{(3N/2)-1}$ is rising very rapidly. The

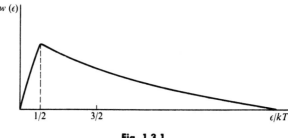

Fig. 1.3.1

result is nearly a delta function, sketched in Fig. 1.3.2. Thus, for one particle, we expect the thermodynamic uncertainty in the energy to be of the same order as the energy itself, whereas for many particles we expect it to be very much less. We shall now work out just what the uncertainty is.

We will need a way to characterize or measure the width of a distribution like $w(\mathscr{E})$—that is, the extent to which thermodynamic quantities fluctuate in equilibrium. However, we are venturing into potentially confusing territory when we speak of the probability that a system, in equilibrium, will have other than the equilibrium values of its thermodynamic variables, so perhaps it will be useful to remind ourselves of precisely the operations we are going to describe. The discussion that follows is generally applicable, not restricted to the ideal gas.

We imagine, as always, that our sample is a part of a much larger system that is itself isolated, with fixed total energy. The overall system, which is in equilibrium, fluctuates about among its allowed states. At some instant in time, we isolate the sample from the medium, so that it has, for example, whatever energy it had at that instant. If either the volume or the number of particles, say, was free to fluctuate as well, we imagine that to be fixed also at its instantaneous value. We now measure the various properties of the sample: temperature, pressure, volume, whatever we are interested in. The result will depend on the particular, say, energy and volume at the instant it

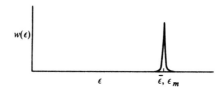

Fig. 1.3.2

1.3 Statistical Mechanics

was isolated, which may or may not be equal to the most probable values, but the body itself, once isolated, is to be thought of as being internally in equilibrium. Thus, T, P, and S, although not necessarily equal to their most probable values, will be completely determined by the energy and volume at the instant of isolation. We repeat this procedure many times and plot for any quantity—E, T, P, and so on—the number of times we get a given result versus the quantity. The result should look like Fig. 1.3.2, giving the probability of any given fluctuation. We must remember, however, that although all the quantities (E, P, V, T, S) fluctuate, they do not fluctuate independently of each other but are instead connected by the equilibrium properties of the body.

Now let us start, for simplicity, by considering a body of fixed N, contained within rigid walls, so that only the energy fluctuates when immersed in the medium. Suppose that at the instant of isolation it is in some state with energy $\mathscr{E}$. The fluctuation in energy is then

$$\Delta E = \mathscr{E} - E \qquad (1.3.150)$$

where E is the average energy as usual;

$$E = \bar{\mathscr{E}} = \int \mathscr{E} w(\mathscr{E})\, d\mathscr{E} \qquad (1.3.151)$$

It follows, then, that

$$\overline{\Delta E} = 0 \qquad (1.3.152)$$

The fluctuations in any quantity always average to zero. However, if we square the fluctuations before averaging, up fluctuations and down fluctuations will both contribute with the same sign, and the result will be nonzero:

$$\overline{(\Delta E)^2} = \overline{(\mathscr{E} - E)^2} = \overline{(\mathscr{E}^2 - 2\mathscr{E}E + E^2)} = \overline{\mathscr{E}^2} - 2\bar{\mathscr{E}}E + E^2 \qquad (1.3.153)$$

or, putting in Eq. (1.3.151),

$$\overline{(\Delta E)^2} = \overline{\mathscr{E}^2} - E^2 \qquad (1.3.154)$$

The quantity $\overline{\mathscr{E}^2}$ is, of course, given by

$$\overline{\mathscr{E}^2} = \int \mathscr{E}^2 w(\mathscr{E})\, d\mathscr{E} \qquad (1.3.155)$$

The mean square fluctuation, $\overline{(\Delta E)^2}$, or its square root, the root mean square, or RMS, fluctuation is a convenient measure of the width of the distribution. Equations analogous to Eqs. (1.3.50) to (1.3.155) can, obviously, be written not only for the energy but also for any fluctuating quantity.

Given $\rho(\mathscr{E})$, and hence $w(\mathscr{E})$, for any system, we can calculate $\overline{(\Delta E)^2}$. However, there is a general way of relating $\overline{(\Delta E)^2}$ to equilibrium properties of the body without knowing $\rho(\mathscr{E})$. The fluctuations are included in the partition function, which was constructed by imagining precisely the same processes that we have discussed here. Written in terms of continuous variables,

$$Z = \int \rho(\mathscr{E}) e^{-\mathscr{E}/kT} \, d\mathscr{E} \qquad (1.3.156)$$

and
$$F = -kT \log Z \qquad (1.3.157)$$

Then

$$\frac{\partial^2 F}{\partial T^2} = \frac{\partial^2}{\partial T^2}(-kT \log Z)$$

$$= \frac{1}{kT^3}\left[\frac{(\int \mathscr{E}\rho(\mathscr{E})e^{-\mathscr{E}/kT} d\mathscr{E})^2}{Z^2} - \frac{1}{Z}\int \mathscr{E}^2 \rho(\mathscr{E})e^{-\mathscr{E}/kT} d\mathscr{E}\right]$$

$$= \frac{\overline{E}^2 - \overline{\mathscr{E}^2}}{kT^3} \qquad (1.3.158)$$

Thus,
$$\overline{(\Delta E)^2} = \overline{\mathscr{E}^2} - \overline{E}^2 = -kT^3 \frac{\partial^2 F}{\partial T^2} = kT^2 C_V \qquad (1.3.159)$$

Equation (1.3.159) gives the mean square fluctuations in the energy of any system in terms of its own heat capacity, provided only that the fluctuations occur at constant volume. To get some feeling for how big these fluctuations actually are, we turn, not unexpectedly, to the ideal gas. From Eq. (1.3.147),

$$E = \overline{\mathscr{E}} = \tfrac{3}{2}NkT \qquad (1.3.160)$$

and since E does not depend on V for fixed N in this case,

$$C_V = \frac{\partial E}{\partial T} = \frac{3}{2}Nk \qquad (1.3.161)$$

Thus, the relative size of the fluctuations is

$$\frac{\sqrt{\overline{(\Delta E)^2}}}{E} = \frac{\sqrt{kT^2 \tfrac{3}{2} Nk}}{\tfrac{3}{2}NkT} = \frac{1}{\sqrt{\tfrac{3}{2}}} \frac{\sqrt{N}}{N} \sim \frac{1}{\sqrt{N}} \qquad (1.3.162)$$

The relative energy fluctuations are of order $1/\sqrt{N}$, a result that will turn out to be typical of extensive quantities. We see that for a single particle, $N = 1$, the fluctuations are of the same order as the energy itself, as asserted above, but if $N \sim 10^{22}$, then the fluctuations introduce an uncertainty of order 1 part in 10^{11}, an imprecision that is not usually considered intolerable.

We can study density fluctuations in an analogous way, by using the

1.3 Statistical Mechanics

formulation in which N is free to vary. Retaining, this time, discrete variables (in order to demonstrate our versatility), we have

$$\left(\frac{\partial^2 \Omega}{\partial \mu^2}\right)_{T,V} = \frac{\partial^2}{\partial \mu^2}\left[-kT \log \sum_\alpha \exp\left(-\frac{E_\alpha - \mu N_\alpha}{kT}\right)\right]$$

$$= -\frac{1/kT \sum_\alpha N_\alpha^2 \exp\left[-(E_\alpha - \mu N_\alpha)/kT\right]}{\sum_\alpha \exp\left[-(E_\alpha - \mu N_\alpha)/kT\right]}$$

$$+ \frac{1/kT\{\sum_\alpha N_\alpha \exp\left[-(E_\alpha - \mu N_\alpha)/kT\right]\}^2}{\{\sum_\alpha \exp\left[-(E_\alpha - \mu N_\alpha)/kT\right]\}^2}$$

$$= -\frac{1}{kT}(\overline{N_\alpha^2} - \overline{N}^2) \qquad (1.3.163)$$

or

$$\overline{(\Delta N)^2} = \overline{N_\alpha^2} - \overline{N}^2 = -kT\left(\frac{\partial^2 \Omega}{\partial \mu^2}\right)_{T,V} \qquad (1.3.164)$$

Here we have allowed the energy and number of particles in a fixed volume to fluctuate simultaneously, holding the temperature and chemical potential of the medium fixed. To put the result in a more convenient form, we recall that

$$\left(\frac{\partial \Omega}{\partial \mu}\right)_{T,V} = -N = -\overline{N_\alpha} \qquad (1.3.165)$$

so that

$$\overline{(\Delta N)^2} = kT\left(\frac{\partial N}{\partial \mu}\right)_{T,V} \qquad (1.3.166)$$

Using Eq. (1.2.35), we find that

$$\left(\frac{\partial \mu}{\partial N}\right)_{T,V} = \frac{V}{N}\left(\frac{\partial P}{\partial N}\right)_{T,V} = \frac{1}{N}\left(\frac{\partial P}{\partial \rho}\right)_{T,V} \qquad (1.3.167)$$

where in the last step we have taken the constant V inside the derivative and defined the density

$$\rho = \frac{N}{V} \qquad (1.3.168)$$

Now, according to Eq. (1.2.99), for a system of constant N,

$$K_T = -\frac{1}{V}\left(\frac{\partial V}{\partial P}\right)_T$$

$$= -\frac{N}{V}\frac{\partial (V/N)}{\partial P}$$

$$= -\rho \frac{\partial (1/\rho)}{\partial P}$$

$$= \frac{1}{\rho}\left(\frac{\partial \rho}{\partial P}\right)_T \qquad (1.3.169)$$

so
$$\left(\frac{\partial \mu}{\partial N}\right)_{T,V} = \frac{1}{N\rho K_T} \quad (1.3.170)$$

$$\overline{(\Delta N)^2} = NkT\rho K_T \quad (1.3.171)$$

where K_T is the isothermal compressibility. Since we are really considering density fluctuations, it is obvious that

$$\frac{\overline{(\Delta N)^2}}{N^2} = \frac{\overline{(\Delta V)^2}}{V^2} = \frac{kT\rho}{N} K_T = \frac{kT}{V} K_T \quad (1.3.172)$$

Once again we can try this out on the ideal gas. In this case,

$$PV = NkT \quad (1.3.173)$$

$$K_T = -\frac{1}{V}\frac{\partial}{\partial P}\left(\frac{NkT}{P}\right) = \frac{NkT}{P^2 V} = \frac{1}{P} \quad (1.3.174)$$

and
$$\frac{\sqrt{\overline{(\Delta N)^2}}}{N} = \frac{\sqrt{kT/PV}}{N} = \frac{1}{\sqrt{N}} \quad (1.3.175)$$

Once again the RMS fluctuations are one part in $\sqrt{N}$.

This result (as well as the one for the energy) depends on the compressibility (or heat capacity) of the ideal gas. However, compressibilities (and heat capacities) do not differ by large factors between the ideal gas and most other substances, and so the relative RMS fluctuations of these quantities are always of order of $N^{-1/2}$. There is, as we shall see later, one important type of exception to this rule: at certain phase transitions, C_V and/or K_T diverge and the fluctuations become very large, ultimately dominating the behavior of the systems in question. The study of that situation is the subject of Chap. 6.

Equations (1.3.159) and (1.3.172) refer, respectively, to energy fluctuations at constant density and to density fluctuations when the energy is allowed to fluctuate as well. We cannot, for example, assume that the energy fluctuations in the latter case are given by Eq. (1.3.159), since the density fluctuations will themselves contribute to the energy fluctuations. However, as long as the density is kept fixed, the fluctuations of all other quantities are determined by those of the energy, as we have said, through the equilibrium properties of the body. For example, the temperature fluctuations are given by

$$\Delta T = \left(\frac{\partial T}{\partial E}\right)_{V,N} \Delta E = \frac{1}{C_V}\Delta E \quad (1.3.176)$$

Squaring and averaging give

$$\overline{(\Delta T)^2} = \frac{1}{C_V^2}\overline{(\Delta E)^2} = \frac{kT^2}{C_V} \quad (1.3.177)$$

a result valid, for the moment, only when the density is held constant.

1.3 Statistical Mechanics

In order to calculate the mean square fluctuations of the various thermodynamic quantities in a general way, it is convenient to write probability functions like Eq. (1.3.140) in the Gaussian form, Eqs. (1.3.55) and (1.3.56). Suppose that we take $\bar{x} = 0$, so x is itself the departure from equilibrium of a quantity; for example, $x = \Delta T$ if we are interested in temperature fluctuations. Then the mean square fluctuations are

$$\overline{x^2} = \sqrt{\frac{\beta}{2\pi}} \int_{-\infty}^{\infty} x^2 \exp\left(-\tfrac{1}{2}\beta x^2\right) dx$$

$$= \frac{1}{\beta} \qquad (1.3.178)$$

To put the probability function $w(x)$ in the proper form, we start with Eq. (1.3.52) in the form

$$w(x) \propto \exp\left(\frac{\Delta S_t}{k}\right) \qquad (1.3.179)$$

where

$$\Delta S_t = S_t(x) - S_t(0) \qquad (1.3.180)$$

defines what we mean by Δ of any quantity and S_t is the combined entropy of the sample plus medium after we have isolated the sample to make our measurement. Then

$$S_t = S + S' \qquad (1.3.181)$$

and since the medium is always taken to be internally in equilibrium,

$$\Delta S' = \frac{\Delta E' + P \Delta V'}{T} = \frac{-\Delta E - P \Delta V}{T} \qquad (1.3.182)$$

In the last step we have used the fact that $V + V'$ and $E + E'$ are constants. Substituting Eqs. (1.3.182), (1.3.181), and (1.3.180) into (1.3.179), we have

$$w \propto \exp\left(-\frac{\Delta E - T \Delta S + P \Delta V}{kT}\right) \qquad (1.3.183)$$

In leading order, the argument of the exponential is zero, corresponding to the fact that w is a maximum at equilibrium [the first member of Eq. (1.3.53)]. We are interested in the curvature about that maximum, which will give the width of the Gaussian curve. In other words, we want to take the exponent of second order. Expanding E about its most probable value, we get

$$\Delta E = \frac{\partial E}{\partial S} \Delta S + \frac{\partial E}{\partial V} \Delta V + \frac{1}{2}\frac{\partial^2 E}{\partial S^2}(\Delta S)^2 + \frac{\partial^2 E}{\partial S\, \partial V} \Delta S\, \Delta V + \frac{1}{2}\frac{\partial^2 E}{\partial V^2}(\Delta V)^2$$

$$(1.3.184)$$

In cutting the series off at second order, we are assuming, in effect, that the sample is large, and the fluctuations will therefore be small. The coefficients are to be evaluated at their most probable values, so that $\partial E/\partial S = T$, $\partial E/\partial V = -P$, and Eq. (1.3.184) reduces to

$$\Delta E - T \Delta S + P \Delta V = \frac{1}{2}\left[\frac{\partial^2 E}{\partial S^2}(\Delta S)^2 + 2\frac{\partial^2 E}{\partial S \partial V}\Delta S \Delta V + \frac{\partial^2 E}{\partial V^2}(\Delta V)^2\right]$$

The term in brackets may be written

$$\Delta S\left(\frac{\partial^2 E}{\partial S^2}\Delta S + \frac{\partial^2 E}{\partial S \partial V}\Delta V\right) + \Delta V\left(\frac{\partial^2 E}{\partial S \partial V}\Delta S + \frac{\partial^2 E}{\partial V^2}\Delta V\right)$$

$$= \Delta S \Delta\left(\frac{\partial E}{\partial S}\right) + \Delta V \Delta\left(\frac{\partial E}{\partial V}\right) = \Delta S \Delta T - \Delta V \Delta P$$

Substituting this result into Eq. (1.3.183), we have

$$w \propto \exp\left(-\frac{\Delta T \Delta S - \Delta P \Delta V}{2kT}\right) \qquad (1.3.185)$$

Let us imagine that after we isolate the sample, we measure its temperature and its volume in order to determine the fluctuations in those quantities. All other properties of the body are definite functions of the temperature and volume, so their fluctuations can be written in terms of those in T and V. Thus, $S = S(T, V)$,

$$\Delta S = \left(\frac{\partial S}{\partial T}\right)_V \Delta T + \left(\frac{\partial S}{\partial V}\right)_T \Delta V = \frac{C_V}{T}\Delta T + \left(\frac{\partial P}{\partial T}\right)_V \Delta V$$

where we have used the Maxwell relation, Eq. (1.2.21). Moreover,

$$\Delta P = \left(\frac{\partial P}{\partial T}\right)_V \Delta T + \left(\frac{\partial P}{\partial V}\right)_T \Delta V = \left(\frac{\partial P}{\partial T}\right)_V \Delta T - \frac{1}{VK_T}\Delta V$$

Multiplying, respectively, by ΔT and ΔV, we get

$$\Delta T \Delta S - \Delta P \Delta V = \frac{C_V}{T}(\Delta T)^2 - \frac{1}{VK_T}(\Delta V)^2 \qquad (1.3.186)$$

Notice that the cross term has dropped out. If we now substitute this result into Eq. (1.3.185), we get

$$w(\Delta T, \Delta V) \propto \exp\left[-\frac{C_V(\Delta T)^2}{2kT^2} + \frac{(\Delta V)^2}{2kTVK_T}\right] \qquad (1.3.187)$$

1.3 Statistical Mechanics

Rather than a simple distribution function for one variable, as in Eq. (1.3.55), we have here the combined probabilities of fluctuations in two variables, with the form

$$w(x, y) \propto \exp\left(-\tfrac{1}{2}\beta_1 x^2 - \tfrac{1}{2}\beta_2 y^2\right) \quad (1.3.188)$$

However, the absence of a cross term allows $w(x, y)$ to split into two independent factors, each of which behaves like Eq. (1.3.55). For example,

$$1 = \iint w(x, y)\, dx\, dy \propto \int \exp\left(-\tfrac{1}{2}\beta_1 x^2\right) dx \int \exp\left(-\tfrac{1}{2}\beta_2 y^2\right) dy$$

so that each is normalized as in Eq. (1.3.56).

$$w(x, y) = \sqrt{\frac{\beta_1}{2\pi}} \exp\left(-\tfrac{1}{2}\beta_1 x^2\right) \sqrt{\frac{\beta_2}{2\pi}} \exp\left(-\tfrac{1}{2}\beta_2 y^2\right) = w(x)w(y) \quad (1.3.189)$$

Then
$$\overline{x^2} = \int x^2 w(x, y)\, dx\, dy = \frac{1}{\beta_1} \int w(y)\, dy = \frac{1}{\beta_1} \quad (1.3.190)$$

as in Eq. (1.3.178), and

$$\overline{y^2} = \frac{1}{\beta_2}$$

If we take $x = \Delta T$ and $y = \Delta V$, we see by extracting β_1 and β_2 from Eq. (1.3.187) that

$$\overline{(\Delta T)^2} = \frac{kT^2}{C_V} \quad \text{and} \quad \overline{(\Delta V)^2} = kTVK_T \quad (1.3.191)$$

in agreement with Eqs. (1.3.177) and (1.3.172). Moreover, we can compute the correlated fluctuations of T and V:

$$\overline{(\Delta T\, \Delta V)} \propto \int \Delta T \exp\left[-\frac{C_V(\Delta T)^2}{2kT^2}\right] dT \int \Delta V \exp\left[-\frac{(\Delta V)^2}{2kTK_T}\right] dV$$

$$\propto \overline{(\Delta T)}\,\overline{(\Delta V)}$$

$$= 0 \quad (1.3.192)$$

Fluctuations in T and V are uncorrelated. ΔT and ΔV are said to be satistically independent variables.

The statistical independence of T and V is due to the fact that the temperature of the medium is not changed by fluctuations in the volume of the sample. Thus, when V fluctuates, say, to a larger than average value, T still fluctuates symmetrically about the value fixed by the medium—there is no net tendency for it to be higher or lower than average. For analogous reasons (as we shall see below), P and S are statistically independent. However, other sets, such as T, S and P, V, are not, since they are connected by the internal properties of any sample.

One consequence of the statistical independence of T and V is that Eq. (1.3.177), which we found for temperature fluctuations at constant V, turns out to be correct [Eq. (1.3.191)] even if V is free to fluctuate as well.

Fluctuations in all other quantities may be deduced from the ones we have already computed. For example, the energy fluctuations when both T and V are free to fluctuate are

$$\Delta E = \left(\frac{\partial E}{\partial T}\right)_V \Delta T + \left(\frac{\partial E}{\partial V}\right)_T \Delta V = C_V \Delta T + \left[T\left(\frac{\partial S}{\partial V}\right)_T - P\right] \Delta V$$

Squaring and averaging give

$$\overline{(\Delta E)^2} = C_V^2 \overline{(\Delta T)^2} + \left[T\left(\frac{\partial S}{\partial V}\right)_T - P\right]^2 \overline{(\Delta V)^2}$$

$$= kT^2 C_V + \left[T\left(\frac{\partial P}{\partial T}\right)_V - P\right]^2 kTVK_T \quad (1.3.193)$$

For the entropy,

$$\overline{(\Delta S)^2} = \left(\frac{C_V}{T}\right)^2 \overline{(\Delta T)^2} + \left(\frac{\partial P}{\partial T}\right)^2 \overline{(\Delta V)^2}$$

$$= kC_V + kTVK_T \left(\frac{\partial P}{\partial T}\right)_V^2 = kC_P \quad (1.3.194)$$

We have used Eq. (1.2.107). The mean square pressure fluctuations are given by

$$\overline{(\Delta P)^2} = \left(\frac{\partial P}{\partial T}\right)_V^2 \frac{kT^2}{C_V} + \frac{1}{(K_T V)^2} kTVK_T$$

$$= \left(\frac{\partial P}{\partial T}\right)_V^2 \frac{kT^2}{C_V} + \frac{kT}{VK_T} \quad (1.3.195)$$

After some juggling this result reduces to

$$\overline{(\Delta P)^2} = -\frac{kT}{V} K_S \quad (1.3.196)$$

We can also multiply together ΔP and ΔS, then average:

$$\overline{\Delta P \, \Delta S} = \left(\frac{\partial P}{\partial T}\right)_V \frac{C_V}{T} \overline{(\Delta T)^2} + \left(\frac{\partial S}{\partial V}\right)_T \left(\frac{\partial P}{\partial V}\right)_T \overline{(\Delta V)^2}$$

$$= \left(\frac{\partial P}{\partial T}\right)_V kT - \left(\frac{\partial S}{\partial V}\right)_T kT = 0 \quad (1.3.197)$$

as promised above. The mean square or correlated fluctuations of all other quantities may be found by similar arguments (see Prob. 1.14).

1.4 REMARKS ON THE FOUNDATIONS OF STATISTICAL MECHANICS

a. Quantum and Thermodynamic Uncertainties

We have developed the formalisms of this chapter, not as an end in themselves, but in order to provide a clear conceptual basis for the studies of the properties of matter that are to follow. In an attempt to keep things relatively simple, we have evaded confronting certain points that are usually dealt with in more traditional treatments. Let us, before closing the chapter, look into some of these questions.

We started out, in Sec. 1.1, with what seemed to be a simple, unambiguous picture: an isolated system of N particles in volume V has a fixed energy E and therefore some definite number, Γ, of ways in which it can divide up the energy among its various particles. With all these ways being equally probable in equilibrium, we could write $E = E(S, V, N)$, where $S = k \log \Gamma$, and such concepts as temperature, $T = \partial E/\partial S$, fell neatly into place. The rest of equilibrium thermodynamics followed essentially from the fact that E is an exact function of two or three variables and has the properties of such a function.

It turned out, however, that quantum mechanically, and even classically, that simple picture of a definite number of states associated with a definite energy could not be retained exactly. In quantum mechanics the difficulty arises because, owing to the necessity of allowing the system to undergo transitions among its allowed states, there is necessarily some uncertainty in the energy of any system and, consequently, some uncertainty in the number of allowed states. The same problem arose in the classical discussion (Sec. 1.3e) when we tried to assign a number of states to the accessible portion of phase space for an isolated system. It was necessary to think of the system as restricted to a volume—a thin shell—rather than a surface in phase space, thus introducing precisely the same uncertainty in energy [when we set $\tau = 2\pi\hbar$ in Eq. (1.3.99) the correspondence became exact].

Operationally, this difficulty was avoided by means of the formalism set up in Sec. 1.3a. By dealing with a body immersed in a medium, rather than an isolated body, we found ways of deducing its statistical properties by doing weighted sums over all its states (or integrals over all its phase space) rather than counting the number of accessible states at fixed energy. To perform the needed operations, it is only necessary to know the body's density of states—the number of states per unit range of energy—rather than the total number of states at any fixed energy. Since the density of states is always well defined, the problem of uncertainty in the energy does not interfere with the application of the formulas derived in Sec. 1.3a. Nevertheless, the derivation of those formulas still rests on the idea that an isolated system has a definite number of states—or at least a definite entropy—in equilibrium. The time has come to look more deeply into this problem.

The resolution of the dilemma lies basically in the fact that the number of states of a macroscopic system is so large that even very substantial uncertainties in it make no appreciable difference in its logarithm, the entropy. This point was alluded to in Sec. 1.1, but we are now in a better position to evaluate its implications. As pointed out there, since the entropy (divided by k) of a typical system not too close to its ground state is of order N, the number of particles, the number of possible states is a number whose logarithm is of order, say, 10^N. Thus, even a factor 10^{23} uncertainty in the number of states yields only an additive uncertainty of log $(10^{23}) \approx 55$ to the entropy, to be compared to its value $N \approx 10^{23}$. These numbers are just made up, however. Let us see how big the errors really are.

We consider an isolated system of energy $\mathscr{E}_0$. We now know that there must be some uncertainty, $\delta\mathscr{E}_0$, in that energy, and we wish to know how much uncertainty is thereby introduced into the entropy. The density of states, $\rho(\mathscr{E})$, is always a monotonically rising function of $\mathscr{E}$, so we can be sure that $\rho(\mathscr{E}_0)$ is larger than the average value of $\rho(\mathscr{E})$ for all lower energies:

$$\rho(\mathscr{E}_0) \geq \frac{1}{\mathscr{E}_0} \int_0^{\mathscr{E}_0} \rho(\mathscr{E}) \, d\mathscr{E} \tag{1.4.1}$$

The entropy of a system in a shell $\delta\mathscr{E}_0$ wide in phase space is

$$S = k \log \left[\rho(\mathscr{E}_0) \, \delta\mathscr{E}_0 \right] \tag{1.4.2}$$

[since the number of states in the shell is $\rho(\mathscr{E}_0) \, \delta\mathscr{E}_0$], which, using Eq. (1.4.1), can be given a lower bound:

$$k \log \left[\rho(\mathscr{E}_0) \, \delta\mathscr{E}_0 \right] \geq k \log \frac{\delta\mathscr{E}_0}{\mathscr{E}_0} + k \log \int_0^{\mathscr{E}_0} \rho(\mathscr{E}) \, d\mathscr{E} \tag{1.4.3}$$

On the other hand, away from the ground state, the number of states at energies up to and including the shell must be greater than the number in the shell alone:

$$\int_0^{\mathscr{E}_0} \rho(\mathscr{E}) \, d\mathscr{E} \geq \rho(\mathscr{E}_0) \, \delta\mathscr{E}_0 \tag{1.4.4}$$

which gives us an upper bound as well:

$$k \log \left[\int_0^{\mathscr{E}_0} \rho(\mathscr{E}) \, d\mathscr{E} \right] \geq S \geq k \log \left[\int_0^{\mathscr{E}_0} \rho(\mathscr{E}) \, d\mathscr{E} \right] - k \log \frac{\mathscr{E}_0}{\delta\mathscr{E}_0} \tag{1.4.5}$$

1.4 Remarks on the Foundations of Statistical Mechanics 85

Since the entropy is rigorously bounded in this way, the largest possible uncertainty in the entropy is $k \log (\mathscr{E}_0/\delta\mathscr{E}_0)$. We can be sure that our formulation is valid if we require that this number be small compared to the entropy itself. First, however, let us see if the form we have found for the uncertainty in the entropy makes sense.

The form seems, at first, paradoxical. The smaller $\delta\mathscr{E}_0$ is, the less certain the entropy. This behavior can be understood in a number of ways. The most direct is from the point of view of phase space. The entropy depends on the number of cells in a thin shell. If the shell gets too thin, the number of cells becomes sensitive to whether cells fall just inside or just outside the shell; there is an edge effect, in other words. A thick shell is less susceptible to these fluctuations; the edges are less important.

We can, of course, imagine a truly isolated system with an exact energy. Such a system would be in a stationary state, never making any transitions at all, and the concept of entropy would therefore have no meaning for it. That is not quite the situation we have envisaged, since in order to speak of the various properties we were concerned with, we have had to think of making measurements from time to time and then averaging over the results of these measurements. For example, we have often spoke of "disconnecting" a sample from the medium, so that the sample was isolated with the energy of the quantum state it happened to be in at that instant. The state of the system at a certain instant cannot be an exact-energy eigenstate—the very process we described introduces uncertainties into the energies of the subsystem and medium. Alternatively, we thought of an isolated system as consisting of perfect gas particles, each of which has energy eigenstates given by Eq. (1.1.1). The state of the system was described in terms of some way of dividing all the energy of the system among the particles. But that could not be an exact description; there had to be some residual interactions in order to allow transitions between the states. The states, then, have finite lifetimes, and the system has a corresponding uncertainty in its energy. If, in fact, the energy of the system is too precisely determined, we must have done so by eliminating some of the interactions (or measurements) that allowed transitions to take place, thus isolating some of the states and increasing the uncertainty in the number still accessible. That is the point of Eq. (1.4.5).

Let us summarize briefly what we have said here. In order to make our statistical arguments, we always imagine dividing our system into subunits of some sort, either a macroscopic sample and a much larger medium or perhaps into individual particles with definite energies in each state (it does not matter whether the particles interact with each other or not). This separation is always approximate; there are always residual interactions or, what is the same thing, occasional measurements, which introduce uncertainty in the system energy but at the same time allow the system to make transitions among its various states. In fact, our original argument, that a given system would have some particular entropy depending on its energy, remains valid provided that the uncertainty in the energy is not too small.

From Eq. (1.4.5) the condition that the entropy be precisely determined is

$$\frac{\mathscr{E}_0}{\delta\mathscr{E}_0} \ll e^{S/k} \tag{1.4.6}$$

The uncertainty $\delta\mathscr{E}_0$ is associated with a time, either the time between measurements in our imaginary sample and medium system or, more realistically for the matter we are going to be studying, the relaxation time for transition between states,

$$\tau \approx \frac{\hbar}{\delta\mathscr{E}_0} \tag{1.4.7}$$

Together with Eq. (1.4.6), this gives

$$\tau \ll \frac{\hbar}{\mathscr{E}_0} e^{S/k} \tag{1.4.8}$$

This is not a severe restriction, since S/k is usually of order 10^{23}.

On the other hand, it is important that $\delta\mathscr{E}_0$ not be too large. For one thing, $\delta\mathscr{E}_0 \ll \mathscr{E}_0$ is a condition for writing Eq. (1.4.4). There is, however, a much more severe restriction. In working out the formalisms of Secs. 1.3a and 1.3f, we imagined a subsystem in a temperature bath to be fluctuating among "definite" energy states, which we now know to be somewhat indefinite. It is obviously necessary for the validity of those arguments that the states among which the sample is fluctuating themselves be at least relatively well defined. To ensure this, we must require that the quantum uncertainty in the energy be small compared to the thermodynamic uncertainty, which we may take to be the RMS fluctuations in the energy:

$$\delta\mathscr{E}_0 \ll \sqrt{\overline{(\Delta E)^2}} \tag{1.4.9}$$

or, for example, using Eq. (1.3.159), which refers to a sample of fixed volume,

$$\delta\mathscr{E}_0 \ll \sqrt{kC_V T}$$

or

$$\frac{\hbar}{T\sqrt{kC_V}} \ll \tau \ll \frac{\hbar}{\mathscr{E}_0} e^{S/k} \tag{1.4.10}$$

(S in the right-hand number is not the entropy at temperature T but rather the entropy when the object in question is isolated with energy $\mathscr{E}_0$.) Equation (1.4.10) gives the outside limits of validity of the formulas we have worked out in this chapter. It will turn out that the restrictions are comfortably satisfied by virtually all macroscopic systems, and even many quite microscopic ones. When we apply these considerations to a small bit of real matter, we must keep in mind that it is never actually isolated from its medium; in other words, the τ is not of our choosing but depends instead on the properties of the stuff we are studying. However, for any macroscopic system there is a wide range of τ's satisfying the conditions (1.4.10).

b. Status of the Equal Probabilities Postulate

The basic assumption we have made is that all allowed states of a system are equally likely. We may, if we wish, allow the justification of this assumption to be the success of the predictions it leads to, which, of course, is very great. It is possible, however, to trace the assumption back to two logically preceding steps, which help to illuminate what we have done and may also be of some use in what follows. The two steps are, first, the ergodic hypothesis and, second, either (quantum mechanically) the principle of detailed balance or (classically) Liouville's theorem.

The ergodic hypothesis states that any system starting out in any (energetically) allowed state will, if we wait long enough, eventually reach any other allowed state. Classically, this supposition is easily refuted by counter-example, and a great deal of time and ingenuity was once expended to find ways of arguing that it was very close to being true in sufficiently complicated systems. Quantum mechanically, it is only necessary to suppose that there exists some way, even if it is only indirect, by way of intermediate states, for the system to make transitions from any one state to any other. That alone is enough to ensure that, starting with the system in any one if its allowed states, the probability of occupation of all other states will eventually become finite.

The second step, both quantum mechanically and classically, is to use a theorem that may be proved directly from the appropriate mechanics. The classical statement, Liouville's theorem, is that the density of points (representing possible states of the system) in phase space is conserved in time. The quantum statement is that in a closed system (what we have called an isolated system) the probability of any transition is equal to the probability of the reverse transition.

Classically, as we have said earlier, if a system occupies a given point in phase space at some instant, its future trajectory is completely determined. If we imagine a cloud of nearby points in phase space, then as time progresses—each point following its own trajectory—the cloud will move through the phase space, changing its shape, but, according to Liouville's theorem, it will not change its volume. We shall not try to prove the theorem here but only note that the proof is applicable only to the volume in a space of coordinates and momenta, not, for example, coordinates and velocities. The two kinds of spaces are equivalent for particles of the kind we have been studying, but they differ if, for instance, we consider charged particles in magnetic fields.

The quantum theorem of detailed balance tells us that if states a and b of a system have the same energy, then if P_{ab} is the probability per unit time of a transition from a to b, and P_{ba} that from b to a,

$$P_{ab} = P_{ba} \qquad (1.4.11)$$

Aside from leading to our basic postulate that all states are equally likely, the statement in the form of Eq. (1.4.11) is often used together with that postulate to analyze dynamical processes that occur in equilibrium. The rate at which transitions take place from a to b is

$$R_{ab} = w_a P_{ab} \qquad (1.4.12)$$

where w_a is the probability that the system be in state a. Since, according to our postulate,

$$w_a = w_b \qquad (1.4.13)$$

when the system is in equilibrium, it follows that

$$R_{ab} = R_{ba} \qquad (1.4.14)$$

in equilibrium. Thus, for example, when a liquid is in equilibrium with its own vapor, not only is the rate of evaporation equal to the rate of condensation, but those rates are also equal in detail—that is, for each direction and velocity of gas atoms incident on the interface. Another celebrated example, from which Einstein deduced the law of spontaneous and stimulated emission: for an atom of matter in equilibrium with a radiation field, the probability of the atom being in the ground state and absorbing a photon of a given frequency must be equal to the probability that it is in the excited state and emits the photon.

Given the ergodic hypothesis and either Liouville's theorem or detailed balance, the equal probability of all states in equilibrium then follows. In equilibrium, the probability of finding the system in any particular state must be independent of time. Classically, only a uniform distribution of probabilities in phase space is consistent with the picture of a cloud of phase points of constant density continually wandering around, getting into every accessible region, without accumulating monotonically somewhere. Quantum mechanically, the rate at which the probability of being in state a changes, $\dot{w}_a$, is given by

$$\dot{w}_a = \sum_b w_b P_{ba} - w_a \sum_b P_{ab} \qquad (1.4.15)$$

the first term on the right being the rate at which transitions are made from all other states into a, the second term the rate at which transitions from a

1.4 Remarks on the Foundations of Statistical Mechanics

into all other states occur. The condition for equilibrium is that $\dot{w}_a = 0$; since $P_{ba} = P_{ab}$, we have

$$0 = \sum_b P_{ab}(w_b - w_a) \qquad (1.4.16)$$

for all a. This is a set of linear homogeneous equations for all the $(w_b - w_a)$ with coefficients P_{ab}. It turns out that the determinant of the matrix P_{ab} is not equal to zero, so that the only solutions of Eq. (1.4.16) are $w_a = w_b$ for all a and b, which is our postulate.

The interesting point here is that the P_{ab} need not all be the same—in fact, some of them may be zero. Suppose that at some instant the system is in state a, with only a very small probability of getting to, say, state z, although it easily makes transitions to states c, d, and so on. We can even suppose that it has little chance of getting from any other state into z. At first glance it looks as if it forever after will be less likely to be found in z than in the other states, but we have just concluded instead that if we wait long enough, z will have the same likelihood as any of the others. The way this situation occurs is that although it is unlikely for it to get into z, once it gets in, it is just as unlikely to get out. The result is that it spends as much time in z as in any other state.

c. Ensembles

In our discussions up to this point, we have made much use of the idea that a given system or subsystem has probabilities of doing various things. When called upon to define that notion more closely, we have imagined sampling the object in question at various times. Thus, the probability of finding a certain state (either macroscopic or microscopic) was exactly the same as the fraction of its time the object spent in that state. There is an alternative way in which we could picture these statistical processes. Instead of thinking of a single system evolving in time, we could think of a collection of many identical systems, distributed in some appropriate way in phase space. The probability of finding a given state is then the fraction of the collection of systems in that state at any single instant rather than the fraction of its time one system spends in that state. Such a collection of systems is called an *ensemble*.

The idea of ensembles was introduced by Gibbs in order to avoid a serious difficulty in classical statistics, the failure of the ergodic hypothesis. There were certain states that could not be reached from other states of what were otherwise perfectly respectable systems. We can imagine, for example, a gas of particles bouncing elastically back and forth in paths perpendicular to walls that are parallel to each other. The system will clearly never get from that region of phase space to any other of the same energy. The idea of ensembles circumvents this kind of difficulty. If we start out, for example, with an ensemble of systems uniformly distributed throughout a shell of constant energy, Liouville's theorem tells us that the uniform distribution will persist forever, and we are then free to apply our statistical arguments, based on that uniform distribution, to the ensemble.

In quantum statistics, the need for an ensemble over which to take our statistical averages is not nearly so acute, and, in fact, an ensemble average is entirely equivalent to a time average on a single system. The question of which conceptual picture to have in the backs of our minds is purely one of taste. The time-average idea was chosen for this chapter only because the initial terminology is simpler, and it seemed possible to get more quickly into intuitive discussions of how statistics works.

Whichever picture we choose to think of, it is useful to know the names of the principal ensembles, as well as what they correspond to in the development given here, in order to understand other books more easily. Very briefly, the three most important are

1. *The Microcanonical Ensemble* This is the ensemble mentioned above: a uniform distribution in a shell of constant energy. It corresponds to our isolated system.

2. *The Canonical Ensemble* This corresponds basically to our subsystem immersed in a temperature bath. Instead of a uniform distribution in a restricted portion of phase space, phase points are distributed over all of phase space, but with a density proportional to $e^{-\mathscr{E}/kT}$ (it does not have to be a temperature bath, and so a parameter other than T could appear in the exponential, but this is the most common case). The phase points all execute trajectories of constant energy, so the density is preserved in time. Given the exponential dependence, the density falls off with increasing energy, but the volume of phase space increases so rapidly with energy (for macroscopic systems) that nearly all the systems in the ensemble have very close to the average energy. The reader should have no difficulty switching this back into the language we have used before.

3. *The Grand Canonical Ensemble* This is basically the same as the canonical ensemble except that it is open instead of closed. For example, instead of only the energy varying from one system to another in the ensemble, the number of particles varies as well, and there is another parameter (μ/kT) in the distribution, which is proportional to $e^{-\mathscr{E}/kT + N\mu/kT}$

APPENDIX A: THERMODYNAMIC MNEMONIC

The sheer feat of memory involved in keeping track of all the thermodynamic definitions, the proper independent variables of each of the energy functions, and so on can become an impediment to the usefulness of thermodynamics. To overcome this problem, there is a well-known mnemonic device for keeping

Appendix A: Thermodynamic Mnemonic

track of all these things. The device is shown in Fig. 1A.1. You may find it necessary to make up your own mnemonic to remember the mnemonic.

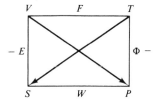

Fig. 1.A.1

We see, first of all, that the four energy functions, each of which occupies an edge of the figure, is each flanked by its proper independent variables at the corners: F by V and T, Φ by T and P, and so on. The energy functions are written in differential form by taking the coefficients from the opposite corners and the differentials from the near corners. If, in going from coefficient to differential the path goes counter to the arrow, the term has a minus sign, as in Fig. 1A.2.

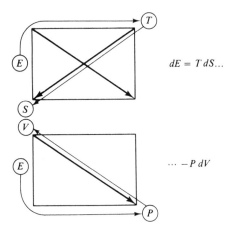

Fig. 1.A.2

Equations of the type $T = (\partial E/\partial S)_V$ follow from these differential forms. In order to get the transformations between the energy functions, start with any one of them. It is equal to the next energy function (in either direction), plus the product of the variable at the next corner times its conjugate across the diagonal, still following the sign convention of the arrows (Fig. 1A.3). The Maxwell relations may also be generated, by going around three corners in order (Fig. 1A.4). In the second step, we go around three corners in the opposite direction, ending on the same side (the paths overlap one side). If there is a minus sign flanking the side where the two paths overlap, there is a minus sign in the result.

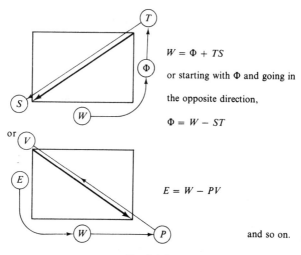

$W = \Phi + TS$

or starting with Φ and going in the opposite direction,

$\Phi = W - ST$

$E = W - PV$

and so on.

Fig. 1.A.3

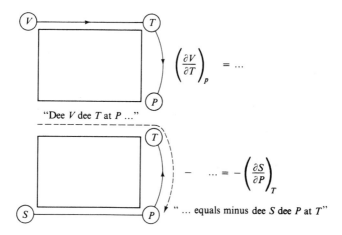

$\left(\dfrac{\partial V}{\partial T}\right)_P = \ldots$

"Dee V dee T at P ..."

$\ldots = -\left(\dfrac{\partial S}{\partial P}\right)_T$

"... equals minus dee S dee P at T"

Fig. 1.A.4

All this sounds more complicated than it is; you will find that after a few minutes of practice generating all the relations, use of the mnemonic becomes easy and automatic.

BIBLIOGRAPHY

The principal reference for this chapter and the next is L. D. Landau and E. M. Lifshitz, *Statistical Physics*, 2nd ed. Translated by J. B. Sykes and M. J.

Kearsley. (Reading, Mass.: Addison-Wesley, 1969.) This is Volume 5 of the Landau and Lifshitz course on theoretical physics. This book, like each of the others in this series, would be a remarkable accomplishment all by itself. Altogether, the series beggars the imagination. The main shortcoming of the book is that it is too elegant. Each sentence seems designed to be chipped in stone and stand as a monument for the ages. The result is that although you can usually be pretty sure that what they have done is correct, you often cannot be sure that you have understood what they have done. It also suffers, inevitably, from being obsolete in places (not incorrect, of course, just superseded)—for example, in the general area of phase transitions. Nevertheless, the book is a landmark in the literature of statistical physics, an endless source of ideas, insights, and techniques.

Wherever possible we have followed the notation of Landau and Lifshitz, as, for example, in the letters for the thermodynamic energy functions. Many of our arguments have been based on theirs, although generally given here in a simplified, or at least elaborated, form. Particular examples are Sec. 1.2d here (variational principles), which should be compared to Sec. 20, Chapter II of Landau and Lifshitz, and Sec. 1.3f (fluctuations), which corresponds to Secs. 112, 113, 114, and 115, Chapter XII. The present chapter will have served a useful purpose if it only has the effect of introducing you to that book.

There is a rich tradition of careful, detailed books by distinguished authors in the field of thermodynamics and statistical mechanics. Probably the first in this line is J. Willard Gibbs, *Elementary Principles in Statistical Mechanics, Developed with Especial Reference to the Rational Foundation of Thermodynamics* (New York: Scribners, 1902), a true classic but a bit dated to be of any real use to study physics from. More to the point is R. C. Tolman, *The Principles of Statistical Mechanics* (Oxford: Clarendon, 1938), which is still the ultimate authority on the most delicate questions of quantum and classical statistics. Also in this tradition are J. Mayer and M. Mayer, *Statistical Mechanics* (New York: Wiley, 1940) and, more recently, K. Huang, *Statistical Mechanics* (New York: Wiley, 1963) and many others. At a more elementary level, M. Zemansky, *Heat and Thermodynamics* (New York: McGraw-Hill, 1957) has become a kind of classic in its own right, and, more recently, there is an excellent treatment by F. Reif, *Statistical Physics* (New York: McGraw-Hill, 1965).

PROBLEMS

1.1 This problem should be done on the basis of Sec. 1.1 only, without using any of the machinery developed later in the chapter.

a. Prove that a perfect gas in equilibrium must have uniform pressure and uniform chemical potential.

b. Show that for a perfect gas, $PV = \frac{2}{3}E$.

c. A system in either mechanical or thermodynamic equilibrium should resist changes. For example, a marble balanced on top of a sphere is not in equilibrium, since a small push will cause its state to change dramatically, and irreversibly, but a marble inside a spherical bowl is in equilibrium at the bottom. We have defined what we mean by equilibrium differently, however. Show that our definition of thermodynamic equilibrium, contained in supposition 1 on page 3, implies stability against small perturbations. Can the connection be made without using supposition 2?

1.2 **a.** For a perfect gas at very low density, we can assume that no two particles have the same quantum numbers ℓ_x, ℓ_y, and ℓ_z. This is the limit in which $PV = NkT$. For this case, using the result of Prob. 1.1b, find the change in entropy and in volume if the masses of the particles are changed at constant P and T.

b. Why could we not have defined the entropy to be Γ itself, instead of $k \log \Gamma$? We would then have had $E = E(\Gamma, N, V)$ and an equation just like Eq. (1.1.6) with S replaced by Γ. Where would we run into trouble?

1.3 Prove that in order for two objects to be in thermodynamic equilibrium with each other, their center of mass velocities must be the same.

1.4 **a.** Show that for a magnetic material at constant T and H, the quantity

$$\varphi = \left(\frac{\partial \Phi}{\partial V}\right)_{T,H}$$

must be uniform in equilibrium. In particular, for a superconductor at critical field, show that the equilibrium condition is

$$\varphi_n(T, H) = \varphi_{sc}(T, H) \tag{1}$$

b. Use general arguments (like additivity of energy) to show that

$$\varphi = \mu\rho - P$$

$$d\varphi = -\frac{S}{V} dT - M\, dH$$

$$\Phi = \varphi V$$

Assume that $\rho = N/V$ = constant and $\int M\, dV = MV$.

c. Construct φ for the superconducting and normal states and show that Eq. (1) above leads to the result we had, Eq. (1.2.93).

$$f_{0,sc} = f_{0,n} - \frac{H_c^2}{8\pi}$$

d. Show that for the superconducting and normal phases taken separately,

$$C_H = C_{eq}$$

where C_H is the heat capacity at constant H and C_{eq} is the heat capacity with $H = H_c(T)$—that is, along phase equilibrium at the same temperature.

e. Our arguments in Sec. 1.2e left V_{sc} undetermined. What does determine V_{sc}?

1.5 It is found that a certain material has heat capacity

$$C_n = aT + bT^3$$

in the normal phase (at all T) and

$$C_{sc} = cT^3$$

in the superconducting phase.

Problems

a. Find the critical temperature in zero field.
b. Find the critical field at zero temperature.

1.6 a. Using the fact (proved in Prob. 1.1h) that $PV = \tfrac{2}{3}E$ in a perfect gas, show that
$$\left(\frac{\partial T}{\partial P}\right)_S = \frac{2}{5}\frac{T}{P}$$
b. Show that
$$\frac{K_S}{K_T} = \frac{C_V}{C_P}$$

1.7 a. Sketch P-V diagrams, including typical isotherms, for the liquid, solid, and gas phases of water and helium.
b. Sketch typical isochores (curves of constant density) crossing the melting and the vapor pressure curves on a P-T diagram for argon. These sketches need not be to scale, nor have correct units; we are interested only in topological features.

1.8 Suppose that in some substance there is a phase transition with a coexistence curve in the P-T plane. Along the curve, the latent heat is zero, and the two phases coexist at the same density, but they have different heat capacities. Find an expression for the slope of the coexistence curve in terms of other measurable quantities.

1.9 We would like to show that particles of the same species (e.g., perfect gas particles with the same mass) must be considered indistinguishable if our formulation of thermodynamics is to be consistent. Do so as follows.
a. Show that
$$\left(\frac{\partial \mu}{\partial T}\right)_N = -\left(\frac{\partial S}{\partial N}\right)_T$$
for a system of fixed volume.
b. For an ideal gas, show that
$$\left(\frac{\partial S}{\partial N}\right)_T = \frac{S}{N} - k$$
but be sure not to use any result that depends on the particles being indistinguishable.
c. Use the additivity property of the entropy to show that the dependence of S/N on the density N/V must be of the form
$$\frac{S}{N} = -k \log\left(\frac{N}{V}\right) + g(T)$$
where $g(T)$ is a function of temperature.
d. Now assume that the particles are distinguishable; use Eq. (1.3.69) without the $N!$ term to show that the entropy is inconsistent with the result of part c.
e. We could, in principle, make a gas of truly distinguishable particles—say, 10^3 particles all using different elements, isotopes, molecules, etc. How does the argument then break down?

1.10 Suppose that $Z = Z(T, V)$ is known explicitly, where Z is the partition function for some system. Find expressions for Φ and W (the Gibbs potential and enthalpy) as a function of Z, T, and V. Show how they then may be written in terms of their proper variables.

1.11 Consider a system that consists of N distinguishable parts, each of which has two possible states, with energies E_0 and $E_0 + \Delta$. Find and sketch the entropy and heat capacity in the high-temperature and low-temperature limits. (What is a natural criterion for high and low temperature in this problem?) Find the energy as a function of temperature.

1.12 There is a model of the thermal behavior of crystalline solids, according to which each of the N atoms of the solid behaves like three independent harmonic oscillators. The $3N$ harmonic oscillators (which are on distinguishable sites) all have the same frequency, ω_0. Their possible energy levels are

$$\varepsilon_n = \hbar\omega_0(n + \tfrac{1}{2}) \qquad (n = 0, 1, 2, \ldots)$$

a. Show that the free energy of the solid is given by

$$F = 3NkT \log\left[1 - \exp(-\hbar\omega_0/kT)\right] + \tfrac{3}{2}N\hbar\omega_0$$

b. Find the heat capacity, C, as a function of temperature. Give the general formula and, to leading order, the low-temperature limit and the high-temperature limit, and sketch a plot of C versus T.

c. Find the average value of n, $\bar{n}(T)$.

d. Suppose that in a fixed volume we wish to add one more atom to the solid without changing the entropy. How much energy is required?

e. There is a contribution to the free energy missing from this picture that will show up if you investigate the conditions for equilibrium between this solid and its own vapor. What is missing from the model?

f. Assume (correctly) that the above deficiency does not affect the entropy of the solid. Then if the model were an accurate one, we could use the calculated entropy, together with the Clausius-Clapeyron equation and measurements of the vapor pressure curve, to measure the absolute entropy of the vapor. Explain why the result would not tell us whether the particles in the vapor are distinguishable or not.

1.13 Compare the form of Eq. (1.3.143)

$$w(E) = \frac{1}{\Gamma(3N/2)}\left(\frac{E}{kT}\right)^{(3N/2)-1}\frac{1}{kT}\exp(-E/kT)$$

to the Gaussian distribution

$$w(x) = \frac{1}{\sqrt{2\pi\bar{x}^2}}\exp(-x^2/2\bar{x}^2)$$

where $x = E - E_m$. This can be done by expanding $\log w(E)$ up to second order in $(E - E_m)$ and comparing the result to $\log w(x)$. Are they the same? Why?

1.14 **a.** In the notation of Sec. 1.3, what is the difference between S_0 and $S_t(0)$? Are they equal to each other? Why not?

b. Find the following quantities: $\overline{(\Delta P \, \Delta V)}$, $\overline{(\Delta T \, \Delta S)}$, $\overline{(\Delta \mu)^2}$, the mean square velocity of a single particle in an ideal gas.

c. Find $\overline{(\Delta W)^2}$, $\overline{(\Delta S \, \Delta V)}$, $\overline{(\Delta T \, \Delta P)}$, $\overline{(\Delta \mu \, \Delta S)}$. See also Prob. 6.2.

1.13 Suppose that in a perfect gas the interactions that cause the system to change its quantum state all take place at the walls of the container. What approximate criteria are necessary for the application of equilibrium thermodynamics? Try to make reasonable numerical estimates of whatever seems important.

TWO

PERFECT GASES

2.1 INTRODUCTION

In this chapter we take the model system so often used as an example in Chap. 1, the perfect gas, and work out its statistical and thermodynamic behavior under a variety of conditions. In Sec. 2.2 we gather together and study in detail all the results in the ideal gas approximation to the perfect gas. As we shall see, that approximation, which is formally that every single-particle quantum state have a low probability of occupation, effectively assumes low densities and high temperatures. Applied to real atoms or molecules, it is consistent with the perfect gas approximation, in that it is under the same conditions that interactions between them may be ignored. The net result is that the formulas arrived at in Sec. 2.2 for the ideal gas correctly describe one of the states of real matter.

As the temperature is reduced and the density increased, the departures from ideal gas behavior observed in real matter are invariably the consequence of the potential energy of interaction between the constituent atoms. However, if there existed a perfect, noninteracting gas in nature, it, too, would eventually depart from ideal behavior, owing to purely quantum mechanical effects. We shall reserve for Chap. 4 the question of how real gases depart from ideality and shall pursue instead in this chapter the apparently academic point of the behavior of perfect gases in the limits of low temperature and high density. Under these limits, when the perfect gas is dominated by quantum mechanical rules, it is said to be degenerate.

2.2 The Ideal Gas

Two types of degenerate behavior are possible, depending on whether the wave function of a state of the many-particle system is symmetric or antisymmetric (i.e., whether it is unchanged or changes sign) under permutation of the particles. The two types of systems form, in the language of Sec. 1.4b, nonergodic sets. The Hamiltonian of the system does not change the symmetry character of the wave function, so that if a system is once symmetric, it is forever after symmetric; there is no way to reach any antisymmetric state. Particles that form symmetric systems are said to obey Bose-Einstein statistics, and those that form antisymmetric systems are said to obey Fermi-Dirac statistics. The necessary distinctions and their basic consequences are worked out in Sec. 2.3.

In Sec. 2.4 we investigate the way in which the two kinds of perfect gas begin to depart from ideality when we start to relax the ideal gas approximation. It turns out that even though the particles are assumed strictly noninteracting, purely quantum statistical effects lead them to behave as if there were forces between them; repulsive forces in the Fermi-Dirac case, attractive in the Bose-Einstein case.

In the last two sections, 2.5 and 2.6, we examine the two kinds of gases in the extreme degenerate limit. The degenerate Fermi-Dirac gas, Sec. 2.5, turns out to be a surprisingly good model for the behavior of conduction electrons in metals, and so we pause in that section to examine some of the reasons why electrons behave so much like perfect gas particles. In the Bose-Einstein case, the apparent attractive forces that we saw start to develop in Sec. 2.4 lead, in the degenerate limit, to a phase transition, a condensation, which must be one of the most spectacular phenomena to occur purely on paper. We treat the condensation as a first-order phase transition and take that occasion to discuss some of the phenomenology of first-order phase transitions in general. Although the Bose condensation never actually occurs in nature, something related to it is responsible for the phenomena superfluidity and superconductivity, to be studied in Chap. 5.

2.2 THE IDEAL GAS

In the course of giving examples for various points made in Chap. 1, the ideal gas approximation to the perfect gas was worked out almost completely. Let us, however, briefly run through the whole thing again here.

The system is a box of many particles, each independently obeying Eqs. (1.1.1) to (1.1.4). We choose as a subsystem all those particles having a particular set of quantum numbers, ℓ_x, ℓ_y, ℓ_z—that is, all the occupants of a particular quantum state, whose wave vector is $\mathbf{q}$. The number of particles in the subsystem is called $n_\mathbf{q}$, so that if $\varepsilon_\mathbf{q}$ is the energy of each particle in that single-particle state, the energy of the subsystem is $n_\mathbf{q}\varepsilon_\mathbf{q}$. $\varepsilon_\mathbf{q}$ and $\mathbf{q}$ are related by Eq. (1.1.1), $\varepsilon_\mathbf{q} = \hbar^2 q^2/2m$.

The Landau potential of the subsystem is Ω_q, related to Ω of the system by Eq. (1.3.77);

$$\Omega = \sum_q \Omega_q \qquad (2.2.1)$$

where the sum is over all the single-particle states. Applying Eq. (1.3.23) to this situation, we obtain, for Ω, Eq. (1.3.76):

$$\Omega_q = -kT \log \sum_{n_q} \exp\left[-\frac{n_q(\varepsilon_q - \mu)}{kT}\right] \qquad (2.2.2)$$

Thus far we have made no approximation of ideality, or classical behavior. The difference between the classical behavior we wish to investigate here and the quantum or degenerate behavior that will concern us in the rest of this chapter lies in the choice of what occupation numbers, n_q, are allowed to contribute to Ω_q in Eq. (2.2.2). To obtain the classical case, we assume that the average value of n_q is small for all $\mathbf{q}$:

$$\overline{n}_q \ll 1 \qquad (2.2.3)$$

As we saw in Sec. 1.3d, this is equivalent to Eq. (1.3.83):

$$e^{\mu/kT} \ll 1 \qquad (2.2.4)$$

which, as we shall see shortly, is valid when the density is low and the temperature high. Each term in the sum in Eq. (2.2.2) has in it a factor $e^{\mu/kT}$ raised to the n_q power. Since the other factor, $\exp(-\varepsilon_q/kT)$, is less than one for all $\mathbf{q}$, the approximation (2.2.4) allows us to drop all terms except $n_q = 0$ and $n_q = 1$:

$$\Omega_q = -kT \log\left[1 + \exp\left(-\frac{\varepsilon_q - \mu}{kT}\right)\right] \qquad (2.2.5)$$

The average occupation number is given by

$$\overline{n}_q = -\frac{\partial \Omega_q}{\partial \mu} = \exp\left(-\frac{\varepsilon_q - \mu}{kT}\right) \qquad (2.2.6)$$

and we see that Eq. (2.2.3) is satisfied for all $\mathbf{q}$ if Eq. (2.2.4) is satisfied.

Substituting (2.2.6) into (2.2.5) gives

$$\Omega_q = -kT \log(1 + \overline{n}_q) = -kT\overline{n}_q \qquad (2.2.7)$$

where we have made use of (2.2.3) to approximate the logarithm. Since $\Omega = -PV$, and

$$N = \sum_q \overline{n}_q \qquad (2.2.8)$$

Eq. (2.2.1) may be applied to give

$$PV = NkT \qquad (2.2.9)$$

2.2 The Ideal Gas

Equation (2.2.9) is the equation of state of the ideal gas, obtained by the same argument earlier.

More information than that is available to us, however. The quantity $\bar{n}_q$ depends on T, V, and μ through Eq. (2.2.6) [recall that ε_q depends on $V = L^3$ from Eqs. (1.1.1) to (1.1.4)]. Thus, Eq. (2.2.7), together with (2.2.1), should give us $\Omega(T, V, \mu)$, which would be everything there is to know. Alternatively, Eq. (2.2.8) gives us $N(T, V, \mu)$, which, together with (2.2.9), can be solved for $\mu(P, T)$. Let us proceed this latter way.

$$N = \sum_q \bar{n}_q = \int \exp\left(-\frac{\varepsilon - \mu}{kT}\right) \rho(\varepsilon)\, d\varepsilon \qquad (2.2.10)$$

Recalling Eq. (1.3.106), we have

$$\rho(\varepsilon) = \frac{4\pi\sqrt{2}\, V m^{3/2}}{(2\pi\hbar)^3} \varepsilon^{1/2} \qquad (2.2.11)$$

Then
$$N = \frac{4\pi\sqrt{2}\, V m^{3/2}}{(2\pi\hbar)^3} e^{\mu/kT} \int e^{-\varepsilon/kT} \varepsilon^{1/2}\, d\varepsilon$$

$$= \frac{V}{\Lambda^3} e^{\mu/kT} \qquad (2.2.12)$$

where, as in Eq. (1.3.126),

$$\Lambda = \frac{2\pi\hbar}{\sqrt{2\pi m kT}} \qquad (2.2.13)$$

and we have used

$$\int x^{1/2} e^{-x}\, dx = \Gamma\left(\frac{3}{2}\right) = \frac{\sqrt{\pi}}{2}$$

Now, using Eq. (2.2.9) to eliminate (N/V) from Eq. (2.2.12), and solving for μ, we find

$$\mu = -kT \log \frac{kT}{P\Lambda^3} \qquad (2.2.14)$$

$$\Phi = N\mu = -kTN \log \frac{kT}{P\Lambda^3} \qquad (2.2.15)$$

We at long last have one of the energy functions in terms of its proper variables. The equation of state is easily retrieved:

$$V = \left(\frac{\partial \Phi}{\partial P}\right)_{T,N} = \frac{kTN}{P} \qquad (2.2.16)$$

and we are now able to find the entropy,

$$S = -\left(\frac{\partial \Phi}{\partial T}\right)_{P,N}$$

$$= -Nk \log P + \frac{5}{2} Nk \log kT + Nk \left(\frac{5}{2} + \frac{3}{2} \log \frac{m}{2\pi \hbar^2}\right) \quad (2.2.17)$$

or using Eq. (2.2.16) to eliminate P in favor of N/V,

$$S = -Nk \log \frac{N}{V} + \frac{3}{2} Nk \log kT + Nk \left(\frac{5}{2} + \frac{3}{2} \log \frac{m}{2\pi \hbar^2}\right) \quad (2.2.18)$$

All other thermodynamic results are easily derived from these formulas. For example, from (2.2.17) and (2.2.18),

$$C_P = T \left(\frac{\partial S}{\partial T}\right)_P = \frac{5}{2} Nk \quad (2.2.19)$$

$$C_V = T \left(\frac{\partial S}{\partial T}\right)_V = \frac{3}{2} Nk \quad (2.2.20)$$

and so on. We may now consider the ideal gas problem solved.

Before going on to nonideal perfect gases, however, let us pause here to make a few observations.

The ideal gas approximation, Eq. (2.2.4), may now be seen, by means of Eq. (2.2.14), to be

$$\frac{kT}{P\Lambda^3} \gg 1 \quad (2.2.21)$$

or

$$\left(\frac{m}{2\pi \hbar^2}\right)^{3/2} \frac{(kT)^{1/2}}{P} \gg 1 \quad (2.2.22)$$

The approximation is thus valid under a combination of high temperature and low pressure. Using (2.2.9) to replace P by NkT/V in Eq. (2.2.21), the condition may also be stated

$$N \frac{\Lambda^3}{V} \ll 1 \quad (2.2.23)$$

We have been imagining the gas to consist of N, a large number, of particles fluctuating about among momentum eigenstates, each particle with an energy, on the average, of order kT. We can, if we wish, think of the same many-particle states among which the gas is fluctuating as consisting of fairly well-localized particles by constructing wave packets out of the available momentum states. Since the particles have linear components of momentum on the order of $(2mkT)^{1/2}$, we can expect the wave packets we construct to

2.2 The Ideal Gas

have linear extent of the order of $\hbar/(2mkT)^{1/2}$—that is, approximately Λ. Thus Λ^3 can be thought of, roughly, as the quantum mechanical size of the particles. Equation (2.3.23) shows that the criterion for the ideal gas approximation is essentially that the density be low enough to ensure that the particles do not overlap quantum mechanically. That being the case, we can picture a rarefied gas of particles with fairly definite positions, whose uncertainties in momentum are no worse than they would have been thermodynamically if they were classical objects. In short, we can use a classical description. As we shall see later, Eq. (2.2.23) may be satisfied, and a classical description therefore valid, even when the system can no longer be thought of as a rarefied gas on other grounds.

We argued in Sec. 1.3d that, for the classical perfect gas, we could obtain the thermodynamic functions from Eq. (1.3.69), which, together with Eq. (1.3.125), may be written

$$F = -NkT \log\left(\frac{V}{\Lambda^3}\right) + kT \log N! \qquad (2.2.24)$$

This should be compared to the free energy constructed from Eq. (2.2.15) with the help of the equation of state,

$$F = \Phi - PV = -kTN \log \frac{V}{N\Lambda^3} - NkT \qquad (2.2.25)$$

The two results are the same provided that

$$\log N! = N \log N - N \qquad (2.2.26)$$

Equation (2.2.26) is called Stirling's formula, and we have arrived at it in a rather roundabout way. A more conventional derivation would begin by noting that

$$N! = 1 \times 2 \times 3 \times \cdots \times N$$

so that

$$\log N! = \log 1 + \log 2 + \cdots + \log N$$

$$= \sum_1^N \log m = \sum_1^N (\log m) \, \Delta m$$

The sum is then approximated by an integral,

$$\sum_1^N (\log m) \, \Delta m \approx \int_1^N \log m \, dm$$

$$= [m \log m - m]_1^N$$

$$\approx N \log N - N \qquad (2.2.27)$$

The approximations become exact in the limit of infinite N. We thus see that it is in that limit that the two formulations we have used, what we have learned to call the canonical and grand canonical ensembles, become exactly the same. That situation is to be expected, since in the canonical ensemble there are exactly N particles in the system, whereas in the grand canonical ensemble the particle number has fluctuations of order $\sqrt{N}$, as we saw in Sec. 1.3f. These latter fluctuations become negligible when N approaches infinity.

In spite of the allegedly classical nature of our results, however, we find $\hbar$ appearing in Eqs. (2.2.17) and (2.2.18). We had actually anticipated that this would occur, in Sec. 1.3e. According to Eqs. (1.3.95) and (1.3.99), we were to expect in every entropy a term of the form

$$-fk \log (2\pi\hbar)$$

where f is the number of degrees of freedom; in this case, $f = 3N$. That is just the term we find in Eqs. (2.2.27) and (2.2.18).

The behavior we have deduced for the ideal gas is to be expected to occur when the low-density approximation we have made is valid. However, that approximation has been applied to what was already only a model: a system of noninteracting particles. That model is, in turn, an approximation to the behavior of real matter under certain conditions. Basically, we can expect the noninteracting model to be a good one to represent real matter when the atoms composing it are far apart and when the kinetic energy of the atoms is large compared to the potential energies of interaction between them. These conditions are met in the same limit as the ideality conditions: high temperature and low density. Real monatomic gases, at sufficiently high temperature and low density, are accurately described by the ideal gas results we have obtained.

The range of conditions under which real gases behave ideally cannot be given in terms of density or temperature alone but require instead a combination of the two, such as Eq. (2.2.23). For example, the vapor of liquid helium satisfies Eq. (2.2.23) down to temperatures below 1°K, and is, in fact, an ideal gas even at those low temperatures, because the density of the gas at the vapor pressure is extremely low.

Suppose that we take some (reasonably large) number of real atoms, N, in a fixed volume V and start to lower the temperature. Regardless of what N and V we start with, it is clear that at some temperature one or the other of the approximations we have made must start to break down. Equation (2.2.18), for example, cannot remain valid to arbitrarily low temperature because the entropy would then diverge instead of vanishing, as required by the Third Law of Thermodynamics. For all real matter, it turns out that the perfect gas, or noninteracting approximation, breaks down first. We thus invariably find departures from ideal behavior when Eq. (2.2.23) is still well satisfied. The criterion necessary for a real gas to obey the ideal gas equations

2.3 Bose-Einstein and Fermi-Dirac Statistics

must depend on the interactions and, in practice, be more stringent than Eq. (2.2.23). We shall return to it in Chap. 4, where we shall study the way in which real gases depart from ideality.

Even if the initial density is quite low (say, much less than the critical value for a given substance), as we further reduce the temperature, still keeping N and V fixed, we reach a point where the gas begins to condense into a liquid; and at lower temperature still (for all substances except helium), the liquid freezes into a solid. These phenomena are, of course, consequences of the interactions. The condensed (liquid or solid) phase eventually comes to contain most of the atoms present, although it occupies only a small fraction of the available volume. The remainder is gas at the vapor pressure. This gas phase may be quite accurately ideal (as in the case of helium mentioned above and even more accurately for other substances), but the number of atoms in the gas phase decreases rapidly as the temperature is reduced. The entropy per gas atom actually does diverge as the temperature approaches zero, but the number of atoms goes to zero, as does the total entropy.

We are left with a question that at first sight seems purely academic. Real substances depart from ideality because the interactions become important; but what of our perfect gas model where there are no interactions? Granted that there are no perfect gases of the kind we have described in nature; nevertheless, our model system ought to obey the laws of thermodynamics. In particular, it should have a nondegenerate ground state, and hence zero entropy at zero degrees. How, then, does our model system behave when the condition (2.2.23) no longer holds?

The question is not purely academic. The answers to it involve a number of physical phenomena on which a substantial part of our understanding of the behavior of real matter is based. The remainder of this chapter will be devoted to investigating how the perfect gas behaves when it is not ideal.

2.3 BOSE-EINSTEIN AND FERMI-DIRAC STATISTICS

In Sec. 1.3d we discussed ways of ensuring that we did not count separately states of a many-particle system that differed from each other only in the matter of which particles had particular quantum numbers. Since the particles are indistinguishable, no new states of the system are to be obtained by interchanging them. In this section we consider some further consequences of the indistinguishability of particles.

It was known before the rise of quantum mechanics that in order to describe the entropy of gases correctly, the particles had to be considered indistinguishable for counting purposes. However, it is only in quantum mechanics that indistinguishability is installed as a fundamental principle. Particles must be indistinguishable in order to obey the uncertainty principle. If we momentarily localize a particle in order to identify it, we can have no idea of its momentum, so that the identification must be lost in the next moment.

As long as the low density and high temperature of a gas ensure that no two particles are likely to have the same quantum numbers, the effects of indistinguishability are purely statistical, as we have seen. However, at low temperatures and higher densities, the kinds of quantum states that the system can be allowed to be in must be restricted in order that there be no way to distinguish between the particles.

To see how this restriction works, suppose that we have a system of two particles that, in a certain state, has wave function $\psi(1, 2)$, or if the particles are interchanged, $\psi(2, 1)$. There must be no way of distinguishing between these two possibilities; thus,

$$|\psi(1, 2)|^2 = |\psi(2, 1)|^2 \qquad (2.3.1)$$

or
$$\psi(1, 2) = \pm \psi(2, 1) \qquad (2.3.2)$$

Interchanging the particles can, at most, change the sign of the wave function. If it does so, the state is said to be antisymmetric; if not, it is symmetric. Now let us suppose that the two particles occupy single-particle states a and b with wave functions ϕ_a and ϕ_b when the system is in the state ψ. If the notation $\phi_a(1)$ means that particle 1 is in state a and so on, the possible ways of constructing two-particle states from the single-particle states ϕ_a and ϕ_b are $\phi_a(1)\phi_b(2)$ and $\phi_a(2)\phi_b(1)$. The physically correct state will be a linear combination of these states that obeys Eq. (2.3.2). Two are possible:

$$\psi_s \propto \phi_a(1)\phi_b(2) + \phi_a(2)\phi_b(1) \qquad (2.3.3)$$

which is symmetric, or

$$\psi_A \propto \phi_a(1)\phi_b(2) - \phi_a(2)\phi_b(1) \qquad (2.3.4)$$

which is antisymmetric.

We see immediately, from Eq. (2.3.4), a remarkable result: if a and b are the same state, so that $\phi_a = \phi_b$, then the antisymmetric version of ψ is zero. For an antisymmetric system of two particles, there is no amplitude for both to occupy the same state.

The procedures for forming Eqs. (2.3.3) and (2.3.4) are easily generalized for systems of many particles. In the symmetric case, the system wave function is the sum over all permutations of the particles of the product of the single-particle functions. The antisymmetric case is formed in the same way, except that each term in the sum is multiplied by a factor (-1) for each

2.3 Bose-Einstein and Fermi-Dirac Statistics

permutation of particles by which it differs from the first term. For example, the antisymmetric wave function for a system of three particles is

$$\psi_A(1, 2, 3) \propto \phi_a(1)\phi_b(2)\phi_c(3) - \phi_a(1)\phi_b(3)\phi_c(2)$$
$$- \phi_a(2)\phi_b(1)\phi_c(3) + \phi_a(2)\phi_b(3)\phi_c(1)$$
$$- \phi_a(3)\phi_b(2)\phi_c(1) + \phi_a(3)\phi_b(1)\phi_c(2) \quad (2.3.5)$$

It is easy to see that if any two of the states are the same (say, $\phi_a = \phi_b$), then $\psi_A = 0$. That is quite generally true. For an antisymmetric system of particles, no more than one can occupy any single-particle state. That restriction is known as the *Pauli exclusion principle*. There is no restriction on the ways in which particles with symmetric wave functions can occupy single-particle states.

Particles whose wave functions are antisymmetric, and which therefore obey the exclusion principle, are said to obey *Fermi-Dirac statistics* and are sometimes called *fermions*. Those with symmetric wave functions obey *Bose-Einstein statistics* and may be called *bosons*. The rules for deciding which type of statistics are to be obeyed by a given particle are simple: particles with odd half-integral spin (spin $\frac{1}{2}$, $\frac{3}{2}$, etc.) are fermions. Those with zero or integer spin are bosons. Notice that if two fermions (or, for that matter, any even number) become bound together, the rules call for us to assign Bose-Einstein statistics to the bound pair. All particles in nature obey either Bose-Einstein or Fermi-Dirac statistics.

For the perfect gas, the states a, b, etc. are those described by Eqs. (1.1.1) to (1.1.4) (with plane-wave functions). The choice of one or the other kind of statistics affects the values that n_q, the occupation number of each single-particle state, is allowed to have. In the classical case, we restricted n_q to either zero or one only because larger values would have negligible probability, ensured by Eq. (2.2.4). We now wish to relax (2.2.4) and allow the choice of statistics to govern the sum Eq. (2.2.2). Let us consider the two cases separately.

Fermi-Dirac statistics

Owing to the exclusion principle, the sum in Eq. (2.2.2) has in it terms only for $n_q = 0, 1$:

$$\Omega_q = -kT \log\left[1 + \exp\left(\frac{\mu - \varepsilon_q}{kT}\right)\right] \quad (2.3.6)$$

The mean occupation number is obtained by using Eq. (1.2.38) in the form

$$\bar{n}_q = -\frac{\partial \Omega_q}{\partial \mu} = \frac{\exp\left[(\mu - \varepsilon_q)/kT\right]}{1 + \exp\left[(\mu - \varepsilon_q)/kT\right]}$$

$$= \frac{1}{\exp\left[(\varepsilon_q - \mu)/kT\right] + 1} \quad (2.3.7)$$

To get Ω and N for the system as a whole, we sum Ω_q and $\bar{n}_q$ over all single-particle states:

$$\Omega = -kT \sum_q \log\left[1 + \exp\left(-\frac{\varepsilon_q - \mu}{kT}\right)\right] \quad (2.3.8)$$

$$N = \sum_q \frac{1}{\exp[(\varepsilon_q - \mu)/kT] + 1} \quad (2.3.9)$$

Bose-Einstein statistics

In this case, there are no restrictions on n_q. The sum in Eq. (2.2.2) runs from zero to infinity:

$$\Omega_q = -kT \log \sum_{n=0}^{\infty} \exp\left[\frac{n(\mu - \varepsilon_q)}{kT}\right] \quad (2.3.10)$$

The sum in Eq. (2.3.10) is of the form

$$\sum_{n=0}^{\infty} x^n \quad (2.3.11)$$

where

$$x = \exp\left(\frac{\mu - \varepsilon_q}{kT}\right) \quad (2.3.12)$$

The condition for the sum, Eq. (2.3.11), to converge is $x < 1$. According to Eq. (2.3.12), we must therefore have $\mu < \varepsilon_q$, and this condition must be met for all q. Thus, for the Bose gas,

$$\mu < \varepsilon_0 \quad (2.3.13)$$

where ε_0 is the single-particle ground state. In our formulation, $\varepsilon_0 = 0$ for the state $q = 0$. Thus, we must require, for a perfect gas of bosons,

$$\mu < 0 \quad (2.3.14)$$

With this restriction, Eq. (2.3.11) may be summed

$$\sum_{n=0}^{\infty} x^n = \frac{1}{1-x} \quad (2.3.15)$$

so that Eq. (2.3.10) reduces to

$$\Omega_q = -kT \log\left[\frac{1}{1 - \exp[-(\varepsilon_q - \mu)/kT]}\right]$$

$$= kT \log\left[1 - \exp\left(-\frac{\varepsilon_q - \mu}{kT}\right)\right] \quad (2.3.16)$$

2.3 Bose-Einstein and Fermi-Dirac Statistics

The mean occupation numbers are

$$\bar{n}_q = -\frac{\partial \Omega_q}{\partial \mu} = \frac{1}{\exp\left[(\varepsilon_q - \mu)/kT\right] - 1} \qquad (2.3.17)$$

and for the system as a whole

$$\Omega = kT \sum_q \log\left[1 - \exp\left(-\frac{\varepsilon_q - \mu}{kT}\right)\right] \qquad (2.3.18)$$

$$N = \sum_q \frac{1}{\exp\left[(\varepsilon_q - \mu)/kT\right] - 1} \qquad (2.3.19)$$

The classical limit

In discussing the ideal gas it was unnecessary to distinguish between fermions and bosons. It follows that, in the proper limit, the formulas for both types of particles should reduce to the ideal values. Let us write together Eqs. (2.3.8) and (2.3.18):

$$\Omega = \mp kT \sum_q \log\left[1 \pm \exp\left(-\frac{\varepsilon_q - \mu}{kT}\right)\right] \qquad (2.3.20)$$

where the upper sign refers to Fermi-Dirac particles and the lower to Bose-Einstein. Using the same convention, Eqs. (2.3.9) and (2.3.19) become

$$N = \sum_q \frac{1}{\exp\left[(\varepsilon_q - \mu)/kT\right] \pm 1} \qquad (2.3.21)$$

Equations (2.3.20) and (2.3.21) emphasize the great similarity between the formulas governing the two types of particles. Yet, as we shall see, they behave in profoundly different ways in the limit of low temperature and high density. On the other hand, in the opposite, classical limit, since $e^{-\mu/kT} \gg 1$, Eq. (2.3.21) reduces to

$$N = \sum_q \exp\left(-\frac{\varepsilon_q - \mu}{kT}\right) \qquad (2.3.22)$$

the same—and equal to the ideal gas result in both cases—and expanding the logarithm by means of

$$\log(1 \pm x) = \pm x - \tfrac{1}{2}x^2 + \cdots \qquad (2.3.23)$$

we find in both cases,

$$\Omega = -kT \sum_q \exp\left(-\frac{\varepsilon_q - \mu}{kT}\right) \qquad (2.3.24)$$

which is, once again, the classical result.

2.4 SLIGHTLY DEGENERATE PERFECT GASES

In the next three sections we shall study the way a perfect gas behaves when its true quantum mechanical nature asserts itself. In doing so, we obviously cannot uncritically make use of classical approximations that we were able to apply earlier. In particular, we must reexamine the procedure for doing sums over states as integrals. We shall start this section with a brief reconsideration of the validity of integrating over discrete states, then go on to consider the leading-order corrections to the ideal gas equation of state as the ideality approximation begins to break down.

The basic criterion for an integral to be an accurate approximation to a sum over discrete states is that the distance between the states be small compared to the smallest units of energy that are otherwise of interest. The sums we wish to perform, basically Eqs. (2.3.20) and (2.3.21), are over single-particle states. It should be obvious that no appreciable error will be introduced by ignoring the discreteness of the states, provided that the distance between adjacent states is much less than the root mean square fluctuations of the energy of a single particle in equilibrium. For a single particle, the energy fluctuations are of order kT, while the spacing between energy states is $(\hbar^2/2m)(2\pi/L)^2$ [see Eqs. (1.1.1) to (1.1.4)]. The criterion for doing integrals is thus

$$kT \gg \frac{\hbar^2}{2m}\left(\frac{2\pi}{L}\right)^2 \qquad (2.4.1)$$

In the rest of this chapter, when we take results in the limit of low temperatures, we shall nevertheless always intend that Eq. (2.4.1) is understood. It is not excessively restrictive. For $L \approx 1$ cm and $m \approx 10^{-24}$ gram—the mass of a hydrogen atom—we find $T \gg 10^{-13}°\text{K}$. If we use the electron mass instead, we must keep $T \gg 10^{-10}°\text{K}$. There is no difficulty in maintaining temperatures above these values; practically speaking, it is impossible to do otherwise.

Equations (2.3.20) and (2.3.21) are to be evaluated, then, by integration over the density of states. There is, however, one minor modification to be made. In distinguishing between Bose-Einstein and Fermi-Dirac particles, we are now taking into account the spin of the particles (the only property we considered in Sec. 1.1 was mass). In the absence of any fields that interact with the spins, a particle with spin S has $2S + 1$ orientations possible, and these are energetically degenerate. This spin degeneracy is superimposed on the momentum degeneracy that we have previously considered; it increases the number of possible states at a given energy by a factor $(2S + 1)$ for each particle and introduces an additive term, $k \log (2S + 1)$, in the entropy for

2.4 Slightly Degenerate Perfect Gases

each particle. All of this will be taken care of properly if we increase the density of states we have been using by a factor $(2S + 1)$. Defining

$$g = 2S + 1 \tag{2.4.2}$$

we rewrite Eq. (1.3.101) for the number of states in $d^3p\, d^3r$ of phase space,

$$g\frac{d^3p\, d^3r}{(2\pi\hbar)^3} \tag{2.4.3}$$

and, correspondingly, rewrite Eq. (1.3.106) as

$$\rho(\varepsilon)\, d\varepsilon = \frac{4\pi\sqrt{2}\, gVm^{3/2}}{(2\pi\hbar)^3}\varepsilon^{1/2}\, d\varepsilon \tag{2.4.4}$$

In this book we shall almost always be concerned with Bose particles of spin zero, $g = 1$ (the value it has had up to now), or Fermi particles with spin $\tfrac{1}{2}$ and $g = 2$.

We are now prepared to investigate the leading-order departures from classical behavior. The classical limit, as we have seen, is obtained from Eq. (2.3.20) by retaining the first-order term in the expansion, Eq. (2.3.23). Higher-order terms are negligible owing to the condition, Eq. (2.2.4),

$$e^{\mu/kT} \ll 1 \tag{2.4.5}$$

We now wish to see what happens when $e^{\mu/kT}$ is still small but has grown large enough so that the next-order term in the expansion of the logarithm can no longer be neglected. We have then

$$\begin{aligned}\Omega &= \mp kT \sum_{\mathbf{q}} \log\left[1 \pm \exp\left(-\frac{\varepsilon_{\mathbf{q}} - \mu}{kT}\right)\right] \\ &= -kT \sum_{\mathbf{q}} \exp\left(-\frac{\varepsilon_{\mathbf{q}} - \mu}{kT}\right) \pm \frac{kT}{2} \sum_{\mathbf{q}} \exp\left[-\frac{2(\varepsilon_{\mathbf{q}} - \mu)}{kT}\right] \\ &= \Omega_{\text{class}} \pm \frac{kT}{2}\exp\left(\frac{2\mu}{kT}\right)\sum_{\mathbf{q}}\exp\left(-\frac{2\varepsilon_{\mathbf{q}}}{kT}\right)\end{aligned} \tag{2.4.6}$$

where Ω_{class} is the classical, or ideal, value of Ω. Notice that the original argument that led to the form of Ω_{class} is somewhat different from the one we have used here. In Sec. 1.3d, and later in Sec. 2.2, we arrived at the form given in Eq. (2.2.5) by arguing that higher-order terms represented multiple occupancy of single-particle states, a condition assumed to be unlikely. We were performing a sum over occupation numbers, $n_{\mathbf{q}}$, inside the logarithm. In this case, the sum over $n_{\mathbf{q}}$ has already been performed, leading to the Bose and Fermi equations of the last section, and so the new term we are considering represents not two particles in a single state (which would be forbidden in the Fermi case) but rather the leading correction as Eq. (2.4.5) begins to fail. We have now to evaluate the sum in Eq. (2.4.6):

$$\sum_q \exp\left(-\frac{2\varepsilon_q}{kT}\right) = \int_0^\infty \rho(\varepsilon) \exp\left(-\frac{2\varepsilon}{kT}\right) d\varepsilon$$

$$= \left(\frac{1}{2}\right)^{3/2} \int_0^\infty \rho(2\varepsilon) \exp\left(-\frac{2\varepsilon}{kT}\right) d(2\varepsilon) \quad (2.4.7)$$

where we have made use of the fact that $\rho(\varepsilon) \propto \varepsilon^{1/2}$. Noting that the 2ε in the integral is a dummy variable, we can write the correction term in Eq. (2.4.6) as

$$\pm \left(\frac{1}{2}\right)^{5/2} kTe^{2\mu/kT} \int_0^\infty \rho(\varepsilon) e^{-\varepsilon/kT} d\varepsilon \quad (2.4.8)$$

The leading-order term is

$$\Omega_{\text{class}} = -kT \exp\left(\frac{\mu}{kT}\right) \sum_q \exp\left(-\frac{\varepsilon_q}{kT}\right)$$

$$= -kT\, e^{\mu/kT} \int_0^\infty \rho(\varepsilon)\, e^{-\varepsilon/kT} d\varepsilon \quad (2.4.9)$$

Comparing Eqs. (2.4.8) and (2.4.9), we see that Eq. (2.4.6) may now be written

$$\Omega = \Omega_{\text{class}} \left[1 \mp \frac{e^{\mu/kT}}{(2)^{5/2}} \right] \quad (2.4.10)$$

or recalling that $\Omega = -PV$ and $\Omega_{\text{class}} = -NkT$,

$$PV = NkT \left[1 \mp \frac{e^{\mu/kT}}{(2)^{5/2}} \right] \quad (2.4.11)$$

The upper sign, remember, is Fermi-Dirac statistics and the lower, Bose-Einstein.

In order to interpret this result, we must think carefully about what we have been doing. Our approach has basically been to calculate Ω in terms of its proper variables, T, V, and μ. We have found that, at a given T, V, and μ, the quantum statistical corrections to the ideal gas equation of state have different signs for Fermi and Bose statistics. The effect of the statistics is evidently to act as a kind of force between the particles, changing the relation between pressure and density at a given temperature. However, we cannot, from Eq. (2.4.11) alone, decide in which case the force is attractive and in which case it is repulsive, since, holding T, V, and μ fixed, we have really found the correction to the quantity P/N. It must be remembered that N is variable in this formulation. What we would like to know is this: What is the leading correction to the pressure of a gas of fixed T, V, and N?

2.4 Slightly Degenerate Perfect Gases

If the pressure is increased, the effective forces are repulsive, and if decreased, they are attractive. We need, then, to compute the pressure as a function of T, V, and N.

To do so, we must construct the free energy, $F(T, V, N)$. In Sec. 1.2b we argued that, owing to the way the transformations among the energy functions were defined, a small change in the system will produce the same change in each of the energy functions, if done with its own proper variables held fixed. In particular, for the correction to the ideal gas free energy, we have from Eq. (1.2.40)

$$(\delta F)_{T,V,N} = (\delta \Omega)_{T,V,\mu} \qquad (2.4.12)$$

where we are using the notation

$$\Omega(T, V, \mu) = \Omega_{\text{class}} + (\delta\Omega)_{T,V,\mu} \qquad (2.4.13)$$

$$F(T, V, N) = F_{\text{class}} + (\delta F)_{T,V,N}$$

Then
$$\delta\Omega = \mp \Omega_{\text{class}} \frac{e^{\mu/kT}}{(2)^{5/2}} = \pm NkT \frac{e^{\mu/kT}}{(2)^{5/2}} \qquad (2.4.14)$$

To find the correction to F, we must eliminate μ from Eq. (2.4.14) in favor of T, V, and N. We can, to sufficient accuracy, replace μ by its classical value in terms of these quantities, using Eqs. (2.2.14) and (2.2.9):

$$\mu_{\text{class}} = -kT \log \frac{V}{N\Lambda^3} \qquad (2.4.15)$$

There will, of course, be corrections to $\mu_{\text{class}}(T, V, N)$, but these corrections, substituted into Eq. (2.4.14), will produce terms of still higher order. With the help of Eqs. (2.4.12) to (2.4.15), we find

$$F = F_{\text{class}} \pm \frac{N^2 kT \Lambda^3}{(2)^{5/2} V} \qquad (2.4.16)$$

Now, $P(T, V, N)$ is given by

$$P = -\frac{\partial F}{\partial V} = P_{\text{class}} \pm \frac{N^2 kT \Lambda^3}{(2)^{5/2} V^2} = \frac{NkT}{V}\left(1 \pm \frac{N\Lambda^3}{(2)^{5/2} V}\right) \qquad (2.4.17)$$

Thus, the magnitude of the corrections depends on the quantum mechanical overlap discussed in the last section. From the signs of the corrections, we see that Fermi statistics, the exclusion of particles from multiple occupancy of single-particle quantum states, has the effect of a repulsive force between the particles, whereas Bose statistics, which is just the absence of any restriction at all, has the effect of an attractive interaction.

In the next two sections we shall look into the ultimate consequences of these effective interactions, studying the Fermi and Bose gases in the extreme degenerate, quantum mechanical limit—that is, at high densities and low temperature.

2.5 THE VERY DEGENERATE FERMI GAS: ELECTRONS IN METALS

At the outset of the nineteenth century, the ageless atomic theory of matter received a solid empirical grounding in the Law of Simple and Multiple Proportions of John Dalton. For the ensuing century, many competent, if conservative, chemists refused to believe in atoms, finding the hypothesis unecessary and inconsistent with their philosophy of how science ought to operate. One of the first to take a stand against Daltonian atomism was Dalton's contemporary compatriot, Sir Humphry Davy, discoverer, among many other things, of the metallic elements sodium and potassium. Davy's position, however, did not derive from the obstinacy of conservatism but rather from the inspiration of a visionary. That is, subsequent events proved his objection to have been essentially correct.

Atoms were the ultimate, indivisible constituents of matter; that is the very meaning of the word. There were 40 known elements at the time, and that, in Dalton's scheme, meant 40 distinct ultimate constituents, each with its own independent, irreducible properties. Of these 40 elements, 26 were metals, sharing in common high electrical and thermal conductivity, surface luster, and other properties. This cannot, said Davy, be an accident that happened 26 times. There must be a single underlying principle of metalization.

Our respect for Davy's insight will only be tarnished if we go into details of what he thought the principle of metalization might be. At the end of the century, the discovery of the electron—that is, the first splitting of Dalton's unsplitable atoms—provided the essential clue. When atoms combine to form a metal, each loses its outermost electron or two, these electrons becoming common property of the system as a whole. It is the behavior of this gas of electrons that gives rise to the properties of metals. Remarkably, an excellent model of the conduction electrons in a metal is the degenerate perfect gas of fermions.

In this section we shall study the degenerate Fermi gas, pausing from time to time to discuss its applicability as a model of electrons in metals. These ideas will then be further refined in Chap. 3, Sec. 3.6, when we look at electron energy bands in solids. We begin by considering the ground state of a perfect gas of fermions and later extend our treatment to finite temperatures.

In the limit of zero temperature, the Fermi gas goes into the lowest energy state allowed by the Pauli exclusion principle, which it is obliged to

2.5 The Very Degenerate Fermi Gas: Electrons in Metals

obey. If we imagine the particles to have spin $\frac{1}{2}$, as electrons do, then two particles can occupy the single-particle ground state, with zero kinetic energy. Subsequent particles must go into the lowest-lying available single-particle states, which means that they must have nonzero kinetic energy. The lowest-lying available states are successively filled, two spin $\frac{1}{2}$ particles in each momentum eigenstate, until all the particles of the gas have been used up. The single-particle phase space will have all its cells filled up, from the origin up to some momentum, called the Fermi momentum, p_F, which will depend on the volume of the box and the number of particles. From the point of view of the single-particle states, we can picture a filled sphere in momentum space, with radius p_F. If we wish to think of many-body phase space, the perfect gas sphere, Eq. (1.3.133),

$$\mathscr{E} = \frac{1}{2m} \sum_1^{3N} p_i^2 \tag{2.5.1}$$

has shrunk to a single point, since there is only one allowed set of the $3N$ values of p_i.

This situation is depicted by the Fermi-Dirac formula for the mean occupation numbers, Eq. (2.3.7),

$$\bar{n}_\mathbf{q} = \frac{1}{\exp\left[(\varepsilon_\mathbf{q} - \mu)/kT\right] + 1} \tag{2.5.2}$$

if we take it in the limit as T goes to zero. In that limit, for states with $\varepsilon_\mathbf{q} < \mu$, the exponent in Eq. (2.5.2) goes to negative infinity, so that $\bar{n}_\mathbf{q}$ becomes equal to one. For states with $\varepsilon_\mathbf{q} > \mu$, the exponent is positive infinity, so $\bar{n}_\mathbf{q}$ is zero. $\bar{n}_\mathbf{q}$ is thus a step function, all states being filled up to energy μ_0, where μ_0 is the value of μ at $T = 0$, as shown in Fig. 2.5.1. The value of μ at $T = 0$ is formally fixed by the equation

$$N = \lim_{T \to 0} \sum_\mathbf{q} \frac{1}{\exp\left[(\varepsilon_\mathbf{q} - \mu)/kT\right] + 1} \tag{2.5.3}$$

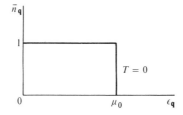

Fig. 2.5.1

We simplify matters greatly if we take the energies, ε_q, to be a continuous distribution, ε, and integrate instead of summing. Our excuse for integrating here is not as it usually is (see Sec. 2.4)—that the thermodynamic fluctuations are large compared to the level spacing—but rather that there are a macroscopic number of single-particle states filled, up to an energy so much greater than the level spacing that the discreteness cannot be important. We may then rewrite Eq. (2.5.3)

$$N = \lim_{T \to 0} \int_0^\infty \frac{\rho(\varepsilon)\, d\varepsilon}{e^{(\varepsilon-\mu)/kT} + 1} = \int_0^{\mu_0} \rho(\varepsilon)\, d\varepsilon \qquad (2.5.4)$$

where we have used the fact that $\bar{n}_q$ becomes a step function at $T = 0$. Instead of turning back to Eq. (2.4.4) to look up $\rho(\varepsilon)$, we can get the result by recalling that N will be the number of cells in single-particle phase space in a sphere of radius p_F:

$$N = \frac{gV(\tfrac{4}{3}\pi p_F^3)}{(2\pi\hbar)^3} \qquad (2.5.5)$$

Solving for p_F, we have

$$p_F = 2\pi\hbar \left(\frac{N}{V}\right)^{1/3} \left(\frac{3}{4\pi g}\right)^{1/3} \qquad (2.5.6)$$

Then μ_0 is given by

$$\mu_0 = \frac{p_F^2}{2m} = \frac{(2\pi\hbar)^2}{2m} \left(\frac{N}{V}\right)^{2/3} \left(\frac{3}{4\pi g}\right)^{2/3} \qquad (2.5.7)$$

Let us take note of the fact that the chemical potential is, in this case, positive. In Sec. 1.2b we argued that μ ought, ordinarily, to be a negative quantity, since, adding a particle to a system, we would usually have to extract energy in order to keep the entropy from increasing. That argument depended, however, on it being possible to add a particle at zero energy. In the system we have here, no particle can be added with energy less than μ_0. The chemical potential is thus positive. μ_0 is sometimes known as the Fermi energy, ε_F. Furthermore, it is customary to define a characteristic temperature, the Fermi temperature, T_F, by

$$kT_F = \varepsilon_F = \mu_0 \qquad (2.5.8)$$

The filled sphere in momentum space is called the Fermi sphere (and sometimes the Fermi sea) and its surface the Fermi surface.

The density of states, given by Eq. (2.4.4), is proportional to $\varepsilon^{1/2}$. One

2.5 The Very Degenerate Fermi Gas: Electrons in Metals 117

consequence is that the average energy per particle is $\frac{3}{5}$ the maximum energy:

$$\bar{\varepsilon} = \frac{\int_0^{\varepsilon_F} \varepsilon \rho(\varepsilon)\, d\varepsilon}{\int_0^{\varepsilon_F} \rho(\varepsilon)\, d\varepsilon}$$

$$= \frac{\int_0^{\varepsilon_F} \varepsilon^{3/2}\, d\varepsilon}{\int_0^{\varepsilon_F} \varepsilon^{1/2}\, d\varepsilon}$$

$$= \frac{\tfrac{2}{5}\varepsilon_F^{5/2}}{\tfrac{2}{3}\varepsilon_F^{3/2}} = \tfrac{3}{5}\varepsilon_F \qquad (2.5.9)$$

The total energy is given by

$$E = \tfrac{3}{5}N\varepsilon_F = \frac{3(2\pi\hbar)^2}{10m}\left(\frac{N}{V}\right)^{2/3}\left(\frac{3}{4\pi g}\right)^{2/3} N \qquad (2.5.10)$$

The entropy is, of course, zero, and the energy is equal to the free energy. The pressure of the system is given by

$$P = -\left(\frac{\partial E}{\partial V}\right)_N = \frac{2}{3}\frac{E}{V} \qquad (2.5.11)$$

This finite pressure at zero temperature is the ultimate consequence of the repulsive effective force between fermions that we saw beginning to develop in Sec. 2.4. We note, in passing, that the result of Eq. (2.5.11), $E = \tfrac{3}{2}PV$, is actually quite generally true in the perfect gas, depending only on the relation $\varepsilon = p^2/2m$. (See Prob. 1.1b.)

Let us now consider qualitatively what we expect to happen as the temperature begins to rise above zero. At sufficiently low temperature, we should be able to think of the system as a perturbed version of what we already have seen at zero temperature. There can be only one criterion for how low a temperature is sufficiently low, since there is only one characteristic temperature in the problem: we shall study the system in the limit

$$T \ll T_F \qquad (2.5.12)$$

We can start by estimating T_F, from

$$T_F = \frac{(2\pi\hbar)^2}{2mk}\left(\frac{3}{4\pi g}\right)^{2/3}\left(\frac{N}{V}\right)^{2/3} \qquad (2.5.13)$$

Since we shall be interested in applying the model to electrons in metals, we use values of the parameters that are appropriate to that case. For the mass, we use the electron mass, 9×10^{-28} gram; we take $g = 2$ corresponding to spin $\frac{1}{2}$. The number density of conduction electrons in most metals is in the range $1 - 10 \times 10^{22}$ cm^{-3}. Taking 8.5×10^{22}, the value for copper, we get

$$T_F = \frac{(6.6 \times 10^{-27})^2}{2 \times (9 \times 10^{-28}) \times (1.4 \times 10^{-16})} \left(\frac{3}{12.6 \times 2}\right)^{2/3} (8.5 \times 10^{22})^{2/3}$$

$$= 8.5 \times 10^{4} \,°\text{K} \quad (2.5.14)$$

This result, 85,000 degrees Kelvin, may be compared, for example, to the melting point of copper, which is on the order of 10^{3}°K. We see that at all temperatures at which copper is a solid, Eq. (2.4.12) is satisfied; the electron gas is in its low-temperature limit.

As the temperature rises above zero, the particles of the gas tend to become excited with an energy of order kT each. However, those electrons deep in the Fermi sea, much more than kT below the Fermi surface, cannot be excited, because there are no available states for them to be excited into. Only those within about kT of the surface, a very small fraction, of order T/T_F of the whole gas, have any chance of being excited. The rest remain unaffected, unchanged from their zero-degree situations. The net result is that the mean occupation number becomes slightly blurred compared to its sharp, step function form at $T = 0$. The width of the blurring is roughly kT, as shown in Fig. 2.5.2. All the thermodynamic behavior of the system takes place, basically, in that narrow band, of width $\sim kT$, about the Fermi surface.

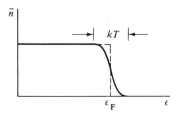

Fig. 2.5.2

We are now in a position to understand the resolution of what was once a deep problem in the theory of metals. The electron, as we started to say earlier, was discovered in 1896 by J. J. Thomson. Thomson quickly made use of his discovery to explain a number of phenomena, including the properties of metals, for which he examined what is basically the model we still have, that the metal is permeated by a gas of free electrons (the model was originally proposed by Drude). He showed some striking successes, among them a derivation of the Wiedemann-Franz law for the universal relation between the electrical and thermal conductivities of metals. However, there was a

2.5 The Very Degenerate Fermi Gas: Electrons in Metals

quite serious failure as well: the electron gas, if it existed, should have had the same heat capacity as any other gas, $\frac{3}{2}k$ per particle. From the known heat capacity of silver, an upper limit could thereby be placed on the number of electrons in the gas. If the number were any larger, silver would need a larger heat capacity. The result was far too small to account for the other electrical properties of the metal.

If the electrons have the properties of the perfect Fermi gas we have been considering, the dilemma can be resolved. All the electrons participate in the conduction of electricity in the metal, since an electric field has the effect of displacing the entire Fermi sphere; the field acts on all the electrons. However, only those electrons within kT of the Fermi surface can be excited thermally, so that for heat capacity purposes there would seem to be a gas consistingly only of the small fraction, T/T_F, of the electrons. If there are N electrons in all, we would expect a heat capacity on the order of $\frac{3}{2}Nk(T/T_F)$. Note, for future reference, that we are predicting a heat capacity linear in T (i.e., proportional to T).

There is, of course, still an important drawback to our use of the perfect Fermi gas as a model for electrons in metals: how can we possibly think of charged particles like electrons as noninteracting? And for that matter, how can we imagine the background of closely spaced atoms through which they travel to be an empty box? Strictly speaking, of course, we cannot do either; and yet, both are, often enough, surprisingly good approximations. The reasons are as follows:

1. The metal as a whole is electrically neutral, the negatively charged electrons traveling through a background of positively charged ions. The positive background has the effect of allowing the electrons to shield themselves from one another. We are dealing, then, not really with free electrons but rather with ones that somehow carry some positive shielding with them; we shall eventually come to modify the model accordingly.

2. The Fermi distribution itself tends to retard collisions between the electrons. Quantum mechanically, the effect of a potential of interaction between the particles, thought of as a perturbation of the perfect gas picture, is to scatter colliding particles into new momentum states. However, at thermal energies, only T/T_F of the particles are close enough to an unfilled state for scattering to be possible. For all the particles deeper in the Fermi sea, there are no states available to scatter into; there simply can be no scattering, hence no collisions.

3. The remaining objection is that even if the particles are nearly noninteracting (with each other), they are not in an empty box, nor (what is the same thing) in a uniform, positively charged background, but instead find themselves in a closely packed lattice of heavy positive ions. It is only in an empty box that we can expect them to have states described by Eqs. (1.1.1) to (1.1.4), specifically, by $\varepsilon_\mathbf{q} = \hbar^2 q^2/2m$. However, with regard to their momen-

tum states, we must think of the electrons as waves, with wavelength $\lambda = 2\pi/q$. The atomic structure of the background cannot be very important for those states with $\lambda \gg a$, where a is the spacing between the atoms; very long wavelength electrons should be insensitive to structure on a much finer scale. We can thus expect that for low-momentum states ($p = \hbar q = 2\pi\hbar/\lambda$) the basic form, $\varepsilon \propto q^2$, will be preserved. At higher momentum, as λ starts to approach a, the relation between ε and p will depart more and more from the perfect gas form. We shall work this process out in more detail in Chap. 3; a typical example of what we shall find is sketched in Fig. 2.5.3. Even at low q, where we preserve the quadratic relation between ε and q, the curve has a different coefficient from the free electron value. We can write

$$\varepsilon = \frac{\hbar^2 q^2}{2m^*} \tag{2.5.15}$$

where m^* is what we shall call the effective mass. This can be thought of as a result of the tendency of the electrons to shield themselves, as if carrying along some of the positive background made the electrons heavier.

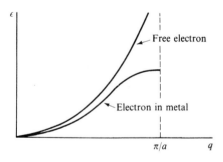

Fig. 2.5.3

If the Fermi energy, ε_F, falls at a value of q where Eq. (2.5.15) is valid (as it does, basically, for many metals), then we need only modify our perfect gas picture by putting *'s on all the m's. If, on the other hand, it falls closer to $q = \pi/a$ (as it does for insulators, semiconductors, and some metals), then we should expect results qualitatively different from perfect gas behavior. In any case, accurate knowledge of $\varepsilon(\mathbf{q})$ will allow us, in principle at least, to find the density of states, $\rho[\varepsilon(\mathbf{q})]$, and do our thermodynamics, just as if we actually had a gas of noninteracting particles in an empty box. The entities, or particles, of this gas are not true particles at all but collective properties of the electron-lattice system. This method of reducing a complicated problem

2.5 The Very Degenerate Fermi Gas: Electrons in Metals

to an effective $\varepsilon(\mathbf{q})$, and hence a density of states, is our first example of a recurring theme in our study of matter: the construction of quasiparticles, imaginary entities that do not interact very much with each other and may thus be studied with our perfect gas formalism.

Let us go back now to the perfect Fermi gas and work out in detail some of its thermal properties. The basic job is to evaluate Eqs. (2.3.8) and (2.3.9). To start with,

$$\Omega = -kT \sum_{\mathbf{q}} \log\left[1 + \exp\left(\frac{\mu - \varepsilon_{\mathbf{q}}}{kT}\right)\right]$$

$$= -kT \int_0^\infty \rho(\varepsilon) \log\left[1 + \exp\left(\frac{\mu - \varepsilon}{kT}\right)\right] d\varepsilon \qquad (2.5.16)$$

We can either integrate (2.5.16) by parts or take a shortcut by noting that for all perfect gases (so long as $\varepsilon \propto p^2$), $E = -\frac{3}{2}\Omega$. Thus,

$$\Omega = -\frac{2}{3} E = -\frac{2}{3} \int_0^\infty \varepsilon \rho(\varepsilon) \bar{n}(\varepsilon) \, d\varepsilon$$

$$= -\frac{2}{3} \frac{4\pi V g \sqrt{2} \, m^{3/2}}{(2\pi\hbar)^3} \int_0^\infty \frac{\varepsilon^{3/2} \, d\varepsilon}{e^{(\varepsilon - \mu)/kT} + 1} \qquad (2.5.17)$$

The job reduces itself to doing the integral in Eq. (2.5.17) or, more generally, integrals of the type

$$I = \int_0^\infty \frac{f(\varepsilon) \, d\varepsilon}{e^{(\varepsilon - \mu)/kT} + 1} \qquad (2.5.18)$$

The factor in the integrand

$$\bar{n}(\varepsilon) = \frac{1}{\exp\left[(\varepsilon - \mu)/kT\right] + 1} \qquad (2.5.19)$$

is sketched in Fig. 2.5.2. With $T \ll T_F$, it is very nearly a step function. Most of the integral I will therefore be given by

$$I_0 = \int_0^\mu f(\varepsilon) \, d\varepsilon \qquad (2.5.20)$$

and we need evaluate only the correction terms, δI, where

$$I = I_0 + \delta I \qquad (2.5.21)$$

I_0 is the value of I at $T = 0$, and all the thermal behavior is included in δI. It is important to remember, however, that we are changing T at constant μ, so that N may have to change. [We could, of course, imagine N constant, but then μ would have to change and I_0, Eq. (2.5.20), would depend on temperature.] Since δI is a small correction, it is natural to seek a way of writing I, in effect, as a Taylor expansion about $T = 0$.

$$I = I_0 + \left(\frac{\partial I}{\partial T}\right)_{T=0} T + \frac{1}{2}\left(\frac{\partial^2 I}{\partial T^2}\right)_{T=0} T^2 + \cdots \quad (2.5.22)$$

The crucial question here is the order of the first nonzero term, since the thermal behavior of the gas at low temperature will be dominated by that term. It will turn out that the linear term in Eq. (2.5.22) vanishes, and the leading-order correction is of order T^2. In deriving the result, let us be careful to see why that happens.

Rather than push blindly ahead with the prescription of Eq. (2.5.22), we can make better progress by some judicious analysis of the task at hand. If, instead of integrating $f(\varepsilon)$ over the real $\bar{n}(\varepsilon)$, Eq. (2.5.19), we integrate it over the step function, we introduce two errors: we put in too much $f(\varepsilon)$ where the step function is bigger than $\bar{n}(\varepsilon)$ just below $\varepsilon = \mu$, and we neglect a piece of the integral just above $\varepsilon = \mu$, where the real $\bar{n}(\varepsilon)$ is finite but the step function is zero. It is because these two errors nearly compensate each other that the term linear in T in the corrections to I_0 vanishes. In Fig. 2.5.4

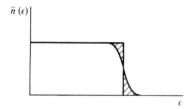

Fig. 2.5.4

the two regions where $\bar{n}(\varepsilon)$ differs from the step function are cross-hatched. Our assertion that the two errors nearly cancel means that the area under the step function is nearly the same as that under $\bar{n}(\varepsilon)$. The number of particles at a given energy is the density of states, $\rho(\varepsilon)$, times $\bar{n}(\varepsilon)$. Since $\rho(\varepsilon)$ changes very little in the region near μ where $\bar{n}(\varepsilon)$ departs from the step functions, the cross-hatched areas are essentially proportional to the number of particles promoted from states below μ, in the one case, and into states above μ, in the other. The near cancellation of these two effects, then, merely reflects the fact that as the temperature is raised, the total number of particles is almost conserved.

In order to proceed in calculating the correction terms, let us first of all shift the origin, defining

$$z = \frac{\varepsilon - \mu}{kT} \quad (2.5.23)$$

2.5 The Very Degenerate Fermi Gas: Electrons in Metals

The correction terms in I are given by integrating $f(\varepsilon)$ over the difference between $\bar{n}(\varepsilon)$ and the step function. This difference can be represented by two functions of z, $g_0(z)$ and $g_1(z)$, which are equal to zero everywhere except close to $z = 0$. They are sketched in Fig. 2.5.5. In terms of these functions, the corrections are given by

$$\delta I = \int_{-\infty}^{\infty} f(\varepsilon)[g_1(z) - g_0(z)] \, d\varepsilon$$

$$= \int_{-\infty}^{\infty} f(\mu + kTz)[g_1(z) - g_0(z)]kT \, dz \qquad (2.5.24)$$

where we have taken the limits of integration from $-\infty$ to $+\infty$, since g_0 and g_1 in the integrand are zero for large $|z|$ in any case. Since all the result will come from z close to zero, we expand $f(\mu + kTz)$ about $z = 0$:

$$f(\mu + kTz) = f(\mu) + kTz \left(\frac{\partial f}{\partial \varepsilon}\right)_{\varepsilon = \mu} + \cdots$$

$$= f(\mu) + kTz f'(\mu) + \cdots \qquad (2.5.25)$$

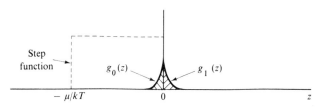

Fig. 2.5.5

Substituting into (2.5.24), we get

$$\delta I = kT f(\mu) \int_{-\infty}^{\infty} [g_1(z) - g_0(z)] \, dz$$

$$+ (kT)^2 f'(\mu) \int_{-\infty}^{\infty} z[g_1(z) - g_0(z)] \, dz \qquad (2.5.26)$$

Examination of the first term on the right shows that it is proportional to the difference in area between $\bar{n}(\varepsilon)$ and the step function, $\int (g_1 - g_0) \, dz$. The second term, however, has $g_1 - g_0$ multiplied by z, which changes sign going through the origin, so that g_1 and g_0 will not cancel but contribute with the same sign. Notice that if the first term drops out, as we have said it will, all odd-order terms will go with it.

The function $g_1(z)$ is just that part of $\bar{n}$ found at $z \geq 0$:

$$g_1(z) = \frac{1}{e^z + 1} \quad (z \geq 0) \tag{2.5.27}$$

while $g_0(z)$ is the step function minus $\bar{n}$ for negative z. That is,

$$g_0(z) = 1 - \frac{1}{e^z + 1} \quad (z < 0) \tag{2.5.28}$$

or

$$g_0(z) = \frac{e^z}{e^z + 1}$$

$$= \frac{1}{e^{-z} + 1} \quad (z < 0) \tag{2.5.29}$$

The first term on the right in Eq. (2.5.26) is then proportional to

$$\int_{-\infty}^{\infty} [g_1(z) - g_0(z)] \, dz = \int_0^{\infty} g_1(z) \, dz - \int_{-\infty}^0 g_0(z) \, dz$$

$$= \int_0^{\infty} \frac{dz}{e^z + 1} - \int_{-\infty}^0 \frac{dz}{e^{-z} + 1}$$

$$= \cdots + \int_{+\infty}^0 \frac{dz}{e^z + 1}$$

$$= \cdots - \int_0^{\infty} \frac{dz}{e^z + 1}$$

$$= 0 \tag{2.5.30}$$

while in the second term we get

$$\int_0^{\infty} z g_1(z) \, dz - \int_{-\infty}^0 z g_0(z) \, dz = 2 \int_0^{\infty} \frac{z \, dz}{e^z + 1} = \frac{\pi^2}{6} \tag{2.5.31}$$

The net result for I, including the next-order term as well, is

$$I = \int_0^{\mu} f(\varepsilon) \, d\varepsilon + \frac{\pi^2}{6} f'(\mu)(kT)^2 + \frac{7\pi^4}{360} f'''(\mu)(kT)^4 + \cdots \tag{2.5.32}$$

Equation (2.5.32) is the basic result needed to work out the thermal properties of the perfect Fermi gas.

If we substitute Eq. (2.5.32) into (2.5.17), we find, up to order T^2, that Ω is given by

$$\Omega = -\frac{2}{3} \frac{4\pi V g \sqrt{2} \, m^{3/2}}{(2\pi\hbar)^3} \left[\frac{2}{5} \mu^{5/2} + \frac{\pi^2}{4} \mu^{1/2} (kT)^2 \right] \tag{2.5.33}$$

2.5 The Very Degenerate Fermi Gas: Electrons in Metals

The number of particles is

$$N = -\frac{\partial \Omega}{\partial \mu} = \frac{2}{3} \frac{4\pi V g \sqrt{2}\, m^{3/2}}{(2\pi\hbar)^3} \left[\mu^{3/2} + \frac{\pi^2}{8} \frac{(kT)^2}{\mu^{1/2}} \right] \quad (2.5.34)$$

As noted above, the number of particles has had to change in order to keep μ fixed, but the change is second order in T. The number at $T = 0$ is the leading term [see Eq. (2.5.7)],

$$N_0 = \frac{2}{3} \frac{4\pi V g \sqrt{2}\, m^{3/2} \mu^{3/2}}{(2\pi\hbar)^3} \quad (2.5.35)$$

so that

$$N = N_0 \left[1 + \frac{\pi^2}{8} \left(\frac{kT}{\mu} \right)^2 \right] \quad (2.5.36)$$

These are results in the grand canonical ensemble, in which we imagine a large "bath" of particles, with constant chemical potential. Many experiments on metals are done using a sample with a fixed number of electrons (although that is not necessarily the case—the sample could be part of an electric circuit), so it is convenient to express our results for constant particle number. In that case, μ will change with T. To allow for that possibility, we once again call μ_0 the zero-temperature value of μ and rewrite Eq. (2.5.36)

$$N = N_0 \left[1 + \frac{\pi^2}{8} \left(\frac{kT}{\mu} \right)^2 \right] \left(\frac{\mu}{\mu_0} \right)^{3/2} \quad (2.5.37)$$

since, from Eq. (2.5.35), $N_0 \propto \mu^{3/2}$. Now, if N is constant, $N = N_0$ but μ changes from μ_0. Solving for μ, and setting $\mu = \mu_0$ in the correction term, we get

$$\mu = \frac{\mu_0}{[1 + \pi^2/8(kT/\mu_0)^2]^{2/3}} = \mu_0 \left[1 - \frac{\pi^2}{12} \left(\frac{kT}{\mu_0} \right)^2 \right] \quad (2.5.38)$$

where, in the last step, we have made a binomial expansion of the denominator, always keeping terms to order T^2. Notice that if we had kept μ instead of μ_0 in the correction term, we could now expand it the same way, and the difference from Eq. (2.5.38) would be of order T^4.

We have found, according to Eq. (2.5.38), that for a gas of a constant number of fermions, μ decreases slightly as T is raised from zero. This decrease in μ was to be expected; after all, we know that in the opposite limit, $T \gg T_F$, the system must reduce to an ideal gas, in which the chemical potential is negative, not positive as it is here. The chemical potential is easily identified from the occupation numbers at any temperature in the range $T \ll T_F$ (sketched in Fig. 2.5.2), since, according to Eq. (2.5.2), at $\varepsilon_q = \mu$, $n = \frac{1}{2}$. μ is thus the point on Fig. 2.5.2 where $\bar{n}$ passed through $\frac{1}{2}$. We see that, to first order in T, it is constant as the temperature rises but that it actually slips slowly to lower values of ε when studied with higher accuracy. The entropy of the gas is calculated from Eq. (2.5.33), using

$$S = -\left(\frac{\partial \Omega}{\partial T}\right)_{V,\mu} = \frac{4\pi V g \sqrt{2} \, m^{3/2}}{(2\pi\hbar)^3} \frac{\pi^2}{3} \mu^{1/2} k^2 T \qquad (2.5.39)$$

We have obtained S in terms of T, V, and μ, rather than $S(T, V, N)$. However, we now know that the change in μ with T if we hold N constant will introduce into Eq. (2.5.39) a correction of order T^3, which we ignore. Using Eq. (2.5.13) to eliminate the constants in favor of T_F, we find

$$S = \frac{\pi^2}{2} Nk \frac{T}{T_F} \qquad (2.5.40)$$

For the heat capacity, we get the same result:

$$C_V = T\left(\frac{\partial S}{\partial T}\right)_V = \frac{\pi^2}{2} Nk \frac{T}{T_F} \qquad (2.5.41)$$

We see, as promised earlier, that the heat capacity is linear and of order $Nk(T/T_F)$.

At temperatures below 1°K, the heat capacity of copper is dominated by the electron contribution. It is found to be linear, as predicted, and, in particular,

$$\frac{C}{NkT} = 0.8 \times 10^{-4} \qquad \text{(experimental)}$$

Putting into Eq. (2.5.41) the copper value [see Eq. (2.5.14)] of $T_F = 8.5 \times 10^{4}$°K, we find

$$\frac{C}{NkT} = 0.6 \times 10^{-4} \qquad \text{(predicted)}$$

Thus, even the magnitude is pretty close, and we even know how to fix up the small difference. Since $C \propto T_F^{-1} \propto m$ [according to Eq. (2.5.13)], we need only assign to electrons in copper an effective mass, $m^* = 1.3m$, to obtain complete agreement.

Things are not really as simple as we have pictured them here. For one thing, metals are not isotropic; they have crystal structure, and so although there is only one Fermi energy (or chemical potential), the Fermi momentum depends on direction with respect to the crystal axes. As a consequence, m^* will depend on direction. Moreover, even if long wavelength electrons do not interact with each other, and are largely undisturbed by a perfect lattice of ions, there are things in any real metal with which they do interact: impurities, imperfections, thermal excitations of the lattice, and so on. Furthermore, as we did mention, the situation does change for electrons of shorter wavelength. Yet, ignoring all those complications temporarily, we have put together a simple, elegant explanation of one of the fundamental properties of matter: the existence of metals. Can we confirm the essential features of this elementary model by any observation other than the linear temperature dependence of the heat capacity? Application of the model has depended on some hand-waving arguments of which we might reasonably be skeptical: do electrons really fail to interact with each other and (under appropriate circumstances) the lattice through which they move? And, do they actually behave as if their masses were other than the free mass, or is m^* just a parameter for fixing up the magnitude of the heat capacity? There is, it turns out, some direct experimental evidence on these points.

It is possible, under certain circumstances, to accelerate electrons in metals in circular paths by the same mechanism that is used to accelerate particles in cyclotrons. The electrons orbit in a plane perpendicular to an applied magnetic field, resonantly accelerated by radio frequency radiation if it is applied at the right frequency,

$$\omega_c = \frac{eH}{m^*c} \tag{2.5.42}$$

where c is the speed of light and e the electron charge. The resonance is detected by loss of radio frequency power to the electron system. The frequency is thus an independent way of measuring the mass. Furthermore, the strength of the signal depends on how far an electron can go in its resonant orbit before being scattered—that is, on the electron mean free path. The experiments are done in very pure metals at very low temperature to maximize the mean free path, precisely by doing away with those complications to our model cited above. If our model is a good one, we should expect it to be possible to reach quite long mean free paths and to find masses in agreement with those deduced from heat capacity measurements.

These expectations are confirmed. Mean free paths as long as 10^8 lattice spacings (~ 1 cm) have been reported. The effective mass does depend on direction with respect to the crystal lattice (we shall return to this kind of detail in Chap. 3), but when properly averaged over direction, the heat capacity is found to be measuring the same kind of mass that enters into Eq. (2.5.42). We may thus feel sufficiently confident of our basic model to consider more sophisticated embellishments. This we shall do in Chap. 3.

2.6 THE BOSE CONDENSATION: A FIRST-ORDER PHASE TRANSITION

In spite of the superficial similarity between the equations governing Bose-Einstein and Fermi-Dirac perfect gases, Eqs. (2.3.20) and (2.3.21), the two kinds of particles behave in profoundly different ways in the low-temperature limit. The essence of Bose-Einstein behavior may be seen by examining the Bose form of Eq. (2.3.21)

$$N = \sum_q \frac{1}{\exp\left[(\varepsilon_q - \mu)/kT\right] - 1} \tag{2.6.1}$$

which we may immediately convert to the integral

$$N = \frac{4\pi V g \sqrt{2}\, m^{3/2}}{(2\pi\hbar)^3} \int_0^\infty \frac{\varepsilon^{1/2}\, d\varepsilon}{\exp\left[(\varepsilon - \mu)/kT\right] - 1} \tag{2.6.2}$$

Equation (2.6.2) relates the variables N, T, V, and μ. As we have already seen in Sec. 2.3, Eq. (2.3.14), in this case, in contrast to the Fermi-Dirac case, the chemical potential is strictly limited to negative values.

Consider a container of fixed volume, held by an external bath at constant temperature, into which we place particles, thus successively increasing N. The remaining variable, μ, must adjust in order to keep Eq. (2.6.2) balanced. The integral on the right will change in the same direction as $e^{\mu/kT}$; that is, as N increases, μ must increase. Since μ is negative, adding particles has the effect of pushing it toward zero. If, at some finite N, it gets pushed asymptotically close to zero, it can no longer change to compensate further additions to the number of particles. We can easily see whether that actually occurs at finite N by solving Eq. (2.6.2) with μ set equal to zero:

$$N = \frac{4\pi V g \sqrt{2}\, m^{3/2}}{(2\pi\hbar)^3} (kT)^{3/2} \int_0^\infty \frac{x^{1/2}\, dx}{e^x - 1} \tag{2.6.3}$$

The definite integral has the value

$$\int_0^\infty \frac{x^{1/2}\, dx}{e^x - 1} = 2.31 = \frac{\sqrt{\pi}}{2} \times 2.612 \tag{2.6.4}$$

We thus run into trouble at a critical density

$$\left(\frac{N}{V}\right)_c = \frac{2.612}{\Lambda^3} \tag{2.6.5}$$

or, conversely, if we imagine reducing the temperature while holding N/V fixed, the problem arises at a critical temperature

$$T_c = \frac{1}{2.31 mk} \left(\frac{N}{V}\right)^{2/3} \frac{(2\pi\hbar)^2}{(4\pi\sqrt{2})^{2/3}} \tag{2.6.6}$$

Equation (2.6.5) is a clear hint that the difficulty is related to the fact that the particles can no longer be localized without overlapping quantum mechanically.

The problem that we have run into here is apparently a serious one.

2.6 The Bose Condensation: A First-Order Phase Transition

There is no physical reason why we should not be able to continue adding particles at fixed T and V, but the total number of particles, as counted by Eq. (2.6.2), can no longer change. What happens to the lost particles? Before examining our mathematical apparatus to find them, let us consider for a moment a more familiar situation in which an analogous phenomenon occurs.

Suppose that we take some common gas, say nitrogen, N_2, and start feeding molecules of it into a fixed volume held at constant temperature. We choose the temperature to be $60°K$, just above the triple point, and assert in advance that throughout the experiment the nitrogen gas will be ideal, obeying

$$N = P \frac{V}{kT} \tag{2.6.7}$$

We count the number of molecules that we have put in and simultaneously measure the pressure, which, according to Eq. (2.6.7), also "counts" N. If we make a plot of the number we have put in versus the measured P, we should, and do, get a straight line with constant slope V/kT. However, we find at some point that Eq. (2.6.7) abruptly breaks down, and the pressure stops changing, as sketched in Fig. 2.6.1. We can describe the Bose gas in exactly the same terms if instead of Eq. (2.6.7), we use Eq. (2.6.2). On the ordinate, instead of P, we plot

$$y = \int_0^\infty \frac{\varepsilon^{1/2} \, d\varepsilon}{\exp\left[(\varepsilon - \mu)/kT\right] - 1} \tag{2.6.8}$$

The initial slope is then $4\pi V g \sqrt{2} \, m^{3/2}/(2\pi\hbar)^3$. At the critical value of N, y abruptly stops changing, just as P does in Fig. 2.6.1.

In the nitrogen experiment, the pressure at which the curve breaks is the vapor pressure. At that point droplets of liquid nitrogen begin to form in

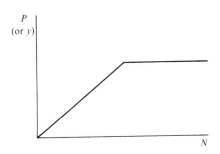

Fig. 2.6.1

the container. As further molecules are added, the liquid droplets grow, but the pressure remains the same; the extra molecules go, on the average, into the denser liquid state. The gas phase, as we have promised, remains very nearly ideal, but the amount in it ceases to change; in fact, as the liquid comes to occupy an appreciable part of the volume, the amount of gas declines. Figure 2.6.1 is evidence that a first-order phase change—the condensation of a vapor—is being observed. We shall find that a very similar phenomenon occurs in the Bose gas.

If Fig. 2.6.2 is taken to describe the Bose case, the break in the curve occurs at the point where μ reaches zero. Beyond that point, the quantity of Eq. (2.6.8) is no longer able to keep track of all the particles. We have made certain approximations in arriving at that integral form, and one of them has evidently broken down. The first thing that comes to mind is the propriety of writing the sum, Eq. (2.6.1), as an integral in the first place. The criterion for doing that, Eq. (2.4.1), is

$$kT \gg \frac{\hbar^2}{2m}\left(\frac{2\pi}{L}\right)^2 \qquad (2.6.9)$$

where $L^2 = V^{2/3}$. To check the validity of the result we have here, we must check that Eq. (2.6.9) is satisfied when $T = T_c$. Substituting in Eq. (2.6.6) for T_c, the criterion becomes

$$N^{2/3} \gg \frac{2.31(4\pi\sqrt{2})^{2/3}}{2} \qquad (2.6.10)$$

The right-hand side is a number of order one. Thus, for any macroscopic number of particles, we are evidently justified in integrating. The difficulty must lie elsewhere.

The resolution of the dilemma is actually quite simple. We have taken the density of states to be

$$\rho(\varepsilon) \propto \varepsilon^{1/2} \qquad (2.6.11)$$

so that the density of states at $\varepsilon = 0$ is zero, where it should, in fact, be one. We have ignored all particles in the ground state. For the ideal gas and for the Fermi gas, that is of no consequence whatever; for the ideal case, the average number of particles in any state, ground state included, is much less than one, out of the 10^{23} or so involved, whereas for the degenerate Fermi gas, there are two particles in the zero-kinetic-energy state. For the Bose gas, however, the mean number in the ground state is

$$\overline{n_0} = \frac{1}{e^{-\mu/kT} - 1} \qquad (2.6.12)$$

In the limit as $\mu \to 0$, the exponential can be expanded, to give

$$\overline{n_0} = -\frac{kT}{\mu} \qquad (2.6.13)$$

2.6 The Bose Condensation: A First-Order Phase Transition

The number of particles in this single state becomes macroscopic as $\mu \to 0$, or, to put it differently, it is into the ground state that the missing particles we have spoken of are disappearing. The phenomenon we are observing is a phase transition: a condensation into the ground state, called the Bose condensation.

The thermodynamics of the partly condensed system is easily worked out once we know where to look for the missing particles. The integral in Eq. (2.6.2) counts correctly the number of particles that are not in the ground state. Let us call this number N^*.

$$N^* = \int_0^\infty \frac{\rho(\varepsilon)\, d\varepsilon}{e^{\varepsilon/kT} - 1} = N\left(\frac{T}{T_c}\right)^{3/2} \quad (2.6.14)$$

The second step follows, since, according to Eq. (2.6.3), what we are now calling N^* is proportional to $T^{3/2}$. All the remaining particles are in the ground state. Let us call this number N_0; then

$$N_0 = N - N^* = N\left[1 - \left(\frac{T}{T_c}\right)^{3/2}\right] \quad (2.6.15)$$

The N^* excited particles have all the energy of the gas. We therefore correctly compute the energy from

$$E = \int_0^\infty \frac{\varepsilon \rho(\varepsilon)\, d\varepsilon}{e^{\varepsilon/kT} - 1}$$

$$= \frac{4\pi V g \sqrt{2}}{(2\pi\hbar)^3}(mkT)^{3/2} kT \int_0^\infty \frac{x^{3/2}\, dx}{e^x - 1} \quad (2.6.16)$$

Integrals of this form [Eq. (2.6.4) is another example] are given by

$$\int_0^\infty \frac{x^n\, dx}{e^x - 1} = \Gamma(n+1)\zeta(n+1) \quad (2.6.17)$$

where ζ is the Riemann zeta function. In this case,

$$\int_0^\infty \frac{x^{3/2}\, dx}{e^x - 1} = \frac{3\sqrt{\pi}}{4} \times 1.341 \quad (2.6.18)$$

We find for the energy,

$$E = 0.770 kT N^* = 0.770 NkT \left(\frac{T}{T_c}\right)^{3/2} \quad (2.6.19)$$

The basic temperature dependence of the energy is $E \propto T^{5/2}$. The heat capacity at constant volume is

$$C_V = \left(\frac{\partial E}{\partial T}\right) = 1.9 Nk \left(\frac{T}{T_c}\right)^{3/2} = 1.9 N^* k \quad (2.6.20)$$

From the point of view of the heat capacity, the behavior of the excited part is not very different from that of an ideal gas, in which $C_V = 1.5Nk$.

At a given density N/V and at sufficiently high T, the Bose gas is, in fact, ideal and has $C_V = 1.5Nk$. As the temperature decreases, $(-\mu/kT)$ becomes smaller, and as we saw in Sec. 2.3, departures from ideal gas behavior start to appear. The effective attraction between bosons, which will eventually result in the condensation, begins to assert itself, and, in addition, the heat capacity starts to rise above its ideal value. It reaches $1.9Nk$ at $T = T_c$; that is, the heat capacity is continuous in passing through T_c but has a cusp at that point. C_V/Nk is plotted over the whole range of T in Fig. 2.6.2.

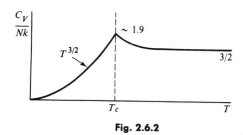

Fig. 2.6.2

It is sometimes said of the Bose condensation that it is a third-order phase transition and that it is a condensation in momentum space, not in real space. We shall argue here that it is actually a first-order phase transition and that condensation takes place in real as well as momentum space, also taking this opportunity for some general discussion of the behavior of matter at first-order phase transitions. Let us begin with a brief discussion of what is meant by the order of a phase transition.

A first-order phase transition is one in which there is a latent heat. As we saw in Sec. 1.2g, two phases of matter separated by a first-order transition can coexist in equilibrium along a curve in the P-T plane, but when they do, they differ in both specific entropy and in density (we also looked into a generalization of these ideas to a magnetic transition). There are also phase transitions in which there is no latent heat; that is, the entropy (and consequently the density) is continuous in passing from one phase to another (recall that the latent heat, L, is just $T \Delta s$, where Δs is the difference in specific entropy between the two phases). There must, however, be discontinuities or singularities of some sort in some thermodynamic function, or we would not call whatever is happening a phase transition. An attempt was once made to

2.6 The Bose Condensation: A First-Order Phase Transition

make an orderly classification of all phase transitions according to where discontinuities showed up. The scheme went like this:

1. In a first-order transition, the entropy is discontinuous.
2. In a second-order transition, the entropy is continuous, but its first derivative, the heat capacity, is discontinuous.
3. In a third-order transition, the entropy and heat capacity are continuous, but the derivative of the heat capacity, $\partial C/\partial T$, is discontinuous.
4. And so on.

Recently this classification scheme has become unfashionable. One reason is that it does not seem to work very well except for first-order transitions; higher-order phase transitions tend to have infinities rather than discontinuities in their thermodynamic functions. Another reason is that it tends to misdirect attention from what is now thought to be the essential sameness of all non-first-order transitions (this second point is the subject of Chap. 6 of this book). Nevertheless, this scheme has entered the language of physics, and we can see from Fig. 2.6.2 that the Bose condensation, in particular, seems to fit the requirements of what we have called a third-order transition above.

Our basic objection to that assignment is that even for first-order transitions the classification scheme can be meaningful only if we apply it in going through transitions at constant pressure, not at constant volume as we have done in Fig. 2.6.2. In order to see the point, let us imagine what would appear to be the heat capacity of some ordinary material if we caused it to pass through what we know to be a first-order transition—say evaporation from a condensed into a gaseous phase—at constant volume. The path we will follow is shown by X's in the P-V and P-T planes in Fig. 2.6.3. At low temperature the material is mostly liquid. As we raise T at constant N and V, evaporation occurs, causing the pressure to rise through the two-phase region (i.e., along the vapor pressure curve) until, finally, at the pressure shown by the dashed line, all the liquid is gone and P departs from the vapor pressure curve, henceforth to follow the pure gas equation of state.

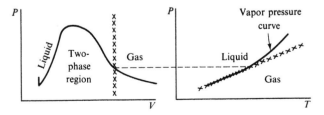

Fig. 2.6.3

Now let us see how the heat capacity behaves through all of this. At low temperature, where the phases coexist, we will have, approximately,

$$\frac{C_V}{N} = xC_g + (1 - x)C_\ell + L\frac{dx}{dT} \qquad (2.6.21)$$

where C_g is the specific heat of the gas, C_ℓ that of the liquid, and x is the fraction of the material in the gas state.

The first two terms represent the amount of heat required, per unit temperature change, to warm up the gas and liquid, respectively. The third term is the amount of heat that goes into causing liquid to evaporate. As the temperature rises, x rises toward one, at which point we depart from the two-phase region. At higher T we have

$$\frac{C_V}{N} = C_g \qquad (2.6.22)$$

In crossing out of the two-phase region, the second term in Eq. (2.6.21) goes continuously to zero, but the third term might vanish discontinuously, depending on the shape of the coexistence curve (L usually does not change rapidly with T; the discontinuity depends on how fast x is changing as the coexistence curve is crossed). Thus, in the entire curve of C_V versus T, the most dramatic evidence we can hope to see of a first-order phase transition is a discontinuity, as sketched in Fig. 2.6.4. The point here is that if we were to use this curve to decide the order of the transition according to our classification scheme, we would (incorrectly) decide it is second order at best, possibly third order.

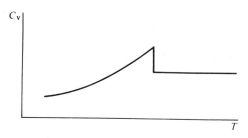

Fig. 2.6.4

The correct way to do the job is to pass through the transition at constant pressure. In that case, the path followed is shown in Fig. 2.6.5. Now the entire two-phase region is traversed at constant temperature; the entire latent heat must be put in with no change in temperature at all. The apparent

2.6 The Bose Condensation: A First-Order Phase Transition

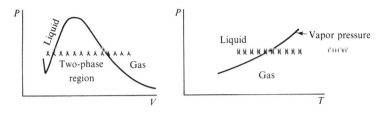

Fig. 2.6.5

heat capacity at that point is a delta-function infinity, the area under the delta function being the latent heat. C_P is sketched in Fig. 2.6.6.

It is, of course, experimentally difficult to measure C_P directly. To do so, we must imagine that the material of interest is in a cylinder with a movable piston, maintained at constant pressure by changing the volume. Nevertheless, Fig. 2.6.6 is the kind of information we need to make a clear identification of a first-order phase transition. Fortunately, we can construct C_P out of the more easily measurable C_V if we know, in addition, the equation of state of the combined system. They are related by Eq. (1.2.107).

$$C_P - C_V = TVK_T \left[\left(\frac{\partial P}{\partial T} \right)_V \right]^2 \quad (2.6.23)$$

In the two-phase region, a change in T at constant V slides P along the vapor pressure curve, so that $(\partial P/\partial T)_V$ is finite. However, the compressibility, K_T, defined by

$$K_T = -\frac{1}{V} \left(\frac{\partial V}{\partial P} \right)_T \quad (2.6.24)$$

is infinite: isotherms in the P-V plane, as shown in Fig. 1.2.6, are horizontal. V changes by a finite amount with no change in P, and T constant as well.

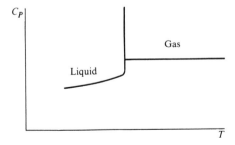

Fig. 2.6.6

Thus, according to Eq. (2.6.23), C_P is infinite when the phases coexist even though C_V is finite. Actually, the horizontal isotherm, meaning that a finite discontinuity in V occurs at constant P and T, is itself sufficient to identify a first-order phase transition.

All this discussion indicates that we must know the equation of state, $P(T, V)$, of the degenerate Bose gas if we are to judge the order of the transition. To find it, we could begin by computing Ω from the Bose form of Eq. (2.3.20) with $\mu = 0$. Then since, in general, $\Omega = F - \mu N$, we have, in this case, $F = \Omega$, and we can find the pressure from $P = -\partial F/\partial V$. It is easier, however, merely to recall that $PV = \frac{2}{3}E$ as usual, so that

$$P = \frac{2}{3}\frac{E}{V} = 0.513\frac{NkT}{V}\left(\frac{T}{T_c}\right)^{3/2} \tag{2.6.25}$$

where we have used Eq. (2.6.19) for E. Eliminating T_c via Eq. (2.6.6), we find

$$P = \frac{4\pi\sqrt{2}\,(2.31m)^{3/2}}{(2\pi\hbar)^3} \times 0.523(kT)^{5/2}$$

$$= 1.2\,\frac{kT}{\Lambda^3} \tag{2.6.26}$$

The pressure thus goes as $T^{5/2}$, but (and here is the important point) it is independent of the volume. At constant temperature, if we change the volume, the pressure stays constant. In other words, the P-V isotherms are horizontal. That was, remember, the unmistakable signature of a first-order phase transition. The compressibility is infinite [since, according to Eq. (2.6.26), $(\partial P/\partial V)_T = 0$], and so it follows from Eq. (2.6.23) that C_P is infinite.

Let us try, now, to make a detailed, consistent description of the degenerate Bose gas as a system in which a first-order phase transition is occurring. When the system becomes degenerate—that is, when $\mu = 0$—we picture it as consisting of two phases in equilibrium: the condensate, which constitutes those particles in the zero-energy state, and the excited part, which, for want of a better term, we can call the Bose vapor. As in other gas-condensate equilibria, the gas has a much higher entropy and occupies a much larger volume per particle than does the condensate, but the Bose case is a bit special in that the condensate has zero entropy and occupies zero volume. The zero volume is made possible because the particles themselves, having been assumed not to have any hard cores (i.e., repulsive interactions at short distance), occupy no volume. We can see that the condensate occupies zero volume by the following argument. Imagine the Bose material to be in a cylinder with a movable piston. According to Eq. (2.6.26), if we push the piston in, decreasing the available volume at constant temperature, the pressure will not change. What happens instead is exactly what would happen if

2.6 The Bose Condensation: A First-Order Phase Transition

we tried the same experiment with a system of water and water vapor, or liquid and gaseous nitrogen: the vapor is forced to condense at constant P and T. For real liquid-vapor systems, however, that process would have an endpoint: once all the vapor had condensed, and the remaining volume was full of liquid, it would take a rapid increase in pressure to compress the stuff further. In our Bose system as well, we can proceed to push in the piston at constant pressure until all the Bose vapor has condensed. The volume at which that point occurs can be found by solving Eq. (2.6.15) for $N_0 = N$, replacing T_c by means of Eq. (2.6.6). The condition is obviously that $T_c = \infty$ or, in other words, $V = 0$. Zero is the volume of the condensate. [Actually, a bit of care must be taken, since our equations will no longer be valid if we violate the quantum condition Eq. (2.4.1). It is fair to say, however, that although the condensate consists of a macroscopic number of particles, it does not occupy a macroscopic volume.] It was on the basis of this argument, incidentally, that we asserted earlier that a condensation takes place in real as well as momentum space.

The reader may be tempted to object that there is, after all, an important difference between the condensation of a real vapor and the condensation of a Bose vapor: a real vapor condenses into a liquid that is physically localized at the bottom of the container, whereas the Bose vapor changes into a condensate whose wave function has constant amplitude everywhere in the container; the vapor and condensate interpenetrate completely. However, the fact is that even in the real system localization of the condensate at the bottom of the container is an artifact of gravity, a perturbation that we have not put into our model. In the absence of gravity, a real liquid, too, would have equal probability of being found anywhere in the container.

Equation (2.6.26) is the equation of the vapor pressure curve of the Bose condensate, sketched in Fig. 2.6.7. Unlike conventional vapor pressure curves, it does not end in a critical point but instead continues on to arbitrarily high P and T. In the P-V plane, the two-phase region occupies the entire lower left-hand corner, as shown in Fig. 2.6.8. Since $P_c \propto T_c^{5/2}$ and

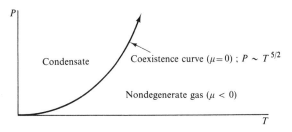

Fig. 2.6.7

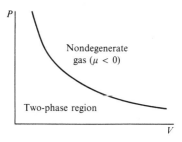

Fig. 2.6.8

$T_c \propto (N/V)^{2/3}$, the curve bounding the two-phase region in Fig. 2.6.8 has the form $P \propto (N/V)^{5/3}$.

Like any first-order phase transition, the Bose condensation must obey the Clausius-Clapeyron equation, Eq. (1.2.128), for the slope of the co-existence curve:

$$\left(\frac{dP}{dT}\right)_{\text{coex}} = \frac{L}{T \, \Delta V} \tag{2.6.27}$$

In the Bose case, the equation of the vapor pressure curve is really

$$\mu(P, T) = 0 \tag{2.6.28}$$

so that, along the curve,

$$0 = d\mu = -\frac{S}{N} dT + \frac{V}{N} dP \tag{2.6.29}$$

The slope of the vapor pressure curve is thus given by

$$\left(\frac{dP}{dT}\right)_{\text{coex}} = \frac{S}{V} \tag{2.6.30}$$

The latent heat, defined as $L = T \Delta S$ is, in this case, just TS, since the condensate has zero entropy, and as we have seen, the difference in volume is just the volume itself, which is also the volume of the gas. Equations (2.6.27) and (2.6.30) thus agree. The entropy at any given volume may be obtained from

$$S = \int_0^T \frac{C_V}{T} dT = \frac{2}{3} C_V = \frac{5}{3} \frac{E}{T} \tag{2.6.31}$$

where we have made use of the fact that $E \propto T^{5/2}$ and $C_V \propto T^{3/2}$ [since $C_V = (\partial E/\partial T)_V$ and $E \propto T^{5/2}$, it follows that $C_V = (\frac{5}{2})E/T$]. The latent heat of the transition is then

$$L = TS = \tfrac{5}{3}E \tag{2.6.32}$$

Substitution into Eq. (2.6.27) gives, for the slope of the vapor pressure curve,

$$\left(\frac{dP}{dT}\right)_{\text{coex}} = \frac{5}{3}\frac{E}{TV} \tag{2.6.33}$$

This result may be checked directly by differentiating Eq. (2.6.26) and finding E in terms of the constants, but perhaps it is easier to recall that $P = \tfrac{2}{3}E/V \propto T^{5/2}$, so that $dP/dT = \tfrac{5}{3}E/TV$ as in Eq. (2.6.33).

Finally, it is amusing to note that since, in general, from Eq. (1.2.34),

$$E = TS - PV + \mu N \tag{2.6.34}$$

and since in this case, $\mu = 0$, we have here

$$E = TS - PV \tag{2.6.35}$$

which may be checked by recalling that $PV = \tfrac{2}{3}E$ and, from Eq. (2.6.32), $TS = \tfrac{5}{3}E$. If we expand the system from zero volume at constant T and P, we absorb from the medium an amount of heat $L = TS$, but this is $\tfrac{5}{3}$ of the energy we wind up with. In the course of performing the expansion, we do work equal to PV on the medium, thus giving back $\tfrac{2}{3}$ of the final energy.

As we have mentioned before, real interactions always dominate the behavior of matter at low temperature, so that the Bose condensation in its pure form is never observed in nature. Nevertheless, the behavior of the Bose degenerate gas underlies our understanding of a number of real phenomena, including superfluidity and superconductivity, to be discussed in Chap. 5.

BIBLIOGRAPHY

The behavior of perfect gases is a traditional part of any course in statistical mechanics, and so the main features of this chapter will be found in any of the statistical mechanics books, from Tolman on, cited in the bibliography of Chap. 1. Our approach to the formalism has particularly followed that of Landau and Lifshitz.

PROBLEMS

2.1 The number of states available to two systems in contact is larger than the number available when the two are isolated from each other with the equilibrium values of their thermodynamic parameters. Thus, for example, if two containers of ideal gas, each with the same N, V, and T, are brought into contact, the entropy of the combined system should increase. Investigate what happens to the entropy of the combined system under all permutations of the following:

    a. The particles in the two containers are/are not identical.

    b. The two gases are/are not allowed to mix when contact is made.

2.2 A two-dimensional perfect gas is formed by adsorbing atoms onto a plane surface with binding energy $-\varepsilon_0$ (atoms are free to move laterally on the surface). This gas is in equilibrium with its own three-dimensional vapor. Find the number adsorbed per unit area as a function of P and T (P is the three-dimensional vapor pressure) in the classical limit. Also find the leading-order corrections for quantum degeneracy in the adsorbed part. Under what circumstances might the three-dimensional gas become degenerate?

2.3 The quantity $(\partial T/\partial P)_W$ (where W is the enthalpy) is called the Joule-Thomson coefficient. If it is positive, a gas can be cooled by expansion through a porous plug.

    a. Find $(\partial T/\partial P)_W$ for a classical ideal gas.

    b. Find the leading-order quantum corrections for Bose-Einstein and Fermi-Dirac statistics.

2.4 For a very degenerate perfect Fermi gas, find the energy as a function of T, V, and N up to fourth order in T.

2.5 N particles of spin $\tfrac{1}{2}$ and mass m in a volume V are able to form bound pairs of spin zero with binding energy $-E_0$ per pair ($E_0 > 0$). The particles are otherwise noninteracting.

    a. Show by means of a variational argument that, in equilibrium,

$$\mu_2 = 2\mu_1$$

where μ_1 is the chemical potential of a particle and μ_2 is the chemical potential of a pair.

    b. Find the number of pairs in equilibrium at $T = 0$.

    c. At what value of μ_1 do the pairs undergo a Bose condensation?

    d. Under what conditions of N, V, and T do the particles become Fermi degenerate?

    e. Find the critical value of (N/V) for a Bose condensation of the pairs when $kT \ll E_0$.

2.6 For perfect gases in two dimensions, find

    a. The Fermi degeneracy temperature, T_F.

    b. The Bose condensation temperature.

    c. The heat capacity at constant area in the low-temperature limit for both the Fermi and Bose cases.

2.7 Below 1°K, a liquid mixture of the isotopes He^3 and He^4 undergoes a spontaneous separation into separate phases, one rich in He^3, the other in He^4. This is a first-order phase transition, with a phase diagram that looks like this:

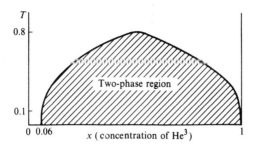

A very good, simple model is as follows: (i) The He^4 is close to its ground state and has very little entropy. (ii) The He^3, in both phases, acts like a degenerate perfect Fermi gas, with effective mass $m^* = 2.5m$, where m is the mass of a He^3 atom. Below about $0.1°K$, the two phases coexist at constant concentration: 6% He^3 in one phase, 100% He^3 in the other.

a. Estimate the difference between the binding energies for a He^3 atom in the two phases at zero degrees.

b. Find the latent heat per He^3 atom below $0.1°K$.

c. Describe, quantitatively where possible, the heat capacity at constant concentration, C_x, as a function of temperature, with $x = 10\%$ He^3.

THREE

SOLIDS

3.1 INTRODUCTION

It seems fair to say that of all the enterprises of physics, the study of the solid state is one of the most successful, whereas the study of the liquid state is one of the least successful. One of our central purposes, in the next two chapters, will be to understand the reasons for this disparity.

To a large extent, our success in understanding solids is a consequence of nature's kindness in organizing them for us. The subject matter of this chapter is chosen to take maximum possible advantage of that natural organization. By the term solid, we shall really always mean crystalline solid, and, moreover, infinite perfect crystalline solid at that, thus ignoring all questions that pertain even to surfaces of perfect crystals, much less to impurities, imperfections, and, worst of all, amorphous solids and glasses. In doing so, we turn our backs on a very deep problem. Our ability to explain the various properties of solids may result from their natural organization—that is, from their periodicities and symmetries—but the question is: To what extent do the properties themselves depend on those qualities? We shall return to this point in the introductory remarks of the next chapter, pointing out that amorphous solids and even liquids have at least some of the properties that appear to result from crystal symmetries.

As we shall see in Chap. 4, there is some difficulty even in defining unambiguously what we mean by a liquid. There is no such difficulty in the

3.1 Introduction

case of a crystalline solid. We can, in the first instance, distinguish between a liquid and a solid by the observation that liquids slosh and solids do not. To say the same thing in language more suitable for a sophisticated textbook such as this, solids resist shear stress and support low-frequency transverse sound while liquids do not. That, however, still leaves us with amorphous solids and even glasses (which slosh but only very, very slowly). The real distinguishing characteristic of the solids we wish to deal with in this chapter is periodicity.

The crystalline solid, as we shall know it, is a regular array of atoms, in a structure that is repeated endlessly through space. The possible geometries of the basic repeating unit, called the *unit cell*, will be discussed in Sec. 3.4. We shall actually find it convenient to think of the crystal as being composed of a finite (though very large) number of atoms, N. We can formally do so, without introducing surfaces that behave differently from interiors, by applying periodic boundary conditions at the surfaces, and this is what we shall always do.

Sections 3.2 and 3.3 deal with various phenomena that arise out of the fact that, at finite temperatures, the atoms in a solid are thermally excited into vibrations about their equilibrium positions. These sections come before the discussion of crystal structures, partly in order to emphasize that the results do not depend on geometric details, although some of the analysis does depend on the fact of periodicity itself.

In Sec. 3.2, in particular, we shall be dealing with the thermal behavior of massive particles. The thermal behavior of the particles in this case differs from that of the perfect gases of Chap. 2 in that the interactions between the particles will now be important and must be taken into account. When we come to count the possible states of such a system, in order to find its entropy, we must, as always, do so in a way that does not distinguish between the particles. This step cannot be done, as in Chap. 2, by counting the occupations of single-particle states, since that procedure depended on the particles being noninteracting. Here, however, the organization provided by nature comes to our rescue. The atomic positions in the periodic array, or lattice, can, in principle, be labeled, and we can speak of the behavior of the atom at a particular site. The fact that the same device cannot be used in the theory of liquids underlies much of the difficulty of that formidable discipline.

In Sec. 3.5 we examine some of the consequences of the fact that all processes occurring in crystals take place in a kind of space that has periodicity but that is not uniformly invariant under translation. As we shall see, one of the results of real space being periodic is that momentum space becomes periodic as well.

Finally, in Sec. 3.6, we shall consider what happens to the degenerate electron gas, Sec. 2.5, when placed in a crystal structure. Among other things, we shall see why some materials are metals while others are insulators or semiconductors. The last two classes are dielectrics, but we shall resist the temptation to call their study dielectrical materialism.

3.2 THE HEAT CAPACITY DILEMMA

When Niels Bohr introduced his new atomic model in 1913, he decided, understandably, to begin his paper with the strongest arguments he could produce to show that unorthodox new directions were needed in physics. Among the failures of classical physics that he cited was its evident inability to account for the heat capacities of solids, a problem that, qualitatively at least, had already been resolved by Einstein in 1907 by applying the embryonic quantum ideas then available. The basic picture of a solid, then as now, was a regular array of atoms, each caused to vibrate harmonically about its equilibrium position by thermal excitation. The classical expectation was that the heat capacity that resulted should be a constant, independent of temperature. That was generally found to be the case at ordinary temperature, but when low-temperature refrigerants (such as liquid nitrogen and oxygen) became available toward the end of the nineteenth century, it became clear that the heat capacities of solids became temperature dependent at low T. A typical heat capacity for a simple, atomic, insulating solid is sketched in Fig. 3.2.1. The decline of C_V at low T was the dilemma referred to by Bohr. We will come to understand both the dilemma and its basic resolution if we construct a workable model of the solid and examine its consequences for the heat capacity.

Rather than try immediately to attack the complex problem of the collective behavior of many densely packed interacting particles, we shall begin by focusing our attention on any one atom and imagine it to interact with the averaged effect of all the other atoms. This is our first example of what we

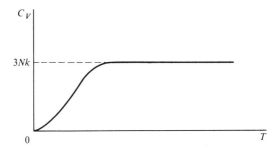

Fig. 3.2.1

3.2 The Heat Capacity Dilemma

shall call a *mean field theory*—others to come later include the van der Waals and Weiss theories of Chap. 6—and we shall find this one typical both in its qualitative success and in certain kinds of quantitative failure.

Each atom is found at a particular site in the lattice. We take **r** to be the displacement of an atom from its equilibrium position and $u(\mathbf{r})$ to be the potential energy of the atom, owing to its interaction with all the other atoms in the crystal. We shall be interested in small, thermally excited displacements, so we expand $u(\mathbf{r})$ about $\mathbf{r} = 0$, assuming, for simplicity, that the potential is isotropic. Since there are no forces on the atom at $\mathbf{r} = 0$, we have at that point

$$\frac{\partial u}{\partial r} = 0 \qquad \frac{\partial^2 u}{\partial r^2} > 0 \qquad (3.2.1)$$

the second condition ensuring that $\mathbf{r} = 0$ is a stable position. The leading terms in the expansion are then

$$u(\mathbf{r}) = u_0 + \frac{\alpha}{2} r^2 \qquad (3.2.2)$$

where $\alpha = (\partial^2 u/\partial r^2)_{r=0} > 0$. This is the potential of a three-dimensional harmonic oscillator whose energy is given by

$$\varepsilon = u_0 + \frac{1}{2m}(p_x^2 + p_y^2 + p_z^2) + \frac{\alpha}{2}(x^2 + y^2 + z^2) \qquad (3.2.3)$$

The term u_0, which is negative, is the binding energy per atom of the crystal. We can immediately write the classical partition function for each atom as

$$Z_1 = \frac{1}{(2\pi\hbar)^3} \int \cdots \int \exp\left[-\frac{\varepsilon(\mathbf{p}, \mathbf{r})}{kT}\right] d^3p \, d^3r$$

$$= \frac{\exp(-u_0/kT)}{(2\pi\hbar)^3} \left[\int \exp\left(-\frac{p^2}{2mkT}\right) dp\right]^3 \left[\int \exp\left(-\frac{\alpha x^2}{2kT}\right) dx\right]^3$$

(3.2.4)

Aside from coefficients, the integrals are of identical form:

$$\int \exp\left(-\frac{a\xi^2}{kT}\right) d\xi = \frac{1}{2}\left(\frac{kT}{a}\right)^{1/2} \int \eta^{-1/2} \exp(-\eta) \, d\eta \qquad (3.2.5)$$

The limits of integration for the momenta are zero to infinity. For the coordinates, we are formally limited to small $\mathbf{r}$, but since the exponential, $\exp(-\alpha r^2/2kT)$, becomes negligible for large $\mathbf{r}$, we take the limits to be $-\infty$ and $+\infty$, and

$$\int_{-\infty}^{\infty} \exp\left(-\frac{\alpha x^2}{2kT}\right) dx = 2 \int_0^{\infty} \exp\left(-\frac{\alpha x^2}{2kT}\right) dx \qquad (3.2.6)$$

since the exponential is an even function of x. Thus, all the integrals in Eq. (3.2.4) are of the form Eq. (3.2.5) with limits zero to infinity. The partition function for each atom becomes

$$Z_1 = A(kT)^3 \exp\left(-\frac{u_0}{kT}\right) \qquad (3.2.7)$$

where all the constants [including six powers of $\int_0^\infty \eta^{-1/2} \exp(-\eta)\, d\eta$] have been gathered into A. The atoms are indistinguishable, but we can distinguish between sites on the lattice and associate Z_1 with whichever atom is on some particular site. The partition function of the N atom crystal is thus

$$Z = Z_1^N = A^N(kT)^{3N} \exp\left(-\frac{Nu_0}{kT}\right) \qquad (3.2.8)$$

The free energy is

$$F = -kT \log Z = Nu_0 - 3NkT \log kT - NkT \log A \qquad (3.2.9)$$

the entropy

$$S = \left(\frac{\partial F}{\partial T}\right)_V = 3Nk \log kT + 3Nk + Nk \log A \qquad (3.2.10)$$

and the heat capacity and energy

$$C_V = T\left(\frac{\partial S}{\partial T}\right)_V = 3Nk \qquad (3.2.11)$$

and $\qquad E = F + TS = 3NkT + Nu_0 \qquad (3.2.12)$

These are the classical results. As one can see from the derivation, for every term in the energy of the system that is quadratic in one of the components of $\mathbf{p}$ or $\mathbf{r}$ of any particle, we wind up with $\frac{1}{2}kT$ of energy and $\frac{1}{2}k$ of heat capacity. Thus, in the ideal gas, each particle had energy $\varepsilon = \frac{1}{2}m \times (p_x^2 + p_y^2 + p_z^2)$, contributing $\frac{3}{2}kT$ to the energy of the system. In the present case, each particle, according to Eq. (3.2.3), has six quadratic terms in its energy, leading to a contribution of $3kT$ to the energy of the system. This result is called the *Law of Equipartition*. Notice that the coefficients of the quadratic terms make no difference: we have $\frac{1}{2}m$ multiplying the p^2 terms and $\frac{1}{2}\alpha$ multiplying the r^2 terms, but each dutifully contributes the same $\frac{1}{2}kT$ to the thermal energy. Details of the model are also unimportant: if the crystal were nonisotropic, so that α had a different value for each direction, or even if we abandoned the mean field aspect and considered cooperative oscillations of the atoms (we shall shortly see how to do that), the classical result would be the same.

As promised, the behavior of real matter, sketched in Fig. 3.2.1, disagrees with this result at low temperature.

a. The Einstein Model

The gross dilemma of why the heat capacity starts to decline at low temperature is resolved as soon as we realize that the oscillations of the crystal must be treated quantum mechanically rather than classically. Let us revert once again to our mean field model. The quantum description corresponding to Eq. (3.2.3) is that each atom behaves like three independent harmonic oscillators, each of frequency ω, with possible energies given by

$$\varepsilon_\omega = (n + \tfrac{1}{2})\hbar\omega + \frac{u_0}{3} = \varepsilon_{0\omega} + n\hbar\omega \qquad (3.2.13)$$

where

$$\varepsilon_{0\omega} = \frac{u_0}{3} + \frac{\hbar\omega}{2} \qquad (3.2.14)$$

is the ground state energy and n is any integer. The partition function for each oscillator is

$$Z_{1\omega} = \exp\left(-\frac{\varepsilon_{0\omega}}{kT}\right) \sum_n \left[\exp\left(-\frac{\hbar\omega}{kT}\right)\right]^n \qquad (3.2.15)$$

$$= \frac{\exp(-\varepsilon_{0\omega}/kT)}{1 - \exp(-\hbar\omega/kT)} \qquad (3.2.16)$$

[We summed the same series earlier, Eq. (2.3.15).] In our mean field theory, the system is just $3N$ identical oscillators associated with distinguishable sites, so we need only find the thermodynamic functions of one of them:

$$F_{1\omega} = -kT \log Z_{1\omega} = \varepsilon_{0\omega} + kT \log\left[1 - \exp\left(-\frac{\hbar\omega}{kT}\right)\right] \qquad (3.2.17)$$

Then

$$S_{1\omega} = -k \log\left[1 - \exp\left(-\frac{\hbar\omega}{kT}\right)\right]$$

$$+ \frac{\hbar\omega}{T} \frac{\exp(-\hbar\omega/kT)}{1 - \exp(-\hbar\omega/kT)} \qquad (3.2.18)$$

$$E_{1\omega} = F_{1\omega} + TS_{1\omega} = \varepsilon_{0\omega} + \bar{n}\hbar\omega \qquad (3.2.19)$$

where

$$\bar{n} = \frac{\exp(-\hbar\omega/kT)}{1 - \exp(-\hbar\omega/kT)} = \frac{1}{\exp(\hbar\omega/kT) - 1} \qquad (3.2.20)$$

Comparison with Eq. (3.2.13) shows that $\bar{n}$ is the mean thermally excited quantum number of the harmonic oscillator. Its form, not accidentally, is the same as that of the number of thermally excited particles in each single-particle state of a degenerate Bose gas.

In order to find the extensive properties of the system, F, S, E, we multiply the single-particle values by $3N$. In particular, for the heat capacity,

$$C_V = \left(\frac{\partial E}{\partial T}\right)_V = 3N \left(\frac{\partial E_{1\omega}}{\partial T}\right)_V$$

$$= 3Nk \left(\frac{\hbar\omega_E}{kT}\right)^2 \frac{\exp(\hbar\omega_E/kT)}{[\exp(\hbar\omega_E/kT) - 1]^2} \quad (3.2.21)$$

All $3N$ oscillators have the same frequency—that is a direct consequence of using an isotropic mean field model. We have labeled the frequency ω_E, and we expect it to have a different value for each material, ω_E being larger for harder, more tightly bound materials. It is customary as well to define a characteristic temperature, Θ_E, the Einstein temperature, by

$$k\Theta_E = \hbar\omega_E \quad (3.2.22)$$

In terms of Θ_E, Eq. (3.2.21) may be rewritten

$$C_V = 3Nk \left(\frac{\Theta_E}{T}\right)^2 \frac{\exp(\Theta_E/T)}{[\exp(\Theta_E/T) - 1]^2} \quad (3.2.23)$$

We can see that Eq. (3.2.21) or (3.2.23) has the general form required by Fig. 3.2.1 by looking at its limiting behavior at high and low temperatures. At high T, which can only mean $kT \gg \hbar\omega_E$,

$$\exp\left(\frac{\hbar\omega_E}{kT}\right) \approx 1 + \frac{\hbar\omega_E}{kT}$$

and we have

$$C_V \to 3Nk \quad \text{(High } T) \quad (3.2.24)$$

the classical result. At low temperature, $kT \ll \hbar\omega_E$,

$$\exp\left(\frac{\hbar\omega_E}{kT}\right) \gg 1$$

in the denominator, and we get

$$C_V \to 3Nk \left(\frac{\hbar\omega_E}{kT}\right)^2 \exp\left(-\frac{\hbar\omega_E}{kT}\right) \quad \text{(Low } T) \quad (3.2.25)$$

Here the temperature dependence is dominated by the exponential, and C_V falls rapidly to zero as $T \to 0$. Thus, C_V goes to zero exponentially at low T, to $3Nk$ at high T, and varies smoothly in between. The net result cannot look very different from Fig. 3.2.1.

The model we have just worked out is the one proposed by Einstein, in 1907, to show how the new quantum ideas would resolve the heat capacity

3.2 The Heat Capacity Dilemma

dilemma. With the advantage, now, of all quantum mechanics at our disposal, we can easily see why it works. The fact is that any quantum mechanical model, regardless of the details, will give $C_V \to 0$ as $T \to 0$. The reason is related to the quantum nature of the Third Law of Thermodynamics, the fact that there is only one ground state for any system, which it occupies at $T = 0$. The number of low-lying states that can be excited thermally decreases to zero as the temperature goes to zero. Consequently, at $T = 0$, $\partial S/\partial T$ must be finite, and the heat capacity, $T(\partial S/\partial T)$, is zero.

Aside from its historic importance in showing the way toward resolving the heat capacity dilemma, we can examine the Einstein model for its applicability to real solids. Broadly speaking, as we have already seen, it predicts a heat capacity not very different from what one actually observes. However, it disagrees with certain details of the behavior of real solids, and these areas of disagreement will turn out to be characteristic of the failures of mean field theories in general. Two particular types of disagreement are worth focusing on.

1. The predicted heat capacity itself is incorrect at low temperature. In contrast to the exponential behavior predicted by Eq. (3.2.25), the lattice heat capacities of all solids at very low temperature are found to obey

$$C_V \propto T^3 \quad (3.2.26)$$

Thus the Einstein heat capacity drops to zero too rapidly as T approaches zero.

2. There are two material-dependent parameters in the model—ω_E, as we have noted, and u_0, the binding energy. It is important to realize that in a self-consistent mean field theory, these quantities cannot depend on the spacing between the atoms or, in other words, on the density of the material. If they did depend on the spacing, the motion of each atom would affect its neighbors, and thus they would not behave independently as we have assumed, for example, in writing Eq. (3.2.21). However, the consequence of this assumption, that the particles act independently, is that the material has absurd elastic properties. The physical reason is obvious: it is indifferent to its density and so does not oppose forces applied to it. Formally, the free energy is $3N$ times the free energy per oscillator; from Eq. (3.2.17), together with Eq. (3.2.14),

$$F = Nu_0 + \tfrac{3}{2}N\hbar\omega_E + 3NkT \log\left[1 - \exp\left(-\frac{\hbar\omega_E}{kT}\right)\right] \quad (3.2.27)$$

It follows that the equilibrium pressure must be zero, since

$$P = -\left(\frac{\partial F}{\partial V}\right)_{T,N} \quad (3.2.28)$$

and the compressibility is infinite:

$$K_T = -\frac{1}{V}\left(\frac{\partial V}{\partial P}\right)_{T,N} = -\left[V\left(\frac{\partial^2 F}{\partial V^2}\right)_{T,N}\right]^{-1} \quad (3.2.29)$$

Both the zero pressure and the infinite compressibility follow from the fact that $F(T, V, N)$ is independent of V.

Obviously, in any realistic model of a solid, the potential energy of an atom must depend on its distance from its neighbors, and, consequently, the atoms must act collectively, not independently. What is not so obvious is that as soon as we fix up point 2, the low-temperature dependence of the heat capacity will come out right as well. We can see the connection between them, however, if we think for a moment about what any real material is likely to do at very low temperature.

At zero degrees (i.e., in its ground state) the system of atoms is in a regular lattice at a spacing that minimizes its energy. The magnitude of the heat capacity in this limit will depend on the lowest energy-excited states of the system because when a bit of energy is put in, thereby raising the temperature, each possible type of excitation, acting like one of the Einstein oscillators we have just studied, will pick up a portion of it proportional to $e^{-\mathscr{E}/kT}$, where $\mathscr{E}$ is the energy of the excitation, and so each will contribute to the heat capacity a term proportional to $e^{-\mathscr{E}/kT}$. Thus, the lower the smallest unit of $\mathscr{E}$, the larger the heat capacity will be. We therefore wish to construct the lowest-lying excited states of the system.

First, imagine moving one atom from its equilibrium position in some direction. Doing so will raise the potential energy of the system because, loosely speaking, the atom will move into the repulsive potential of a neighbor on one side, and out of the attractive potential of a neighbor on the other. The amount by which the potential changes is proportional to α of Eq. (3.2.2), and since that is the effective spring constant, it is proportional to ω^2 of the vibrations that will result from this motion. However, we can construct a configuration of lower potential energy, and hence lower vibrational frequency, if we allow the two neighboring atoms to move together with the one we are considering, and their neighbors, in turn, to move together with them, and so on. We cannot allow that process to go too far because if all the atoms move together, we simply have a displacement of the center of mass of the crystal as a whole, which is not the kind of excitation we want to consider. Instead, we can imagine all the atoms moving nearly together, with very little relative displacement of neighbors, but an absolute displacement from equilibrium that gradually grows to a maximum as we proceed over a macroscopic distance through the crystal, then diminishes back to zero again. The result will be a kind of motion that, having very little potential energy, will have a very low frequency. Since the energy to excite this motion comes

3.2 The Heat Capacity Dilemma

in units of $\hbar\omega$, this type of motion will have very low energy, and it is therefore the low-lying excitation we seek. It is a long wavelength sound wave. We may thus conclude that the low-temperature heat capacity of a solid will be dominated by collective rather than independent motions of the atoms. Just as in the Einstein model, we can think of the system as a collection of quantum mechanical harmonic oscillators but now they will not all have the same frequency. Instead, there will be many types of collective or cooperative motions, each with some characteristic frequency. We will call each a *collective mode*, or *normal mode* of the system. The magnitude of the contribution each mode makes to the free energy, and to the heat capacity, will depend, as we have argued, on the frequency of the mode, but the form will be the same as in the Einstein model. We have already found the thermal behavior of any single oscillator, or mode, if only we know its frequency. If we label each frequency ω_α, then Eqs. (3.2.17), (3.2.19), and (3.2.20) may immediately be adopted for the general case

$$F = E_0 + kT \sum_\alpha \log\left[1 - \exp\left(-\frac{\hbar\omega_\alpha}{kT}\right)\right] \quad (3.2.30)$$

$$E = E_0 + \sum_\alpha \bar{n}_\alpha \hbar\omega_\alpha \quad (3.2.31)$$

and

$$\bar{n}_\alpha = \frac{1}{\exp(\hbar\omega_\alpha/kT) - 1} \quad (3.2.32)$$

There remains only the job of enumerating the possible modes of vibration of a real solid.

The enumeration of the normal modes of an harmonically bound crystal is the subject of Sec. 3.3. However, we can understand the essential nature of the heat capacities if we can answer immediately two questions: What is the distribution of the lowest-lying modes (for they will govern the low-temperature behavior) and what is the total number of modes (that will be needed for normalization)? Take the second question first. There will obviously be some maximum frequency of vibration of the crystal, corresponding to the motion with the highest potential energy (it is a mode in which neighbors move in opposite directions). Let us call this maximum frequency ω_m. At sufficiently high temperatures (i.e., $kT \gg \hbar\omega_m$), we expect the law of equipartition to apply—after all, real crystals do obey equipartition at high temperature—and so the thermal energy of the crystal must be $3NkT$. Under these conditions, $\hbar\omega_\alpha/kT \ll 1$ for all α, so we can expand the exponential in Eq. (3.2.32), giving

$$\bar{n}_\alpha = \frac{kT}{\hbar\omega_\alpha} \quad (3.2.33)$$

which simply restates the Law of Equipartition; the excitation energy of each mode is $\bar{n}_\alpha \hbar\omega_\alpha = kT$. Since the total excitation energy is $3NkT$, there must always be $3N$ modes. That was true in the Einstein model (where all $3N$ modes had the same frequency, ω_E), and it will be true in all other models as well.

b. Long Wavelength Compressional Modes

The frequencies of the long wavelength sound modes will be given by the formula

$$\omega = cq \tag{3.2.34}$$

where c is the speed of sound and $q = 2\pi/\lambda$ is the magnitude of a wave vector, $\mathbf{q}$, which has discrete values fixed [as in Eq. (1.1.2)] by periodic boundary conditions. Solids support both transverse and longitudinal sound waves, with different velocities, c_t and c_ℓ. However, in order to get a feeling for how Eq. (3.2.34) comes about in the simplest possible way, let us temporarily ignore the existence of transverse waves and find the lowest-lying frequencies of a medium that has only longitudinal sound. The result will be applicable, actually, only to liquids.

We wish to consider only long wavelength modes, so the underlying atomic nature of the medium cannot be important, and we may think of it as a continuum. A small volume element of the medium has a pressure P, density ρ, and a velocity v (we will take all motion to be along the x direction). The force per unit volume on the volume element is in the direction opposite to the gradient of the pressure, so the Newton's law may be written

$$\frac{\partial P}{\partial x} + \rho \frac{\partial v}{\partial t} = 0 \tag{3.2.35}$$

A second equation arises from the conservation of mass

$$\frac{\partial \rho}{\partial t} + \frac{\partial}{\partial x}(\rho v) = 0 \tag{3.2.36}$$

The waves we are interested in are very small disturbances from the average values of P, ρ, and v, so we write

$$P = P_0 + P'$$
$$\rho = \rho_0 + \rho' \tag{3.2.37}$$
$$v = v'$$

where P_0 and ρ_0 are constants and the primed quantities are the small disturbances. We now linearize Eqs. (3.2.35) and (3.2.36) by discarding any term in which more than one primed quantity is found:

$$\frac{\partial P'}{\partial x} + \rho_0 \frac{\partial v'}{\partial t} = 0$$
$$\frac{\partial \rho'}{\partial t} + \rho_0 \frac{\partial v'}{\partial x} = 0 \tag{3.2.38}$$

3.2 The Heat Capacity Dilemma

At this point we have three unknowns, P', ρ', and v', and only two equations. To remedy the situation, we take the density to be a function only of the pressure, $\rho = \rho(P)$. Then

$$\rho = \rho(P_0) + \frac{\partial \rho}{\partial P} P' \tag{3.2.39}$$

or

$$\rho' = \frac{\partial \rho}{\partial P} P' = \frac{1}{c^2} P' \tag{3.2.40}$$

where we have defined

$$c^2 = \frac{\partial P}{\partial \rho} = \frac{1}{K\rho} \tag{3.2.41}$$

In the full equation of state, the density should depend both on pressure and on temperature [as in Eq. (2.2.9) for the ideal gas]. Ignoring the temperature dependence of ρ is equivalent to ignoring the difference between C_P and C_V, or K_T and K_S, a good approximation for condensed media where the compressibilities are small. The K appearing in Eq. (3.2.41) is thus either K_T or K_S. Substituting Eq. (3.2.40) into (3.2.38), we get

$$c^2 \frac{\partial \rho'}{\partial x} + \rho_0 \frac{\partial v'}{\partial t} = 0$$
$$\frac{\partial \rho'}{\partial t} + \rho_0 \frac{\partial v'}{\partial x} = 0 \tag{3.2.42}$$

If we now take $\partial/\partial x$ of the first of these equations, and $\partial/\partial t$ of the second, and subtract, we have

$$\frac{\partial^2 \rho'}{\partial x^2} - \frac{1}{c^2} \frac{\partial^2 \rho'}{\partial t^2} = 0 \tag{3.2.43}$$

which is the equation for density waves traveling at velocity c. However, the information we want is to be found from Eqs. (3.2.42) by putting in trial solutions

$$\rho' = a \exp\left[i(\omega t - qx)\right]$$
$$v' = b \exp\left[i(\omega t - qx)\right] \tag{3.2.44}$$

The results are

$$iqc^2 a - i\omega \rho_0 b = 0$$
$$-i\omega a + iq\rho_0 b = 0 \tag{3.2.45}$$

These are two linear, homogeneous equations in two unknowns, a and b.

The condition that they have nontrivial solutions (i.e., other solutions than $a = b = 0$) is that the determinant of the coefficients of a and b vanish:

$$0 = \begin{vmatrix} iqc^2 & -i\omega\rho_0 \\ -i\omega & iq\rho_0 \end{vmatrix}$$

$$= -q^2 c^2 \rho_0 + \omega^2 \rho_0$$

or
$$\omega = \pm cq \tag{3.2.46}$$

where the two solutions are for wave vectors in the positive and negative x directions. Equation (3.2.46) is the desired result. The method that we have used to get it parallels closely the techniques we shall use in the next section to enumerate the normal modes of a crystal. Here we have enumerated the low-frequency, long wavelength modes only. ω actually has a series of discrete values given by the discrete values of q. These values, in turn, are specified by requiring that the disturbances be periodic. For example, if in Eq. (3.2.44) we are to have

$$\rho'(x = 0) = \rho'(x = L) \tag{3.2.47}$$

where L is the size of the medium, then

$$q = \frac{2\pi}{L} \ell \tag{3.2.48}$$

where ℓ is any (not too large) integer. Although we have not yet proved it, we shall assume that an equation of the form (3.2.46) holds for transverse sound waves as well, when they exist.

c. The Debye Model

With the knowledge that the lowest energy excitations of the solid are sound waves described by Eq. (3.2.46), we should, and shall, be able to demonstrate that these sound waves are responsible for the T^3 dependence of the low-temperature heat capacity. Our picture is that the solid consists of $3N$ quantum mechanical harmonic oscillators, which, however, are associated not with individual atoms but rather are collective properties of the crystal as a whole. Each oscillator makes a contribution of the same form to the thermal behavior of the system, as does each oscillator in the Einstein model, and therefore all we need to know is the list, or enumeration of the frequencies, in order to be able to perform the operations indicated in Eqs. (3.2.30) to (3.2.32).

As mentioned above, the long wavelength modes of a solid have three polarizations, two transverse and one longitudinal, each with an equation of the form Eq. (3.2.46) but distinguished by the fact that they have different sound speeds, c, since the various oscillations have different restoring forces. Let us call the speed for one polarization c_i, and consider them one at a time.

3.2 The Heat Capacity Dilemma

Since there will be a very large number of modes, of order N, in each polarization, we shall want to do the sums indicated in Eqs. (3.2.30) to (3.2.32) as integrals. We therefore wish to put the enumeration in the form of a number of modes per unit range of frequencies—that is, a density of modes. That step is easily done. Rewriting Eq. (3.2.48) for the three-dimensional case, we have

$$\mathbf{q} = \left(\frac{2\pi}{L}\right)\boldsymbol{\ell} \tag{3.2.49}$$

where
$$\boldsymbol{\ell} = \ell_1 \hat{x} + \ell_2 \hat{y} + \ell_3 \hat{z} \tag{3.2.50}$$

and we are imagining a crystal of dimension L on each side, so that $L^3 = V$. Since the integers ℓ_1, ℓ_2, ℓ_3 simply count modes, the number of modes between $\boldsymbol{\ell}$ and $\boldsymbol{\ell} + d\boldsymbol{\ell}$ is just $d\boldsymbol{\ell}$, which is a three-dimensional differential, $d\boldsymbol{\ell} = d^3\ell$. Thus, the number between $\mathbf{q}$ and $\mathbf{q} + d\mathbf{q}$ is

$$d^3\ell = \frac{V}{(2\pi)^3} d^3q \tag{3.2.51}$$

or for an isotropic solid,

$$\frac{4\pi V q^2\, dq}{(2\pi)^3} \tag{3.2.52}$$

These equations have, of course, a familiar ring about them. Since each polarization has frequencies given by

$$\omega = c_i q \tag{3.2.53}$$

the density of modes for each polarization (i.e., the number of modes between ω and $\omega + d\omega$) is given by

$$\rho_i(\omega)\, d\omega = \frac{4\pi V}{(2\pi)^3} q^2(\omega)\, dq(\omega) = \frac{4\pi V}{(2\pi c_i)^3} \omega^2\, d\omega \tag{3.2.54}$$

The density of modes here plays precisely the same role as the density of single-particle states in the perfect gas problem of Chap. 2. Both are basically just the number of complete waves of a given wavelength that can be fitted into a box of a given size, and so they are identical in q space. Equation (3.2.54) differs from Eq. (1.3.106) only in the relation between ε and p or, equivalently, ω and q. Just as the single-particle states of the perfect gas were each to be filled with some average number of particles, depending on the temperature, we shall now populate the modes of the harmonic crystal with an average degree of excitation, an average number of units of energy $\hbar\omega$, given by Eq. (3.2.32). The units of excitation are thus closely analogous to particles in a perfect gas. We think of them, in fact, as quasiparticles and call them phonons. This follows the convention of giving quantized entities names ending in –ons: photons, electrons, protons, and so on. Even human populations have quantized units called persons.

The total density of modes is given by summing Eq. (3.2.54) over the three polarizations. If we assume that the crystal is isotropic, the two transverse modes will, by symmetry, have the same speed, c_t, and we can call the longitudinal speed c_ℓ. The density of modes for low frequency will then be given by

$$\rho(\omega)\, d\omega = \frac{4\pi V}{(2\pi)^3} \left(\frac{1}{c_\ell^3} + \frac{2}{c_t^3} \right) \omega^2 \, d\omega \qquad (3.2.55)$$

We can save some writing by defining an average speed, $\bar{c}$, by

$$\frac{3}{\bar{c}^3} = \frac{1}{c_\ell^3} + \frac{2}{c_t^3} \qquad (3.2.56)$$

and then

$$\rho(\omega) = \frac{12\pi V}{(2\pi \bar{c})^3} \omega^2 \qquad (3.2.57)$$

Equation (3.2.57) is valid only for low-frequency, long wavelength modes. However, we expect that only these modes will be excited at very low temperature, and so we should now be able to find, say, the energy and heat capacity of a solid in that limit. If the temperature is low enough so that Eq. (3.2.57) is valid up to frequencies such that

$$\hbar\omega \gg kT \qquad (3.2.58)$$

then the higher-frequency modes, for which Eq. (3.2.57) is not valid, will have $\bar{n}$, given by Eq. (3.2.32), vanishingly small, and the thermal energy will be given by

$$E_{\text{th}} = \int_0^\infty \bar{n}(\omega)\rho(\omega)\hbar\omega \, d\omega$$

$$= \frac{12\pi V \hbar}{(2\pi \bar{c})^3} \int_0^\infty \frac{\omega^3 \, d\omega}{\exp(\hbar\omega/kT) - 1}$$

$$= \frac{12\pi V}{(2\pi \hbar \bar{c})^3} (kT)^4 \int_0^\infty \frac{x^3 \, dx}{e^x - 1} \qquad (3.2.59)$$

The definite integral in the last step of Eq. (3.2.59) is simply a number

$$\int_0^\infty \frac{x^3 \, dx}{e^x - 1} = \frac{\pi^4}{15} \qquad (3.2.60)$$

and so the energy has the dependence

$$E_{\text{th}} \propto T^4 \qquad (3.2.61)$$

3.2 The Heat Capacity Dilemma

and the heat capacity, as required, is

$$C_V = \left(\frac{\partial F}{\partial T}\right)_V \propto T^3 \qquad (3.2.62)$$

We note, in passing, that the discussion of phonons in a solid at low temperature is, up to this point, formally identical to the behavior of photons in a cavity (i.e., blackbody radiation) except that photons do not have a longitudinal polarization. The photon formulas may thus be produced by replacing

$$\frac{3}{\bar{c}^3} \to \frac{2}{c^3} \qquad (3.2.63)$$

where c, in the photon case, is the speed of light.

We have not yet taken into account the higher-frequency modes, which cannot be expected to obey Eq. (3.2.46). However, it is not difficult to see that these modes will not have much effect on the heat capacity. The high-temperature heat capacity, as we saw in the arguments in the vicinity of Eq. (3.2.33), depends only on the total number of modes in the crystal, not on their distribution. The low-temperature heat capacity, as we have just argued, depends only on the low-frequency modes, whose distribution we already know. Any smooth form that connects the high- and low-temperature behavior simply cannot be very different from the form sketched in Fig. 3.2.1. It follows, then, that an adequate form for the heat capacity will result if we simply extend Eq. (3.2.57) beyond its range of validity, up to a cutoff frequency chosen to give the right number of modes. That is, we assume that $\rho(\omega) \propto \omega^2$ up to some maximum frequency ω_m such that

$$3N = \int_0^{\omega_m} \rho(\omega) \, d\omega \qquad (3.2.64)$$

This construction, really an interpolation formula between the high- and low-temperature heat capacities, is called the *Debye model*. The maximum frequency ω_m, like the Einstein frequency ω_E of Eq. (3.2.21), is a property of the material, depending on the elastic properties by way of the average sound speed; substituting Eq. (3.2.57) into (3.2.64), we get

$$\omega_m = 2\pi\bar{c}\left(\frac{3N}{4\pi V}\right)^{1/3} \qquad (3.2.65)$$

We could have cut off the density of states in other ways—for example, by choosing a different cutoff for each polarization, to ensure that the shortest wavelength allowed is twice the lattice spacing. Doing so, as we shall see later, also gives the right number of modes, but, for our purposes, which require only a simple interpolation formula, it is an unnecessary complication.

As in the Einstein case, Eq. (3.2.22), it is customary to define a characteristic temperature, the Debye temperature, Θ_D, by

$$\hbar\omega_m = k\Theta_D \tag{3.2.66}$$

If we then repeat Fig. 3.2.1, plotting C_V/Nk versus T/Θ_D, we get a universal curve for the lattice heat capacity of all substances, as in Fig. 3.2.2. Debye temperatures are typically in the range 10^2 to $10^{3\,\circ}$K, increasing with the hardness of the material. For example, for copper, $\Theta_D \sim 340^\circ$K, while for diamond, $\Theta_D \sim 2000^\circ$K.

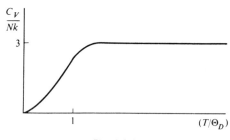

Fig. 3.2.2

We have already drawn attention to the fact that the thermal behavior of the lattice could be ascribed to the excitation of a perfect gas of quasiparticles called phonons. Let us work out the details of the Debye model, using that language, in order to underline the point that we have, in effect, reduced the thermal behavior of a solid to that of a kind of perfect gas.

The energy of each phonon

$$\varepsilon = \hbar\omega \tag{3.2.67}$$

and its momentum

$$p = \hbar q \tag{3.2.68}$$

are related by

$$\varepsilon = \bar{c}p \tag{3.2.69}$$

up to the maximum energy, $k\Theta_D$. The density of single-particle states of the gas is thus

$$\rho(\varepsilon)\,d\varepsilon = 3\,\frac{4\pi V}{(2\pi\hbar)^3}\,p^2\,dp \tag{3.2.70}$$

where the factor 3 is inserted for the three polarizations, just as, for the perfect

3.2 The Heat Capacity Dilemma

gas in Chap. 2, we inserted a factor g for spin orientations. With Eq. (3.2.69), this becomes

$$\rho(\varepsilon) = \frac{12\pi V}{(2\pi\hbar c)^3} \varepsilon^2 \tag{3.2.71}$$

[which differs by the constant factor $\hbar^{-3}$ from the density of modes, $\rho(\omega)$]. There is no limit to the number of phonons occupying each state (mode), so phonons obey Bose statistics. The total number of phonons at any T and V is not fixed but instead adjusts itself to minimize the free energy. If n_{ph} is the total number of phonons, then

$$\left(\frac{\partial F}{\partial n_{\text{ph}}}\right)_{T,V} = 0 \tag{3.2.72}$$

But $(\partial F/\partial n_{\text{ph}})_{T,V}$ is just μ_{ph}, the chemical potential of the phonon gas. Thus,

$$\mu_{\text{ph}} = 0 \tag{3.2.73}$$

The phonons therefore behave like the uncondensed part of a degenerate Bose gas. It follows that the average number of phonons in any state (mode) is

$$\bar{n} = \frac{1}{\exp(\varepsilon/kT) - 1} = \frac{1}{\exp(\hbar\omega/kT) - 1} \tag{3.2.74}$$

The first form of Eq. (3.2.74) is the Bose distribution with zero chemical potential [see Eq. (2.3.17)]. The second is identical to Eq. (3.2.32).

We may now use perfect gas formulas to compute, say, the energy and heat capacity of the solid, but it is convenient first to write the density of states in terms of the cutoff energy instead of the speed of sound as in Eq. (3.2.71). To produce the Debye model, in fact, all we need to remember is that there are $3N$ states with $\rho(\varepsilon) \propto \varepsilon^2$. Then, writing $\rho(\varepsilon) = A\varepsilon^2$, we have

$$3N = A \int_0^{k\Theta_D} \varepsilon^2 \, d\varepsilon \tag{3.2.75}$$

and we may solve for A; $\rho(\varepsilon)$ becomes

$$\rho(\varepsilon) = \frac{9N}{(k\Theta_D)^3} \varepsilon^2 \tag{3.2.76}$$

Then the thermal energy is

$$E_{\text{th}} = \int_0^{k\Theta_D} \varepsilon \rho(\varepsilon) \bar{n}(\varepsilon) \, d\varepsilon$$

$$= 9NkT \left(\frac{T}{\Theta_D}\right)^3 \int_0^{\Theta_D/T} \frac{x^3 \, dx}{e^x - 1} \tag{3.2.77}$$

The integral in Eq. (3.2.77) depends on T through its upper limit and cannot be done analytically in general (it can be done numerically). At low T, in particular, for $T \ll \Theta_D$, the upper limit may be replaced by infinity, and with the help of Eq. (3.2.60), we get

$$E_{th} = \frac{3}{5}\pi^4 NkT \left(\frac{T}{\Theta_D}\right)^3 \quad \text{(low } T\text{)} \tag{3.2.78}$$

and

$$C_V = \frac{12}{5}\pi^4 Nk \left(\frac{T}{\Theta_D}\right)^3 = 234 Nk \left(\frac{T}{\Theta_D}\right)^3 \tag{3.2.79}$$

The Debye temperature of a substance may thus be determined from measurements of its low-temperature heat capacity. At high temperature, $\Theta_D/T \ll 1$, we may expand $e^x - 1 \approx x$ in the denominator of the integral in Eq. (3.2.77), and the energy becomes

$$E_{th} = 3NkT$$

We have, you will recall, ensured this result by choosing the number of modes to be $3N$.

3.3 NORMAL MODES

We saw in the last section that the $3N$ harmonic oscillators that a mean field approach would lead us to expect somehow turn into $3N$ collective modes, with $3N$ distinct frequencies, which are properties of the crystal as a whole. In this section we shall see how that situation occurs, examining the physical principles that underlie the way the modes evolve out of the forces between the atoms.

We shall wind up by actually solving for all the modes only in certain simple cases, chosen to illustrate the important principles. However, let us begin by trying to set up the general problem of the modes of a real crystal. Even here we must make an approximation at the outset: we shall assume that the forces binding each atom to its equilibrium position are harmonic— that is, that the restoring force acting on the atom is proportional to its displacement. This approximation, basically the same one we made in cutting off the series in Eq. (3.2.2), is a good one, but not exact, for the small-amplitude, thermally excited oscillations we wish to study. It has the important mathematical consequence of giving a set of exact (or eigen) frequencies for the crystal vibrations. Since it is a good approximation, the effects of anharmonic forces, which are present in real crystals, can be treated later as perturbations to the results that emerge now.

In contrast to the mean field approach of Eq. (3.2.2), we must consider not only the displacement of an atom from its equilibrium position in

3.3 Normal Modes

absolute space but also the instantaneous displacements of all the other atoms that exert forces on it. Let us temporarily call the possible independent displacements of individual atoms that can take place in a crystal U_i, where i is an index running from 1 to $3N$. In other words, three of the U_i belong to the three independent displacements of each atom. Each displacement, U_j, that occurs anywhere in the crystal may, in principle, exert a force on any of the $3N$ possible motions that can occur; the only restriction is that, being harmonic, the force—and hence the acceleration, $\ddot{U}_i$—will be proportional to U_j. We can represent this by writing

$$\ddot{U}_i = -\sum_{j=1}^{3N} \lambda_{ij} U_j \tag{3.3.1}$$

The coefficients λ_{ij} depend on the nature of the forces acting (or the model we use to represent the forces) in the crystal. For example, if the forces between atoms are short ranged, all the λ_{ij} will be zero except those representing displacements i and j on nearly atoms. Each λ_{ij} also has the mass of the particle with displacement i in its denominator. There is an equation of the form Eq. (3.3.1) for each value of i, $3N$ of them in all.

We now try to solve these $3N$ differential equations (bear in mind that $N \approx 10^{23}$) by putting in oscillatory trial solutions,

$$U_i = U_{i0} \exp(i\omega t) \tag{3.3.2}$$

Substituting Eq. (3.3.2) into (3.3.1), we get $3N$ equations of the form

$$\sum_j (\lambda_{ij} - \omega^2 \delta_{ij}) U_{j0} = 0 \tag{3.3.3}$$

We now have $3N$ linear, homogeneous equations for the $3N$ unknown U_i's, which is, of course, just enough. Just as in Eq. (3.2.45), where we had two linear homogeneous equations in two unknowns, the condition that there be nontrivial solutions is that the determinant of the coefficients of U_{j0}'s vanish:

$$\text{Det } |(\lambda_{ij} - \omega^2 \delta_{ij})| = 0 \tag{3.3.4}$$

This is a $3N \times 3N$ determinant whose off-diagonal terms are λ_{ij} and whose diagonal terms are $(\lambda_{ii} - \omega^2)$. Even if we assume short-range forces, so that many of the λ_{ij} are zero, we still have a $3N \times 3N$ determinant to multiply out. With $N \approx 10^{23}$, the job of multiplying it out may prove tedious and require a great deal of paper, but the result will inevitably be an equation of order $3N$ in ω^2. There will thus be $3N$ distinct solutions for ω^2 in terms of the λ_{ij}, and these solutions are the normal modes of the crystal.

We shall, of course, find means to simplify the problem, and not ever have to confront a 3×10^{23}-order determinant. However, we have already made an important point: it is no accident that the crystal has just exactly $3N$ modes. We argued in Sec. 3.2 that $3N$ was the required number to give the known heat capacity at high temperature, and here we have seen how that $3N$ comes about.

The enormous simplification, which changes the problem from being absurd to being merely very difficult, arises from the fact that we are dealing with a crystal, whose physical properties are periodic in space. Take, by way of illustration, the simplest possible three-dimensional example, an infinite crystal of a single kind of atom arranged on a regular lattice. For, say, the x displacement of a particular atom, we have an equation like Eq. (3.3.1). But for the x displacement of every other atom, we have exactly the same equation; all the x displacements act under the same forces and must therefore behave in the same way. They must execute all the same oscillations although not necessarily at the same time. What we do, then, is write down three equations for the three types of displacement, instead of $3N$ for the individual displacements. Then instead of $3N$ trial solutions like Eq. (3.3.2), we use three trial solutions, one for each type of displacement, differing from one lattice point to another only in the phase (or time) at which the motion occurs. In other words, we try running waves,

$$U_x = U_{x0} \exp\left[i(\omega t - \mathbf{q} \cdot \mathbf{r}_s)\right] \qquad (3.3.5)$$

where $\mathbf{r}_s$ is a discrete vector whose possible values take us from one atom to any other on the lattice. This is really the same as Eq. (3.3.2) except that N values of U_{i0} are replaced by

$$U_{x0} \exp(-i\mathbf{q} \cdot \mathbf{r}_s) \qquad (3.3.6)$$

where $\mathbf{r}_s$ takes on N values to reach all lattice points. We still have the same number of solutions, but we have arranged them in a more convenient way. We now have three equations, three types of solution, a 3×3 determinant to solve. There will thus be three solutions for ω^2, but now each is a function of the variable $\mathbf{q}$; for each value of $\mathbf{q}$, each of the solutions gives a mode. We have already seen, in the last section, that at least for small $\mathbf{q}$, we expect three equations of the form Eq. (3.2.46), $\omega = cq$. These are for the two transverse and one longitudinal polarizations.

Rather than work the whole thing out for one kind of atom only in the crystal, let us set it up in a way that can handle more complex situations. The *sine qua non* of the crystalline solid is repeatability. There is some smallest building block, which can be imagined to be a parallelepiped called the unit cell, and the solid is constructed by repeating that cell in space in all directions. The unit cell can have in it only one atom, as in the example just above, or it can be extremely complex, involving many different kinds of atoms. Either way, all the different types of displacement that can be found in the crystal are found in each unit cell. We need only write as many equations for the displacements as there are independent degrees of freedom within one cell. Suppose that there are r atoms in the unit cell. We label

3.3 Normal Modes

the independent displacements U_n^s, where n is now an index that runs from 1 to $3r$ and s locates the unit cell; from some fixed origin on the corner of a unit cell, the vector $\mathbf{r}_s$ goes to the same corner of the sth unit cell. We can now rewrite Eq. (3.3.1) as

$$\ddot{U}_n^s = -\sum_{n's'} \lambda_{nn'}^{s-s'} U_{n'}^{s'} \tag{3.3.7}$$

This is really the same as Eq. (3.3.1) except that we have reorganized things. Instead of counting i over degrees of freedom from 1 to $3N$, we count n' over those in one cell, from 1 to $3r$, and s' over unit cells, from 1 to N_σ where N_σ is the number of unit cells in the crystal. The number of terms in each equation like Eq. (3.3.7) is $3rN_\sigma = 3N$ just as in Eq. (3.3.1). The notation $s - s'$ in the superscript of the coefficients $\lambda_{nn'}^{s-s'}$ is meant to indicate that the coefficients can only depend on how far apart cells s and s' are, not on their absolute positions in space. We now try running wave solutions.

$$U_n^s = U_n \exp\left[-i(\mathbf{q} \cdot \mathbf{r}_s - \omega t)\right] \tag{3.3.8}$$

The crucial point, once again, is that U_n does not depend on s; it is the same for the nth degree of freedom in every unit cell. At any instant, the amplitude of the wave in any other cell, say, s_0, differs from that in the sth cell only by the phase factor $\exp\left[-i\mathbf{q} \cdot (\mathbf{r}_s - \mathbf{r}_{s0})\right]$. Substituting Eq. (3.3.8) into (3.3.7) and multiplying through by $\exp(i \cdot \mathbf{q} \cdot \mathbf{r}_s)$, we get

$$-\omega^2 U_n = -\sum_{n's'} \lambda_{nn'}^{s-s'} U_{n'} \exp\left[i\mathbf{q} \cdot (\mathbf{r}_s - \mathbf{r}_{s'})\right]$$

$$= -\sum_{n'\sigma} \lambda_{nn'}^\sigma U_{n'} \exp(i\mathbf{q} \cdot \mathbf{r}_\sigma) \tag{3.3.9}$$

where, in the last step, we have made the obvious change of notation, $\mathbf{r}_s - \mathbf{r}_{s'} = \mathbf{r}_\sigma$ and $s - s' = \sigma$, since nothing can depend on the absolute values of s and s'. (We are, in fact, considering an infinite crystal, or a finite one with periodic boundary conditions, in all these discussions. Otherwise the position relative to the surface could be important, not only the distance between cells.) Equation (3.3.9) is like (3.3.3). The condition that it have solutions is

$$\mathrm{Det}\left|\sum_\sigma \lambda_{nn'}^\sigma \exp(i\mathbf{q} \cdot \mathbf{r}_\sigma) - \omega^2 \delta_{nn'}\right| = 0 \tag{3.3.10}$$

In contrast to Eq. (3.3.4), however, this is a $3r \times 3r$ determinant rather than $3N \times 3N$.

We have, then, solutions for $3r$ values of ω^2 in terms of the quantities $\sum_\sigma \lambda_{nn'}^\sigma \exp(i\mathbf{q} \cdot \mathbf{r}_\sigma)$. These latter terms are just Fourier sums with coefficients that depend on the forces and masses of the atoms. $\mathbf{r}_\sigma$ is the dummy variable in these sums, so each sum is just some function of $\mathbf{q}$. The net result is $3r$ functional solutions $\omega = \omega(\mathbf{q})$. As we have already seen, if $r = 1$ (one atom per unit cell), these three solutions reduce for low q to the two transverse and one longitudinal sound polarizations. We shall see shortly, by example, what happens when there is more than one atom per cell.

The functional dependence of ω on $\mathbf{q}$ is called the *dispersion relation*. For sound in crystals, the dispersion relation has $3r$ branches. On each branch there is a normal mode of definite frequency for each value of $\mathbf{q}$; since we know (by our earlier arguments) that there will be $3N$ normal modes in all, there must be just N_σ allowed values of $\mathbf{q}$ (recall $rN_\sigma = N$). Those values are determined by the boundary conditions on the waves, as we shall see shortly by example.

Let us consider some specific examples of systems that have normal modes —that is, vibrational modes that are collective properties of the system. For simplicity, we shall consider only one-dimensional cases, since nearly all the principles involved are illustrated by them, and, of course, they are much more tractable than three-dimensional cases. Furthermore, we shall restrict ourselves to nearest-neighbor forces. As a matter of fact, since the forces are harmonic and act only between neighbors, we may as well replace the atomic forces altogether by imaginary springs of spring constant k.

The simplest problem of this kind is the linear triatomic molecule—that is, three masses connected by two springs in one dimension (Fig. 3.3.1).

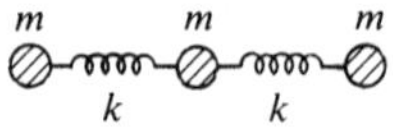

Fig. 3.3.1

Since there are three degrees of freedom, there are three solutions:

$$\omega^2 = 0, \quad \frac{k}{m}, \quad \frac{3k}{m} \tag{3.3.11}$$

(see or, better, do Prob. 3.3). There is no dependence on q—no running waves—since this problem has no repeatable features. Each atom has different forces acting on it. The only way to approach the problem is by Eq. (3.3.1). The zero-frequency solution corresponds to the same displacement in all three atoms, which has no restoring force. There is always one zero-frequency mode for each macroscopic degree of freedom of the system as a whole, reserved for telling us where, in space, the system can be found. For a three-dimensional crystal, there are actually only $3N - 6$ nonzero frequencies, six coordinates being needed to fix the position and orientation of the crystal. However, since $N \approx 10^{23}$, the six modes are not badly missed.

The next easiest problem to do is an infinite chain of identical masses

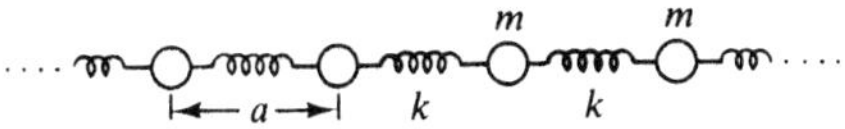

3.3 Normal Modes

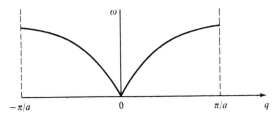

Fig. 3.3.2

(actually N masses, length L, periodic boundary conditions) connected by identical springs, as in Fig. 3.3.2. In this case, we have repeatability and running waves. There is, however, only one degree of freedom per unit cell, a 1×1 determinant to solve, and only one branch of the dispersion curve (Prob. 3.4), given by

$$\omega^2 = 4\frac{k}{m}\sin^2\left(\frac{qa}{2}\right) \qquad (3.3.12)$$

where $\qquad q = \left(\frac{2\pi}{L}\right)\ell \quad \left(\ell = 0, \pm 1, \ldots, \pm \frac{L}{2a}\right) \qquad (3.3.13)$

A plot of ω versus q is shown in Fig. 3.3.3. Notice that as $q \to 0$, $\sin qa/2 \to qa/2$, and the dispersion relation becomes

$$\omega = \pm \left(\frac{k}{m}a^2\right)^{1/2} q \qquad (3.3.14)$$

which is the same form that we got for the long wavelength, low q modes in the continuum hydrodynamic case, Eq. (3.2.46). Let us defer detailed analysis of the results in this case, since we shall now do explicitly the next least complicated problem, the linear chain with two different masses. That problem will reduce to the one above when we allow the masses to become the same.

Fig. 3.3.3

The problem and some of its notation are sketched in Fig. 3.3.4. The particles have masses m_1 and m_2; let us arbitrarily choose $m_1 > m_2$. The basic repeating unit, the unit cell, has in it two atoms and, being one dimensional, two degrees of freedom. The index n can be 1 or 2 for mass m_1 or m_2, respectively, and the unit cells are labeled $s = \ell - 1, \ell, \ell + 1, \ldots$. The equilibrium spacing between atoms is a, so that the distance between equiv-

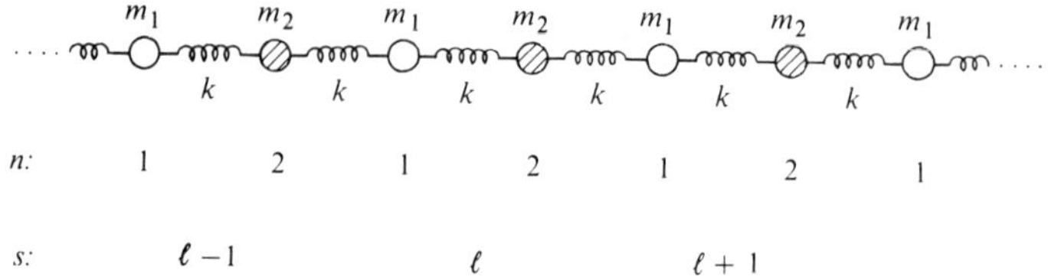

Fig. 3.3.4

alent points in adjacent unit cells is $2a$. The vector $\mathbf{r}_s$ thus comes in units of $2a$:

$$r_s = X_0 + 2as = X_0 + [2a(\ell - 1), 2a\ell, \ldots] \tag{3.3.15}$$

where X_0 is the arbitrary origin of r_s (a scalar in this one-dimensional problem). Let the equilibrium positions of the particles be labeled X_{0n}^s, so that

$$X_{02}^s - X_{01}^s = a \tag{3.3.16}$$

If we call the instantaneous positions X_n^s and the displacements from equilibrium U_n^s, we have

$$U_n^s = X_n^s - X_{0n}^s \tag{3.3.17}$$

We may now start writing equations of motion. The force to the right on the atom at X_1^ℓ is

$$m_1 \ddot{U}_1^\ell = k(X_2^\ell - X_1^\ell - a) + k(X_2^{\ell-1} - X_1^\ell + a) \tag{3.3.18}$$

the forces just being due to the stretching or contracting of springs on either side. With the use of Eqs. (3.3.16) and (3.3.17), this becomes

$$\ddot{U}_1^\ell = \frac{k}{m_1}(U_2^\ell + U_2^{\ell-1} - 2U_1^\ell) \tag{3.3.19}$$

The same arguments for the atom at X_2^ℓ give

$$\ddot{U}_2^\ell = \frac{k}{m_2}(U_1^{\ell+1} + U_1^\ell - 2U_2^\ell) \tag{3.3.20}$$

Comparing with Eq. (3.3.7), we see that our model for the forces gives the coefficients $\lambda_{nn'}^{s-s'}$,

$$\lambda_{12}^{0} = -\frac{k}{m_1}, \quad \lambda_{12}^{-1} = -\frac{k}{m_1}, \quad \lambda_{11}^{0} = 2\frac{k}{m_1}$$
$$\lambda_{21}^{1} = -\frac{k}{m_2}, \quad \lambda_{21}^{0} = -\frac{k}{m_2}, \quad \lambda_{22}^{0} = 2\frac{k}{m_2} \tag{3.3.21}$$

3.3 Normal Modes

All other λ's are zero. With these identifications, Eqs. (3.3.18) and (3.3.19) are the two distinct forms of Eq. (3.3.7).

The next step is to try solutions of the form Eq. (3.3.8):

$$U_1^{\ell} = U_1 \exp\{-i[q(2\ell a) - \omega t]\} \tag{3.3.22}$$

$$U_2^{\ell} = U_2 \exp\{-i[q(a + 2\ell a) - \omega t]\} \tag{3.3.23}$$

In writing Eqs. (3.3.21) and (3.3.22), we have taken $X_0 = 0$ for $n = 1$ and $X_0 = a$ for $n = 2$. With those starting points [see Eq. (3.3.15)], r_s reaches, respectively, every $n = 1$ position and every $n = 2$ position when all values of ℓ are substituted in. We now substitute Eqs. (3.3.22) and (3.3.23) into (3.3.18) and (3.3.19), getting

$$-\omega^2 U_1 = \frac{k}{m_1}(U_2 e^{-iqa} + U_2 e^{iqa} - 2U_1)$$

$$-\omega^2 U_2 = \frac{k}{m_2}(U_1 e^{-iqa} + U_1 e^{iqa} - 2U_2)$$

or since $e^{iqa} + e^{-iqa} = 2\cos qa$,

$$\left(2\frac{k}{m_1} - \omega^2\right)U_1 - 2\frac{k}{m_1}(\cos qa)U_2 = 0$$

$$-2\frac{k}{m_2}\cos qa + \left(2\frac{k}{m_2} - \omega^2\right)U_2 = 0 \tag{3.3.24}$$

This is the result for Eq. (3.3.9). We see that the sum of the Fourier components over unit cells has given us functions of q in the off-diagonal terms of the matrix of coefficients of U_n—namely, $\cos qa$. Applying the condition, Eq. (3.3.10), we get

$$\left(2\frac{k}{m_2} - \omega^2\right)\left(2\frac{k}{m_1} - \omega^2\right) - 4\frac{k^2}{m_1 m_2}\cos^2 qa = 0 \tag{3.3.25}$$

a quadratic equation in ω^2, as promised, with solutions

$$\omega^2 = k\left(\frac{1}{m_1} + \frac{1}{m_2}\right) \pm k\left[\left(\frac{1}{m_1} + \frac{1}{m_2}\right)^2 - \frac{4\sin^2 qa}{m_1 m_2}\right]^{1/2} \tag{3.3.26}$$

We can simplify the writing if we define a reduced mass, μ,

$$\frac{1}{\mu} = \frac{1}{m_1} + \frac{1}{m_2} \tag{3.3.27}$$

whereupon Eq. (3.3.26) becomes

$$\omega^2 = \frac{k}{\mu}\left[1 \pm \left(1 - \frac{4\mu^2 \sin^2 qa}{m_1 m_2}\right)^{1/2}\right] \tag{3.3.28}$$

There are, as expected, two solutions for ω as a function of q—that is, two branches of the dispersion relation corresponding to the two degrees of freedom in each unit cell. Notice that no new solutions appear for values of $|qa| > \pi/2$; we get all the frequencies if we limit $|q|$ to run from 0 to $\pi/2a$. For a wave with $q = \pm \pi/2a$, we have from Eq. (3.3.22) or (3.3.23) the ratio of displacements in adjacent unit cells:

$$\frac{U_n^\ell}{U_n^{\ell+1}} = \exp(\pm i\pi) = -1 \tag{3.3.29}$$

Atoms in adjacent cells are exactly out of phase; that is the shortest wavelength that can be constructed. Thus all possible wavelengths are accounted for by writing

$$q = \frac{2\pi}{L}\left[0, \pm 1, \pm 2, \ldots, \pm \frac{1}{2}\left(\frac{L}{2a}\right)\right] \tag{3.3.30}$$

where we are imagining N atoms and $N_\sigma = N/2$ unit cells in a chain of length L†. ($L/2a$) is just N_σ, the number of unit cells, so on each branch we have one value of q for each unit cell; between both branches there will be one solution, ω, for each degree of freedom. The number of normal modes, as always, is equal to the number of degrees of freedom.

We wish now to sketch the general form of Eq. (3.3.28). We can get a pretty good idea of how it goes by evaluating its limiting behavior, as $qa \to 0$ and as $qa \to \pi/2$ (negative values of q, which correspond to waves running in the opposite direction, will give exactly the same results as positive values). When $qa \to 0$, $\sin^2 qa \to (qa)^2 \to 0$, and

We get
$$\left(1 - \frac{4\mu^2 \sin^2 qa}{m_1 m_2}\right)^{1/2} \to 1 - \frac{2\mu^2 q^2 a^2}{m_1 m_2} \tag{3.3.31}$$

$$\lim_{qa \to 0} \omega^2 = 0, 2\frac{k}{\mu} \tag{3.3.32}$$

We thus get one solution at zero frequency, which locates the chain in space, and $2N_\sigma - 1 = N - 1$ nonzero solutions, including one finite frequency solution at $q = 0$—that is, with infinite wavelength. We shall see later what is physically vibrating at that frequency. For small but finite q, the two solutions are, respectively,

$$\omega = \pm\left(\frac{2k\mu}{m_1 m_2}\right)^{1/2} aq \tag{3.3.33}$$

† More precisely, L is the length over which we apply periodic boundary conditions on each kind of wave. It is therefore the distance between atom 1 in the first cell and atom 1 in the last. The length of the chain is really $L + 2a$, and $N_\sigma = (L + 2a)/2a$. Then $q = 2\pi/L$ $[0, \pm 1, \ldots, \pm\frac{1}{2}(N_\sigma - 1)]$. There are *exactly* N_σ discrete values of q on each branch.

3.3 Normal Modes

which is of the form of Eq. (3.2.46)—that is, an acoustic wave—and

$$\omega = \pm \left(2\frac{k}{\mu}\right)^{1/2}\left[1 - \frac{2\mu^2(qa)^2}{m_1 m_2}\right] \qquad (3.3.34)$$

These solutions are called, respectively, the acoustic and optical branches. Notice that the optical branch, Eq. (3.3.34), comes into $q = 0$ at zero slope. That is,

$$\lim_{q \to 0} \frac{\partial \omega}{\partial q} = 0 \qquad \text{(optical branch)} \qquad (3.3.35)$$

We may now begin to construct the dispersion relation as it appears at low q (see Fig. 3.3.5).

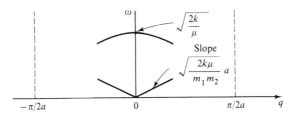

Fig. 3.3.5

Close to $qa = \pi/2$, we define

$$qa = \frac{\pi}{2} - \delta a \qquad (3.3.36)$$

Then, for $\delta a \ll 1$, we have, expanding about $qa = \pi/2$,

$$\sin qa = \sin\left(\frac{\pi}{2} - \delta a\right)$$

$$= \sin\left(\frac{\pi}{2}\right) - \frac{(\delta a)^2}{2}\sin\left(\frac{\pi}{2}\right)$$

$$= 1 - \frac{(\delta a)^2}{2}$$

$$\sin^2 qa = \left[1 - \frac{(\delta a)^2}{2}\right]^2 = 1 - (\delta a)^2 \qquad (3.3.37)$$

so that Eq. (3.3.28) becomes

$$\omega^2 = \frac{k}{\mu}\left\{1 \pm \left[1 - \frac{4\mu^2}{m_1 m_2} + \frac{4\mu^2}{m_1 m_2}(\delta a)^2\right]^{1/2}\right\} \qquad (3.3.38)$$

At $\delta a = 0$, we get the two solutions

$$\omega^2 = \frac{k}{\mu}\left[1 \pm \left(1 - \frac{4\mu^2}{m_1 m_2}\right)^{1/2}\right] \qquad (3.3.39)$$

but

$$1 - \frac{4\mu^2}{m_1 m_2} = \frac{(m_1 - m_2)^2}{(m_1 + m_2)^2}$$

so that

$$\omega^2 = \frac{2k}{m_1}, \frac{2k}{m_2} \qquad (3.3.40)$$

Moreover, we may see from Eq. (3.3.38) that as $\delta a \to 0$, then $\partial \omega / \partial \delta \to 0$ as well. With all these clues, the complete picture may be sketched (Fig. 3.3.6).

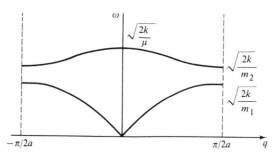

Fig. 3.3.6

All the modes of the linear crystal are shown in Fig. 3.3.6. They fall on two curves, the acoustic and optical branches, but are restricted to a range of q space just big enough to accommodate one mode for every degree of freedom; there are just N solutions in Fig. 3.3.6, half of them, or N_σ, on each branch. The region of q space in which all the solutions are found is our first example of a *Brillouin zone*, a concept we shall generalize later. The dispersion relation comes in to the boundaries of the zone with zero slope, a behavior we shall later see to be characteristic of all kinds of waves in crystals, including electrons. Some of the significance of this behavior may be seen if we recall that the group velocity of a wave is

$$v_g = \frac{\partial \omega}{\partial q} = \frac{\partial \varepsilon}{\partial p} \qquad (3.3.41)$$

Thus, at the zone boundary the solutions are not running waves at all but standing waves, with zero group velocity. Moreover, there are two standing waves with the same q—and therefore the same wavelength—but different

3.3 Normal Modes

frequencies at the zone boundary. There are no modes in the crystal with frequencies (or, what is the same thing, phonon energies) between these two values.

Let us investigate what kinds of physical oscillations are associated with the various modes seen in Fig. 3.3.6. After writing Eqs. (3.3.24) for the amplitudes of the two kinds of atoms, we have since directed our attention to the conditions that those equations have solutions, rather than the solutions themselves. We shall now consider what the solutions are like at various ω and q where we have found that solutions are to be expected. The first of Eqs. (3.3.24) may be rewritten

$$\frac{U_1}{U_2} = \frac{2(k/m_1)\cos qa}{(2k/m_1) - \omega^2} \qquad (3.3.42)$$

Recall that, in each mode, the heavier atoms are moving with amplitude proportional to U_1 and the lighter to U_2; U_1/U_2 is the ratio of the amplitudes in each mode. Looking first at the long wavelength modes, $qa \to 0$, and, in the acoustic branch, $\omega^2 \to 0$, so that $U_1/U_2 \to 1$. Thus, in the acoustic branch, at low q, both kinds of atom move together with the same amplitude. These modes are just the long wavelength sound waves we are already familiar with. They are basically macroscopic in nature, quite insensitive to such fine details as the fact that the medium is made up of two kinds of atoms. In the optical branch, $\omega^2 \to 2k/\mu$, so

$$\frac{U_1}{U_2} \to \frac{2k/m_1}{2k/m_1 - 2k/\mu} = \frac{1}{1 - [(m_1 + m_2)/m_2]} = -\frac{m_2}{m_1} \qquad (3.3.43)$$

Thus, in these modes, the two kinds of atoms oscillate in opposite directions but with stationary center of mass. The mode found at zero wave vector (i.e., infinite wavelength) but finite frequency has all the atoms of one kind moving together, but in the opposite direction to all the atoms of the other kind. It involves the largest potential energy, has the largest restoring force, and the highest frequency and energy of any mode in the crystal.

Near the zone boundary, we once again define δ by means of Eq. (3.3.36) and find

$$\cos qa \to \delta a \qquad (3.3.44)$$

so that

$$\frac{U_1}{U_2} \to \frac{(2k/m_1)\,\delta a}{(2k/m_1) - \omega^2} \qquad (3.3.45)$$

For the acoustic branch, taking the $(-)$ solution of Eq. (3.3.38) and recalling that the limit will be $\omega^2 \to 2k/m_1$, we see that the dispersion relation will have the form

$$\omega^2 = \frac{2k}{m_1} - \alpha_0(\delta a)^2 \qquad (3.3.46)$$

where α_0 depends on the masses. Substituting into Eq. (3.3.45) and letting $\delta a \to 0$, we find for the acoustic branch

$$\frac{U_1}{U_2} \to \infty \qquad (3.3.47)$$

at the zone boundary. In the optical branch, $\omega^2 \to 2k/m_2$. Substituting that into Eq. (3.3.45) and letting $\delta a \to 0$, we find

$$\frac{U_1}{U_2} \to 0^- \qquad (3.3.48)$$

Using all these results, the ratio of the heavy-to-light amplitudes, U_1/U_2, is sketched as a function of q in Fig. 3.3.7.

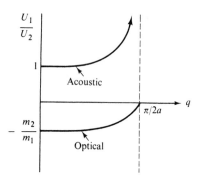

Fig. 3.3.7

Since all our solutions are for small amplitude, the only way we can have $U_1/U_2 \to \infty$ is if $U_2 \to 0$. Thus, the mode found in the acoustic branch at the zone boundary has all of the m_2 (i.e., light atoms) standing still while the heavy atoms oscillate in a standing wave with the shortest possible wavelength, $\lambda = 2\pi/q = 4a$, or twice the length of a unit cell. At the same q in the optical branch, $U_1 = 0$, only the light atoms oscillate while the heavy ones stand still.

Guided by Fig. 3.3.7, we can now describe the oscillations corresponding to all parts of the dispersion relation, Fig. 3.3.6. Starting in the acoustic branch at low q, all the atoms, heavy and light, move together in long wavelength, ordinary sound waves. At higher q, shorter wavelength, on the acoustic branch, more and more of the motion is carried by the heavy atoms, until, at the zone boundary, the light atoms stand still and a standing wave of wavelength $4a$ appears in the heavy atoms. At the same wavelength, still on the zone boundary, there is a higher-frequency standing wave, in which only

3.3 Normal Modes

the light atoms oscillate, and that mode is the endpoint of the optical branch. Moving to smaller q along the optical branch, motion of the heavy atoms comes increasingly into play, but now, in contrast to the acoustic branch, neighboring atoms, light and heavy, tend to move in opposite directions. Finally, at $q = 0$, there is a standing wave of infinite wavelength, all heavy atoms moving together in one direction while all light atoms go in the other.

As we have already seen, the dispersion relation gives the possible modes or states into which phonons may be excited, each with energy $\hbar\omega$ in the crystal. The density of modes, $\rho(\omega)$, or, what is the same thing, the density of states for phonons, has in it an important feature that we shall also find in the single-particle electron states in crystals: there is a band of frequencies, between the acoustic and optical branches, within which no propagating modes exist—in other words, a band of forbidden energies for phonons. There will thus be a gap in the density of states. Moreover, the states on each side of the forbidden band are standing waves at the zone boundary, with zero group velocity v_g. Now, the density of states is always given by

$$\rho(\omega)\, d\omega \propto dq(\omega) \tag{3.3.49}$$

and therefore

$$\rho(\omega) \propto \frac{dq}{d\omega} = \frac{1}{v_g} \tag{3.3.50}$$

It follows that the density of states, which is zero inside the gap, is infinite at its edges. The density of states near the gap is sketched in Fig. 3.3.8.

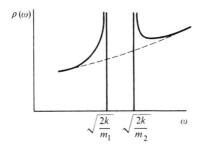

Fig. 3.3.8

The total number of states in the crystal (the area under the curve) is just the same as if the two masses were the same, there were no forbidden band, and $\rho(\omega)$ followed the dotted line in Fig. 3.3.8. The states that would have been in the gap get pushed out to the edges, thereby causing infinities there. The areas under the solid and dotted curves are the same.

It is interesting to look at this problem in the limit $m_1 \to m_2$. In that case, it should reduce to the problem of the linear chain with one kind of mass (Prob. 3.4), whose solutions are given in Eq. (3.3.12) and Fig. 3.3.3. Of the differences between those results and the ones we have here, it is obvious that one will vanish: the gap in frequencies between $\sqrt{2k/m_1}$ and $\sqrt{2k/m_2}$ goes away when $m_1 \to m_2$. It looks at first as if we are left with an unwanted optical branch in this limit, but the reason is that when $m_1 = m_2$, the basic periodicity of the lattice changes from $2a$ to a, so that the Brillouin zone extends not to $\pi/2a$ but all the way to π/a. The optical branch must thus be folded out into the region between $\pi/2a$ and π/a. Further details of the comparison are left to Prob. 3.5.

If we consider a chain with, say, three different masses instead of two, the dispersion relation will, as we have seen, have three branches. The question arises as to whether the third branch will be acoustic or optical in character—that is, whether it will come into $q = 0$ with zero or finite frequency. The answer can be seen by recalling the solution to the triatomic linear molecule, Eq. (3.3.11). In that case, the atoms differed not in mass but in the forces acting on them. Nevertheless, the triatomic molecule is qualitatively similar to one unit cell of the chain with three masses. There are no running wave solutions, but instead we may expect the three solutions to correspond to the three $q = 0$ modes of the long triatomic chain. Thus, at $q = 0$, we expect two finite-frequency (optical) and one zero-frequency mode. In fact, no matter how complicated the unit cell, we expect one zero-frequency mode, reserved for the position of the center of mass, and $N - 1$ finite-frequency modes for any one-dimensional chain. It follows that there will be one acoustic branch, and all the rest will be optical.

Although we have only solved a one-dimensional problem, all the features of our results are to be found in real three-dimensional crystals. For example, in order to represent the phonon spectrum of a crystal, we can still make a sketch like Fig. 3.3.3 (for a monatomic crystal) or Fig. 3.3.6 (for a diatomic crystal), but we must now understand that q plotted on the abscissa is along some particular direction in the crystal and that details of the curve (slopes, intercepts) will differ in different directions. These features will introduce new complexity into, for example, the density of states, in which all directions are lumped together and we only count the number of states at a given frequency. There will also be transverse branches in addition to the longitudinal ones we have studied here, but their behavior, insofar as $\omega(\mathbf{q})$ is concerned, is not different from what we have seen here, and they will make similar contributions to the density of states.

Details of the kind of complications that arise when the zone boundary is encountered at different energies in different directions are far more important for electrons in crystals than for phonons, and so we shall defer discussion of them to the section on electrons in this chapter. However, broadly speaking, we may expect $\rho(\omega)$, even for a monatomic crystal, to have a number of peaks in it corresponding to zone boundaries in various directions, as in Fig. 3.3.9. Here we have sketched the real density of states of a hypothetical crystal, and superimposed on it, the dotted line, is the density of states of the Debye model. The Debye $\rho(\omega)$ shares two features with the real one: $\rho(\omega) \propto \omega^2$ for low ω, and the total area under $\rho(\omega)$ is $3N$. These features, as we have seen, are quite enough to give a good account of the heat capacity of the crystal. The Einstein model could be plotted on Fig. 3.3.9 as well. Its density of states would have the form of a delta function at frequency ω_E. The area under the delta function would still be $3N$.

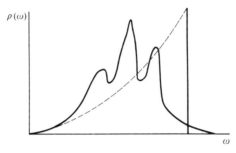

Fig. 3.3.9

3.4 CRYSTAL STRUCTURES

In order to understand the central common features of the solid state, it is sufficient to know that a crystal is a periodic structure in space, with a basic repeating unit called the unit cell. On the other hand, the rich variety of phenomena that present themselves in solids, the differences between different solids, depend on the geometrical structure of the unit cell, and the nature and arrangement of the atoms within it. In this section we shall outline briefly the principles by which complex crystal structures are organized into categories with common geometrical properties.

The first step of the job of organizing things we have already done by mentally separating the crystal into unit cells, each with some arrangement of atoms within it. The usual nomenclature is to speak of a *lattice*, and a *basis*. The lattice is an imaginary array of points in space, while the basis is the atom or group of atoms that must be placed at each lattice point to form the crystal. The unit cell is the repeating structural unit of the lattice. For example, the salt cesium chloride, CsCl, has a simple cubic lattice—that is, the unit cell is a cube, the lattice being the points at the corners of the cubes. The basis is a cesium chloride molecule, with, say, the cesium at the corner and the chlorine at the body center inside the cube. An alternative way to describe the same arrangement is to speak of two interpenetrating cubic lattices, one with cesium atoms at all the corners, the other with chlorine atoms at the corners. That way does not, however, follow the convention we are setting up.

Given any lattice of points in space, there is always more than one way to choose a unit cell. For example, if we have a simple cubic lattice, we could choose two adjacent cubes to be the unit cell; this unit cell, like the simple cube itself, is a repeating unit which fills all of space. There is clearly a distinction to be made between any old unit cell and the smallest possible unit cell for a given lattice. The smallest possible choice is called the *primitive unit cell*. One should not be mislead, however, into thinking that there is some primitive unit cell for each lattice and that all other unit cells are somehow incorrect. As we shall see below, sometimes there are real advantages in using, for descriptive purposes, a choice other than the primitive unit cell of certain lattices. Moreover, even the choice of primitive unit cell is not unique in any lattice.

Suppose that we have a lattice for which we have chosen a convenient unit cell to deal with. Although it is not the only way to make the choice, we will imagine that the unit cell has lattice points at its corners. In that case, it will be some sort of parallelepiped. Choose one corner of the parallelepiped, and let the three edges joining at that corner be vectors, **a**, **b**, **c**, as shown in Fig. 3.4.1. Notice that **a**, **b**, and **c** are not necessarily orthogonal. We now construct the lattice translation vector

$$\mathbf{r}_s = n_1 \mathbf{a} + n_2 \mathbf{b} + n_3 \mathbf{c} \qquad (3.4.1)$$

where the n_1, n_2, and n_3 may be any positive or negative integer. Some vector with length and direction given by Eq. (3.4.1) will get us from the lattice point at the corner of any unit cell to the lattice point at the corner of any other. For that matter, it will take us from any position within the cell, say, the position of a certain atom in the basis, to the same position in every other cell. We have already used this property of $\mathbf{r}_s$ in constructing Eqs. (3.3.8), (3.3.9), and (3.3.10).

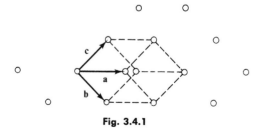

Fig. 3.4.1

3.4 Crystal Structures

We now make the following distinction: if the vectors r_s, starting from one lattice point, enable us to reach every other lattice point, the cell that we have chosen is primitive. It is easy to see that if the unit cell we have chosen to define **a**, **b**, and **c** is primitive, the entire lattice may be generated by putting points at the ends of all vectors produced by all possible choices of positive and negative integers for n_1, n_2, and n_3. There is one lattice point for every choice (n_1, n_2, n_3). Since we also generate another unit cell every time we change n_1, n_2, or n_3 by one unit and place **a**, **b**, and **c** at that point, it also follows that there is one primitive unit cell for each lattice point. That is really the essence of the primitive unit cell: there is just one lattice point for each of them. The volume of the unit cell, primitive or otherwise, is given by the geometric formula

$$V_0 = \mathbf{a} \cdot \mathbf{b} \times \mathbf{c} \qquad (3.4.2)$$

If the cell is primitive, this volume is also equal to the volume of the entire lattice divided by the number of lattice points.

It is possible to construct a primitive unit cell in such a way that the lattice point is at its center rather than its corner. This construction is done by drawing lines to all other lattice points, forming the plane that is the perpendicular bisector of each of those lines, and taking as the unit cell the smallest volume enclosed by these planes. The cell so constructed is called the Wigner-Seitz unit cell. It will not generally have the same shape as the ordinary primitive cell, but, having one lattice point per cell, it must have the same volume. An example in a two-dimensional lattice is shown in Fig. 3.4.2.

All the possible lattice configurations that can fill all of space are categorized into types according to their symmetry properties. One symmetry property is common to all lattices: every lattice point is a center of symmetry. This statement means that, starting from one point, if another point is found a certain distance in one direction, another point must be found the same distance in the opposite direction. This property is obvious from Eq. (3.4.1): for the point at (n_1, n_2, n_3) there is another at

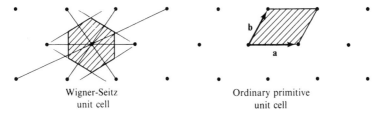

Wigner-Seitz unit cell

Ordinary primitive unit cell

Fig. 3.4.2

$(-n_1, -n_2, -n_3)$. The symmetries that distinguish between different lattice types are of two kinds: rotation about an axis and reflection through a plane. If there is some axis through a crystal, and a rotation of the entire lattice through an angle of $2\pi/n$ about that axis will simply reproduce the original lattice, the lattice is said to have an n-fold axis, and that is one of its symmetry properties (let us defer examples until we get to the classification scheme itself). Similarly, if the lattice can reproduce itself by reflecting through some plane, it is said to have a symmetry plane, and that again is one of its symmetry properties.

Classifying the possible lattices according to these symmetry properties has a number of advantages. For example, as we shall see in Sec. 3.5a, a crystal acts as a three-dimensional diffraction grating for X rays. Thus, if a monochromatic beam of X rays is sent into a crystal, it will emerge diffracted in various directions, to be recorded on a photographic plate, as sketched in Fig. 3.4.3. Now, suppose that the incident beam happened to be

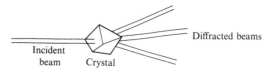

Fig. 3.4.3

sent in along an n-fold symmetry axis. Rotation of the crystal through $2\pi/n$ about that axis would bring it back into itself. It follows that, whatever physical processes cause the beam to be diffracted, rotation of the photographic plate through $2\pi/n$ must bring the pattern on it back into itself. The symmetries of the crystal show up in the symmetries of the diffraction pattern, which, of course, will help the crystallographer to classify the crystal from the patterns he finds. There are other advantages in that the symmetries of the crystal may be used to simplify the tensors that represent the directional response of the crystal—for example, to various stresses, to electric fields, or to magnetic fields. It should be kept in mind that the crystal may have less symmetry than its lattice type, since the basis may not be properly reproduced under some of the symmetry operations of the lattice.

We shall list the various systems, also called *Bravais lattices*, into which the possible lattices are categorized and give a brief description of each. Afterwards we shall give some examples of the symmetry operations that distinguish the various categories and discuss some special cases.

The *triclinic* Bravais lattice is the most general type, with no symmetries other than the necessary fact that each atom is a center of symmetry. The unit cell is a parallelepiped with unequal edges, $|\mathbf{a}| \neq |\mathbf{b}| \neq |\mathbf{c}|$, and unequal angles between $(\mathbf{a}, \mathbf{b})$, $(\mathbf{b}, \mathbf{c})$, and $(\mathbf{c}, \mathbf{a})$.

3.4 Crystal Structures

The *monoclinic* system differs from the triclinic in that the angle between **c** and the plane **a, b** is 90°. Unit cells of the triclinic and monoclinic systems are sketched in Fig. 3.4.4.

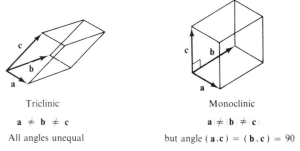

Triclinic
a ≠ **b** ≠ **c**
All angles unequal

Monoclinic
a ≠ **b** ≠ **c**
but angle (**a**,**c**) = (**b**,**c**) = 90

Fig. 3.4.4

In the *orthorhombic* system all the angles are 90°, but the edges are still unequal. Its unit cell is thus a rectangular parallelepiped.

The *tetragonal* system has all right angles and two equal sides: |**a**| = |**b**| ≠ |**c**|. The cell is a rectangular parallelepiped with a square base. The orthorhombic and tetragonal cells are sketched in Fig. 3.4.5.

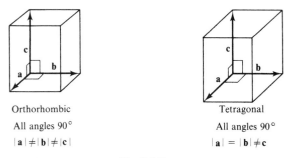

Orthorhombic
All angles 90°
|**a**| ≠ |**b**| ≠ |**c**|

Tetragonal
All angles 90°
|**a**| = |**b**| ≠ **c**

Fig. 3.4.5

In the *trigonal* system all the edges and angles are equal, but the angles are not 90°. The trigonal unit cell is a cube that has been stretched or compressed along a body diagonal. This system is also known as *rhombohedral*. It will be noticed that the sequence triclinic, monoclinic, orthorhombic, and tetragonal was in an obviously increasing order of symmetry approaching the cube. The trigonal (rhombohedral) system departs from that sequence. We now start increasing symmetry again but along a different path.

In the *hexagonal* system |**a**| = |**b**| and the angle (**a**, **b**) is 120°. **c** is

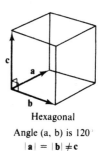

Trigonal (rhombohedral)
Angles and edges equal

Hexagonal
Angle (a, b) is 120
$|a| = |b| \neq c$

Fig. 3.4.6

perpendicular to the plane **a**, **b**. The trigonal and hexagonal cells are sketched in Fig. 3.4.6.

The final system is, of course, *cubic*, with $|a| = |b| = |c|$ and all right angles.

Figure 3.4.6 is actually somewhat misleading in that it obscures the basic symmetries of the hexagonal system and also fails to reveal the close similarity between the trigonal and hexagonal systems, and even between both of those and the cubic. We can get some insight into these symmetries and similarities by considering the symmetry operations that distinguish these systems. First consider the trigonal system, imagining the **a**, **b**, and **c** vectors arranged as shown in Fig. 3.4.7. The endpoints of the vectors form an equilateral triangle in the horizontal plane indicated by the cross-hatched circle. The vertical dotted line, perpendicular to the plane, is a threefold rotation axis. In the horizontal plane passing through the origin, there are three twofold axes, two of which are shown by dotted lines. They are the projections of **a**, **b**, and **c** in that plane. Furthermore, there are three vertical reflection planes, passing through the vertical dotted line, midway between the horizontal

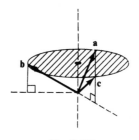

Fig. 3.4.7

3.4 *Crystal Structures* 181

rotation axes. The basic structure of the lattice is of planes of points arranged in equilateral triangles; these are the horizontal planes in Fig. 3.4.7. Seen from above, the lattice points in adjacent planes are displaced, as the origin is from the endpoints of **a**, **b**, **c**, the lattice points in each plane falling above the centers of the triangles in the plane below.

The hexagonal system is also formed by planes of equilateral triangles, but in this case the lattice points in successive planes fall directly above one another. Although it is not the primitive unit cell, it is easiest to imagine this lattice to be formed of right hexagonal prisms. The way in which hexagons are formed in the horizontal planes from the bases of the primitive cell of Fig. 3.4.6 is sketched in Fig. 3.4.8. Here we see the bases of four unit cells (dotted lines) and the hexagon picked out from among them. The axis through any lattice point in the vertical direction—that is, perpendicular to the (horizontal) plane of Fig. 3.4.8—is a sixfold axis of rotation. The horizontal plane itself is a reflection plane. There are six twofold axes in the horizontal plane; these axes are any of the dotted lines through a lattice point, then another through the same point every 30°. Finally, there are six vertical reflection planes, one through each of the horizontal axes.

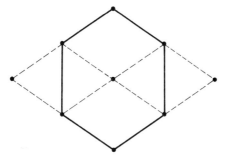

Fig. 3.4.8

Rather than make a detailed listing of the many symmetries of the cubic system, let us call attention to the similarity between the cubic and hexagonal systems in the following way: in Fig. 3.4.9 we have reproduced the heavy outline of the hexagon from Fig. 3.4.8. The picture we have drawn, however, is that of a cube.

In some of the systems we have described it is possible to have more than one lattice point per unit cell without losing any of the symmetry operations that distinguish the system. For example, in the monoclinic system (Fig. 3.4.4) the **c** vector is a twofold axis and the **a**, **b** plane is a reflection plane; these are the only symmetries of the system. Neither operation is lost if we place

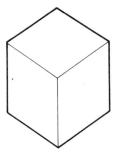

Fig. 3.4.9

additional lattice points at the center of the base (i.e., in the **a**, **b** plane) of each unit cell. Thus, the monoclinic system is said to have both primitive- and base-centered unit cells. By contrast, the trigonal and hexagonal systems have only primitive unit cells. Remarkably, in spite of the many symmetry operations that distinguish the cubic system, all are retained for cubic unit cells with an additional lattice point at the center of each face (called face-centered cubic, or just fcc), or an additional lattice point where the body diagonals cross (called body-centered cubic, or bcc). To be sure, the lattices described by the nonprimitive cells in any system can also be described by an appropriate choice of primitive cell, although perhaps at a considerable descriptive sacrifice. For example, the bcc lattice actually has a primitive unit cell that is trigonal, with the three equal angles being 109°28′. Although that description of the lattice is correct, it does not lead to the easy visualization, nor does it make obvious the many symmetries that immediately come to mind when we call the lattice bcc.

The structures that are formed simply by packing spheres together are of special interest, for they are often found in nature. In any one plane, the closest packing is achieved by arranging the centers of the spheres on equilateral triangles. If successive layers have their equilateral triangles directly above one another, we have the hexagonal lattice, but that is not the closest packing. For closest packing, the lattice points in each plane must lie above the centers of the triangles below. There are still two ways of doing that, however. The third plane can lie directly above the first plane, or there is still an alternative position it can take. On the triangular array sketched in Fig. 3.4.10, we have indicated A, B, and C positions, each forming a triangular array. Close-packed arrays of spheres, with the same average density, can be built up by successive layers in positions ABABAB..., or ABCABCABC.... As we have just noted, the arrangement AAAA... belongs to the hexagonal system. It is obvious, by comparison to the discussion of Fig. 3.4.7, that ABABAB... is a special case of the trigonal system; nevertheless, it is

3.5 Crystal Space: Phonons, Photons, and the Reciprocal Lattice

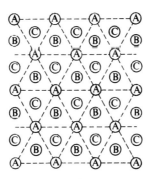

Fig. 3.4.10

called hexagonal close packed, or hcp. We can assign a crystal with this structure to the hexagonal system if we say it has a basis of two atoms, one at the A positions (the corners of the unit cells) and one at the B positions (midway between the planes of the hexagonal system). In that case, we would say that the basis had lowered the symmetry of the Bravais lattice type. However we choose to assign it, the hcp lattice does not have as high a symmetry as the hexagonal system. The ABCABC... arrangements can, by means of a significant feat of spatial imagination, be seen to be in an fcc lattice. A triangular array, when layered AAA..., or ABAB..., or ABCABC..., belongs, respectively, to the hexagonal, trigonal, and cubic systems, thus once again indicating the basic similarities between these systems.

A final comment about the types of symmetries possible. When we speak of an n-fold axis, n may be 2, 3, 4, or 6 (see Prob. 3.9). The absence of five-fold symmetry is related to the fact that one cannot cover a floor with pentagonal tiles (a proof based on that observation was first given by Kepler).

3.5 CRYSTAL SPACE: PHONONS, PHOTONS, AND THE RECIPROCAL LATTICE

The law of conservation of momentum is an expression of the fact that free space is homogeneous. It may be deduced in quantum mechanics from the requirement that the Hamiltonian be invariant under a differential displacement of an entire system. The converse statement is obvious: if the Hamiltonian changes under displacement, it must include a spatially varying potential, hence a force, which can cause the momentum to change.

Phenomena that take place inside a crystal occur in a space that is not invariant under differential displacement. The atoms or ions that constitute the crystal have potentials and exert forces that can change the momentum of entities found inside, entities such as electrons, neutrons, photons, even phonons, which, as we have seen, are composed of disturbances of the crystal itself. Nevertheless, there is, in the space inside a crystal, a restricted kind of translational invariance. Although the Hamiltonian changes on infinitesmal translation (with respect to the fixed lattice), it is restored by translation through any lattice vector, r_s. The operation of translation through any r_s, starting from any point in a crystal (not necessarily a lattice point), brings us to another place that is physically the same as our starting point. We might expect that this special property will lead to some restricted form of momentum conservation, which, indeed, turns out to be the case.

Needless to say, any compromise of the absolute conservation of momentum is fiction, because of the fact that we specify that the lattice be fixed in space but ignore the forces required to keep it there when momentum is transferred to it. This is not so much an omission as an approximation. The center of mass of the crystal as a whole acts as a reservoir for any momentum, lost or gained, in our problems. If a microscopic entity, such as an X-ray photon, gives a momentum, **p**, of microscopic magnitude to the center of mass of a crystal, the crystal is set in motion with a velocity

$$\mathbf{v} = \frac{\mathbf{p}}{M} \tag{3.5.1}$$

where M is the mass of the entire crystal—that is, a macroscopic quantity which, for most purposes, may formally be imagined to be infinite. It is the negligible velocity **v** that we are basically ignoring. More to the point, in any process in which momentum is to be lost or gained, energy must nevertheless be conserved. This statement means that we are ignoring the amount by which the energy of the system (excluding the center of mass of the crystal) must actually change:

$$\Delta E = \tfrac{1}{2} M v^2 = \frac{p^2}{2M} \tag{3.5.2}$$

The approximation we will use—that the constituents of our system can gain or lose momentum under certain circumstances while still conserving energy —will be valid as long as the energies in the problem are large compared to ΔE of Eq. (3.5.2). ΔE is the recoil energy of the crystal.

Although we can think of momentum as being not strictly conserved in the crystal, there are, as we have mentioned, restrictions on its changes that arise from the periodicity of the crystal—that is, the fact that the Hamiltonian is invariant under certain kinds of translation operations. In more physical terms, the restrictions occur because the process by which momentum is gained or lost must happen uniformly in every unit cell of the crystal. If there is more local momentum transfer in some regions of the crystal than in others, relative displacement of the atoms occurs, so that a phonon will develop, carrying away with it not only momentum but energy (large compared to ΔE) as well. Only if every unit cell receives the same velocity $\mathbf{v}$ [of Eq. (3.5.1)] can we imagine that the crystal has been left undisturbed and the momentum transferred to the center of mass.

a. Diffraction of X rays

The values of momentum transfer to which we are thus restricted are most easily found from an argument commonly given to find the conditions for elastic diffraction of X rays, a phenomenon in which the X rays change their directions without changing their frequencies (hence energies). In other words, X-ray diffraction is an example of precisely the type of phenomenon we have been discussing. Imagine a line of atoms of the same kind along the $\mathbf{a}$ axis of some crystal, spaced a distance $a = |\mathbf{a}|$ apart [$\mathbf{a}$ is one of the fundamental constituent vectors of $\mathbf{r}_s$, as in Eq. (3.4.1)]. An X-ray beam with a definite wave vector $\mathbf{q}_0$ (that is, a plane wave) impinges on the line from an angle α_0, gives the same push to each atom, and this beam comes off with wave vector $\mathbf{q}$ in direction α, as shown in Fig. 3.5.1. The condition that the beam come off in the same direction from each scattering center is

$$a(\cos \alpha - \cos \alpha_0) = \ell \lambda \tag{3.5.3}$$

In more wavelike language, this is the condition that the scattered waves interfere constructively in the $\mathbf{q}$ direction. Thus, λ is the wavelength of the X rays and ℓ is an integer. Since the recoil energy goes into the center of mass of the crystal, we are assuming that M is sufficiently large that

$$\hbar \omega = \frac{2\pi \hbar c}{\lambda} \gg \Delta E \tag{3.5.4}$$

Therefore, the wavelength is unchanged in the process, and

$$|\mathbf{q}| = |\mathbf{q}_0| = \frac{2\pi}{\lambda} \tag{3.5.5}$$

Then if we write

$$\mathbf{q} = |\mathbf{q}|\hat{\mathbf{s}} \quad \text{and} \quad \mathbf{q}_0 = |\mathbf{q}_0|\hat{\mathbf{s}}_0 \tag{3.5.6}$$

we may rewrite Eq. (3.5.3)

$$\mathbf{a} \cdot (\hat{\mathbf{s}} - \hat{\mathbf{s}}_0) = \ell \lambda \tag{3.5.7}$$

But

$$\hat{\mathbf{s}} - \hat{\mathbf{s}}_0 = \frac{\mathbf{q} - \mathbf{q}_0}{|\mathbf{q}|} = \frac{\lambda}{2\pi} \Delta \mathbf{q} \tag{3.5.8}$$

Substituting Eq. (3.5.8) into (3.5.7), we get

$$\mathbf{a} \cdot \Delta \mathbf{q} = 2\pi \ell \tag{3.5.9}$$

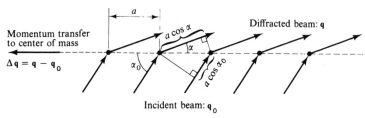

Fig. 3.5.1

Clearly, the same condition must hold in the **b** and **c** directions as well, so we can write the general condition on the momentum transfer $\Delta\mathbf{q}$ as

$$\mathbf{a} \cdot \Delta\mathbf{q} = 2\pi\ell_1 \qquad \mathbf{b} \cdot \Delta\mathbf{q} = 2\pi\ell_2$$

$$\mathbf{c} \cdot \Delta\mathbf{q} = 2\pi\ell_3 \tag{3.5.10}$$

where ℓ_1, ℓ_2, and ℓ_3 are integers. Equations (3.5.10) are called the von Laue conditions for X-ray diffraction.

A beam of X rays sent into a crystal will therefore come out scattered coherently, elastically, into various directions, each of which represents a momentum transfer satisfying Eq. (3.5.10). The X-ray diffraction pattern is thus a map in q space of a discrete lattice of vectors $\Delta\mathbf{q}$ that represent those values of the momentum which may be transferred to (or from) the crystal as a whole without any corresponding transfer of energy. The argument did not depend on any particular property of electromagnetic radiation; other plane waves, particularly electrons, neutrons, and phonons, will suffer the same diffractions (as we shall see in more detail below).

b. The Reciprocal Lattice and Brillouin Zones

The set of vectors $\Delta\mathbf{q}$ satisfying Eq. (3.5.10) obviously play an important role in the physics of crystals and deserve a name of their own. The lattice of points in **q** space that may be generated by the vectors $\Delta\mathbf{q}$ is called the *reciprocal lattice*. The vectors have the property that, for any translation vector of the original lattice in real space, $\mathbf{r}_s$,

$$\Delta\mathbf{q} \cdot \mathbf{r}_s = \Delta\mathbf{q} \cdot (n_1 \mathbf{a} + n_2 \mathbf{b} + n_3 \mathbf{c})$$

$$= 2\pi(n_1 \ell_1 + n_2 \ell_2 + n_3 \ell_3)$$

$$= 2\pi n \tag{3.5.11}$$

where n is any integer. We shall henceforth refer to these vectors $\Delta\mathbf{q}$ by the symbol **G**. **G** is called the reciprocal lattice translation vector (just as $\mathbf{r}_s$ is the ordinary lattice translation vector) and may be constructed from funda-

3.5 Crystal Space: Phonons, Photons, and the Reciprocal Lattice

mental vectors **A**, **B**, and **C**, the edges of the unit cell in q space of the reciprocal lattice, by

$$\mathbf{G} = m_1\mathbf{A} + m_2\mathbf{B} + m_3\mathbf{C} \qquad (3.5.12)$$

where m_1, m_2, and m_3 are positive or negative integers. **A**, **B**, and **C** depend on the real space lattice and its fundamental vectors **a**, **b**, **c** and, like the latter, are not necessarily orthogonal. It is easy to verify that **A**, **B**, and **C** may be constructed from **a**, **b**, and **c** by means of the formulas

$$\mathbf{A} = 2\pi \frac{\mathbf{b} \times \mathbf{c}}{V_0}$$

$$\mathbf{B} = 2\pi \frac{\mathbf{c} \times \mathbf{a}}{V_0}$$

$$\mathbf{C} = 2\pi \frac{\mathbf{a} \times \mathbf{b}}{V_0} \qquad (3.5.13)$$

where V_0 is the volume of the real unit cell, Eq. (3.4.2),

$$V_0 = \mathbf{a} \cdot \mathbf{b} \times \mathbf{c} \qquad (3.5.14)$$

The von Laue conditions for diffraction of X rays and other waves may then be described by

$$\mathbf{G} = \Delta\mathbf{q}; \quad \mathbf{G} \cdot \mathbf{r}_s = 2\pi n \qquad (3.5.15)$$

Let us see how lattice vibrations, or phonons, may be represented in the reciprocal lattice. The spatial variations in the amplitude of vibrations in any particular mode at a given instant are given by Eq. (3.3.8),

$$U_n^s(\mathbf{q}) = U_n \exp(-i\mathbf{q} \cdot \mathbf{r}_s) \qquad (3.5.16)$$

where we have chosen the instant $t = 0$. The quantity U_n refers to a particular degree of freedom of the atom at a definite position in the unit cell, but it is the same in every unit cell. The quantity U_n^s is not, however, the same in every unit cell, changing by the phase factor $\exp(-i\mathbf{q} \cdot \mathbf{r}_s)$ from cell to cell (i.e., for different values of $\mathbf{r}_s$). We identify **q** with the momentum of the phonon **p**, by $\mathbf{p} = \hbar\mathbf{q}$. Now suppose that we try to construct a mode in the same set of degrees of freedom (the nth in each unit cell) that differs from this one by a reciprocal lattice vector—that is, with wave vector $\mathbf{q} + \mathbf{G}$. The displacements in this new mode are given by

$$U_n^s(\mathbf{q} + \mathbf{G}) = U_n \exp[-i(\mathbf{q} + \mathbf{G}) \cdot \mathbf{r}_s]$$

$$= U_n \exp(-i\mathbf{q} \cdot \mathbf{r}_s) \exp(-i\mathbf{G} \cdot \mathbf{r}_s)$$

$$= U_n \exp(-i\mathbf{q} \cdot \mathbf{r}_s) \exp(-i2\pi n)$$

$$= U_n^s(\mathbf{q}) \qquad (3.5.17)$$

In other words, we have only succeeded in reproducing the original mode. Although the amplitude of a phonon changes from cell to cell in real space, the phonon spectrum is perfectly periodic in q space, with the periodicity of the reciprocal lattice. It makes no difference in which cell of reciprocal (or q) space we wish to represent it. To be sure, all the phonon modes may be represented in any one cell in reciprocal space, as noted in Sec. 3.3, just as the complete physical arrangement of the atoms in a crystal may be represented in any one cell in real space. Nevertheless, we can, if we wish, represent the phonon spectrum (i.e., the dispersion curve) as periodic. The periodic version of Fig. 3.3.6 (in one dimension) is sketched in Fig. 3.5.2. It must be remembered that, in this case, the lattice spacing is $2a$. The reciprocal lattice vector, given by the one-dimensional version of Eq. (3.5.15), comes in units of π/a rather than the usual $2\pi/a$. The periodic representation is produced by redrawing each piece of the curve at intervals of π/a. Thus, for example, the region marked I, which appears also in Fig. 3.3.6, shows up in Fig. 3.5.2 in region II, III, as well as $-$II, and so on.

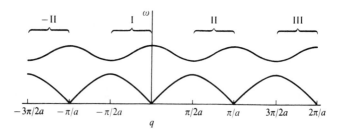

Fig. 3.5.2

It is important to realize that the situation we have just described does not mean that all phonons are free to change their wave vectors by amounts equal to **G**, but rather that the same phonon mode may be represented again a distance **G** away in reciprocal space. Only under restricted conditions can a phonon change its wave vector by **G**—that is, be reflected by the lattice, give momentum to the center of mass, and go off in a different direction, but with the same energy. The restrictions are the same as those that apply to photons. We are speaking here of a transition in which, say, a photon changes from wave vector **q** to **q**′, where

$$\mathbf{q} + \mathbf{G} = \mathbf{q}' \tag{3.5.18}$$

but $|\mathbf{q}| = |\mathbf{q}'|$ to conserve energy; that is,

$$q^2 = q'^2 \tag{3.5.19}$$

3.5 Crystal Space: Phonons, Photons, and the Reciprocal Lattice

This ensures that **q** and **q'** have the same magnitude. Squaring both sides of Eq. (3.5.18), we get the condition

$$2\mathbf{q} \cdot \mathbf{G} + G^2 = 0 \tag{3.5.20}$$

Only photons or phonons with values of **q** satisfying this relation may be coherently reflected by the crystal.

It is easy to find, by construction, the places in reciprocal space where vectors **q** satisfying Eq. (3.5.20) are located. Take any reciprocal lattice vector **G**, and construct the plane that is its perpendicular bisector, as shown in Fig. 3.5.3. It is obvious from the sketch that any vector **q** from the origin to the plane obeys the equation

$$|q| \cos \theta = \frac{G}{2} \tag{3.5.21}$$

which is just a way of writing Eq. (3.5.20). The smallest set of vectors satisfying this condition are those that fall on the bisectors nearest the origin. The construction we are describing is exactly that of the Wigner-Seitz unit cell, sketched in Fig. 3.4.2. The planes on which the **q** vectors fall are the boundaries of the Wigner-Seitz cell. The volume of the cell so constructed, as pointed out in Sec. 3.4, must be the same as that of the original unit cell in reciprocal space:

$$\mathbf{A} \cdot \mathbf{B} \times \mathbf{C} = \frac{(2\pi)^3}{V_0} \tag{3.5.22}$$

Equation (3.5.22) gives the result of integrating $d\mathbf{q} = d^3q$ from the origin to the boundaries of the cell. In other words,

$$\int_V \int_{\text{cell}} \frac{d^3q \, d^3r}{(2\pi)^3} = \frac{V}{V_0} = N_\sigma$$

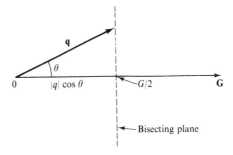

Fig. 3.5.3

Thus the amount of phase space encompassed by the volume of the cell in q space, together with the volume of the crystal in real space, is just enough to accommodate one state for each unit cell or, in the language of Sec. 3.3, one mode on each of the $3r$ branches of the dispersion relation, for each unit cell. In other words, all the phonon modes are to be found in the cell; it is just what we earlier called the *first Brillouin zone*. To repeat: the Wigner-Seitz cell about the origin in reciprocal space is the first Brillouin zone.

As we noted in Fig. 3.3.6, the phonon modes become standing waves as they approach the zone boundary. Although we have proved the point only for photons, it is just at the zone boundary where phonons (and other waves) satisfy the condition, Eq. (3.5.20), that elastic reflection by the crystal is possible. We may thus imagine that phonons, trying to propogate with wave vector at the zone boundary, are reflected back and forth, setting up the standing waves found here. This picture will be useful later in understanding the very similar behavior of electrons at the zone boundary. It also helps us to see that standing waves will be found (for electrons as well as phonons) not only on the boundaries of the first zone but everywhere on the planes in reciprocal space that satisfy Eq. (3.5.20) as well. These planes, the bisectors of all reciprocal lattice vectors, are everywhere zone boundaries, bounding higher-order Brillouin zones as we move away from the origin. Each zone— the second Brillouin zone, the third, and so on—has the same volume and may be folded back into the first zone by means of displacements of reciprocal lattice vectors. For a monatomic chain in one dimension with lattice spacing a, reciprocal lattice vectors equal to $2\pi n/a$, the first three zones are shown in Fig. 3.5.4. The vectors at the bottom of the figure show how to displace each

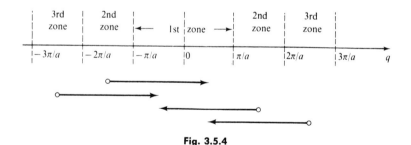

Fig. 3.5.4

segment of the higher zones in order to fold it into the first. The length of each vector is a reciprocal lattice vector. For a square lattice in two dimensions, which has a square reciprocal lattice, the first three zones are shown in Fig. 3.5.5. The dotted lines in the figure on the left show the reciprocal lattice vectors that were bisected to get the boundaries shown. The figure on the right shows how pieces of the third zone are folded back to fill the first zone.

3.5 Crystal Space: Phonons, Photons, and the Reciprocal Lattice

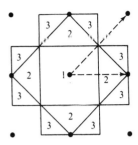 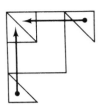

Fig. 3.5.5

As we shall see, in Sec. 3.6, constructions of this type are important in determining certain properties of metals.

c. Umklapp Processes

The ability of a phonon gas to change its momentum in units of $\hbar \mathbf{G}$ under appropriate circumstances has an interesting effect on the thermal conductivity of insulating materials. Ignoring the factor $\hbar$, the total momentum of a phonon gas is given by

$$\mathbf{J} = \sum_{\mathbf{q}} n_{\mathbf{q}} \mathbf{q} \tag{3.5.23}$$

where $n_{\mathbf{q}}$ is the number of phonons in the mode with wave vector $\mathbf{q}$. If the material is in thermal equilibrium, then $n_{\mathbf{q}} = \bar{n}$, where $\bar{n}$ is given by Eq. (3.2.74) with $\omega = \omega(\mathbf{q})$; the result, then, is that $\mathbf{J} = 0$, since there are equal numbers of phonons going in each direction in each mode. However, suppose that we make one end of the material hotter than the other, so that there is a net flux of energy, carried by the phonons, toward the cooler end. The question, now, is this: If we isolate the material from the heat source, how does the phonon gas come back to equilibrium? The gas must lose its net momentum, $\mathbf{J}$.

Up to this point we have not considered any mechanism for interactions that can change the states of phonons or, what is the same thing, create and destroy phonons. In this sense, the situation is quite the same as that of the perfect gas of Sec. 1.1; we describe the energies of the single-particle states (the frequencies of the phonon modes), ignoring interactions. However, by applying statistical mechanics to find equilibrium distributions of particles (phonons) among states (modes), we necessarily imply that the interactions exist. In the case of the phonons, the interactions take place by way of the higher-order terms in the forces between the atoms, neglected, for example, in writing Eq. (3.3.7). These higher-order, or anharmonic, terms have the effect of degrading slightly our precise solutions for distinct frequencies at which vibrations in the crystal may take place. To the extent that the anharmonicity is small, it leaves our phonon picture a reasonably good one, but we must now associate with each phonon a small spread in frequencies, $\delta\omega$. The spread means that the phonon has a finite lifetime τ related to $\delta\omega$ by the uncertainty principle,

$$\tau = \frac{\hbar}{\hbar\,\delta\omega} = \frac{1}{\delta\omega} \qquad (3.5.24)$$

This lifetime can be thought of as the mean time between collisions that change the states of phonons. Clearly, the mean free path, or mean distance traveled by a phonon, is given by $\langle \ell \rangle = c/\tau$, where c is the speed of sound or the group velocity of the mode.

We may thus imagine collisions between phonons in which two phonons with q_1 and q_2 give rise to two others with q_3 and q_4 such that

$$q_1 + q_2 = q_3 + q_4 \qquad (3.5.25)$$

It should be obvious, however, that this kind of momentum-conserving collision, which we have only just seen to be possible in the first place, still cannot do the job of bringing a crystal to equilibrium. Remember, we have a phonon gas with net momentum J which must be extracted. The net momentum of the gas, Eq. (3.5.23), is not changed by an event of the type in Eq. (3.5.25).

All the same arguments could be made for the perfect gas of Sec. 1.1. A net momentum flux in the gas could not be relaxed, even by interactions among the particles, unless there were some mechanism for transmitting momentum to the center of mass of the box containing the gas. The box containing the phonon gas is the crystal itself, and the necessary mechanism arises from the ability of the phonon system to exchange momentum with the center of mass in units of G.

The most important interactions leading to changes in J are of the type

$$q_1 + q_2 = q_3 + G \qquad (3.5.26)$$

with the auxiliary condition of conservation of energy,

$$\omega_1 + \omega_2 = \omega_3 \qquad (3.5.27)$$

Collisions of this type are called *umklapp* processes, in contrast to those of Eq. (3.5.25), which are called normal processes.

The best empirical supporting evidence for umklapp processes comes from the thermal conductivity, κ, of crystals pure enough so that phonons collide mainly with other phonons. For the phonon gas, as for any gas, the thermal conductivity is proportional to the mean free path

$$\kappa \propto \langle \ell \rangle \qquad (3.5.28)$$

3.5 Crystal Space: Phonons, Photons, and the Reciprocal Lattice

where we expect $\langle \ell \rangle$ to be the mean free path between umklapp processes, since only they can relax **J** leading to thermal resistance. In order for Eq. (3.5.27) to hold with $\mathbf{q}_3$ in the first Brillouin zone, $\mathbf{q}_1 + \mathbf{q}_2$ must add up to a magnitude outside of the first zone (so that **G** can bring the result back in). This means that $|\mathbf{q}_1|$ and $|\mathbf{q}_2|$ must be roughly of order of halfway out to the zone boundary or more. The frequency corresponding to **q** at the zone boundary is always comparable to the highest in the crystal and may be estimated by the Debye frequency. Phonons capable of suffering umklapp collisions thus have roughly half that energy or about $(k\Theta_D/2)$. At high temperature $(T \gg \Theta_D)$, all modes are excited, the number of phonons in each mode being

$$\bar{n} = \frac{1}{\exp(\hbar\omega/kT) - 1} \approx \frac{kT}{\hbar\omega} \propto T \tag{3.5.29}$$

Since $\langle \ell \rangle$ is inversely proportional to the number of phonons capable of scattering,

$$\langle \ell \rangle \propto \bar{n}^{-1} \propto T^{-1} \quad \text{(high } T\text{)} \tag{3.5.30}$$

At low T, the number of phonons with energy around $k\Theta_D/2$ is

$$\bar{n} = \frac{1}{\exp(\Theta_D/2T) - 1} \approx \exp\left(-\frac{\Theta_D}{2T}\right) \tag{3.5.31}$$

and so

$$\langle \ell \rangle \propto \exp\left(\frac{\Theta_D}{2T}\right) \tag{3.5.32}$$

We thus expect

$$\kappa \propto T^{-1} \quad \text{(high } T\text{)}$$
$$\kappa \propto \exp\left(\frac{\Theta_D}{2T}\right) \quad \text{(low } T\text{)} \tag{3.5.33}$$

That is just the kind of behavior observed.

Let us summarize briefly what we have seen in this section. Space inside a crystal has periodicity but is not uniformly invariant. Consequently, momentum is conditionally rather than absolutely conserved for entities that interact with the crystal. Those events that conserve energy but not momentum are restricted to values of momentum transfer that may be given directly to the center of mass of the crystal, thereby conserving overall momentum. The particular values of momentum that can be exchanged in this way are just $\hbar$ times the translation vectors of the reciprocal lattice. Processes in which this occurs, involving phonons, and, for that matter, electrons as well, are called umklapp processes. When the same thing happens to photons, they are said to be diffracted. Among the phenomena that may be seen to be governed by these processes, aside from the thermal conductivity of phonons and the diffraction of X rays, is the shape of the dispersion relation of the phonons, with standing waves, $\partial\omega/\partial q = 0$, at the boundaries of the Brillouin zones. As we shall see in the next section, very much the same thing happens to the energy-momentum relation for electrons in metals, standing waves and gaps in the energy, like those in Fig. 3.3.6, occurring at the zone boundaries.

d. Some Orders of Magnitude

Let us end this section with a brief comparison of the orders of magnitude of the energies and momenta of the types of particles whose interactions in the crystal we are considering. Since it is only orders of magnitude that we are interested in, it will be adequate for our purposes to compare the simplest possible models: free electrons, as in Sec. 2.5, Debye phonons, Sec. 3.2, and photons in free space. In real crystals, these quasiparticles will wind up having energies differing from (usually lower than) our estimates by factors of between one and two or so. The frequency versus q, or energy versus momentum, of phonons, electrons, and photons is sketched on a single, highly distorted plot in Fig. 3.5.6. For the phonons, q runs up to the zone boundary, $q \approx \pi/a \approx 1\text{Å}^{-1}$. The slope of the curve, $\bar{c}$, is the (averaged) speed of sound, $\bar{c} \approx 10^2 - 10^3$ m/sec. The maximum energy, in temperature units, is $\Theta_D \approx 10^2 - 10^{3}\,^\circ\text{K}$. For the electrons, the interesting physics occurs at the Fermi surface, where the characteristic energy is $T_F \approx 10^4 - 10^{5}\,^\circ\text{K}$. As we shall see in the next section, the Fermi surface falls, in momentum space, at a value of q of the same order as the zone boundary. Thus, comparing phonons and electrons, they characteristically have roughly the same order of momenta, but the electrons have far more energy. Consequently, in a collision between an electron and a phonon,

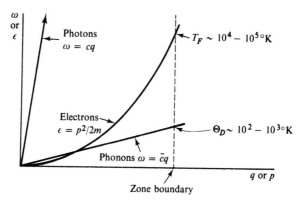

Fig. 3.5.6

the electron can suffer a serious change in direction (a change in momentum of order of the momentum itself), but its energy can hardly be affected at all (recall, from Sec. 2.5, that only electrons close to the Fermi surface can have collisions). The photons, like the phonons, have a dispersion relation of the form $\omega = cq$, but here c is the speed of light, $c = 3 \times 10^8$ m/sec, in contrast to the much smaller value, the speed of sound, for the phonons. At roughly optical frequencies, the photons have virtually zero wave vector ($q = 2\pi/\lambda \approx 10^{-3}\text{Å}^{-1}$) or momentum. A glance at Fig. 3.3.6 shows that if the phonon spectrum has an optical branch, that branch will cross the photon curve. At that point, phonons and photons at the same ω and q exist, so it is possible, for example, for a photon to be absorbed, thereby creating a single phonon while conserving energy and momentum. Strong absorption of light results at the crossover frequency, resulting in the name of the optical branch. Finally, where the photon curve crosses the zone boundary, those photons have very high energy compared to either electrons or phonons in the crystal, but they can satisfy Eq. (3.5.20) and so may be diffracted. These are the X-ray photons.

3.6 ELECTRONS IN CRYSTALS

As we discussed in Sec. 2.4, a metal may be thought of as a lattice occupied by positive ions, interpenetrated by and bound together by a gas of perfect degenerate fermions, perfect, that is, in the sense that they barely interact with each other. They do, however, interact with the lattice, and so the single-particle states are not quite those of a perfect gas of particles in an otherwise empty box. In this section we wish to investigate those electron-lattice interactions. We shall do so by first discussing qualitatively what we expect the single-electron states to be like, then making a detailed analysis of a particular model, the Kronig-Penney model, which has many of the important features of the states in real metals. The behavior of an electron in an ionic lattice is yet another example of an entity interacting in periodic crystal space, and the results bear strong similarities to those for phonons and photons found in the last section. Nevertheless, the consequences of this behavior in the case of electrons are so far-reaching that electrons deserve a special section of their own.

a. A Semiqualitative Discussion

The principal features of the electron states are simply those summarized for phonons and photons at the end of the last section. The momentum of an electron in a particular state is not a unique, conserved quantity, but instead a state with momentum **q** (ignoring the factor $\hbar$) may just as well be represented with momentum **q** + **G**, where **G** is any reciprocal lattice vector. Moreover, electrons with **q** at a zone boundary are subject to umklapping, and, consequently, there are no propagating modes at the zone boundary. Instead, standing waves are formed on the zone boundary, and the electrons are left with gaps, reminiscent of Fig. 3.3.8, in their density of states. Our first job is to obtain some idea of how these phenomena occur.

The wave functions for an electron in an empty box the size of the crystal are of the plane wave type

$$\psi_\mathbf{q} = A \exp(i\mathbf{q} \cdot \mathbf{r}) \qquad (3.6.1)$$

where A is a normalization constant we need not worry about, and the index **q**, which appears in the phase, we identify with the momentum. Now suppose that we begin to "turn on" a very weak potential with the periodicity of a crystal lattice. We can imagine the wave function of each single-particle state modifying itself, taking best possible advantage of the potential in order to lower its energy. It tries not to change too rapidly, since doing so would increase the kinetic energy of the state, which goes as $\nabla^2 \psi$, but it does wish to concentrate as much of its amplitude as possible near the attractive parts of the potential. For a weak periodic potential, it seems reasonable to guess that the result will be to modulate the amplitude of Eq. (3.6.1) in a periodic way

$$\psi_\mathbf{q} = \exp(i\mathbf{q} \cdot \mathbf{r}) U_\mathbf{q}(\mathbf{r}) \qquad (3.6.2)$$

where $U_\mathbf{q}(\mathbf{r})$ is some function having the periodicity of the lattice

$$U_\mathbf{q}(\mathbf{r}) = U_\mathbf{q}(\mathbf{r} + \mathbf{r}_s) \qquad (3.6.3)$$

where $\mathbf{r}_s$ is any lattice translation vector. The constant A of Eq. (3.6.1) has been absorbed into $U_\mathbf{q}(\mathbf{r})$. The remarkable fact is that the solutions of the Schrödinger equation remain of the form Eq. (3.6.2), not only for weak potentials but also no matter how strong the potential becomes, as long as it is periodic. This result is known in physics as Bloch's theorem and in mathematics as Floquet's theorem. We shall not formally prove the Bloch/Floquet theorem, although we shall come rather close to doing so. The proof of the theorem is similar to the analysis of the Kronig-Penney model given below, and we shall see there, for a particular choice of periodic potential in one dimension, that the form Eq. (3.6.2) remains valid regardless of the strength of the potential.

The electron wave functions must satisfy the Schrödinger equation

$$-\frac{\hbar^2}{2m} \nabla^2 \psi(\mathbf{r}) + V(\mathbf{r})\psi(\mathbf{r}) = \varepsilon\psi(\mathbf{r}) \qquad (3.6.4)$$

where, in the crystal, the potential is periodic

$$V(\mathbf{r}) = V(\mathbf{r} + \mathbf{r}_s) \qquad (3.6.5)$$

3.6 Electrons in Crystals

Only in the special case, $V = $ constant, can we expect the solutions to be of the form Eq. (3.6.1)—that is, momentum eigenstates, associated with a single wave vector $\mathbf{q}$. When the periodic part of the potential becomes nonzero, we expect to obtain new wave functions, which, however, may be constructed out of linear combinations of the old ones:

$$\psi(r) = \sum_{\mathbf{q}} C_{\mathbf{q}} \exp{(i\mathbf{q} \cdot \mathbf{r})} \qquad (3.6.6)$$

The potential, in other words, serves to mix together the $\mathbf{q}$ eigenstates, and we should not expect a definite phase factor, $\exp{(i\mathbf{q} \cdot \mathbf{r})}$, to emerge, particularly when the potential is strong so that many of the $C_\mathbf{q}$ in Eq. (3.6.6) may be expected to be nonzero. This is the reason why the Bloch theorem is so remarkable. It tells us that we always wind up with states that can be identified with a definite $\mathbf{q}$—that is, a definite momentum. The states thus always retain their correspondence with those of a particle in an empty box, which is one of the main reasons why the arguments of Sec. 2.5 are successful in accounting for metal properties. For the low $\mathbf{q}$ states, the modulations on a scale much shorter than the wavelength will change all the energies in a smooth way, the dispersion relation still retaining the form $\varepsilon \propto q^2$. We can, therefore, still apply the arguments of Sec. 2.5, only, as noted there, replacing the mass by an effective mass, m^*.

The momentum $\mathbf{q}$ identified with each state of the form Eq. (3.6.2), although definite, is, however, not unique. We can rewrite the state in Eq. (3.6.2)

$$\psi_\mathbf{q} = \exp{[i(\mathbf{q} + \mathbf{G}) \cdot \mathbf{r}]}[\exp{(-i\mathbf{G} \cdot \mathbf{r})}U_\mathbf{q}(\mathbf{r})]$$

$$= \exp{(i\mathbf{Q} \cdot \mathbf{r})}U_\mathbf{Q}(\mathbf{r}) \qquad (3.6.7)$$

where we now have the wave vector $\mathbf{Q} = \mathbf{q} + \mathbf{G}$ in the phase factor, $\mathbf{G}$ being a reciprocal lattice vector, and

$$U_\mathbf{Q}(\mathbf{r}) = \exp{(-i\mathbf{G} \cdot \mathbf{r})}U_\mathbf{q}(\mathbf{r}) \qquad (3.6.8)$$

The function $U_\mathbf{Q}(\mathbf{r})$ has the same periodicity property of $U_\mathbf{q}(\mathbf{r})$, Eq. (3.6.3), as we can see by direct substitution:

$$U_\mathbf{Q}(\mathbf{r} + \mathbf{r}_s) = \exp{[-i\mathbf{G} \cdot (\mathbf{r} + \mathbf{r}_s)]}U_\mathbf{q}(\mathbf{r} + \mathbf{r}_s)$$

$$= \exp{(-\mathbf{G} \cdot \mathbf{r})}U_\mathbf{q}(\mathbf{r}) = U_\mathbf{Q}(\mathbf{r}) \qquad (3.6.9)$$

where we have made use of Eq. (3.6.3), plus Eq. (3.4.16), $\mathbf{G} \cdot \mathbf{r}_s = 2\pi n$, so that $\exp{(-i\mathbf{G} \cdot \mathbf{r}_s)} = 1$. We could not have changed the energy of the state merely by rewriting the wave function in Eq. (3.6.7), but it is once again of the Bloch form, and associated with a new momentum, $\mathbf{Q}$; in fact, there are a number of new momenta, since $\mathbf{Q}$ can differ from $\mathbf{q}$ by any reciprocal lattice vector.

We thus see that the electron states have the same property as the phonon modes discussed previously: they are periodic in reciprocal space, or, alternatively, all states of all energies may be represented in the first Brillouin zone. Let us defer sketching the states according to these schemes, however, until we see a little more about what the states are like at higher **q**.

It should be possible to construct states of the Bloch form out of the pure plane wave states by an appropriate choice of the $C_\mathbf{q}$ in Eq. (3.6.6), and indeed it is. We have already seen the basic trick—it is just that $\exp(i\mathbf{G}\cdot\mathbf{r})$ is always periodic in $\mathbf{r}_s$—and so any function expanded in those factors will be periodic. So we choose as nonzero only $C_\mathbf{q}$ and all the $C_\mathbf{Q}$, where $\mathbf{Q} = \mathbf{q} + \mathbf{G}$. The sum is then over the **G**'s.

$$\psi_\mathbf{q}(\mathbf{r}) = \sum_\mathbf{G} C_{\mathbf{q}+\mathbf{G}} \exp\left[i(\mathbf{q}+\mathbf{G})\cdot\mathbf{r}\right]$$

$$= \exp(i\mathbf{q}\cdot\mathbf{r}) \sum_\mathbf{G} C_{\mathbf{q}+\mathbf{G}} \exp(i\mathbf{G}\cdot\mathbf{r}) \qquad (3.6.10)$$

It is easy to see that

$$U_q(\mathbf{r}) = \sum_\mathbf{G} C_{\mathbf{q}+\mathbf{G}} \exp(i\mathbf{G}\cdot\mathbf{r}) \qquad (3.6.11)$$

has the form Eq. (3.6.3). Thus, just as we said when we wrote Eq. (3.6.6), the crystal potential has the effect of mixing other plane wave states into each electron state, but we now see that it mixes in only states that differ from the free state by reciprocal lattice vectors.

We can use all this discussion, together with our experience with the phonons and photons, to make an informed guess as to the form of the dispersion relation near the zone boundary. Like the other waves, single electrons at the zone boundary may suffer umklapp processes, being reflected back and forth. So we can expect, just as in the phonon case, that there will be no propagating states at the zone boundary but only standing waves instead, with $\partial\varepsilon/\partial\mathbf{q} = 0$. If the dispersion relation, $\varepsilon(\mathbf{q})$, is to have $\varepsilon \propto q^2$ for low q, and $\partial\varepsilon/\partial\mathbf{q} = 0$ at the zone boundary, it cannot be very different from that sketched in Fig. 3.6.1. In sketching this figure we have assumed that the potential acting on the electrons is small in magnitude, so that it is negligible at high energies, and the energy will once again take on the free electron form, $\varepsilon \propto q^2$. Thus, the departure from free electron behavior is principally at the zone boundaries.

Figure 3.6.1 has features closely similar to Fig. 3.3.6; we can make it look even more similar, and we shall later, by folding the second zone back into the first. Like the phonon modes of Fig. 3.3.6, the electron states form two standing waves, of the same **q**, at the zone boundary, with an energy gap between them. Just as we did for the phonons in Sec. 3.3, we can see what the standing waves are and why they differ in energy.

Consider, again, the electron states in a weak potential. In order to

3.6 Electrons in Crystals

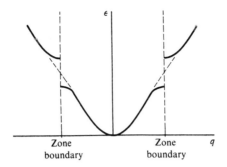

Fig. 3.6.1

simplify the arguments, take the case of one dimension and imagine that the potential is produced by a small net positive charge at $x = 0$, $\pm a$, $\pm 2a$, ... etc. The reciprocal lattice vectors are $G_n = 2\pi n/a$, and we can index the C's with n:

$$\psi_q(x) = e^{iqx} \sum_n C_n e^{2\pi inx/a} \qquad (3.6.12)$$

Far from the zone boundary, for a sufficiently weak potential, the unperturbed wave function is a sufficiently good solution (e.g., for computing the energy to first order in the strength of the potential), and so $\psi_q(x)$ can be formed by letting all the C_n be zero except C_0. Near the zone boundary, however, that will not be adequate, and we can use our experience with phonons to see why. The zone boundary is simply the point where we can satisfy the one-dimensional form of Eq. (3.5.20)

$$q = -\tfrac{1}{2}G_{-1} \qquad (3.6.13)$$

where G_{-1} is the reciprocal lattice vector with $n = -1$—that is, $G_{-1} = -2\pi/a$. An electron trying to propagate through the crystal with q given by Eq. (3.6.13) merely satisfies the condition for reflection into the state given by Eq. (3.5.18),

$$q' = q + G_{-1} = -\frac{\pi}{a} \qquad (3.6.14)$$

whereupon it just satisfies the condition for reflection back into q, and so on. Thus, as we consider states approaching the zone boundary at $G_1/2$, even in a weak potential, we must start to mix into the wave function some amplitude for states G_{-1} away; in Eq. (3.6.12) C_0 diminishes slightly, and C_{-1} starts to grow. Then at the zone boundary itself, where the electrons are freely reflected back and forth, the magnitudes of the coefficients C_0 and

C_{-1} will be equal: $C_{-1} = \pm C_0$. Letting the magnitude be just C_0, the two states that are constructed this way are

$$\psi_{q+} = C_0 e^{iqx}(1 + e^{-2\pi ix/a})$$
$$\psi_{q-} = C_0 e^{iqx}(1 - e^{-2\pi ix/a})$$

(3.6.15)

Since $q = \pi/a$, these equations reduce to

$$\psi_{q+} = C_0(e^{i\pi x/a} + e^{-i\pi x/a}) = 2C_0 \cos\left(\frac{\pi x}{a}\right)$$

$$\psi_{q-} = C_0(e^{i\pi x/a} - e^{-i\pi x/a}) = 2iC_0 \sin\left(\frac{\pi x}{a}\right)$$

(3.6.16)

It is easy to see that exactly the same two states will be produced at $q' = -\pi/a$. The wave functions in Eq. (3.6.16) are not exact solutions of the Schrödinger equation, but, for weak potentials, they should be good enough to compute the energy to leading order, just as the plane waves are far from the zone boundary.

The two states given by Eq. (3.6.16) are, as expected, standing waves, of wavelength $2a$. The difference in energy between the two states arises from the difference in potential energy of interaction with the lattice. In any state the charge density of the electron is given by $-e|\psi|^2$, where $-e$ is the electron charge. Plane wave states, Eq. (3.6.1), have charge densities that are uniform in space, but the states in Eq. (3.6.16) have charge densities given by

$$|\psi_{q+}|^2 \propto \cos^2 \frac{\pi x}{a}$$

$$|\psi_{q-}|^2 \propto \sin^2 \frac{\pi x}{a}$$

(3.6.17)

The charge densities of the two states are sketched in Fig. 3.6.2. In the solid curve, representing $|\psi_{q+}|^2$, the negative charge density is concentrated on the positive ions, whereas the dotted curve, $|\psi_{q-}|^2$, has the negative charge between the ions. Consequently, the state ψ_{q+} has lower potential energy than ψ_{q-}. These two states are the ones that appear at both zone boundaries in Fig. 3.6.1, the energy gap being the difference in energy between them.

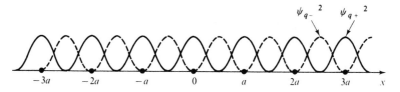

Fig. 3.6.2

3.6 Electrons in Crystals

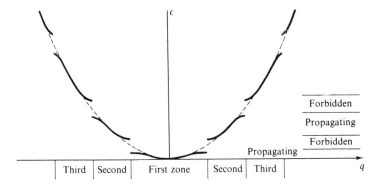

Fig. 3.6.3

The same story may be told at all zone boundaries in the reciprocal lattice. In order to get an idea of what the dispersion relation for electrons looks like in one dimension, or along any particular direction in two or three dimensions, we can simply sketch the free electron parabola, and flatten it out as we approach each zone boundary, as in Fig. 3.6.3. This figure is called the extended zone scheme. The whole picture can be shifted into the first zone by the operation shown in Fig. 3.6.4. The arrow shows a typical operation transfering one segment. This is called the reduced zone scheme. Finally, the whole thing can be represented periodically in reciprocal lattice space, and as in Fig. 3.6.5, which is called the periodic zone scheme.

The choice of which to use depends on the problem that one wishes to solve. For example, as we saw earlier, a photon with frequency far below the X ray has very little momentum, even if its energy is comparable to the scales

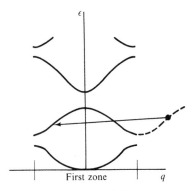

Fig. 3.6.4

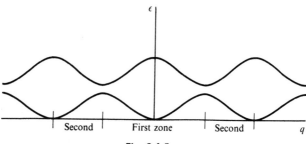

Fig. 3.6.5

of these figures. Thus, if Fig. 3.6.3 told the whole story, a single electron could not absorb a photon, for there are no available states for it at higher energy but roughly the same momentum. However, using the scheme of Fig. 3.6.4, we see that it is possible to absorb the photon, making transitions of the kind shown by the arrow in Fig. 3.6.6. The transition requires that the lower state be occupied (i.e., fall below the Fermi level) and that the upper state be unoccupied (i.e., fall above the Fermi level). These transitions may be detected by studying the absorption of electromagnetic radiation of solids.

The question of where the Fermi level falls is of considerable interest, since it determines the difference between metals and nonmetals in many cases. However, we shall defer that discussion until after we have examined the Kronig-Penney model.

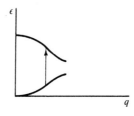

Fig. 3.6.6

b. The Kronig-Penney Model

In the discussion up to this point we have seen how we may understand the existence of energy bands of propagating states for electrons, separated by energy gaps. The arguments have, however, been largely qualitative and have also been based on a single point of view: we imagine a preexisting crystal order, into which we place our electrons and on which we impose a potential. There is another way of looking at it. Imagine two neutral atoms, far apart. Each may be thought of as an ion, which forms a

3.6 *Electrons in Crystals*   203

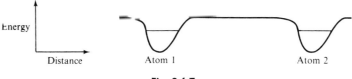

Fig. 3.6.7

potential well with bound states for its outermost electron. This picture of the two atoms, showing the ground state of the outermost electron, is sketched in Fig. 3.6.7. The electron states in the two ions are separated by a broad energy barrier. Now suppose that the atoms are brought close together. The area of the barrier becomes much smaller, and the probability grows that an electron in the bound state of atom 1 will tunnel into that of atom 2 and vice versa. Consequently, an electron in either state now has a finite lifetime, τ, after which it will hop to the other atom. This means that the states themselves, which were previously quite sharp in energy, must develop some width, $\Delta\varepsilon$, given by

$$\Delta\varepsilon \simeq \frac{\hbar}{\tau} \qquad (3.6.18)$$

The new picture is sketched in Fig. 3.6.8. As we bring up more atoms, eventually forming a metal, the lifetime on any one atom gets shorter, the

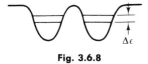

Fig. 3.6.8

state's width gets correspondingly larger, and it develops into one of the bands of propagating energy states in Fig. 3.6.3, with one state in the band for each atom in the crystal. The next band develops from the next higher state of the free atoms and so on. The Kronig-Penney model is a simple, one-dimensional model in which we shall be able to see, quantitatively, all the features discussed qualitatively in the previous subsection and, in addition, follow the development of the energy bands all the way from discrete states in the free atoms to the free electron picture of Sec. 2.5, with all the energy band phenomena in between.

The potential, $V(x)$, of the model is sketched in Fig. 3.6.9. It is a series of potential square wells of width a, separated by energy barriers of width b and height V_0. Later we shall be able to go continuously from free atoms to free electrons by letting the area of the barrier, bV_0, vary from infinity to

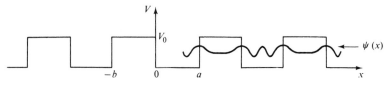

Fig. 3.6.9

zero. The potential in the wells is taken, for convenience, to be zero. Part of a typical wave function, oscillatory in the well and decaying exponentially in the barrier, is also sketched in the figure.

We may immediately write down the Schrödinger equation for this problem. In the wells,

$$\frac{d^2\psi}{dx^2} + \frac{2m}{\hbar^2}\varepsilon\psi = 0 \qquad (0 < x < a) \qquad (3.6.19)$$

and in the barriers,

$$\frac{d^2\psi}{dx^2} + \frac{2m}{\hbar^2}(\varepsilon - V_0)\psi = 0 \qquad (-b < x < 0) \qquad (3.6.20)$$

The wells and barriers, and hence these equations, are, of course, repeated periodically for all x. For ease in writing, we define two quantities, α and β, which are real and have dimensions of wave vectors, (length)$^{-1}$:

$$\alpha^2 = \frac{2m\varepsilon}{\hbar^2} \quad \text{and} \quad \beta^2 = \frac{2m(V_0 - \varepsilon)}{\hbar^2} \qquad (3.6.21)$$

Then Eqs. (3.6.19) and (3.6.20) become

$$\frac{d^2\psi}{dx^2} + \alpha^2\psi = 0 \qquad \text{(wells)} \qquad (3.6.22)$$

$$\frac{d^2\psi}{dx^2} - \beta^2\psi = 0 \qquad \text{(barriers)} \qquad (3.6.23)$$

The solutions of these equations will be of the form, respectively,

$$\psi_1 = Ae^{i\alpha x} + Be^{-i\alpha x} \qquad \text{(wells)} \qquad (3.6.24)$$

$$\psi_2 = Ce^{\beta x} + De^{-\beta x} \qquad \text{(barriers)} \qquad (3.6.25)$$

that is, oscillatory in the wells and decaying exponentially in the barriers. However, the overall potential being periodic, we are clever enough to seek solutions of both equations together of the Bloch form, Eq. (3.6.2):

$$\psi_q = e^{iqx}U_q(x) \qquad (3.6.26)$$

3.6 Electrons in Crystals

We shall see that solutions of this form are applicable regardless of the area of the barriers, thus, in effect, proving the Bloch/Floquet theorem for this special case. Substituting Eq. (3.6.26) into (3.6.22) and (3.6.23), we find differential equations for the function $U_q(x)$:

$$\frac{d^2 U_q}{dx^2} + 2iq \frac{dU_q}{dx} + (\alpha^2 - q^2)U_q = 0 \quad \text{(wells)} \quad (3.6.27)$$

$$\frac{d^2 U_q}{dx^2} + 2iq \frac{dU_q}{dx} - (\beta^2 + q^2)U_q = 0 \quad \text{(barriers)} \quad (3.6.28)$$

We now form trial solutions for U_q, U_1 in the wells and U_2 in the barriers, which reduce exactly to Eqs. (3.6.24) and (3.6.25) written in the form of Eq. (3.6.26):

$$\begin{aligned} U_1 &= A \exp\left[i(\alpha - q)x\right] + B \exp\left[-i(\alpha + q)x\right] \quad \text{(wells)} \\ U_2 &= C \exp\left[(\beta - iq)x\right] + D \exp\left[-(\beta + iq)x\right] \quad \text{(barriers)} \end{aligned} \quad (3.6.29)$$

It is easy to see that multiplying U_1 and U_2 by e^{iqx} recovers ψ_1 and ψ_2.

The trial solutions have four unknowns, A, B, C, and D, for the determination of which we need four boundary conditions. Two conditions are that ψ and $d\psi/dx$ be continuous in passing from well to barrier—that is, at $x = 0$. The other two arise from requiring that $U_q(x)$ have the periodicity property, Eq. (3.6.3), where $r_s = n(a + b)$ in this case. We apply the condition both to $U_q(x)$ and to its derivative, dU_q/dx. In particular, we have

$$U_1(0) = U_2(0) \quad (\psi \text{ continuous}) \quad (3.6.30)$$

$$\left(\frac{dU_1}{dx}\right)_0 = \left(\frac{dU_2}{dx}\right)_0 \quad \left(\frac{d\psi}{dx} \text{ continuous}\right) \quad (3.6.31)$$

$$U_1(a) = U_2(-b) \quad (U_q \text{ periodic}) \quad (3.6.32)$$

$$\left(\frac{dU_1}{dx}\right)_a = \left(\frac{dU_2}{dx}\right)_{-b} \quad \left(\frac{dU_q}{dx} \text{ periodic}\right) \quad (3.6.33)$$

Applying these conditions respectively to the solutions, Eqs. (3.6.29), we find

$$A + B - C - D = 0 \quad (3.6.34)$$

$$i(\alpha - q)A - i(\alpha + q)B - (\beta - iq)C + (\beta + iq)D = 0 \quad (3.6.35)$$

$$\exp\left[i(\alpha - q)a\right]A + \exp\left[-i(\alpha + q)a\right]B - \exp\left[-(\beta - iq)b\right]C$$
$$- \exp\left[(\beta + iq)b\right]D = 0 \quad (3.6.36)$$

$$i(\alpha - q)\exp\left[i(\alpha - q)a\right]A - i(\alpha + q)\exp\left[-i(\alpha + q)a\right]B$$
$$- (\beta - iq)\exp\left[-(\beta - iq)b\right]C + (\beta + iq)\exp\left[(\beta + iq)b\right]D = 0$$
$$(3.6.37)$$

These are four linear, homogeneous equations in the four unknowns. As usual, we are more interested in the conditions that solutions exist than in the solutions themselves. These will be conditions on the values of α and β as a function of q; since α and β depend on the energy ε, the conditions that solutions exist will give us the dispersion relation of the electron states in the Kronig-Penney potential. The required condition is that the determinant of the coefficients of A, B, C, and D in Eqs. (3.6.34) to (3.6.37) be equal to zero:

$$0 = \begin{vmatrix} 1 & 1 & -1 & -1 \\ i(\alpha - q) & -i(\alpha + q) & -(\beta - iq) & +(\beta + iq) \\ \exp[i(\alpha - q)a] & \exp[-i(\alpha + q)a] & -\exp[-(\beta + iq)b] & -\exp[(\beta + iq)b] \\ i(\alpha - q)\exp[i(\alpha - q)a] & -i(\alpha + q)\exp[-i(\alpha + q)a] & -(\beta - iq)\exp[-(\beta - iq)b] & (\beta + iq)\exp[(\beta + iq)b] \end{vmatrix} \quad (3.6.38)$$

After a bit of algebra, Eq. (3.6.38) reduces to

$$\frac{\beta^2 - \alpha^2}{2\alpha\beta} \sinh \beta b \sin \alpha a + \cosh \beta b \cos \alpha a = \cos q(a + b) \quad (3.6.39)$$

Equation (3.6.39) is the dispersion relation we seek, but it is, unfortunately, somewhat less than entirely transparent in this form. Even so, certain general features may be seen directly. The condition that there be propagating solutions (i.e., with real q) is that the right-hand side, $\cos q(a + b)$, fall between $+1$ and -1. There will, in general, be values of α and β—in effect, values of ε—for which there are no such solutions. These will be the forbidden bands, the energy gaps discussed in the previous subsection. Moreover, these bands will begin at values of q for which $\cos q(a + b) = \pm 1$—that is, at

$$q(a + b) = n\pi \quad (n = 0, \pm 1, \pm 2, \ldots) \quad (3.6.40)$$

These values of q, $q = n\pi/(a + b)$, are just the zone boundaries for the lattice whose periodicity is $a + b$.

In order to make further progress analyzing Eq. (3.6.39), we will now make a simplifying assumption. We will let $V_0 \to \infty$ and $b \to 0$ in such a way that the product $V_0 b$ remains finite and adjustable. The barriers will then be δ functions with finite, variable area, and the lattice periodicity will reduce simply to a. Let us see what this does to each of the terms in Eq. (3.6.39).

3.6 Electrons in Crystals

Using the definitions of β and α, Eqs. (3.6.21), we find

$$\frac{\beta^2 - \alpha^2}{2\alpha\beta} = \frac{2m}{2\hbar^2} \frac{V_0 - 2\varepsilon}{\alpha\beta} \to \frac{mV_0}{\hbar^2\alpha\beta} \qquad (3.6.41)$$

$$\sinh \beta b = \frac{e^{\beta b} - e^{-\beta b}}{2} = \beta b + \frac{(\beta b)^3}{3!} + \cdots \to \beta b \qquad (3.6.42)$$

$$\cosh \beta b = \frac{e^{\beta b} + e^{-\beta b}}{2} = 1 + \frac{(\beta b)^2}{2!} + \cdots \to 1 \qquad (3.6.43)$$

Putting (3.6.41) and (3.6.42) together gives

$$\frac{\beta^2 - \alpha^2}{2\alpha\beta} \sinh \beta b \to \frac{mV_0}{\hbar^2\alpha\beta} \beta b = \frac{m}{\hbar^2\alpha} V_0 b \qquad (3.6.44)$$

which is the combination we seek. We can now define a dimensionless parameter, P, proportional to the area of the barrier:

$$P = \frac{ma}{\hbar^2} V_0 b \qquad (3.6.45)$$

Then

$$\frac{\beta^2 - \alpha^2}{2\alpha\beta} \sinh \beta b \to \frac{P}{\alpha a} \qquad (3.6.46)$$

Substituting Eqs. (3.6.43) and (3.6.46) into (3.6.39), we find for the dispersion relation

$$P \frac{\sin \alpha a}{\alpha a} + \cos \alpha a = \cos qa \qquad (3.6.47)$$

This is an equation for α as a function of q with P as a parameter. Recalling from Eq. (3.6.21) that α is proportional to $\varepsilon^{1/2}$, we can see more clearly that this is basically the relation, $\varepsilon = \varepsilon(q)$, that we seek.

Let us consider what happens to Eq. (3.6.47) in some special limits. First, take the case $P = 0$. This choice eliminates the barriers altogether, and Eq. (3.6.47) reduces to

$$\cos \alpha a = \cos qa$$

or

$$\alpha a = qa + 2\pi n \qquad (3.6.48)$$

Using Eq. (3.6.21), this becomes

$$\varepsilon = \frac{\hbar^2[q + (2\pi n/a)]^2}{2m} \qquad (3.6.49)$$

For $n = 0$, we have the familiar free electron result. The term $2\pi n/a$ is a residue of the fact that we have solved the problem in a periodic crystal space: we can add any reciprocal lattice vector to q.

Next let us seek zero-energy solutions: $\varepsilon \propto \alpha^2 = 0$. Then

$$\cos \alpha a \to 1$$

$$\frac{\sin \alpha a}{\alpha a} \to 1$$

and Eq. (3.6.47) reduces to

$$P + 1 = \cos qa \qquad (3.6.50)$$

A solution of this equation can exist only if $P = 0$—that is, in the free electron case. In all other cases, the lowest-lying electron state has zero-point energy.

Finally, consider the opposite limiting case, $P \to \infty$. This choice will correspond to the limit of free atoms, in which our model of the free atom is an infinite square well with bound states for electrons. For this case, Eq. (3.6.47) can have solutions only for the zeros of $(\sin \alpha a)/\alpha a$. As we have just pointed out, $(\sin \alpha a)/\alpha a \to 1$ for $\alpha \to 0$, but it will be zero for all the other zeros of $\sin \alpha a$. Thus, the solutions come at

$$\alpha a = n\pi \qquad [n = \pm 1, \pm 2, \ldots \text{(not zero)}] \qquad (3.6.51)$$

Substitution into Eq. (3.6.21) gives for the allowed energies

$$\varepsilon_n = \frac{n^2(\pi\hbar)^2}{2ma^2} \qquad (3.6.52)$$

These values are simply the discrete energy levels for electrons in our model of the free atom.

For finite values of P, between the limits we have looked at, there will be allowed values of α whenever the left-hand side of Eq. (3.6.47) falls between $+1$ and -1. If we make a plot of the left-hand side versus αa, allowed solutions occur wherever the curve falls in the region between $+1$ and -1, as shown in Fig. 3.6.10. The allowed bands of αa are indicated at the bottom of the figure. The curve starts, with $\alpha a = 0$, at $1 + P$ and undergoes oscillations with period 2π in αa. In each period, $\cos \alpha a$ varies between $+1$ and -1, as does $\sin \alpha a$, but $(\sin \alpha a)/\alpha a$ gets smaller as αa increases. So the oscillations gradually decrease in amplitude until at very large αa they remain between $+1$ and -1 for all αa. Consequently, as αa increases, the widths of the allowed bands increase and those of the forbidden bands diminish. The widths of the first few allowed bands depend on P, getting smaller as P increases. We can depict the widths of the bands,

3.6 Electrons in Crystals

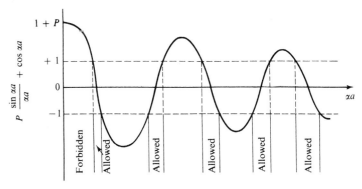

Fig. 3.6.10

over the entire range of P, from zero to infinity by plotting them against P from 0 to 1, then against $1/P$ from 1 back to 0, as in Fig. 3.6.11. Here the regions of allowed, propagating states have been cross-hatched, and the principal quantum number, n, in Eq. (3.6.52) of the free atom state from which each band develops is marked on the right-hand axis.

The magnitude of the parameter P, as we have defined it, depends on a combination of two features of the problem: how far apart the atoms are (i.e., b) and how tightly the electrons are bound to the atoms (i.e., V_0). If we imagine the atoms always to be bound together in a crystal, then P is simply a measure of how tightly the electrons are bound. According to Fig. 3.6.11, then, bands of propagating states for electrons exist for all finite binding energies. If it is the existence of these bands that leads to metallic behavior in solids, why then, the question arises, are not all solids metals? Part of the answer, as we shall see in the next section, depends on just where in these bands the Fermi level falls. That is, however, not the whole story.

At the outset of this subsection we said that we expected this picture

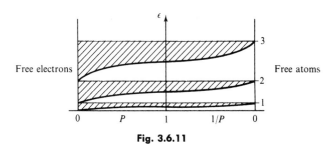

Fig. 3.6.11

to develop energy bands whose widths would be related to the time τ that an electron would spend bound to any single atom,

$$\tau \simeq \frac{\hbar}{\Delta\varepsilon} \quad (3.6.53)$$

where $\Delta\varepsilon$, the width of the band, as we can now see, depends on both P and the principal quantum number n. Obviously it is nonsense for us to speak of electrons more or less freely propagating through the crystal if τ is very long. Thus, we cannot expect metallic behavior if the atoms of the material have their outermost electrons fall into shells that are very tightly bound. A reasonable criterion for metallic behavior is that τ be very short compared to the natural time scale for thermal behavior of the material, say, the inverse of the Debye frequency, ω_D^{-1}, which is the period of the lattice vibrations discussed in Secs. 3.2 and 3.3. Accordingly, we require for metallic behavior

$$\tau \ll \omega_D^{-1} \quad (3.6.54)$$

Typical values of ω_D^{-1} are in the range 10^{-12} to 10^{-14} sec. Thus, one way that a solid can be an insulator is if Eq. (3.6.54) is not satisfied—that is, if its outermost electrons are very tightly bound.

It is obvious even from the crude model that we have used here that the width of the band will depend not only on the structure of the free atom but also on the effects of bringing the atoms together into a condensed material, and even then on the density of the material. Thus, for example, solid hydrogen is known to be an insulator, but it is thought that it might become a metal under ultrahigh pressures ($\sim$ millions of atmospheres), say, at the core of the planet Jupiter. Such changes, from insulating to metallic behavior, may, however, involve more subtle effects not evident in our model. The potential exerted by each positive ion on any single electron is modified by the existence of the other free electrons if the material is a metal, since the free electrons tend to arrange their charge densities in a way that shields the ions. Although we alluded to this behavior in constructing Eq. (3.6.2), the feedback effect that can result from it has been left out entirely from our model. As the material becomes, say, less dense, so that P in our model grows, the binding of electrons to single atoms increases. Consequently, the shielding effect that the electrons had contributed to other atoms is reduced, thereby causing the binding to grow still more. At some point the process may run away, causing an abrupt transition from metallic to insulating behavior. This phenomenon (whose existence has not been unambiguously demonstrated in real materials) is known as a *Mott transition*.

Let us return now to our model in order to bring out a few final important features of the behavior of electrons in metals. Each band of allowed states

3.6 Electrons in Crystals

can, as we have seen, be represented as having values of q that fall in a single Brillouin zone—that is, between $-\pi/a$ and π/a. So far we have treated q as a continuous variable, but it is really composed of discrete states that may be fixed by imposing on the wave function, Eq. (3.6.26), in addition to the conditions, Eqs. (3.6.30) to (3.6.33), the further condition that it be periodic at the endpoints of the macroscopic crystal—that is, at $x = \pm Na/2$, where N is the number of atoms in the crystal. With this condition, applied exactly as discussed in connection with Eq. (3.3.30), we find that q takes on precisely N values in the Brillouin zone, just as it did for phonons. Therefore, we have one single electron translational or kinetic energy state in each zone for each atom in the material. This statement is correct not only in the one-dimensional case we have studied here but in two and three dimensions as well. It must be remembered, however, that the Pauli exclusion principle permits each of these states to accommodate two electrons, one for each orientation of spin $\frac{1}{2}$. It thus requires two electrons per atom to fill up any single zone. This point is central to determining the difference between metals and semiconductors, and we shall return to it in the next subsection.

Let us make one final point to illustrate the correspondence between the results of the Kronig-Penney model in this subsection and the qualitative arguments of the previous subsection. We were led to believe in Sec. 3.2a that the states to be found at the zone boundaries would be standing wave states, and we saw that the energy gaps could be understood in terms of these states. We can test whether these standing waves arise in our model by computing the quantity $\partial\varepsilon/\partial q$ or, more easily, $\partial\alpha/\partial q$, which will be zero for a standing wave. To do so, we need to take derivatives (at constant P and a) of Eq. (3.6.47):

$$\cos qa = P \frac{\sin \alpha a}{\alpha a} + \cos \alpha a$$

giving

$$-(\sin qa)a \, dq = P\left[\frac{\cos \alpha a}{\alpha a} - \frac{\sin \alpha a}{(\alpha a)^2}\right] a \, d\alpha \qquad (3.6.55)$$

or

$$\frac{d\alpha}{dq} = \frac{\sin qa}{\sin \alpha a \{1 + [P/(\alpha a)^2]\} - P[(\cos \alpha a)/\alpha a]} \qquad (3.6.56)$$

Clearly, at the zone boundaries where $\sin qa = 0$, we get standing waves.

Thus, if we examine the Kronig-Penney model for small P—that is, weak binding where the qualitative arguments of Sec. 3.6a were valid—we see that the results approach Eq. (3.6.49), modified by the standing waves at the zone boundary found for $P > 0$. The pictures we would draw to represent these results are just the same that we deduced earlier, Figs. 3.6.3 to 3.6.5. We may consider the qualitative arguments of the previous subsection to have been validated by this quantitative model.

c. The Fermi Surface: Metal or Nonmetal?

Whether a material is a metal or not really comes down to the question of how its electrons respond to an applied electric field. If all the electrons are tightly held, say, by individual atoms or by covalently bonded molecules, an electric field will have only minor effects, and the material will be an insulator. If, on the other hand, large numbers of electrons are free to move in response to a field, the material will have a shiny surface because it reflects light, will freely conduct electricity and heat, and will, in short, be a metal. The oscillatory electric field of incident light will excite corresponding oscillations in the electrons, which, in turn, give rise to outgoing or reflected light; electric charge and thermal energy are both easily transported by free electrons in a metal. As we have seen, free electrons may exist if the material develops sufficiently broad bands of propagating states out of the outer electron shells of the constituent material. That is, however, not a sufficient condition for metallic behavior. It is also necessary that the Fermi level, in reciprocal space, not coincide with zone boundaries. The reason may be seen from a brief discussion of how conduction works in a Fermi degenerate gas of electrons.

To begin the discussion, let us suppose that the Fermi level falls sufficiently far within the first zone so that the dispersion relation is little altered from the free electron case, or, at any rate, so that we can write it as

$$\varepsilon = \frac{p^2}{2m^*} \tag{3.6.57}$$

as discussed in Sec. 2.5. The electrons fill a Fermi sphere of states in q space. Electrons in each of the states have kinetic energies and velocities, but in the absence of an applied field equal numbers move in all directions and there are no net currents. Now suppose that an electric field $\mathscr{E}$ is applied in, say, the negative x direction. Each of the electrons is initially accelerated by an amount

$$\ddot{x} = -\frac{e\mathscr{E}}{m^*} \tag{3.6.58}$$

where x is the coordinate of any electron. Since $\dot{x} = \hbar q_x/m^*$, this corresponds to a rate of change of the x component of each wave vector

$$\dot{q}_x = \frac{m^*}{\hbar} \ddot{x} = -\frac{e\mathscr{E}}{\hbar} \tag{3.6.59}$$

Since all the q's change at this same rate, we see that a uniform acceleration of all the electrons in real space is described by a motion in the same direction but with constant velocity, $\dot{q}_x$, of the entire Fermi sphere in q space.

This acceleration of the electrons does not continue unimpeded, however.

3.6 Electrons in Crystals

Phonons and various impurities and imperfections in the material will scatter the electrons, thus breaking up the uniform acceleration. But although these scattering processes occur everywhere in real space, it is easy to see from our prior considerations that they can occur only at the surface of the Fermi sphere in q space. For definiteness, let us consider the scattering of electrons by phonons (the dominant mechanism in reasonably pure metals at room temperature) and recall the relative magnitudes of energy and momentum sketched in Fig. 3.5.6. The Fermi level falls at a q of order of the zone boundary (we have supposed here that it is less than that), comparable to typical phonon q's, but an energy (in temperature units) of order 10^4 to $10^{5\circ}$K, much greater than the phonon value, $\Theta_D \sim 10^2$ to $10^{3\circ}$K. An electron more than Θ_D in energy below the Fermi surface thus cannot be scattered at all by a phonon, because there is simply not enough energy available to promote it into an unoccupied state. The electrons near the Fermi surface may be scattered, but not to very different energies compared to the Fermi energy. However, enough q is available to change the momentum of the electron substantially compared to its own initial value. In the kind of collisions we expect, then, only electrons in states near the Fermi surface are affected, and they will suffer serious changes in momentum with hardly any change in energy; that is just what happens when a particle is reflected, changing its direction at essentially constant speed.

The whole process can be described in q space as follows: an applied electric field initially sets the entire Fermi sphere into uniform translational motion. However, the leading edge of the sphere soon starts to get preferentially eaten away by collisions with phonons that scatter those electrons into states still on the surface, but toward the trailing edge. Eventually a steady state is reached (in constant field) in which electrons are constantly being scattered from the front to the rear surface of the sphere, being promoted by the field through the sphere until they reach the leading surface, then getting scattered again. The net effect is a constant displacement, δq_x, of the entire sphere. The displacement is typically very small compared to q_F, the radius of the sphere; however, an exaggerated version is sketched in Fig. 3.6.12.

If each electron, on the average, has its q changed by a fixed amount, δq_x, then, in real space, the entire electron gas has taken on a net drift velocity, v_x:

$$v_x = \frac{\hbar}{m^*} \delta q_x \qquad (3.6.60)$$

This velocity is given just by the acceleration, Eq. (3.6.58), times the mean time between collisions for an electron at the Fermi surface, τ_c:

$$v_x = -\frac{e\mathscr{E}}{m^*} \tau_c \qquad (3.6.61)$$

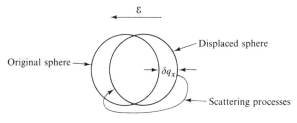

Fig. 3.6.12

The electric current density carried by the material, j, is given by

$$j = -\rho_e e v_x \qquad (3.6.62)$$

where ρ_e is the number density of participating electrons and $-e$ the electron charge. Notice that although only those electrons at the Fermi surface are scattered, all the electrons in the band are accelerated by the field and hence contribute to the current. Substituting in Eq. (3.6.61) for v_x, we get

$$j = \sigma \mathscr{E} \qquad (3.6.63)$$

where

$$\sigma = \frac{\rho_e e^2 \tau_c}{m^*} \qquad (3.6.64)$$

Equation (3.6.63) is Ohm's law, the empirical equation obeyed by currents and fields, and σ, given by Eq. (3.6.64), is the electrical conductivity of metal.

The displacement of the Fermi sphere in an electric field occurs because all the electrons in a band of propagating states tend to be promoted by the field into states of successively higher q in the direction of the force, that process continuing until they are scattered back from the leading side to a state of negative q—going in the opposite direction—and have to start all over. The process is thus a dynamic one, with a net current resulting only because, at any instant, the sphere of filled states is slightly displaced from the zero-field sphere. The next question we must ask is: How is this picture affected if instead of a Fermi sphere well within the first Brillouin zone, we have, rather, all states filled with electrons right up to the zone boundary?

Consider the behavior of an electron, in a state close to the zone boundary, under the influence of a field pushing it toward the boundary. Suppose, in particular, that it starts in the state in position A of Fig. 3.6.13. An electric field, giving rise to a force on the electron in the positive q direction, promotes it toward state B at the zone boundary. Even in state B the electron is under the influence of a force that makes it wish to continue to increase its value of q. This it is able to do because, as we now know, state B and state C are the *same state*; an electron in state B has the form of a standing wave, such as ψ_{q+} of Eq. (3.6.16), which may equally well be assigned to

3.6 Electrons in Crystals   215

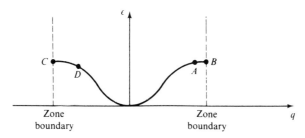

Fig. 3.6.13

state C. The continued application of the force now drives the electron from state C toward state D. The acceleration the electron is undergoing continues to be in the same direction, toward larger, more positive values of q, but its velocity is now in the opposite direction. Another description of the same process is to say that an electron, driven to the zone boundary, is umklapped back to the opposite boundary, whereupon it proceeds to traverse the entire dispersion curve and gets umklapped again. Each electron is accelerated up to the zone boundary, elastically scattered to the same velocity in the opposite direction, decelerated to rest, reaccelerated, and so on. All of this is rather reminiscent of what we have just said about Ohm's law conduction, but it differs from that case in two important ways. For one thing, the scattering in the present case is elastic. In the Ohm's law case, energy from the electric field (small amounts by electron standards but appreciable for the phonons) is constantly fed into the phonon gas, thereby heating the material. No energy is extracted from the source of the electric field by the present kind of elastic umklapp scattering. Second, and here is the crucial point, no electric current is produced by the process we are presently describing. In the Ohm's law case, the displacement of the Fermi sphere means that each electron spends a bit more of its time traveling in the forward direction than in the reverse, leading to a net current. In the present case, with all states filled, there can be no displacement of the Fermi sea; each electron spends equal time moving in each direction. We therefore reach the following important conclusion.

Any completely filled band of electron states makes no contribution to the electric currents carried by a material. Conversely (since no empty band can make a contribution either), in order to be a metal, a material must have one or more partially filled bands.

There are thus two distinct conditions that a material must meet in order to be a metal. First, it must have electrons sufficiently loosely bound to go into sensibly wide bands of allowed, propagating states; and, second, it must manage to populate a band or two without filling them. We come down, then, to the question of what decides the populations of the states when bands exist.

As we have already seen, each allowed band—that is, each Brillouin zone—has in it, for each atom in the crystal, one translational state, with capacity for two electrons (spin up and spin down). The question therefore becomes: How many loosely held electrons per atom are there in the material? If the number is odd, say, one conduction electron per atom, the zone will be only half filled—that is, the states that are occupied will fill half the volume of the zone in q space—and the material will be a good conductor, with the basically spherical Fermi surface we used earlier in discussing ohmic conduction. That is the case, for example, in metals like copper, silver, and gold. If, on the other hand, an even number of electrons per atom goes into bands, it is possible that the uppermost occupied band will be completely filled. A material in which that is true is called a semiconductor.

It is possible, however, for a material with an even number of electrons per atom to contribute to its bands, such as bismuth, indium, and lead, to be a metal. The way in which this situation can come about is not evident from the one-dimensional pictures we have considered so far, for it is basically a geometric effect, depending on the fact that a material can have different properties along different crystal directions. The available electrons in a material will fill up the lowest energy states presented to them, without regard to which band, or zone, they fall in. If it happens, say, that the lowest-lying states in the second Brillouin zone have lower energies than the uppermost states in the first zone, those states in the second zone will be populated before the first zone is filled. If the material in question has two electrons per atom, it will wind up with some unoccupied states in the first zone and some occupied states in the second. It thus has two partially filled bands, both contributing to metallic behavior. Materials of that kind are, nevertheless, not as metallic, say, as copper, silver, and gold—they have lower electrical conductivities, σ, for example—and are sometimes called semimetals. In order to understand these points more clearly, we shall have to consider in greater detail the geometry of the Brillouin zones, as well as the Fermi surfaces within them.

The principles we wish to illustrate can be seen adequately in two dimensions or, what is exactly the same thing, in any single plane through a three-dimensional crystal. Since the states to be filled in q space are chosen strictly for their low energies, the Fermi surface, regardless of its shape, will always be a contour (or surface) of constant energy in q space. We should therefore begin by obtaining some idea of what such contours of constant energy typically look like.

Consider, for simplicity, a material with a square (or cubic) lattice, so that it has a square (or cubic) reciprocal lattice. The first Brillouin zone, together with the dispersion relation along the q_x direction, is sketched in Fig. 3.6.14.

3.6 Electrons in Crystals

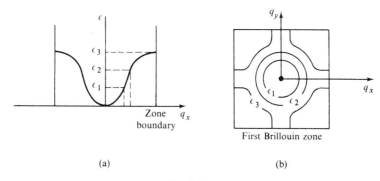

Fig. 3.6.14

In the first zone we have sketched three contours of constant energy, equal energies apart, at ε_1, ε_2, and ε_3. At low energies the contours are circular, as they would be in free space. But as the zone boundary is approached, the dispersion relation bends over toward the horizontal, and, consequently, the contours get farther apart. The net result of the fact that the dispersion relation always comes into the zone boundary horizontally is that the contours of constant energy always come in perpendicular to the zone boundary if they meet it at all.

The Fermi level will fall on one of these contours. To see its general shape, we can begin by sketching what it would look like, compared to the first zone, if there were no interactions between the electrons and the ions. This is done, for the case of one electron per atom, in the left-hand member of Fig. 3.6.15. Half the states in the zone are filled up, so the circle on the left has half the area of the square zone. The electron ion interactions will cause the parts of the circle near the boundary to reach out toward it, with suitable modifications elsewhere to conserve area, thereby resulting in the Fermi contour shown on the right. Details of the shape depend, of course, on the

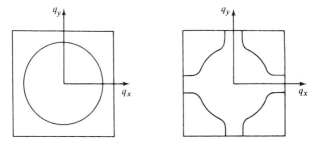

Fig. 3.6.15

strength of the interactions, but there will be a tendency to develop necks out toward the boundary. In all cases, the Fermi level falls on the contour (or surface) of constant energy whose area (volume) in q space is precisely enough to accommodate all the available electrons.

Now let us suppose that there are, say, two electrons per atom (the average number of electrons need not be an integer in, for example, alloys, but it commonly is an integer in simple atomic materials). With two electrons per atom to accommodate, the circle representing the noninteracting case has the same area as the square first zone, and so parts of it fall outside the first zone, into the second, as sketched in Fig. 3.6.16. The second zone, as constructed in Fig. 3.5.5, is also shown. Now the details of the contour, dictated by the interactions, become crucial. If the interactions are weak, so that the gaps in the dispersion relation are small, Fig. 3.6.16 will be altered only slightly, and both the first and second zones will be partly filled, leading to metallic behavior. On the other hand, if the gaps are large, the corners of the first zone will fill before any states become occupied in the second zone, resulting in a semiconductor.

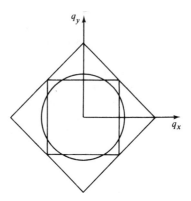

Fig. 3.6.16

In Fig. 3.6.17 we have dispersion relations drawn along the α direction to the right of the origin and the β direction to the left. The α and β directions through the zone are indicated at the bottom of the figure. The left-hand dispersion relation shows a small gap, the right hand a large gap. The lowest-lying state in the second zone falls at the zone boundary in the α direction, whereas the highest state in the first zone is at the corner, where the boundary is encountered in the β direction. If these energies overlap, as in the left-hand picture, the result is a metal; if they do not overlap, as in the right-hand picture, the result is a semiconductor. The Fermi level, up to which all states are filled, is shown by the dashed line in each case.

3.6 Electrons in Crystals 219

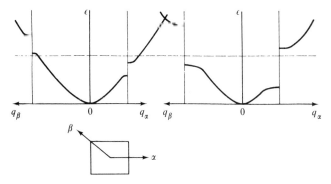

Fig. 3.6.17

In the semiconducting case, the Fermi level (which is, to be precise, the energy at which the mean occupation number for each spin orientation is equal to one half) falls within the energy gap separating states in the two zones. The contour followed by the Fermi level in q space is just the square boundaries of the first zone.

For the case of band overlap and metallic behavior, the Fermi level may be constructed following the rules for constant energy contours outlined above, as sketched in Fig. 3.6.18, where the occupied states are cross-hatched. The diagram shows the Fermi surface in what corresponds to the extended zone scheme, Fig. 3.6.3. We can often get a better idea of the phenomena to be observed by considering instead the periodic zone scheme, Fig. 3.6.5. This we can now construct by repeating the first zone in Fig. 3.6.18 through q space, as done in Fig. 3.6.19, and by using the technique of Fig. 3.5.5, folding the second zone into the first and repeating the second zone, as in

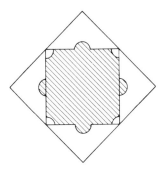

Fig. 3.6.18

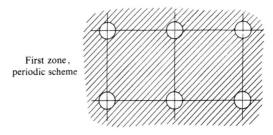

First zone, periodic scheme

Fig. 3.6.19

Fig. 3.6.20. Evidently the second zone is characterized by isolated groups of filled states and the first zone by isolated groups of empty states. Just as the second zone, in our example here, may be thought of as a basically empty zone (or band) with a few electrons in it, the first zone may be thought of as a filled band with a few holes in it. Since, as we have seen, a completely filled band makes no contribution to the electrical properties of the material, everything that goes on can be described in terms of the behavior of the holes, which can be thought of as another kind of quasiparticle. The behavior of holes in electric and magnetic fields (really just the result of the behavior of all the electrons in the filled states) will be consistently described if we assign to holes the following dynamical characteristics: compared to the properties of an electron occupying the same cell in q space, the hole has the opposite charge, mass of the opposite sign, and $\mathbf{q}$ of the opposite sign, while the magnitudes of the charge, mass, and $\mathbf{q}$ are the same as those of the electron. The mass we speak of here is the effective mass, first encountered in Sec. 2.5 and further discussed below. We shall leave the working out of these dynamical assignments to Prob. 3.13.

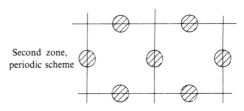

Second zone, periodic scheme

Fig. 3.6.20

An electron (or for that matter, a hole) responds to applied forces in a way that depends on which state it is in. Imagine, for example, an electron in a state near a zone boundary, as shown in Fig. 3.6.21. Suppose that there is an electric field in the $-q_x$ direction, so that the force acting on the electron

3.6 Electrons in Crystals

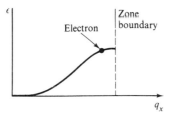

Fig. 3.6.21

tends to push it in the $+q_x$ direction. The acceleration produced (in real space) results from a displacement to higher q_x, just as in Eq. (3.6.59):

$$-\frac{e\mathscr{E}}{\hbar} = \dot{q}_x = \frac{m^*\ddot{x}}{\hbar}$$

In each state, however, the group velocity of the particle is

$$\dot{x} = \frac{1}{\hbar}\frac{\partial \varepsilon}{\partial q_x} \qquad (3.6.65)$$

It is obvious from Fig. 3.6.21 that as q_x increases in this case, $\dot{x}$, proportional to the slope of the dispersion curve, will decrease. Thus, the force is actually serving to decelerate the electron: in other words, the effective mass is negative. The negative effective mass near the zone boundary is an effect we have already described in other words; as the boundary is approached, the electron is approaching a standing wave state with zero velocity. Elsewhere on the dispersion curve, the effective mass will also differ from the free electron mass, although usually not so dramatically. The acceleration produced by a force need not even be collinear with the force; the effective mass is, in general, a tensor. The reason for all this peculiar behavior is, of course, that the electron is responding not only to the externally applied force but to forces from the positive ions as well. For the simple one-dimensional case, we can see how m^* depends on the curvature of the dispersion relation by rewriting Eqs. (3.6.59) and (3.6.65).

$$\frac{1}{m^*} = \frac{\partial \dot{x}/\partial t}{\hbar\, \partial q_x/\partial t} = \frac{1}{\hbar^2}\frac{\partial^2 \varepsilon}{\partial q_x^2} \qquad (3.6.66)$$

Thus, at low q, where $\varepsilon \propto q_x^2$, in Fig. 3.6.21, m^* is a constant, usually not too different from the free electron mass. Higher on the curve, m^* becomes infinite, where there is a point of inflection in $\varepsilon(q_x)$, and, finally, it becomes negative at the zone boundary. A hole, remember, has an effective mass of the opposite sign in each of these states. The electron effective mass that enters into the heat capacity, Eq. (2.5.15), is the average value at the Fermi surface.

Finally, there is a point that should be mentioned if only for its enormous technological importance. A material that forms a natural semiconductor, with bands that are all either full or empty (e.g., silicon, germanium), can nevertheless be made to have some mobile electrons and holes available for conducting electricity. In fact, there are always a few at finite temperature due to thermal excitation of electrons across the energy gap. A semiconductor with only thermally produced electrons and holes has the same number of each and is said to be an instrinsic semiconductor. Its electrical conductivity, proportional to the number of charge carriers (electrons plus holes), is a very strong, exponential function of temperature. More important technologically, however, the material can be made to have permanent charge carriers by putting into it impurities whose chemistry (i.e., whose numbers of outermost electrons) is such that they either contribute or soak up electrons in the bands, producing, respectively, mobile electrons or holes, or, again respectively, negative and positive carriers. The Fermi level, in the middle of the gap in the intrinsic semiconductor, is pushed up or down by the impurities. When pieces of n-type material (i.e., with net negative carriers) and p-type material (net positive carriers) are put together, their Fermi levels must be the same (the Fermi level is the chemical potential for electrons) so that their band structures are pushed up or down relative to one another. A small electric potential, in the right direction, can allow electrons from a filled band to pour into an empty one, but the reverse potential produces no current. Upon this principle, in its various manifestations, is based the transistor, and thus a growing part of the world's economy.

BIBLIOGRAPHY

The subject matter of this chapter, like that of Chaps. 1 and 2, falls squarely within the traditional physics curriculum and may therefore be studied in textbooks; for the remaining chapters, we will have to rely increasingly on monographs and even research articles.

One of the most widely used introductory texts in solid-state physics is C. Kittel's *Introduction to Solid State Physics*, 3rd ed. (New York: Wiley, 1967.) It is encyclopedic, covering many subjects rather superficially. Generally it is a good starting point for finding out about some particular phenomenon in solids. Another rather good book at roughly the same level is A. J. Dekker, *Solid State Physics* (Englewood Cliffs, N.J.: Prentice-Hall, 1957). The field is not distinguished by imaginative book titles.

Many statistical mechanics books discuss at least some of the subjects in this chapter. For example, much of what has been said here may be found in one place or another in Landau and Lifshitz, *Statistical Physics* (see bibliography, Chap. 1).

Among many solid state books at a more advanced level, two particularly well-known examples are J. M. Ziman, *Principles of the Theory of Solids* (London:

Cambridge Press, 1972) and C. Kittel, *Quantum Theory of Solids* (New York: Wiley, 1963). A very good recent contribution is W. A. Harrison, *Solid State Theory* (New York: McGraw-Hill, 1970).

Finally, there is a lovely classic to which we must pay homage: *Theory of the Properties of Metals and Alloys*, by N. F. Mott and H. Jones (Oxford: Clarendon Press, 1936; New York: Dover, 1958). There are very few physics books written as long ago as 1936 that can still be read with profit, but this one certainly can be.

PROBLEMS

3.1 Find the high- and low-temperature limits of the heat capacity of a Debye solid in two dimensions.

3.2 There is an empirical formula relating the Debye temperature and the melting temperature, T_m, of solids:

$$\Theta_D = C \left(\frac{T_m}{Mv^{2/3}}\right)^{1/2}$$

where M is the atomic mass (in amu), v is the volume per atom (in Å^3), and C is a constant, roughly the same for all solids.

a. Estimate the magnitude of C from whatever information you have in your head about the properties of some particular solid. If you lack some necessary datum, make an order of magnitude guess. The formula was first proposed by Lindemann in 1910. According to the textbooks, it has no theoretical explanation. In parts *b* and *c*, you will provide an explanation.

b. Recalling that the energy of an oscillator is twice its potential energy, calculate the mean square displacement $\langle x^2 \rangle$ of an atom in a Debye solid. (*Hint*: For the kth mode, you can show that the energy per atom is $E_k = M\omega_k^2 \langle x_k^2 \rangle$, where ω_k is the frequency and $\langle x_k^2 \rangle$ is the contribution to the mean square displacement of the kth mode. Since the modes are independent, $\langle x^2 \rangle = \sum_k \langle x_k^2 \rangle$. Recall that melting occurs at high temperature.)

c. Assuming a crystal melts when the root mean square displacement is some reasonable fraction (say, 0.1) of the interatomic spacing, estimate C in the Lindemann formula.

d. By the same kind of arguments, find the melting temperature of a Debye solid in two dimensions.

3.3 Consider the linear triatomic molecule in one dimension (three identical masses connected by two identical springs). Show that the normal modes have frequencies 0, $(k/m)^{1/2}$, and $(3k/m)^{1/2}$, where k is the spring constant and m the atomic mass. Identify the modes (what's vibrating?).

3.4 Find and sketch the complete dispersion relation for an infinite linear chain of identical masses and springs in one dimension.

$$\text{Answer: } \omega = \pm 2 \left(\frac{k}{m}\right)^{1/2} \sin\left(\frac{qa}{2}\right)$$

where q is the wave vector and a the equilibrium interatomic spacing.

3.5 For the diatomic linear chain done in the text, show that both branches have zero-group velocity at the zone boundary, $q = \pm \pi/2a$. What happens to the group velocity at this point when $m_1 \to m_2$?

3.6 It is claimed on page 164 of the text that a real crystal in three dimensions will have $3N - 6$ nonzero frequency modes. We know that the solution of the normal mode problem should give exactly $3N$ modes in all, so six of them should be $\omega^2 = 0$. Of these six, three come from the $q = 0$ limit for the three acoustic branches. There must be three others. Discuss how proper counting of modes will leave three others at zero frequency.

3.7 Consider a linear chain of lattice spacing a, with $N - 1$ atoms of mass M and force constant C, which has at some lattice site inside an impurity atom of mass M' and the same force constant, C.

a. Write down the two distinct equations of motion for the two kinds of atom in the crystal (it is convenient to place the impurity atoms at the origin). There is one vibrational mode for each atom in the crystal. A mode is a wave with a particular frequency and wave vector. Since there is one special atom, we can expect to find one special mode, called the impurity mode. To find it, we must construct trial solutions to the equations of motion. Answers to the following questions should help in guessing at a trial solution.

b. Will the mode be a running wave or a standing wave? (Ask yourself: what is the physical basis of running waves in crystals?)

c. Should the solution have any particular spatial symmetry built into it?

d. How big should the amplitude of the mode be very far from the site of the impurity?

You should now be prepared to construct a trial solution. For example, if you're looking for a running wave, then the displacement at the sth position will be proportional to exp $(iska)$, whereas the oscillating spatial part of a standing wave may be represented by an amplitude proportional to $(-1)^s$ (if the impurity is at the origin, s counts the number of sites from the origin).

e. Write the trial solution dictated by your answers to parts b, c, and d.

f. Assuming $M' < M$, what is the frequency of the impurity mode?

g. How does it compare to the frequencies of the pure crystal?

h. What is the ratio of the amplitude of the impurity mode at the impurity to the amplitude s sites away?

i. What becomes of the mode as $M' \to M$?

j. What is the mode like if $M' > M$?

3.8 It has recently been reported that, for a certain material, the following empirical relationship is obeyed:

$$\left(\frac{\partial V}{\partial n_q}\right)_P = \left(\frac{\partial \varepsilon_q}{\partial P}\right)_T$$

where V is the volume, P the pressure, T the temperature, ε_q the energy of an elementary excitation (such as a phonon) in mode q, and n_q the number of excitations in the mode. The equation has the appearance of a thermodynamic law, and the question has arisen of whether the result is, in fact, a necessary consequence of thermodynamics. We wish to show that it is not. The idea is to show that it is not correct for the particular instance of a perfectly harmonic solid (which we shall call a phs). For orientation, show that the following is true for a phs:

   a. It will not expand if heated at constant pressure.
   b. The quantity

$$\lambda = \frac{V}{\Theta_D}\left(\frac{\partial \Theta_D}{\partial V}\right)_T$$

is called the Grüneisen constant; here Θ_D is the Debye temperature, taken from the low-temperature heat capacity. For the phs, $\lambda = 0$.

   c. $K_T = K_S$, where K_T and K_S are, respectively, the isothermal and adiabatic compressibilities. It may be helpful to you to remember that the linear chain problems done earlier are just phs's in one dimension.

   d. Now to prove our point. Show first that, for the phs,

$$\left(\frac{\partial V}{\partial n_q}\right)_P = 0$$

It is adequate to show this for any q; it might be convenient to choose the long wavelength longitudinal phonons.

   e. For the same mode,

$$\left(\frac{\partial \varepsilon_q}{\partial P}\right)_T \neq 0$$

   f. Why does this argument, for a particular mode of the phs, show that the equation is not a thermodynamic result for some other substance? Of what importance is this for the scientist trying to understand the significance of his observation?

   g. Now that you have done that, show that $(\partial \varepsilon/\partial P)_T$ for a harmonic solid *is* in fact equal to zero if by ε we mean the energy of a particular mode rather than the energy at a particular wave vector. Showing this for the one-dimensional chain will be enough to illustrate the principle.

3.9 For an n-fold rotation axis, show that the only allowed values of n are 2, 3, 4, and 6.

3.10 Show that the maximum portion of the available volume that may be filled by hard spheres arranged on various lattices is: simple cubic, 0.52; body-centered cubic, 0.68; face-centered cubic, 0.74.

3.11 **a.** In the Kronig-Penny model, show that the energy at the top of each allowed band is independent of the barrier parameter P.
   b. Discuss how the width of the band depends on P for various different bands (identified by principal quantum number n).

3.12 Consider a two-dimensional crystal whose reciprocal lattice is hexagonal:

$$A = B, \varphi = 120°.$$

   a. What is the lattice in real space?
   b. Draw the first three Brillouin zones, using the extended zone scheme. Assuming that there are two conduction electrons per unit cell,

c. Sketch the Fermi surface in the first three zones, using the periodic zone scheme and weak interactions.

d. How large a band gap (compared to Fermi energy) would one need to depopulate the third zone? The second zone? Would the crystal conduct in these cases?

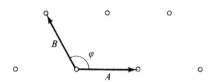

3.13 The force $\mathscr{F}$ on a particle of charge Q in electric field $\mathscr{E}$ and magnetic field **H** is

$$\mathscr{F} = Q\left(\mathscr{E} + \frac{1}{c}\mathbf{v} \times \mathbf{H}\right)$$

where **v** is the particle velocity. In a particular state in a metal, an electron would have charge $Q = -e$, effective mass m^*, wave vector **q**. Suppose that a band is completely filled except for that state. Show that the behavior of all the electrons in the band under the influence of electric and magnetic fields is just what it would be if the entire band were empty except for a single charge carrier of charge $Q = +e$, mass $(-m^*)$, and wave vector $-\mathbf{q}$.

3.14 You are given a piece of solid of unknown composition. Your laboratory is equipped for the following kinds of measurements:

> Optical observation (you can look at the thing)
> Specific heat
> Thermal conductivity
> Electrical conductivity
> Thermal expansion

Each can be done in any desired temperature range. Discuss what you could learn about the material from these observations and how you would analyze the data to look for each feature.

FOUR

LIQUIDS AND INTERACTING GASES

4.1 INTRODUCTION

At the outset of Chap. 1 we posed the question: Given a box containing a large number of noninteracting particles, what would its pressure be at a given temperature? In other words, we wanted a description of the behavior of the system as a macroscopic body. By the end of Chap. 2 we had solved that problem in very considerable detail. In this chapter we pose the same question again, except that the particles do interact as real atoms do. No dependable general solution will be found, not even for the simplest of realistic cases.

Certain portions of the answer are, of course, well known. At high densities and low temperatures, the equilibrium state is the crystalline solid, studied in Chap. 3. In the opposite limit of low density and high temperature, the interactions become unimportant, and the ideal gas is an adequate description. Now we turn to the intermediate case, liquids and interacting gases, where we can make use neither of crystalline order nor of independent particle behavior to simplify our analysis. The problem becomes very sticky.

Let us review the phase diagram of a simple substance in order to see more clearly what it is we are trying to describe. We show here in Fig. 4.1.1 the P-V and P-T planes, repeating the diagrams that were discussed in Sec. 1.2g. As pointed out then, it is possible, by manipulating T and V, to go from any point in the liquid region to any point in the gas region without ever undergoing a phase transition—that is, without ever passing through a

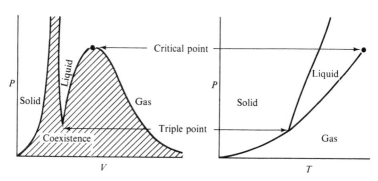

Fig. 4.1.1

point at which one could say: here is where we cease to be a liquid and start to be a gas. It is for this reason that liquids and interacting gases are treated here in a single chapter. By contrast, there is no way to enter the solid region without undergoing a phase transition.

There is just one instance in which there is a clear and unambiguous distinction between liquid and gas, and that is when they coexist with each other. When the two are seen together, with an interface between them, one has no trouble identifying which is the liquid. This is, of course, just the circumstance in which a phase transition separates the two.

Precisely what do we mean by the term liquid? Asking what is a liquid is a little like asking what is life; we usually know it when we see it, but the existence of some doubtful cases makes it hard to define precisely. If we examine our intuitive feeling for the word liquid, it is almost invariable tied up with the existence of a free interface, which we can see slosh about. If asked, do we believe the substance is still a liquid when compressed slightly so that there is no interface (say, just below the surface of a liquid under gravity), we must admit that we do and yet from there, as we have seen, it can change continuously into a gas. We could insist on using a single term, like fluid, for both liquid and gas, but doing so would be pedantic; we know very well that there are liquids and gases and that they are not quite the same things. We shall use all three terms—fluid, liquid, and gas—hoping that in each instance either the meaning is clear or the distinction is not important. If pressed for a definition, we can take a liquid to be a fluid that is basically self-condensed—that is, it will not expand smoothly without limit if the pressure is reduced isothermally.

If it is difficult to distinguish clearly between homogeneous liquid and homogeneous gas, it is not so difficult to distinguish between those states on the one hand and other possible states on the other. As we have seen, a crystalline solid is always separated from either type of fluid by a first-order

4.1 Introduction

phase transition. Moreover, a crystal may always be indentified by its long-range order, operationally, by its X-ray diffraction pattern. A more subtle problem is to distinguish between a liquid and an amorphous or disordered solid, or a glass. It is sometimes said that the distinction here lies in the order of magnitude of the viscosity, since, for example, glasses do flow, very slowly, and may therefore be thought of as very viscous liquids. However, there is, at least in principle, a clear difference between these states and a liquid. The liquid is an equilibrium state of matter, whereas the others are not. If an amorphous solid is heated and allowed to cool slowly, it will not return to precisely the same state; in fact, if the annealing is done slowly enough, it will crystallize. To be sure, disordered solids can be very stable, as can other non-equilibrium states (e.g., diamond is not the equilibrium state of carbon at ordinary temperature and pressure). Moreover, amorphous solids and glasses do behave in many ways just like liquids and may be of great interest in their own right. The distinction we are making, however, is a real one, since, for example, it means we cannot expect to be able to study these substances from the point of view of equilibrium statistical mechanics, not, at least, without special assumptions.

Now that we know what we mean by liquid and interacting gas, we can turn our attention to the question of how to study them. A number of approaches are used. In this chapter we shall adopt a point of view that becomes increasingly accurate at low density, finally reducing to the ideal gas. Other approaches tend to see the liquid as a perturbation of the solid state—it is an imperfect solid, one that flows. The starting point is of atoms in a regular lattice. Imperfections may be introduced into the lattice to give it more liquidlike properties. For example, there may be vacancies to produce a lower density than the solid, and there may be other defects as well. In lattice theories the atoms usually are not confined to lattice sites but are free to move around in a cell, with the cells being arranged in a lattice. For this reason, they are often referred to as cell theories. Cell theories succeed in accounting for certain properties of liquids (they must; otherwise we would not have heard of them), especially near the melting curve.

The pivotal idea in the line of reasoning that we will follow is that of the structure of a liquid. An ideal gas has no structure at all, the average density being perfectly uniform. When interactions become important, however, a kind of structure does appear. It is not a structure that can be described from the point of view of absolute space, as in a crystal, but it does show up from the point of view of any particular atom in the fluid. In Sec. 4.2 we shall see how the structure may be determined by X-ray diffraction measurements, just as in a solid. We then examine how the equation of state and other thermodynamic properties are related to that structure. The structure, which is isotropic, takes the form of correlations of the mean positions of other atoms with respect to the central one we have chosen. The rest of the chapter examines attempts to deduce the structure from the law of interactions between the atoms.

There seems at first to be a basic disparity between this line of argument and the one we followed to deduce the thermal behavior of solids in Chap. 3. For solids, we took the structure to be essentially a given fact, and the thermodynamic behavior arose not from the structure itself but from departures from the structure, which we resolved into normal modes. Why, then, the question arises, does the thermal behavior of liquids depend on the structure, whereas the thermal behavior of solids does not? In other words, if we can obtain the heat capacity and equation of state of a liquid from its microscopic structure, as we do in Eqs. (4.2.58), (4.3.24), and (4.3.36) below, why can we not do the same for a solid?

The answer is that we can do the same for a solid (and the reader is asked to try it in Prob. 4.2). The structure that enters this analysis is not the mere geometric arrangement of a lattice; rather, it includes the average effect of the thermal motions of the atoms. However, although the formalism we shall develop here for liquids is perfectly applicable to solids, we shall see that it is certainly cumbersome and unenlightening compared to the elegant techniques of Chap. 3.

The essential difference between the two approaches is that the formalism of this chapter is worked out in terms of the spatial correlations between the positions of atoms, averaged over time, whereas in treating solids we considered correlations not only in space but in time as well. The trial solution to the equations of motion, Eq. (3.3.8), was a guess at just what those space-time correlations were, and it turned out to be right. The elegant formalism we used, in which the solid could finally be replaced by a perfect gas of phonons, could be applied because the fluctuations of the atoms themselves were correlated in space and time in an extremely simple way.

It must surely be true that the motions of atoms in a liquid are correlated in time as well as in space. We certainly know that fluids transmit ordinary sound; thus, the low-frequency, long wavelength correlations are just like those in solids, at least for longitudinal modes. At short distances and short times, the motion of any atom must force others to get out of the way, thereby introducing another type of correlation. A great deal of effort in the theory of liquids is devoted to studying these dynamic effects, plus the correlations they produce. There is actually one spectacular example of a liquid whose thermal fluctuations can be resolved into good normal modes, just like a solid. That liquid is superfluid helium, and we shall study it separately in Chap. 5. For other, more ordinary liquids, there is some indication that the correlations may be basically similar to those in helium, but they are not as distinct and so cannot be treated as long-lived quasiparticles suitable for analysis as a perfect gas. In any case, the study of the dynamical behavior of liquids has so far been more successful in illuminating their transport and other nonequilibrium properties than in describing their thermodynamic behavior.

The basic approach that we shall take throughout this chapter will be toward making what progress we can in understanding the equilibrium properties of the simplest imaginable liquids, say, argon or xenon. In thus narrowing the scope of our inquiries, we will necessarily be unable to investigate a number of questions of great interest and importance. For example, in Chap. 3 we saw that certain materials are metallic for reasons that depended essentially on the geometric structure of the lattice; for instance, lead, which has four valence electrons, is a metal due to band overlap (Sec. 3.6c). Do we expect, then, that when lead melts and the geometric reasons for its metalization are gone, lead will be an insulator? Quite the contrary; anyone who has any experience soldering knows that liquid lead is a metal.

No, it is evidently not the band overlap that goes away when lead melts but rather the band itself. The simplest way to understand the point is to recall that the energy gap was attributed to the existence of standing wave states for the electrons whose periodicity matched that of the lattice: no periodic lattice, no gap. By extension of this line of reasoning, then, we reach a different conclusion: those materials that are nonmetallic in the solid state because of the band gap should become metals when they melt. That is, all semiconductors should be metals in the liquid state. Looking into the matter, we find that germanium, which is a semiconductor when solid, becomes a metal when it melts. Let us not, however, be too hasty in congratulating ourselves on this triumph of pure reason. The fact is that there do exist liquid semiconductors, and, what is just as relevant to this argument, amorphous semiconductors as well. Moreover, the metalization of germanium when it melts seems more a result of its short-range structure than its loss of long-range order: in crystalline germanium and in amorphous germanium, which is also a semiconductor, each atom has four nearest neighbors, whereas in liquid germanium each atom has eight.

In any case, the existence of liquid and amorphous materials with gaps in their electron spectra is simply not well understood. Their existence suggests that it is not the gap itself but our mathematical solution to the problem of the possible electron states that depends on periodicity. Whether that is true remains to be resolved.

Let us return, if reluctantly, to the question of simple liquids like argon and xenon.

4.2 THE STRUCTURE OF A FLUID

We have followed the hallowed custom of pointing out that all theories of the liquid state treat liquids either as imperfect solids or as dense

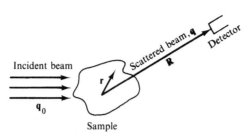

Fig. 4.2.1

gases. In this chapter we shall adopt the dense gases picture because, as we have seen in the previous section, there is no real distinction between liquids and dense gases, whereas one can, presumably, always distinguish between a liquid and a solid. Nevertheless, the central theme of the entire body of language we shall build up for discussing liquids is one that is most naturally associated with solids rather than gases: it is the idea of structure.

a. Measuring the Structure

The principal experimental means of examining the structure of solids is X-ray diffraction, discussed in Sec. 3.5a. Let us take as our starting point the consequences of applying the same technique to a liquid. The experiment to be performed is sketched in Fig. 4.2.1. Incident upon the sample there is a plane wave beam of X rays, with wave vector q_0. With respect to an origin in the sample, a detector at **R** receives those X rays that come out with wave vector q. The sample can be a liquid, a solid, or, for that matter, a gas. The thermal energies available for scattering X rays are of the same order of magnitude in all cases, all of them (recalling Fig. 3.5.6) very small compared to the energy of an X-ray photon. Thus, for scattering that leaves reasonably undisturbed the structure that we wish to investigate, the X rays come off with the same energy they come in with:

$$|q| \cong |q_0| \quad (4.2.1)$$

The momentum transfer

$$\Delta q = q - q_0 \quad (4.2.2)$$

depends only on the angle, θ, between the incident beam and the detector (Fig. 4.2.2). The dependence is

$$\Delta q = 2q_0 \sin \frac{\theta}{2} \quad (4.2.3)$$

as shown in the sketch. It is thus possible to study the scattered X ray as a function of the momentum transfer, Δq, by changing the angle of the detector relative to the beam.

4.2 The Structure of a Fluid

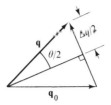

Fig. 4.2.2

The incident plane wave, whose amplitude at $\mathbf{r}_j$ is proportional to $\exp(i\mathbf{q}_0 \cdot \mathbf{r}_j)$, is scattered by the atom at that position into an outgoing spherical wave centered at $\mathbf{r}_j$. Thus, the amplitude of the scattered beam at the detector is proportional to

$$\exp(i\mathbf{q}_0 \cdot \mathbf{r}_j) \frac{\exp(iq|\mathbf{R} - \mathbf{r}_j|)}{|\mathbf{R} - \mathbf{r}_j|} \tag{4.2.4}$$

But since $\mathbf{R} \gg \mathbf{r}_j$,

$$|\mathbf{R} - \mathbf{r}_j| = R - \hat{R} \cdot \mathbf{r}_j$$

The second term on the right is unimportant in the denominator of the amplitude, but in the phase

$$q|\mathbf{R} - \mathbf{r}_j| = qR - \mathbf{q} \cdot \mathbf{r}_j$$

since $\mathbf{q}$ is in the same direction as the unit vector $\hat{R}$. Hence the amplitude at the detector is proportional to

$$\frac{\exp(iqR)}{R} \exp[-i(\mathbf{q} - \mathbf{q}_0) \cdot \mathbf{r}_j] = \frac{\exp(iqR)}{R} \exp(-i\Delta\mathbf{q} \cdot \mathbf{r}_j)$$

If we consider X rays scattered at the same (prior) instant everywhere in the sample, the total amplitude reaching the detector is proportional to

$$\frac{\exp(iqR)}{R} \sum_j \exp(-i\Delta\mathbf{q} \cdot \mathbf{r}_j) \tag{4.2.5}$$

In addition to Eq. (4.2.5), the amplitude at the detector depends on the efficiency of the electron clouds at each atom in scattering the X rays. We can write the total amplitude at the detector, A, as a product of a factor that is the same for each atom, A_0, and one that depends on the atomic positions, the $\mathbf{r}_j$:

$$A = A_0 \sum_j \exp(-i\Delta\mathbf{q} \cdot \mathbf{r}_j) \tag{4.2.6}$$

This separation is convenient, for the whole idea is to try to deduce from the experiment information about the structure of the sample—that is, all of the $\mathbf{r}_j$.

The instantaneous intensity at the detector is just $|A|^2$. However, in all experiments the data are collected, and therefore averaged, over a time that is long compared to the time scale of thermodynamic fluctuations in the sample (say, 10^{-12} to 10^{-14} sec, as we have discussed earlier), so that the experimental results will be given by

$$I = \langle |A|^2 \rangle = |A_0|^2 \left\langle \left| \sum_j \exp(-i\Delta\mathbf{q} \cdot \mathbf{r}_j) \right|^2 \right\rangle \quad (4.2.7)$$

where the angular brackets, $\langle \ \rangle$, have precisely the meaning of the bar over the f in Eq. (1.3.45); it is the thermodynamic average value. The sum in Eq. (4.2.7) is to be multiplied by its complex conjugate, giving

$$I = |A_0|^2 \left\langle \sum_{i,j} \exp[-i\Delta\mathbf{q} \cdot (\mathbf{r}_j - \mathbf{r}_i)] \right\rangle \quad (4.2.8)$$

which may be rewritten

$$I = |A_0|^2 \left\langle \int d^3r \sum_{i,j} \exp(i\Delta\mathbf{q} \cdot \mathbf{r}) \, \delta(\mathbf{r} + \mathbf{r}_i - \mathbf{r}_j) \right\rangle \quad (4.2.9)$$

where we are letting the δ function pick out just those terms that belong in Eq. (4.2.8). The exponential factor can now be taken out of the sum, and, moreover, only the δ function itself will change in the fluctuations between states that are to be averaged over. Thus,

$$I = |A_0|^2 \int d^3r \exp(i\Delta\mathbf{q} \cdot \mathbf{r}) \left\langle \sum_{i,j} \delta(\mathbf{r} + \mathbf{r}_i - \mathbf{r}_j) \right\rangle \quad (4.2.10)$$

We see, then, that the intensity, $I(\Delta q)$, is actually the Fourier transform of the quantity $\langle \sum_{i,j} \delta(\mathbf{r} + \mathbf{r}_i - \mathbf{r}_j) \rangle$.

Let us separate out from the double sum those terms for which $i = j$:

$$\left\langle \sum_{i,j} \delta(\mathbf{r} + \mathbf{r}_i - \mathbf{r}_j) \right\rangle = \left\langle \sum_{i \neq j} \delta(\mathbf{r} + \mathbf{r}_i - \mathbf{r}_j) \right\rangle + N\,\delta(r) \quad (4.2.11)$$

since there are N such terms, each a δ function at $\mathbf{r} = 0$. The first term on the right is a quantity that will be of central importance to us throughout the rest of this chapter. We define, provisionally, the *radial distribution function*, $g(r)$, by

$$\rho g(r) = \frac{1}{N} \left\langle \sum_{i \neq j} \delta(\mathbf{r} + \mathbf{r}_i - \mathbf{r}_j) \right\rangle \quad (4.2.12)$$

where $\rho = N/V$ is the density. Substituting Eqs. (4.2.11) and (4.2.12) into (4.2.10), we find for the scattered intensity

$$I(Q) = |A_0|^2 N \left[1 + \rho \int \exp(i\mathbf{Q} \cdot \mathbf{r}) g(r) \, d^3r \right] \quad (4.2.13)$$

4.2 The Structure of a Fluid

where we have replaced Δq by Q. In the particular case of $Q = 0$—that is, in the forward direction—there is no way to distinguish between scattered and unscattered X rays, and one always finds a strong peak due to the transmitted (i.e., unscattered) part of the beam. We therefore cannot expect $I(Q)$ to represent the data at $Q = 0$, and it makes no difference if we change our formula at that point. It is conventional to subtract off from the expression in Eq. (4.2.13) a δ function at $Q = 0$,

$$\rho \, \delta(Q) = \rho \int \exp\left(i\mathbf{Q} \cdot \mathbf{r}\right) d^3r \qquad (4.2.14)$$

With this change,

$$I(Q) = |A_0|^2 N \left\{ 1 + \rho \int \exp\left(i\mathbf{Q} \cdot \mathbf{r}\right)[g(r) - 1] \, d^3r \right\} \qquad (4.2.15)$$

The δ function has had the effect of shifting the zero of $g(r)$. The function $h(r)$, defined by

$$h(r) = g(r) - 1 \qquad (4.2.16)$$

is called the *total correlation function*.

Let us consider for a moment the function $g(r)$ as it is defined in Eq. (4.2.12). We are told to perform the following operation: we consider the vector $\mathbf{r}$ which has both a length, r, and a direction in space. In each configuration of the atoms of the sample, we record a δ function whenever any two atoms anywhere in the sample are separated by $\mathbf{r}$. This step is done for each possible configuration, and a weighted average of the results is taken in the usual thermodynamic way. The result is made dimensionless by dividing by $N\rho$, giving the value of g at $\mathbf{r}$. If the sample is a crystal, the result will actually depend on both magnitude and direction, and we should write $g(\mathbf{r})$. For a liquid or a gas, however, the result will be isotropic, so that it is enough to write $g(r)$. In any case, the question we are basically asking is: If we know that there is an atom at the origin, is it more or less likely than average that there will be an atom at $\mathbf{r}$? For example, if we compute $g(\mathbf{r})$ for a solid, taking $\mathbf{r}$ in the direction of one of the edges of the unit cell, we find a narrow Gaussian distribution about each lattice site (See Fig. 4.2.3.). The opposite case is the perfect gas, where an atom at the origin has no effect at all on the uniform distribution of the other atoms, so that quite simply

$$g(r) = 1 \quad \text{(perfect gas)} \qquad (4.2.17)$$

(we shall prove this result more formally later on). Liquids are neither as structured as solids nor as structureless as perfect gases. A typical $g(r)$ for a liquid is sketched in Fig. 4.2.4. The main features are a region at small r where $g(r) = 0$ owing to repulsion of other atoms by the one at the origin, a bump representing a shell of neighbors in the attractive part of the central

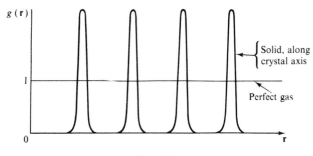

Fig. 4.2.3

atom's potential, a few further wiggles representing their neighbors and so on, and an asymptotic value of $g(r) \to 1$ far away. The function $h(r)$ differs only in that it starts at -1 and goes to zero far away.

Now consider the X-ray scattering intensity, $I(Q)$, produced in these cases. For the perfect gas, substituting Eq. (4.2.17) into (4.2.15), we find

$$I_{\text{P.G.}} = |A_0|^2 N \qquad (4.2.18)$$

This result offers a natural normalization procedure. We can rid ourselves of A_0 by defining

$$S(Q) = \frac{I(Q)}{I_{\text{P.G.}}} = 1 + \rho \int \exp(i\mathbf{Q}\cdot\mathbf{r}) h(r)\, d^3r \qquad (4.2.19)$$

$S(Q)$ is called the *structure factor*. The structure factor for a crystal is formed by substituting Fig. 4.2.3 into Eq. (4.2.19). To simplify matters, let us ignore the thermal broadening into Gaussians about the lattice sites for the moment and represent the crystal by the form

$$h(r) = g(r) - 1 = \frac{V}{N^2} \sum_{\mathbf{r}_s \neq 0} \delta(\mathbf{r} - \mathbf{r}_s) - 1 \qquad (4.2.20)$$

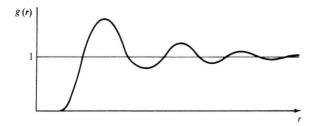

Fig. 4.2.4

4.2 The Structure of a Fluid

where the r_s are the lattice translation vectors, given by Eq. (3.4.1). This is a double sum, as in Eq. (4.2.12), because we take every r_s starting from every atomic position. Then substituting into Eq. (4.2.19), we have

$$S(Q) = 1 + \sum_{r_s} \frac{1}{N} \int \exp(i\mathbf{Q} \cdot \mathbf{r}) \delta(\mathbf{r} - \mathbf{r}_s) d^3r - \rho \delta(Q)$$

$$= 1 + \frac{1}{N} \sum_{r_s} \exp(i\mathbf{Q} \cdot \mathbf{r}_s) \qquad (4.2.21)$$

where the term $-\rho \delta(Q)$, arbitrarily added earlier, has now arbitrarily been dropped. Substituting Eq. (3.4.1) into (4.2.21), we have for the scattered intensity

$$S(Q) = 1 + \frac{1}{N} \sum_{m_1,m_2,m_3} \exp[i\mathbf{Q} \cdot (m_1\mathbf{a} + m_2\mathbf{b} + m_3\mathbf{c})] \qquad (4.2.22)$$

The sums over the integers m_1, m_2, m_3 cause the exponentials to average to zero except for those values of Q such that

$$\mathbf{Q} \cdot \mathbf{a} = 2\pi \ell_1$$
$$\mathbf{Q} \cdot \mathbf{b} = 2\pi \ell_2 \qquad (4.2.23)$$
$$\mathbf{Q} \cdot \mathbf{c} = 2\pi \ell_3$$

at which points we get δ function peaks in the intensity. ℓ_1, ℓ_2, and ℓ_3 can be any integer. Equations (4.2.23) with $\mathbf{Q} = \Delta\mathbf{q}$ are just the von Laue conditions, Eq. (3.5.10). If we replace our model Eq. (4.2.20) by more realistic Gaussian forms instead of the δ functions, the Fourier transforms of the Gaussian will again be Gaussians, with peaks at those values of $\mathbf{Q}$ satisfying Eq. (4.2.23)—that is, wherever $\mathbf{Q}$ is a reciprocal lattice vector (see Prob. 4.1).

Thus, X-ray scattering from a solid, in this description, turns out to be just the same as what we expected in Chap. 3, but we have now generalized our point of view to be able to discuss liquids as well. The structure factor of a liquid is typically like the one sketched in Fig. 4.2.5. A measurement of $S(Q)$ for a liquid can be Fourier transformed numerically to give $h(r)$ or $g(r)$ by means of Eq. (4.2.19). Our next task is to see how the result is related to other properties.

b. The Statistical Mechanics of Structure

The basic results needed to study the statistical mechanics of an interacting system of particles were worked out at the end of Sec. 1.3e. The essential point was that the kinetic energies of the particles could be dealt

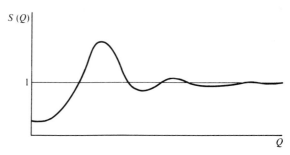

Fig. 4.2.5

with separately, leaving only the configurations to work out. The partition function for N particles could be written [Eq. (1.3.123)]

$$Z_N = \frac{Z_{IG}}{V^N} Q_N \qquad (4.2.24)$$

where Z_{IG} is the partition function of an N-particle ideal gas [Eq. (1.3.125)]

$$Z_{IG} = \frac{V^N}{N! \Lambda^{3N}} \qquad (4.2.25)$$

Λ is the thermal wavelength [Eq. (1.3.126)]

$$\Lambda = \frac{2\pi\hbar}{\sqrt{2\pi m k T}} \qquad (4.2.26)$$

and Q_N is the configurational integral [Eq. (1.3.122)]

$$Q_N = \int \cdots \int \exp\left(-\frac{U\{N\}}{kT}\right) d\{N\} \qquad (4.2.27)$$

The notation we are using here is that $\{N\}$ stands for the coordinates of the N particles, $\mathbf{r}_1, \mathbf{r}_2, \ldots, \mathbf{r}_N$, and $d\{N\}$ stands for $d^3r_1 \, d^3r_2 \ldots d^3r_N$. The dimensions of Q_N are those of V^N. The grand partition function is [Eq. (1.3.132)]

$$\mathscr{L} = \sum_N \frac{z^N Q_N}{N! \Lambda^{3N}} \qquad (4.2.28)$$

where [Eq. (1.3.131)]

$$z = e^{\mu/kT} \qquad (4.2.29)$$

The thermodynamic averaging indicated by the $\langle \ \rangle$ in Eq. (4.2.12) may be performed either in the fixed N or variable N formulations (i.e., the canonical or grand canonical ensembles). The results, however, will not be precisely the same in the two cases, and we shall distinguish between them

4.2 The Structure of a Fluid

by defining $g(r)$ and $h(r)$ according to Eqs. (4.2.12) and (4.2.16), and specifying that the variable N formulation is to be used. In the fixed N formulation, we define an analogous quantity, $g_N(r)$, by

$$g_N(r) = \frac{V}{N(N-1)} \left\langle \sum_{i \ne j} \delta(\mathbf{r} + \mathbf{r}_i - \mathbf{r}_j) \right\rangle_N \qquad (4.2.30)$$

The difference between the two cases arises because of the fluctuations in the value of N in the variable N formulation. For example, consider

$$\int g_N(r) \, d^3r = \frac{V}{N(N-1)} \int \left\langle \sum_{i \ne j} \delta(\mathbf{r} + \mathbf{r}_i - \mathbf{r}_j) \right\rangle_N d^3r$$

$$= \frac{V}{N(N-1)} \left\langle \int \sum_{i \ne j} \delta(\mathbf{r} + \mathbf{r}_i - \mathbf{r}_j) \, d^3r \right\rangle_N \qquad (4.2.31)$$

In each configuration to be averaged over, there will be *precisely* $N(N-1)$ δ functions in the double sum, each pair of atoms being counted twice in summing independently over i and j. Thus,

$$\int h_N(r) \, d^3r \equiv \int [g_N(r) - 1] \, d^3r = 0 \qquad (4.2.32)$$

is an exact result. On the other hand, the same argument cannot be applied to find $\int h(r) \, d^3r$, since the number of pairs will vary, as well as the configurations in the states of the subsystem that are to be averaged over when μ rather than N is held fixed. As we shall see below, $\int h(r) \, d^3r$ will depend on the compressibility, for that factor governs the fluctuations in the number of particles. It must be remembered, however, that $g(r)$ and $g_N(r)$ are measurable quantities and cannot depend on whether N or μ is held fixed, unless the subsystem is so small that the fluctuations do indeed become important. The two quantities, $g(r)$ and $g_N(r)$, must be the same in the formal limit of very large systems:

$$\lim_{\substack{N \to \infty \\ V \to \infty \\ \rho = \text{constant}}} g_N(r) = g(r) \qquad (4.2.33)$$

Let us formally write $g_N(r)$ in terms of our statistical distributions of Chap. 1. The averaging indicated by $\langle \ \rangle$ in Eq. (4.2.30) is to be performed according to Eq. (1.3.8) in conjunction with (1.3.27) and (1.3.28). Once these equations have been transformed by means of the operations indicated in Sec. 1.3e for interacting particles, we find that

$$\left\langle \sum_{i \ne j} \delta(\mathbf{r} + \mathbf{r}_i - \mathbf{r}_j) \right\rangle$$

$$= \frac{1}{Q_N} \int \cdots \int \sum_{i \ne j} \delta(\mathbf{r} + \mathbf{r}_i - \mathbf{r}_j) \exp\left(-\frac{U\{N\}}{kT}\right) d\{N\} \qquad (4.2.34)$$

The quantity to be averaged depends only on the configurations and is averaged only over them, the kinetic energy parts canceling top and bottom. Consider the first term in the double sum on the right-hand side of Eq. (4.2.34). It may be taken out as a separate integral,

$$\frac{1}{Q_N} \int \cdots \int \delta(\mathbf{r} + \mathbf{r}_1 - \mathbf{r}_2) \exp\left(-\frac{U\{N\}}{kT}\right) d^3r_1 \, d^3r_2 \, d^3r_3 \cdots d^3r_N$$

(4.2.35)

Our instructions here are to hold one vector, $\mathbf{r} = \mathbf{r}_2 - \mathbf{r}_1$, fixed and integrate over the remaining $N - 1$ coordinates. The function $U\{N\}$ in the integrand depends only on the distances between the atoms, not on their absolute positions in space, so that one volume integration can always be performed in integrals of this type (including Q_N itself) by choosing an origin and integrating it over space. In other words, $U\{N\}$ depends on $\mathbf{r}_2 - \mathbf{r}_1$, $\mathbf{r}_3 - \mathbf{r}_1, \ldots, \mathbf{r}_N - \mathbf{r}_1$, but not on $\mathbf{r}_1$ itself. Then

$$\begin{aligned}
Q_N &= \int \cdots \int \exp\left(-\frac{U\{N\}}{kT}\right) d\{N\} \\
&= \int \cdots \int \exp\left(-\frac{U\{N\}}{kT}\right) d^3r_1 \, d^3(|\mathbf{r}_2 - \mathbf{r}_1|) \cdots d^3(|\mathbf{r}_N - \mathbf{r}_1|) \\
&= V \int \cdots \int \exp\left(-\frac{U\{N\}}{kT}\right) d^3(|\mathbf{r}_2 - \mathbf{r}_1|) \cdots d^3(|\mathbf{r}_N - \mathbf{r}_1|) \\
&= V \int \cdots \int \exp\left(-\frac{U\{N\}}{kT}\right) d\{N - 1\}
\end{aligned}$$

(4.2.36)

where the last step introduces the notation we shall use instead of the more cumbersome notation of the second and third steps. In the term isolated in Eq. (4.2.35), we similarly take out the integration over $\mathbf{r}_1$, $\mathbf{r}_2$ is held fixed, and we get

$$\frac{V}{Q_N} \int \cdots \int \exp\left(-\frac{U\{N\}}{kT}\right) d\{N - 2\}$$

(4.2.37)

This is the first term in Eq. (4.2.34). All subsequent terms are exactly the same except for the irrelevant choice of which is the ith atom and which the jth. In each term there are N identical choices for the ith atom and $N - 1$ remaining choices for the jth atom, so there will be $N(N - 1)$ terms in the sum. The result for Eq. (4.2.34) is therefore

$$\left\langle \sum_{i \neq j} \delta(\mathbf{r} + \mathbf{r}_i - \mathbf{r}_j) \right\rangle$$

$$= \frac{N(N-1)}{Q_N} V \int \cdots \int \exp\left(-\frac{U\{N\}}{kT}\right) d\{N - 2\} \quad (4.2.38)$$

4.2 The Structure of a Fluid

or substituting into Eq. (4.2.30),

$$g_N(r) = \frac{V^2}{Q_N} \int \cdots \int \exp\left(-\frac{U\{N\}}{kT}\right) d\{N-2\} \qquad (4.2.39)$$

The radial distribution function is thus basically the statistical weight given to those configurations in which atoms 1 and 2 are separated by $\mathbf{r} = \mathbf{r}_2 - \mathbf{r}_1$, and it is a function of r (or of $\mathbf{r}$ for a crystal). Proper account is taken of the fact that 1 and 2 can be any two atoms. Earlier we said, in effect, that it is proportional to the probability of finding any two atoms separated by $\mathbf{r}$. Yet another way of writing it (valid for $r \neq 0$) is

$$\rho g(r) = \frac{1}{\rho} \langle \rho(0)\rho(r) \rangle \qquad (4.2.40)$$

Equation (4.2.39) is easily generalized to give expressions for the statistical weight of particular configurations of 3, 4, or n particles. Let us call these n-particle distribution functions, $g_N^{(n)}\{n\}$:

$$g_N^{(n)}\{n\} = g_N^{(n)}(\mathbf{r}_{12}, \mathbf{r}_{13}, \ldots, \mathbf{r}_{1n}) \qquad (4.2.41)$$

where

$$\mathbf{r}_{12} = \mathbf{r}_2 - \mathbf{r}_1 \qquad (4.2.42)$$

and so on. Each correlation function thus depends on $n-1$ vectors. Generalizing Eq. (4.2.30) gives

$$g_N^{(n)} = \frac{V^{n-1}(N-n)!}{N!} \left\langle \sum_{\substack{i \neq j \neq k \\ \neq \cdots}} \delta(\mathbf{r}_{12} + \mathbf{r}_i - \mathbf{r}_j)\, \delta(\mathbf{r}_{13} + \mathbf{r}_i - \mathbf{r}_k) \cdots \right\rangle_N$$

$$(4.2.43)$$

where $N!/(N-n)! = N(N-1)\cdots(N-n+1)$ is the number of terms in the multiple sum in each configuration, so that, by analogy to Eq. (4.2.32),

$$\int [g_N^{(n)} - 1]\, d\{n-1\} = 0 \qquad (4.2.44)$$

The notation used in the right-hand side of Eq. (4.2.43) is rather unwieldy; let us define n-particle correlation functions

$$W_N^{(n)} = \frac{1}{V} \left\langle \sum_{\substack{i \neq j \neq k \\ \neq \cdots}} \delta(\mathbf{r}_{12} + \mathbf{r}_i - \mathbf{r}_j)\, \delta(\mathbf{r}_{13} + \mathbf{r}_i - \mathbf{r}_k) \cdots \right\rangle_N$$

$$= \frac{N!}{V^n (N-n)!} g_N^{(n)} \qquad (4.2.45)$$

Performing the operations indicated by $\langle \ \rangle_N$ just as in Eqs. (4.2.35) to (4.2.40), we find

$$W_N^{(n)} = \frac{N!}{(N-n)!\, Q_N} \int \cdots \int \exp\left(-\frac{U\{N\}}{kT}\right) d\{N - n + 1\} \quad (4.2.46)$$

Let us write down for later reference the rather obvious general relation

$$W_N^{(n-1)} = \frac{1}{N - n + 1} \int W_N^{(n)}\, d^3 r_n \quad (4.2.47)$$

We see that there is a hierarchy of equations relating each correlation function to the next higher-order one. This hierarchy is generally quite useless, for nothing much is known about the form of any of the W's beyond $W_N^{(2)}$. However, one of the theories to be discussed in Sec. 4.5 cuts off the hierarchy by making an approximation for $W_N^{(3)}$ in terms of $W_N^{(2)}$. We shall then have occasion to make use of

$$W_N^{(2)} = \frac{1}{N - 1} \int W_N^{(3)}\, d^3 r_3 \quad (4.2.48)$$

or

$$g_N^{(2)} = g_N(r) = \frac{1}{V} \int g_N^{(3)}(\mathbf{r}_{12}, \mathbf{r}_{13})\, d^3 r_3 \quad (4.2.49)$$

The correlation functions and distribution functions can be written in the variable N formulation (the grand canonical ensemble) by performing the necessary averages over the configurations of subsystems of all N for fixed V and μ. In the formal limit $N \to \infty$, $V \to \infty$ with ρ constant, the results will be the same as those of the fixed N formulation, as in Eq. (4.2.33). However, the difference between the two for a finite system depends on the fluctuations in the number of particles, and we can take advantage of this difference to obtain a result that will be of much use in what follows, an exact equation of state in terms of $g(r)$. In order to perform the average indicated in Eq. (4.2.12), we write

$$\left\langle \sum_{i \neq j} \delta(\mathbf{r} + \mathbf{r}_i - \mathbf{r}_j) \right\rangle$$

$$= \frac{1}{\mathscr{L}} \sum_N \frac{z^N}{N!\, \Lambda^{3N}} \int \cdots \int \sum_{i \neq j} \delta(\mathbf{r} + \mathbf{r}_i - \mathbf{r}_j) \exp\left(-\frac{U\{N\}}{kT}\right) d\{N\}$$

$$= \frac{1}{\mathscr{L}} \sum_N \frac{z^N N(N-1)V}{N!\, \Lambda^{3N}} \int \cdots \int \exp\left(-\frac{U\{N\}}{kT}\right) d\{N - 2\}$$

$$(4.2.50)$$

4.2 The Structure of a Fluid

where we have followed the steps leading from Eq. (4.2.34) to (4.2.39). We therefore have for $g(r)$, Eq. (4.2.12),

$$g(r) = \frac{1}{\rho^2 \mathscr{L}} \sum_N \frac{z^N N(N-1)}{N! \Lambda^{3N}} \int \cdots \int \exp\left(-\frac{U\{N\}}{kT}\right) d\{N-2\} \quad (4.2.51)$$

The n-particle distribution functions and correlation functions may be written in this formulation, but care must be taken because the combinatorial factors, such as $N!/(N-n)!$ in Eq. (4.2.45), now change with the numbers of particles in the subsystem. We can keep things straight if we can define the correlation functions by

$$W^{(n)} = \frac{1}{V} \left\langle \sum_{\substack{i \neq j \neq k \\ \neq \cdots}} \delta(\mathbf{r} + \mathbf{r}_i - \mathbf{r}_j)\, \delta(\mathbf{r} + \mathbf{r}_i - \mathbf{r}_k) \cdots \right\rangle$$

$$= \frac{1}{\mathscr{L}} \sum_{N \geq n} \frac{z^N}{N! \Lambda^{3N}} \frac{N!}{(N-n)!}$$

$$\times \int \cdots \int \exp\left(-\frac{U\{N\}}{kT}\right) d\{N - n + 1\} \quad (4.2.52)$$

in analogy to the first member of Eq. (4.2.45), then write the distribution function as

$$g^{(n)} = \frac{1}{\rho^n} W^{(n)} \quad (4.2.53)$$

The $W^{(n)}$'s each have dimension ρ^n, and the $g^{(n)}$'s are dimensionless. In any case, the equation of state we seek is obtained by integrating $g(r)$ over $\mathbf{r}$:

$$\int g(r)\, d^3r = \frac{1}{\rho^2 \mathscr{L}} \sum_N \frac{z^N N(N-1)}{N! \Lambda^{3N}} \int \cdots \int \exp\left(-\frac{U\{N\}}{kT}\right) d\{N-1\}$$

$$= \frac{1}{\rho^2 V \mathscr{L}} \sum_N \left[\frac{z^N Q_N}{N! \Lambda^{3N}} N(N-1)\right] \quad (4.2.54)$$

where, in the last step, we have used Eq. (4.2.36). Thus, $\int g(r)\, d^3r$ is just proportional to the quantity

$$\frac{1}{\mathscr{L}} \sum_N N(N-1) \frac{z^N Q_N}{N! \Lambda^{3N}} = \langle N(N-1) \rangle$$

$$= \langle N^2 - N \rangle$$

$$= \langle N^2 \rangle - N^2 + N^2 - N$$

$$= \overline{(\Delta N)^2} + N^2 - N \quad (4.2.55)$$

where we have switched, in the last step, to the notation for average value that we used in Chap. 1, particularly

$$\overline{(\Delta N)^2} = \overline{N^2} - N^2 \qquad (4.2.56)$$

which is an application of Eq. (1.3.154). We could write $\overline{N}$ for N here, but we always understand quantities like E, T, and, in the variable N formalism, N, to be properly averaged without writing bars above them. From Eqs. (4.2.54) and (4.2.55), we have

$$\rho \int g(r)\, d^3r = \frac{\overline{(\Delta N)^2} + N^2 - N}{\rho V} = \frac{\overline{(\Delta N)^2}}{N} + N - 1 \qquad (4.2.57)$$

or using Eq. (1.3.171) and absorbing $N = \rho \int d^3r$ into the integral,

$$\rho \int [g(r) - 1]\, d^3r + 1 = \rho k T K_T \qquad (4.2.58)$$

where K_T is the isothermal compressibility. This is the equation of state we have been seeking. We shall refer to it as the *fluctuation equation of state*.

In order to get some feeling for the relative magnitudes of the various terms in Eq. (4.2.58), let us look at some examples. For an ideal gas, $h(r) = g(r) - 1 = 0$, and $\rho k T K_T = 1$. Thus, any nonzero contribution of the integral in Eq. (4.2.58) can be thought of as a correction to the ideal gas compressibility. On the other hand, consider the model expressed by Eq. (4.2.20), which is really the classical picture of a solid at $T = 0$. We have in that case

$$\rho \int g(r)\, d^3r = \frac{1}{N} \sum_{r_s \neq 0} \int \delta(\mathbf{r} - \mathbf{r}_s)\, d^3r$$

$$= N - 1 \qquad (4.2.59)$$

where the (-1) comes from explicitly excluding the atom at the origin—that is, at the lattice site $\mathbf{r}_s = 0$. Substituting into Eq. (4.2.58)

$$\rho \int [g(r) - 1]\, d^3r = -1 \qquad (4.2.60)$$

we get
$$\rho k T K_T = 0 \qquad (4.2.61)$$

In this model there are no fluctuations, not even quantum fluctuations. However, it is important to realize that the entire difference between this result and the ideal gas result occurs because of the missing atom at the origin. In a self-condensed medium, such as a solid or a liquid near the triple point,

4.2 The Structure of a Fluid

the compressibility is usually much less than that of an ideal gas (i.e., such a medium is nearly incompressible),

$$0 \le \rho k T K_T \ll 1 \qquad (4.2.62)$$

where the left-hand inequality is given by the rigorous condition, Eq. (1.2.115). It follows that

$$\rho \int [g(r) - 1] \, d^3r \approx -1 \qquad (4.2.63)$$

This result comes about from a $g(r)$ that looks qualitatively like the one sketched in Fig. 4.2.4. By analogy to the argument leading to Eq. (4.2.60), we can see that most of the magnitude of this integral arises from the region close to $r = 0$ where $g(r) = 0$—that is, from the repulsion of other atoms by the core of the atom at the origin. It makes eminently good sense that this should be so; after all, the essential incompressibility of condensed media does, in fact, arise from the reluctance of atoms to penetrate each other's cores. However, it leaves us with the realization that the small difference between Eq. (4.2.32) and Eq. (4.2.58) is the result of a near cancellation in the integral over the wiggles in $g(r)$ farther from the origin. If our program in the study of liquids involves measuring or computing $g(r)$ as a function of temperature and density, and then trying to compare the results to the measured equation of state by means of Eq. (4.2.58), we see that our attempt will be doomed to failure unless the job can be done with great precision. That program is followed, as we shall see, and the needed precision is lacking.

We cannot, then, expect to be able to measure $g(r)$ for, say, a liquid near the triple point by means of the X-ray scattering experiments of the previous subsection [obtaining it from a Fourier transform of $S(Q)$] and use the result to predict the equation of state via Eq. (4.2.58). However, that equation does enter into the scattering results in a more direct manner. Taking Eq. (4.2.19) in the limit as $Q \to 0$, we see that

$$S(0) = 1 + \rho \int h(r) \, d^3r = \rho k T K_T \qquad (4.2.64)$$

Thus, Eq. (4.2.58) does serve to tell us that $\rho k T K_T$ will be the limiting value of $S(Q)$ for small Q in Fig. 4.2.5.

There is a case in which the connection between $g(r)$ and K_T by means of Eq. (4.2.58) does become important. At the critical point of a gas-liquid phase transition, as we shall see in Chap. 6, a P-V isotherm has an inflection point where $(\partial P / \partial V)_T = 0$, so that

$$K_T = -\frac{1}{V}\left(\frac{\partial V}{\partial P}\right)_T = \infty \quad \text{(critical point)} \qquad (4.2.65)$$

At the critical point, then, the left-hand side of Eq. (4.2.58), instead of nearly canceling to zero, actually diverges:

$$\int [g(r) - 1] \, d^3r = \int h(r) \, d^3r$$

$$= 4\pi \int h(r) r^2 \, dr$$

$$\to \infty \qquad (4.2.66)$$

The total correlation function, $h(r)$, in a fluid, even one at the critical point, always falls to zero for large r. However, we see from Eq. (4.2.66) that unless it falls off faster than r^{-3}, the integral will diverge (for an infinite volume system). That is just what happens; we shall return to study this point at length in Chap. 6.

c. Applicability of This Approach

We have at this point begun to develop a language in which we can discuss the properties of a liquid, such as its equation of state, in terms of its structure, embodied in $g(r)$. The language is, of course, mathematical and, up to now, rigorous, at least for classical systems. We should not, however, be mislead into thinking that it necessarily follows that the language is useful. To paraphrase Galileo, *Il libro della natura e scritto nella lingua matematica*. (Galileo had a disconcerting habit of making his celebrated statements in Italian.) The book of nature is written in the language of mathematics. "*Si*", we might answer, "*ma deve essere letto nella lingua fisica*." Yes, but it must be read in the language of physics. Given that our purpose is to gain insight into how liquids behave and why, the formalism we have been following, albeit rigorous and even formidable, has some quite serious drawbacks.

For one thing, the principal result we have obtained so far—the fluctuation equation of state—is, as we argued, not really a very powerful or useful tool for studying condensed matter. We shall, in the next section, develop a more useful equation of state based on $g(r)$, but doing so will require making an approximation on the potential energy that may not be a good one at liquid densities. There is, moreover, a problem that may be even more fundamental and that seems worth pointing out. The body of knowledge that we are pursuing in this chapter revolves around the radial distribution function. We shall, on the one hand, try to find ways to relate it to the interatomic potential (in Sec. 4.5) and, on the other hand (as we have already begun to do), to the macroscopic properties of liquids. In other words, $g(r)$ is the pivotal entity in a program designed to deduce the properties of liquids from the interactions between atoms. This is an ambitious and certainly an

4.2 The Structure of a Fluid

important program, but it must be kept in mind that even were it to succeed entirely, we have given up all hope of obtaining certain kinds of fundamental information merely by focusing our attention on $g(r)$.

Ordinarily, in applying statistical mechanics to a problem, our immediate goal is to find either Z or $\mathscr{L}$, as we did for perfect gases and solids. From either of these functions, a complete thermodynamic description of the system follows. Even then, to be sure, we lack any understanding of dynamical behavior—that is, of nonequilibrium properties such as conductivities—but at least we have a thorough description of the equilibrium behavior of the system. However, and this is the point, even if we know $g(r)$ exactly for all densities and temperatures in the liquid range, we would still be left with an incomplete description of the system. We could obtain the equation of state; we could relate it to the statistical properties of the system through Eq. (4.2.51), although, of course, we cannot deduce $\mathscr{L}$ from $g(r)$. We might even, with the help of approximations (as in the next section), relate $g(r)$ to the heat capacity and thus to changes in entropy, when the density and temperature are changed, but normally there is no way to get the absolute entropy of the system from $g(r)$. Knowing $g(r)$ is rather similar to knowing the energy, E, as a function of T and V; it is useful but incomplete information. This situation is distressing because the absolute entropy—that is, the disorder of the system—is very close to the crux of the problem of liquids. The configurational entropy of an ideal gas is complete—any atom can be anywhere in the gas. That of a solid is minimal—there is an atom executing small oscillations about each lattice site. But a liquid has a kind of hindered communal entropy, and $g(r)$ has only partial knowledge of it. We cannot, from $g(r)$ alone, tell when a liquid will freeze or evaporate, or whether two liquids will mix spontaneously. In other words, we cannot tell whether a liquid is stable with respect to some other state. To give another example, the $g(r)$ of an amorphous solid, always a thermodynamically unstable state, is quite the same as that of an equilibrium liquid. Given only $g(r)$, there is no way to tell the difference.

Nevertheless, our program is, as we have said, a worthwhile one, and we shall forge ahead with it. Before examining the potential energy of the interacting system in the next section, let us pause a moment to examine one further question: What is the range of validity of our use of classical statistics?

Two basic errors are introduced by our use of classical mechanics and statistics to study liquids. One is that we may be incorrect in computing the possible energy states of the many-body system, which we do by adding the classical kinetic and potential energies of the particles. The other is the effect of quantum statistics in dictating the configurations that are allowed. It is not difficult to guess that the basic condition for these errors to be small is, as for perfect gases, that the particles have sufficient thermal momentum so that we can think of them as localized and therefore interacting classically. The condition is thus just Eq. (2.2.23):

$$N \frac{\Lambda^3}{V} \ll 1 \tag{4.2.67}$$

This is a more difficult condition to satisfy for liquids than for gases, however, since we must now be prepared to deal with much higher values of N/V. Let us consider the problem numerically. We may be confident of our approximation if

$$\Lambda \ll 2 \times 10^{-8} \text{ cm} \tag{4.2.68}$$

since 2×10^{-8} cm is roughly the size of an atom. Substituting numerical values into Eq. (4.2.26), we find

$$\Lambda \approx \frac{2 \times 10^{-7}}{(M_A T)^{1/2}} \tag{4.2.69}$$

where M_A is the atomic mass number. For argon, then, $M_A \simeq 40$, and the criterion is

$$T \gg 2.5°\text{K} \qquad \text{(argon)} \tag{4.2.70}$$

Since the triple point for argon is $\sim 80°\text{K}$, we seem quite safe in treating liquid argon classically. On the other hand, for helium, $M_A = 4$, so we require

$$T \gg 25°\text{K} \qquad \text{(helium)} \tag{4.2.71}$$

This condition is never satisfied for liquid helium, whose critical temperature is $5.2°\text{K}$. Below $2.17°\text{K}$, helium becomes a purely quantum fluid, which we shall study separately in Chap. 5. However, except for helium and certain intermediate cases (neon, hydrogen), our classical approach is valid for most liquids.

4.3 THE POTENTIAL ENERGY

The energy of any system of interacting particles can be written, using Eqs. (1.3.58) and (4.2.24), as

$$E = \frac{kT^2}{Z_N} \frac{\partial Z_N}{\partial T} = \frac{3}{2} NkT + \frac{kT^2}{Q_N} \frac{\partial Q_N}{\partial T} \tag{4.3.1}$$

where the first term, equal to the energy of the ideal gas, is also the kinetic energy of a system with interactions. We are working here in the fixed N

4.3 The Potential Energy

formalism. The mean potential energy $\bar{U}_N$ is obtained from Eq. (4.3.1), together with Eq. (4.2.27),

$$\bar{U}_N = \frac{1}{Q_N} \int \cdots \int U\{N\} \exp\left(-\frac{U\{N\}}{kT}\right) d\{N\} \quad (4.3.2)$$

This result is sufficiently obvious that we might just have started with it.

Another energy function, a kind of potential in the system, will be of considerable interest to us. We define

$$\varphi_N(r) = -kT \log g_N(r) \quad (4.3.3)$$

The physical significance of $\varphi_N(r)$ may be seen by computing

$$-\frac{\partial \varphi_N(r)}{\partial \mathbf{r}_1} = \frac{kT}{g_N(r)} \frac{\partial g_N(r)}{\partial \mathbf{r}_1}$$

$$= -\frac{\int \cdots \int (\partial U\{N\}/\partial \mathbf{r}_1) \exp(-U\{N\}/kT) \, d\{N-2\}}{\int \cdots \int \exp(-U\{N\}/kT) \, d\{N-2\}} \quad (4.3.4)$$

where the notation $\partial/\partial \mathbf{r}_1$ means the gradient with respect to the coordinates of particle 1 and where we have used Eq. (4.2.39). The quantity on the right may be interpreted as the force on particle 1 if we hold particle 2 fixed and average over the configurations of all the other particles. $\varphi_N(r)$ is the potential for that force and is called the *potential of mean force*. It is generally a considerably more complicated object than the mere potential of interaction between 1 and 2, since it involves the effects of particles 1 and 2 on the configurations of the other particles. A large part of the remainder of this chapter will be devoted to examining the differences between $\varphi_N(r)$ and the direct interaction, or pair potential, operating between 1 and 2. The analogous quantity in the variable N formulation,

$$\varphi(r) = -kT \log g(r) \quad (4.3.5)$$

also has the property of being a potential of mean force:

$$\frac{\partial \varphi(r)}{\partial \mathbf{r}_1}$$

$$= \frac{\sum_N \left[z^N/(N-2)! \, \Lambda^{3N}\right] \int \cdots \int (\partial U\{N\}/\partial \mathbf{r}_1) \exp(-U\{N\}/kT) \, d\{N-2\}}{\sum_N \left[z^N/(N-2)! \, \Lambda^{3N}\right] \int \cdots \int \exp(-U\{N\}/kT) \, d\{N-2\}} \quad (4.3.6)$$

We shall also wish to use potentials of mean force on particle 1 holding $n-1$ other particles fixed:

$$\varphi_N^{(n)} = -kT \log g_N^{(n)} \quad (4.3.7)$$

and

$$\varphi^{(n)} = -kT \log g^{(n)} \quad (4.3.8)$$

These formal definitions avail us little, of course, without some physical model that allows us to calculate $U\{N\}$. Let us start to seek one by considering the interaction between two atoms that are otherwise isolated.

a. The Pair Potential

The potential energy of interaction, $u(r)$, between two simple, spherically symmetric atoms, say for example, argon atoms, is sketched in Fig. 4.3.1. Briefly, the interaction is repulsive at small r because of the unwillingness of the electron clouds to overlap, and it is attractive at long distances due to the effects of mutually induced dipole moments. The potential passes through zero at $r = \sigma$, where σ is roughly twice what we shall call the hard-core radius (strong repulsion begins when the two nuclei are separated by σ). There is a minimum in the potential at $r = r_0$, of depth ε_0, where the short-range repulsive and long-range attractive parts just balance. We shall call ε_0 the well depth.

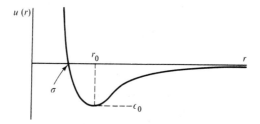

Fig. 4.3.1

The potential $u(r)$ is basically a coulomb effect, resulting from the distribution of charges within the atom. For a spherically symmetric, neutral atom, the electric potential outside the charge distribution is zero, but when two spherically symmetric atoms interact, they find that they can lower their combined potential energies by shifting their charge distributions a bit, as sketched in Fig. 4.3.2. Quantum mechanically, each atom perturbs the other, mixing asymmetric excited states into the ground-state wave function. The situation is adequately described classically, however. Each atom induces a dipole moment in the other, and the force between them is the force between electric dipoles. The energy, u, of an electric dipole of moment $\mathbf{p}$ in an electric field $\mathscr{E}$ is given by

$$u = -\mathbf{p} \cdot \mathscr{E} \qquad (4.3.9)$$

The dipole moment induced in an atom by the field $\mathscr{E}$ is

$$\mathbf{p} = \alpha \mathscr{E} \qquad (4.3.10)$$

4.3 The Potential Energy

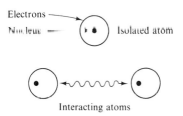

Fig. 4.3.2

where α, the polarizability, is just the susceptibility of the atom to polarization by the field. Equation (4.3.10) may be thought of as an empirical equation (much like, say, Ohm's law) defining α, although α may be computed quantum mechanically as well. The interaction energy, Eq. (4.3.9), is therefore

$$u = -\alpha \mathscr{E}^2$$

The field $\mathscr{E}$ is due to the dipole induced on the other atom, so that at sufficiently large r it has the dipole form

$$\mathscr{E} \propto \frac{1}{r^3} \tag{4.3.11}$$

Thus, the attractive part of the pair potential at long distances has the general form

$$u(r) \propto -\frac{1}{r^6} \tag{4.3.12}$$

It is possible that at very large distances the power law changes, perhaps to $u(r) \propto -r^{-7}$, due to retarded potential effects in communicating back and forth between the atoms, but that refinement is unimportant for our purposes.

There is no reason to expect that the sharp rise in the potential when the hard cores begin to overlap will have the form of a power law or, for that matter, any simple analytic form. It is, however, extremely useful to have an analytic form that fits the potential reasonably well over the entire range of r. Many such forms have been proposed, and they are more or less useful for various particular purposes. Let us record a particularly popular one here as an example. It is the Lennard-Jones potential

$$u(r) = 4\varepsilon_0 \left[\left(\frac{\sigma}{r}\right)^{12} - \left(\frac{\sigma}{r}\right)^6 \right] \tag{4.3.13}$$

This form fits the potential curve sketched in Fig. 4.3.1 with two parameters—σ, the point where $u(r) = 0$, and ε_0, the well depth. It is easy to verify that $r_0 = 2^{1/6}\sigma$. The values of ε_0 and σ for various kinds of atoms are measured by a variety of techniques, including atomic scattering experiments as well as some that will be discussed in Sec. 4.4 below. Since the form Eq. (4.3.13) is not exact, the values of σ and ε_0 that result from analysis of data will depend somewhat on the type of measurement used, plus the portion of the potential to which it is most sensitive. For our purposes, however, when an explicit form for $u(r)$ is needed, Eq. (4.3.13) will serve.

b. The Approximation of Pairwise Additivity

We now come to an approximation that lies at the heart of much of what we shall do throughout the rest of this chapter. We shall assume that the potential energy of the system, in any configuration, is composed of the sum of the potentials of all the atoms interacting two by two with the pair potential we have just discussed. In other words, if $u(r_{ij})$ is the pair potential that would act between the ith and jth atoms if they were isolated and separated by a distance r_{ij}, we assume that the many-body potential of the system as a whole is given by

$$U\{N\} = \sum_{i<j} u(r_{ij}) \qquad (4.3.14)$$

or
$$U\{N\} = \frac{1}{2} \sum_{i \neq j} u(r_{ij}) \qquad (4.3.15)$$

In the second of these equations we sum separately over i and j, a procedure that counts each pair twice, then divide by two. In either case, there are $N(N-1)/2$ pairs to be accounted for.

Let us pause for a moment and consider the nature of the approximation that we are making. The problem can be seen in a system of just three particles, for which we have assumed that

$$U\{3\} = u(r_{12}) + u(r_{13}) + u(r_{23}) \qquad (4.3.16)$$

Let us suppose that we have atoms 1 and 2 separated by r_{12}, with atom 3 far away. The potential of the system is $u(r_{12})$. We now bring up particle 3. Just as particles 1 and 2 have induced each other to have dipole moments, as sketched in Fig. 4.3.2, particle 3 now causes the charge distribution in particles 1 and 2 to shift again in order best to take advantage of its own fields. This rearrangement causes a change in the forces acting between 1 and 2 alone. It is this last change that we ignore. In writing Eq. (4.3.16) we take the potential between 1 and 2, $u(r_{12})$, to be the same as if 3 were not present. In the more general case, Eqs. (4.3.14) or (4.3.15), we take the direct potential between 1 and 2 to be the same as if none of the other atoms was present.

The creative part of theoretical physics lies mainly in making inspired approximations, for it is at this point that we decide which parts of the problem at hand are important and which parts may be neglected. The rest

4.3 The Potential Energy

of the work is largely mechanical. The approximations we make, however, must be justified, and basically there are three ways of doing so. The first is to make an estimate of the magnitude of the errors introduced and to argue that they are negligible, at least under some specified range of circumstances. Equation (4.2.67), specifying the range of validity of the classical formulation to the problem of liquids, is such an argument. There seems no way to make a similar argument in the present case of the approximation of pairwise additivity of potentials. A second possibility is to make the approximation, work out its consequences, and show that they agree with experimental reality. This alternative is probably the intent of most people who use pairwise additivity, but it has not been possible to carry out this program, for reasons that will become clear as we go along. The third possibility is to give up the expectation of quantitative accuracy and make an approximation with the hope that it will be mathematically tractable and contain the essential qualitative physical features of the problem, thereby lending insight into whatever is important in the problem. In this last case, the approximation is usually called a model, thus relieving its inventor of the burden of producing quantitatively accurate predictions. It is possible that pairwise additivity actually falls into this class, although most of the practitioners in the field do not seem to regard it that way explicitly. In any case, we cannot justify the approximation at this point but instead shall make use of it and continue on, watching, as we do, the interplay between attempts at justifications of the second and third kinds. Perhaps it should be mentioned that we made implicit use of this approximation, without much agonizing evaluation, in Chap. 3—for example, Eq. (3.3.18). It was clear there that we were constructing a model.

With the help of Eq. (4.3.15), the potential energy of the system, Eq. (4.3.2), may be written

$$\bar{U}_N = \frac{1}{Q_N} \int \cdots \int \frac{1}{2} \sum_{i \neq j} u(r_{ij}) \exp\left(-\frac{U\{N\}}{kT}\right) d\{N\} \quad (4.3.17)$$

This is really a sum of $N(N-1)$ integrals, each over a single pair potential, $u(r_{ij})$. Since all the coordinates are to be integrated in each integral, the integrals each give the same numerical result, and we may write

$$\bar{U}_N = \frac{N(N-1)}{2Q_N} \int \cdots \int u(r_{12}) \exp\left(-\frac{U\{N\}}{kT}\right) d\{N\}$$

$$= \frac{N(N-1)}{2Q_N} \int \cdots \int u(r_{12}) \exp\left(-\frac{U\{N\}}{kT}\right) d\{N-2\} \, d^3r_1 \, d^3r_2$$

$$(4.3.18)$$

The vector $\mathbf{r}_{12}$ is to be held fixed in the integration over $\{N-2\}$, and so $u(r_{12})$ is a constant for those operations:

$$\bar{U}_N = \frac{N(N-1)}{2Q_N} V \int \left[\int \cdots \int \exp\left(-\frac{U\{N\}}{kT}\right) d\{N-2\}\right] u(r_{12}) \, d^3 r_{12}$$

$$= \frac{N(N-1)}{2V} \int g_N(r) u(r) \, d^3 r \tag{4.3.19}$$

where we have extracted one volume integral as usual and used Eq. (4.2.39). The mean potential energy per particle, a quantity that should not depend on the size of the system, may be found in the variable N formulation by using this formula in the limit $N, V \to \infty$ at constant ρ:

$$\bar{u} = \frac{\bar{U}}{N} = \frac{1}{2} \rho \int u(r) g(r) \, d^3 r \tag{4.3.20}$$

Equations (4.3.19) and (4.3.20) are examples of the considerable simplification of the general problem introduced by the pairwise additivity approximation.

As we have already seen, the radial distribution function, $g(r)$, approaches 1 for large r in a liquid or gas (the argument we shall make here may be extended trivially to solids as well). Let us suppose that the long-range part of $u(r)$ falls off as r^{-m}:

$$u(r) \propto r^{-m} \quad \text{(large } r\text{)} \tag{4.3.21}$$

Then the contribution to $\bar{u}$ from large r will be

$$\bar{u} \sim \frac{4\pi\rho}{2} \int r^{-m} r^2 \, dr \tag{4.3.22}$$

In the formal thermodynamic limit $N, V \to \infty$, the upper limit of the integral is ∞. The condition that $\bar{u}$ be finite is then

$$m > 3 \tag{4.3.23}$$

As we saw earlier [Eq. (4.3.12)], we actually expect $m = 6$ or more for mutually induced dipole forces (and, it turns out, other common forces between neutral atoms and molecules as well), so that the condition is well satisfied for forces of the type that we have been considering.

Real atoms do, however, interact gravitationally as well as electrically. The gravitational forces between atoms are normally much smaller (very, very much smaller) than those we have been discussing, but they do come to dominate at very large distances. Moreover, they are really pairwise additive (as far as one knows), and so our formalism should apply to them. For gravity, however, Eq. (4.3.23) is violated, since $m = 1$. As matter of any kind grows toward infinite size at constant density (the formal thermodynamic limit), the potential energy per particle becomes infinite; the free

4.3 The Potential Energy

energy grows and the uniform density in whatever state it is in ultimately becomes unstable merely because of its size. The result is called gravitational collapse, a phenomenon that causes $g(r)$ to go to zero instead of 1 at large r. The nature of matter after gravitational collapse (black holes, not to be confused with the holes of Sec. 3.6) is still somewhat speculative and will not be treated here. We do, however, note in passing that it is, if not important, at least amusing to remember that whenever the formal thermodynamic limit is taken, gravity must be carefully neglected.

The energy of a system cannot be measured directly, but its heat capacity can. We have

$$C_V = \left(\frac{\partial E}{\partial T}\right)_V = \frac{3}{2} Nk + \frac{1}{2} N\rho \int u(r) \frac{\partial g(r)}{\partial T} d^3r \qquad (4.3.24)$$

where we have used Eq. (4.3.1), together with (4.3.20), for the potential energy. Equation (4.3.24) may be regarded as a verifiable prediction of the approximation of pairwise additivity, since C_V, N, ρ, $u(r)$, and $g(r)$ as a function of T may all be measured independently in principle. There are formidable difficulties, however, in measuring all these quantities, particularly $g(r)$ as a function of T, with sufficient precision to make a useful test.

Equations (4.3.20) and (4.3.24) are applicable to solids (at high temperature) if we remember that the radial distribution function will not be isotropic in a crystal. Some insight into the point of view we are applying (or trying to apply) to fluids may be gained by looking at solids in the same way. The heat capacity of a solid when these equations are valid is given by Eq. (3.2.11) as

$$C_V = 3Nk \qquad (4.3.25)$$

Just half of this value is given by the first term on the right in Eq. (4.3.24), in accordance with the now familiar Law of Equipartition. That term is the consequence of the kinetic energy of the atoms. The second term, which, as we have just seen, is the temperature derivative of the potential energy, has the same value, $\frac{3}{2}Nk$, in a solid. That result followed inexorably in Chap. 3 from the model that the atoms execute harmonic motion, whether collectively as in the Debye model or independently as in the Einstein model. We are now trying, however, to look at things in a quite different way. In Chap. 3 we imagined each atom to be bound harmonically to its equilibrium site under the combined influence of all other atoms. Equation (4.2.24), on the other hand, relates the heat capacity to the direct potential acting between a single pair of atoms, $u(r)$. It is important to understand that $u(r)$ in a solid is not a periodic potential harmonic about equilibrium positions; rather it is just that same pair potential sketched in Fig. 4.3.1. The thermal behavior of the system is all contained in $g(\mathbf{r})$, which is sketched in Fig. 4.2.3. The product of $g(\mathbf{r})$ and $u(r)$ integrated over all space must contribute $\tfrac{3}{2}NkT$ to the energy above the ground state.

It is easy to see that most of the contribution comes from the first peak in each direction in $g(\mathbf{r})$, since $u(r)$ is very small farther away. It is the temperature dependence of the width of the first peak that governs the heat capacity (the area under the peaks simply counts the number of nearest neighbors and is thus independent of T). The broadening of the first peak with T is sketched in Fig. 4.3.3. We shall leave the quantitative details,

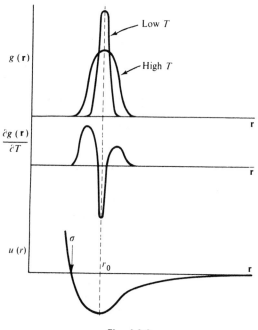

Fig. 4.3.3

showing that all this leads to Eq. (4.3.25) in a solid, to the problems. The arguments are generally similar in a liquid. The product $u(r)\,\partial g(r)/\partial T$ is predominantly positive, with positive contributions at $r < \sigma$ and $r \approx r_0$. The point to remember is that if the large-scale collective properties of the liquid are to be reflected in the energy and heat capacity, they must somehow show up in the very short-range structure, where $u(r)$ is finite. Otherwise they will not show up in Eqs. (4.3.20) and (4.3.24).

An additional equation connects the structure to macroscopic properties.

4.3 The Potential Energy

It is an equation of state, which, unlike Eq. (4.2.58), depends on the pairwise additivity approximation. The equation of state in the fixed N formulation is

$$P = -\left(\frac{\partial F}{\partial V}\right)_T$$

$$= kT \frac{\partial \log Z_N}{\partial V}$$

$$= \rho kT + \frac{kTV^N}{Q_N} \frac{\partial}{\partial V}\left(\frac{Q_N}{V^N}\right) \qquad (4.3.26)$$

where we have used Eqs. (1.2.9), (1.3.32), (4.2.24), and the ideal gas equation of state. The configurational integral, Q_N, Eq. (4.2.27), depends on the volume both through the limits of integration over $\{N\}$ and through the coordinate dependences of $U\{N\}$. In order to perform the derivative in (4.3.26), we can temporarily extract the volume dependence of the limits of integration by rewriting all the r_i's in terms of dimensionless r_i''s:

$$r_i = V^{1/3} r_i' \qquad (4.3.27)$$

so that
$$Q_N = \int_V \cdots \int \exp\left(-\frac{U\{N\}}{kT}\right) d\{N\}$$

$$= V^N \int_0^1 \cdots \int_0^1 \exp\left(-\frac{U\{N\}}{kT}\right) d\{N'\} \qquad (4.3.28)$$

The point here is that the integrand depends only on the distances between particles, and each of these distances will scale with $V^{1/3}$ when V is changed to take the derivative. We thus have

$$\frac{\partial}{\partial V} \frac{Q_N}{V^N} = \int_0^1 \cdots \int_0^1 \frac{\partial}{\partial V} \exp\left(-\frac{U\{N\}}{kT}\right) d\{N'\}$$

$$= -\frac{1}{kT} \int_0^1 \cdots \int_0^1 \exp\left(-\frac{U\{N\}}{kT}\right) \frac{\partial U\{N\}}{\partial V} d\{N'\} \qquad (4.3.29)$$

At this point we put in the pairwise additive approximation, Eq. (4.3.15). Then

$$\frac{\partial U\{N\}}{\partial V} = \frac{1}{2} \sum_{i \neq j} \frac{\partial u(r_{ij})}{\partial V}$$

$$= \frac{1}{2} \sum_{i \neq j} \frac{du(r_{ij})}{dr_{ij}} \frac{\partial r_{ij}}{\partial V} \qquad (4.3.30)$$

Now use Eq. (4.3.27) once more:

$$\frac{\partial r_{ij}}{\partial V} = \frac{1}{3} \frac{r_{ij}'}{V^{2/3}} = \frac{r_{ij}}{3V} \qquad (4.3.31)$$

which again merely states that each r_{ij} scales with $V^{1/3}$. Equations (4.3.30) and (4.3.31) go into (4.3.29) to give

$$\frac{\partial}{\partial V} \frac{Q_N}{V^N} = -\frac{1}{6VkT} \int_0^1 \cdots \int_0^1 \sum_{i \neq j} r_{ij} \frac{du(r_{ij})}{dr_{ij}} \exp\left(-\frac{U\{N\}}{kT}\right) d\{N'\}$$

$$= -\frac{1}{6kTV^{N+1}} \int_V \cdots \int \sum_{i \neq j} r_{ij} \frac{du(r_{ij})}{dr_{ij}} \exp\left(-\frac{U\{N\}}{kT}\right) d\{N\}$$
(4.3.32)

where we have reverted back to integrals over volume in the last step. The summation may be taken outside the multiple integral, and we then have $N(N-1)$ integrals, all numerically the same, for it cannot matter which are the ith and jth atoms. Thus,

$$\frac{\partial}{\partial V} \frac{Q_N}{V^N} = -\frac{N(N-1)}{6kTV^{N+1}} \int \cdots \int r_{12} \frac{du(r_{12})}{dr_{12}} \exp\left(-\frac{U\{N\}}{kT}\right) d\{N\}$$

$$= -\frac{N(N-1)}{6kTV^{N+1}} \iint r_{12} \frac{du(r_{12})}{dr_{12}}$$

$$\times \left[\int \cdots \int \exp\left(-\frac{U\{N\}}{kT}\right) d\{N-2\}\right] d^3 r_1 \, d^3 r_2 \quad (4.3.33)$$

We have separated out of the multiple integral the factor in the integrand that depends only on $\mathbf{r}_1$ and $\mathbf{r}_2$. As usual, only $\mathbf{r}_2 - \mathbf{r}_1$ matters, so one volume integral comes out as V, and we use Eq. (4.2.39) to replace the factor in brackets, giving

$$\frac{\partial}{\partial V} \frac{Q_N}{V^N} = -\frac{N(N-1)}{6kTV^N} \frac{Q_N}{V^2} \int r \frac{du(r)}{dr} g_N(r) \, d^3 r \quad (4.3.34)$$

Substitute this result into Eq. (4.3.26) to give

$$P = \rho kT - \frac{N(N-1)}{6V^2} \int r \frac{du(r)}{dr} g_N(r) \, d^3 r \quad (4.3.35)$$

Finally, we take the formal thermodynamic limit, replacing $N(N-1)/V^2$ by ρ^2 and $g_N(r)$ by $g(r)$, with the result

$$P = \rho kT \left[1 - \frac{\rho}{6kT} \int r \frac{du(r)}{dr} g(r) \, d^3 r\right] \quad (4.3.36)$$

Equation (4.3.36) is the new equation of state.

Once again let us try to get some feeling for the usefulness of this equation by looking at some special cases. Obviously the form is best suited to considering perturbations from ideal gas behavior, since that is the first term

4.3 The Potential Energy

on the right. The ideal gas itself is retrieved by setting $u(r) = 0$, $g(r) = 1$. In the next approximation, with a realistic $u(r)$, there should be some suitable $g(r)$ for low-density gases that will give the leading correction to the ideal gas due to interactions. That leading correction will be examined in Sec. 4.4, the interacting gas. On the other hand, consider the situation in a liquid near the triple point. In this case, the interactions have had the effect of making the matter self-condense, causing the pressure to be very much lower than it would be for an ideal gas at the same density. We might therefore expect that the two terms in the brackets in Eq. (4.3.36) will be very nearly equal and opposite. It is easy to see, in fact, that such is exactly the case. Near the triple point, on the vapor pressure curve, the liquid finds itself in equilibrium with a gas at the same pressure and temperature but a much lower density, ρ_g. The vapor usually obeys the ideal gas equation quite accurately, so we may write

$$P = \rho_g kT \tag{4.3.37}$$

Eliminating P between Eqs. (4.3.36) and (4.3.37), we have

$$1 - \frac{\rho}{6kT}\int r\frac{du(r)}{dr}g(r)\,d^3r = \frac{\rho_g}{\rho} \tag{4.3.38}$$

But ρ_g/ρ is normally very small near the triple point (in argon, $\rho_g/\rho \approx 2 \times 10^{-3}$ at the triple point), so that the two terms on the left in Eq. (4.3.38) do indeed nearly cancel. The result is that even fairly accurate forms of $u(r)$ and $g(r)$ may be expected to give poor results for P in this region. Recall that this is also exactly the region where the fluctuation equation of state, Eq. (4.2.58), becomes difficult to use for much the same reason [see Eqs. (4.2.62) and (4.2.63)].

In a sense, it is just near the triple point that we would most like to understand liquids, for it is here that liquid solid and gas coexist and can most clearly be observed as distinct states of matter. The unsuitability of the equations of state, Eqs. (4.2.58) and (4.3.36), in that region might lead us to conclude that the whole enterprise of discussing liquids in terms of structure has the exasperating feature of being least useful precisely when it is most interesting. On the other hand, we can take the point of view that Eqs. (4.2.63) and (4.3.38) give us two good approximations for integrals over $g(r)$ that should help us to find accurate values of that function just where it is most interesting. The two approximate equations, valid near the triple point, are, respectively [from (4.2.63) and (4.3.38)],

$$\rho\int [g(r) - 1]\,d^3r \approx -1 \tag{4.3.39}$$

and

$$\frac{\rho}{6kT}\int r\frac{du(r)}{dr}g(r)\,d^3r \approx 1 \tag{4.3.40}$$

These two equations are limiting forms for the structure of a liquid when it is strongly self-condensed (pressure and compressibility both near zero). Under all other circumstances, the theory of liquids is simply a theory of dense gases.

4.4 INTERACTING GASES

In Chap. 2 we discussed departures from ideal gas behavior due purely to quantum statistics in a noninteracting gas. In real gases, however, the interactions always become important before quantum effects do. In this section we shall see how departures from ideality due to interactions may be taken into account.

Real gases behave ideally in the limit of low density. It is natural, therefore, to represent the equation of state as a power series in the density, with the ideal equation of state as the leading term:

$$P = \rho kT[1 + \rho B(T) + \rho^2 C(T) + \cdots] \qquad (4.4.1)$$

Equation (4.4.1) is called the *virial equation of state*. In principle, it is exact, but the rate at which it converges depends, of course, on the density and it is most useful at low densities. $B(T)$, $C(T)$, and so on are called the second virial coefficient, third virial coefficient, etc. and depend, as we shall see, on two-particle collisions, three-particle collisions, and so forth. Our first job will be to consider the leading-order correction to the ideal gas, $B(T)$.

a. Cluster Expansions and the Dilute Gas

The equation of state of what we shall call the dilute gas is

$$P = \rho kT[1 + \rho B(T)] \qquad (4.4.2)$$

with the obvious condition for applicability,

$$\rho B(T) \ll 1 \qquad (4.4.3)$$

This is a low-density approximation, generally more restrictive than the condition, Eq. (4.2.67),

$$\rho \Lambda^3 \ll 1 \qquad (4.4.4)$$

Equation (4.4.4) is the condition for ignoring quantum effects. At some density we must take into account binary collisions, or interactions between pairs of atoms, but the density is still low enough so that the probability of three or more atoms colliding simultaneously is negligible. Under these circumstances pairwise additivity of interatomic potentials is not a further

4.4 Interacting Gases

approximation but rather a consequence of the low density. If we substitute Eq. (4.3.14) into (4.2.27), we have

$$Q_N = \int \cdots \int \exp\left[-\sum_{i<j} \frac{u(r_{ij})}{kT}\right] d\{N\}$$

$$= \int \cdots \int \prod_{i<j} \exp\left[-\frac{u(r_{ij})}{kT}\right] d\{N\} \quad (4.4.5)$$

At this point we use an elementary trick that lies at the basis of much of the remaining mathematics of this chapter. We define the quantity

$$f_{ij} = \exp\left[-\frac{u(r_{ij})}{kT}\right] - 1 \quad (4.4.6)$$

Here f_{ij} has the important property of being equal to zero unless r_{ij} is sufficiently small so that $u(r_{ij})$ is not negligible compared to kT. In other words, f_{ij} is zero unless atoms i and j are close together. Using Eq. (4.4.6), Eq. (4.4.5) may be written

$$Q_N = \int \cdots \int \prod_{i<j} (1 + f_{ij}) \, d\{N\} \quad (4.4.7)$$

The integrand is a product of $N(N - 1)/2$ factors of the form $(1 + f_{ij})$. Let us multiply the product out and collect the resulting terms as follows: the first term is 1, picking out the 1 from each of the $(1 + f_{ij})$. Next we pick out one factor f_{ij} and 1's from all the others, then products of the form $f_{ij}f_{jk}$, and so on. We get

$$\prod_{i<j} (1 + f_{ij}) = 1$$
$$+ f_{12} + f_{13} + f_{23} + f_{14} + \cdots$$
$$+ f_{12}f_{23} + f_{13}f_{34} + f_{12}f_{24} + \cdots$$
$$+ \text{etc.} \quad (4.4.8)$$

Here the second line on the right consists of terms that are nonzero when the pair in the subscripts is close together. There are $N(N - 1)/2$ terms on that line. Terms in the third line are nonzero when the three particles referred to are simultaneously close together—the number of such terms is $N(N - 1)(N - 2)/3!$—and so on. In the dilute gas only the first and second lines survive, so we may write

$$Q_N = \int \cdots \int \left[1 + \sum_{i<j} f_{ij}\right] d\{N\}$$
$$= V^N + V^{N-2} \sum_{i<j} \iint f_{ij} \, d^3r_i \, d^3r_j$$
$$= V^N \left[1 + \frac{N(N-1)}{2V} \int f_{12} \, d^3r_{12}\right] \quad (4.4.9)$$

The manipulations in Eq. (4.4.9) should be entirely familiar by now.

Products of the f_{ij}'s are called *clusters*, and integrals over them are called *cluster integrals*. The arrangement of terms in Eq. (4.4.7) in clusters obviously facilitates picking out contributions from various numbers of atoms coming simultaneously together, and that step, as we shall see, leads naturally to expansions in powers of the density. At the moment we are interested in the term of order ρ^2 in the equation of state, Eq. (4.4.2), and that term arises from two particle clusters, represented in Eq. (4.4.9).

We have made a rather subtle assumption, however, that bears comment before we proceed. In arriving at Eq. (4.4.9) we have not only ignored triplet and higher-order collisions but we have also thrown away clusters of the form

$$f_{12}f_{34} + f_{12}f_{35} + \cdots \qquad (4.4.10)$$

These are terms representing independent binary collisions between two sets of pairs in a single configuration but at different places in the system. We are able to do so by a strictly formal trick. We assume that the total quantity of gas is small, so that collisions are restricted not only to pairs but also to one pair at a time. The argument will lead to a free energy density that does not depend on the quantity N, and we then argue that, by the additivity property of the energy functions discussed in Chap. 1, the result applies to systems of any size, not only to small quantities. The same trick is used in each order of approximation—that is, for higher densities as well.

When we use Eqs. (4.4.9) and (4.4.6), the partition function Z_N, Eq. (4.2.24), becomes

$$Z_N = Z_{IG} \left\{ 1 + \frac{N(N-1)}{2V} \int \left[\exp\left(-\frac{u(r)}{kT}\right) - 1 \right] d^3r \right\} \qquad (4.4.11)$$

The integrand in Eq. (4.4.11) is nonzero only for small r, so the integral, although formally taken over the volume of the system, does not, in fact, depend on the volume. Z_N is thus the ideal gas value plus a correction term multiplied by N^2/V. For the correction to be small, not only N/V but N as well must be small, as we have just pointed out. The free energy is given by

$$F = -kT \log Z_N$$

$$= F_{IG} - kT \log \left\{ 1 + \frac{N(N-1)}{2V} \int \left[\exp\left(-\frac{u(r)}{kT}\right) - 1 \right] d^3r \right\}$$

$$= F_{IG} - \frac{kTN(N-1)}{2V} \int \left[\exp\left(-\frac{u(r)}{kT}\right) - 1 \right] d^3r \qquad (4.4.12)$$

In the last step we have taken the leading-order expansion of the logarithm, assuming the correction term to be small. Although N is small, it is much bigger than 1; we replace $N - 1$ by N and have for the free energy density

$$\frac{F}{N} = \frac{F_{IG}}{N} - \frac{\rho kT}{2} \int \left[\exp\left(-\frac{u(r)}{kT}\right) - 1 \right] d^3r \qquad (4.4.13)$$

4.4 Interacting Gases

The free energy density, as promised, does not depend on N. We can construct a system of any size by putting together many systems small enough to satisfy the assumptions that we have made—this is just the additivity property mentioned above—so we now assume that Eq. (4.4.13) is applicable to any size system, even to the thermodynamic limit. For the pressure we find

$$P = -\frac{\partial F}{\partial V} = \rho k T - \frac{kT}{2}\rho^2 \int \left[\exp\left(-\frac{u(r)}{kT}\right) - 1\right] d^3r \quad (4.4.14)$$

Comparing to Eq. (4.4.2), we see that the second virial coefficient is given by

$$B(T) = -\frac{1}{2}\int\left[\exp\left(-\frac{u(r)}{kT}\right) - 1\right] d^3r \quad (4.4.15)$$

or

$$B(T) = -\frac{1}{2}\int f_{12}\, d^3r_{12} \quad (4.4.16)$$

b. Behavior and Structure of a Dilute Gas

The net effect of the pair interaction leading to $B(T)$ can be either attractive (decreasing the pressure relative to the ideal gas value) or repulsive (increasing the pressure), depending on the temperature. We can get some feeling for what goes into $B(T)$ by considering the situation when $kT \gg \varepsilon_0$, where ε_0 is the well depth in Fig. 4.3.1 [this argument does not depend on the particular equation, (4.3.13), for the potential but does depend on the form shown in the figure]. Since ε_0/k is usually of the order of the critical temperature at which liquid-gas phase separation begins to occur, these arguments will be applicable well above that temperature.

For separation $r < \sigma$ (we continue to refer to Fig. 4.3.1), $u(r)$ rapidly becomes large and positive, so that we can approximate

$$f_{12} = -1 \quad r < \sigma$$

When $r > \sigma$, $u(r) < 0$, and its largest magnitude is ε_0, which is small compared to kT. Thus,

$$f_{12} = \left[\exp\left(-\frac{u(r)}{kT}\right) - 1\right] \approx -\frac{u(r)}{kT} \quad (r > \sigma)$$

We therefore find for $B(T)$,

$$B(T) = \frac{1}{2}\int_0^\sigma d^3r + \frac{1}{2kT}\int_0^\infty u(r)\, d^3r$$

$$= b - \frac{a}{kT} \quad (4.4.17)$$

where

$$b = \frac{2}{3}\pi\sigma^3 \quad (4.4.18)$$

and

$$a = -\frac{1}{2}\int_\sigma^\infty u(r)\, d^3r \quad (4.4.19)$$

a and b are both positive, since $u(r)$ is negative for r bigger than σ. Recalling that σ is twice the hard-core radius of the atom, we see that b is four times the atomic hard-core volume. We shall encounter the same a and b in Chap. 6, when we study the van der Waals equation of state, Eq. (6.3.1).

From Eq. (4.4.19) we see that $B(T)$ approaches a positive constant at high temperature. The thermal kinetic energies of the atoms are then so large that the weak attractive well in the potential has little effect, and the dominant interaction is hard-core repulsion. At lower T, however, the attractions between molecules exert their influence and the pressure is reduced, ultimately falling below the ideal gas value when $B(T)$ becomes negative. Anticipating Eq. (6.3.12), $B(T)$ given by Eq. (4.4.17) will change sign at a temperature about three or four times higher than the critical temperature. A rather important consequence of the interactions at high T will be discussed in the next subsection, 4.4c.

If we compare the dilute gas equation, (4.4.2), to the pressure equation of the last section, Eq. (4.3.36), we see that in the dilute gas $g(r)$ has a definite form needed to satisfy

$$\int \left[\exp\left(-\frac{u(r)}{kT}\right) - 1 \right] d^3r = \frac{1}{3kT} \int r \frac{du(r)}{dr} g(r) \, d^3r \qquad (4.4.20)$$

In order to obey Eq. (4.4.2), $g(r)$ must have a form depending only on the temperature, not the density, since we know that $B(T)$ is independent of density. The exact form can be found by integrating the left-hand side by parts:

$$\int \left[\exp\left(-\frac{u(r)}{kT}\right) - 1 \right] d^3r = 4\pi \int \left[\exp\left(-\frac{u(r)}{kT}\right) - 1 \right] r^2 \, dr$$

$$= 4\pi r^3 \left[\exp\left(-\frac{u(r)}{kT}\right) - 1 \right]\Big|_0^\infty$$

$$- 4\pi \int r \, d \left\{ r^2 \left[\exp\left(-\frac{u(r)}{kT}\right) - 1 \right] \right\}$$

The term to be evaluated at zero and infinity is zero at both limits. At the upper limit we can expand the exponential, and the term becomes

$$-4\pi r^3 u(r)$$

which is zero, since $u(r)$ always goes to zero faster than r^{-3}, as mentioned in the last section. At the lower limit the term is just

$$-4\pi r^3$$

4.4 Interacting Gases

because when $r \to 0$, $e^{-u(r)/kT} \to 0$ as well. We are left with

$$\int \left[\exp\left(-\frac{u(r)}{kT}\right) - 1 \right] d^3r$$

$$= -4\pi \int r^3 \, d\left[\exp\left(-\frac{u(r)}{kT}\right) \right]$$

$$\quad -4\pi \int \left[\exp\left(-\frac{u(r)}{kT}\right) - 1 \right] 2r^2 \, dr$$

$$= -4\pi \int r^3 \, d\left[\exp\left(-\frac{u(r)}{kT}\right) \right] - 2 \int \left[\exp\left(-\frac{u(r)}{kT}\right) - 1 \right] d^3r$$

where we have absorbed $4\pi r^2 \, dr$ back into d^3r in the last term. The integral in that term is now the one we are trying to find. Bringing it over to the left, we have

$$\int \left[\exp\left(-\frac{u(r)}{kT}\right) - 1 \right] d^3r = -\frac{4\pi}{3} \int r^3 \, d\left[\exp\left(-\frac{u(r)}{kT}\right) \right]$$

$$= -\frac{4\pi}{3} \int r^3 \left[-\frac{1}{kT} \frac{du(r)}{dr} \exp\left(-\frac{u(r)}{kT}\right) \right] dr$$

$$= \frac{1}{3kT} \int r \frac{du(r)}{dr} \exp\left(-\frac{u(r)}{kT}\right) d^3r$$

Comparing to Eq. (4.4.19), we see that

$$g(r) = \exp\left(-\frac{u(r)}{kT}\right) \qquad (4.4.21)$$

Equation (4.4.21) gives the radial distribution function of a dilute gas. It is easy to see that with $u(r)$ given by Fig. 4.3.1, $g(r)$ will have the form shown in Fig. 4.4.1. The qualitative form is obvious. The particle at the

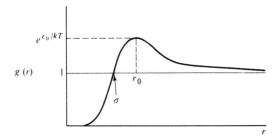

Fig. 4.4.1

origin excludes others out to σ by hard-core repulsion and causes some accumulation in its attractive well.

We might have guessed at the equation for $g(r)$ by realizing that, under the dilute gas approximation, the potential of mean force, $\varphi(r)$ in Eq. (4.3.5) would simply be the pair potential $u(r)$. This fact follows because there are only binary interactions in the dilute gas, the other atoms having no effect on the forces between any two. Thus,

$$\varphi(r) = u(r) \qquad (4.4.22)$$

and Eq. (4.4.21) then follows from (4.3.5).

In any case, Eq. (4.4.21) is the leading-order correction to the ideal gas value, $g(r) = 1$. Comparing Fig. 4.4.1 to Fig. 4.2.4, we see that the essential features of $g(r)$ in a liquid out to the first big bump are already present when binary collisions alone are considered. The wiggles farther out will, of course, require more complicated interactions. In Sec. 4.5 we will seek closed-form expressions for the corrections to Eq. (4.4.21) to all orders in the density, by trying to sum the appropriate cluster expansion. We will write

$$\log g(r) + \frac{u(r)}{kT} = W(r) \qquad (4.4.23)$$

In the ideal gas $W(r) = u(r)/kT = 0$ and $g(r) = 1$. For the dilute gas $W(r) = 0$ and we have Eq. (4.4.21). The problem will be to find $W(r)$ for all densities.

c. The Liquefaction of Gases and the Joule-Thomson Process

One day in 1823 a certain Dr. Paris was watching a young chemist named Michael Faraday perform an experiment that involved generating chlorine from a heated solution inside a sealed glass tube. Paris noticed, and pointed out, that an oily substance seemed to be accumulating in the tube. Faraday filed open the tube to investigate, whereupon it promptly exploded. Nothing daunted, he was able to send Paris a note the next day saying that the oily substance was liquid chlorine.

When Faraday published an announcement of the liquefaction of chlorine, he must have received criticism for falsely claiming to be the first to liquefy a gas, because a little later there appeared an article by Faraday analyzing all previously reported experiments he could find in the literature in which liquefaction might have occurred, whether or not the original authors were aware of the possibility. It is now generally agreed that the first man knowingly to liquefy a substance that is a gas under common conditions of temperature and pressure was van Marum, a Dutch chemist who partially liquefied ammonia while trying to test Boyle's law, late in the eighteenth century. Faraday, however, grasped the wider significance of what

4.4 Interacting Gases

had happened in his tube and immediately undertook further experiments, designed to liquefy other gases, by generating them at high pressure in sealed tubes.

It was perilous work. "I met with another explosion on Saturday evening," he wrote to a friend in that same year of 1823, "which again has laid up my eyes. It was from one of my tubes, and was so powerful as to drive the pieces of glass like pistol-shot through a window. However, I am getting better and expect to see as well as ever in a few days."† In the experiments of 1823, which, happily, Faraday and his eyes both survived, he established the principle that under proper conditions of temperature and pressure, the same matter could exist in all phases—solid, liquid, and gas.

At about the same time that Faraday was liquefying chlorine, Cagniard de la Tour was busy vaporizing ether. He found that at about 160°C and 37 atmospheres pressure, liquid ether turned to gas at the same density, the interface between the two states vanishing. This phenomenon, first called the critical point in 1869 by Thomas Andrews, was referred to somewhat cumbersomely by Faraday as the Cagniard de la Tour state. Nevertheless, he saw the significance of that, too. It meant that some of the gases he wanted to liquefy, particularly nitrogen and hydrogen, which he suspected might be metals, could not be liquefied at any pressure, no matter how high, at ordinary temperatures. He would need a combination of low temperature and high pressure in order to succeed.

The necessary low-temperature bath was provided by Thilorier, who in 1835 devised an apparatus for producing large quantities of liquid carbon dioxide (which liquefies at room temperature at about 30 atm). When the liquid was ejected into air, it produced a snow at $-78°C$, which temperature could be reduced further by mixing with ether, and further still (to about $-110°C$, or about $160°K$) by pumping away its vapor. In 1844 Faraday constructed an apparatus in which he could compress various gases to about 40 atm and cool them with "Thilorier's beautiful bath." He reported the results in 1845. He had succeeded in liquefying all the known permanent gases except, he sadly noted, those he really wanted. Three compounds— carbon monoxide, nitric oxide, and methane—and three elements—oxygen, nitrogen, and hydrogen—had failed to succumb to his techniques.

We shall pick up the thread of our story again in Sec. 6.3. Nitrogen and oxygen eventually fell to a cascade process using a series of refrigerants of successively lower temperature, each used to liquefy the next. However, there were no intervening liquids between nitrogen (which freezes at about $55°K$) and hydrogen (whose critical point is about $33°K$). The liquefaction of hydrogen, and even the last stage of cooling of nitrogen and oxygen, was

† Letter to T. Huxtable, March 25, 1823. L. Pearce Williams, *The Selected Correspondence of Michael Faraday* (London: Cambridge University Press, 1971), Vol. 1, p. 141.

actually accomplished by what at first sight appears to be a quite different technique.

Starting in 1845 and culminating in 1862, Joule, and later Joule and Thomson (the same Thomson who later became Lord Kelvin), carried out a series of experiments that showed that under certain circumstances the expansion of a gas into a chamber held at constant pressure would produce cooling. Seen properly, the physical reason turns out to be the same one that causes a liquid to cool when it evaporates, but Joule and Thomson were thinking along quite different lines; they were testing the interchangeability of work and heat. In any case, it is the cooling effect of the Joule-Thomson expansion that provided (and still provides) the last step in the liquefaction of cryogenic fluids. Let us analyze the Joule-Thomson process.

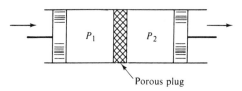

Fig. 4.4.2

For purposes of formal analysis, we can imagine a gas expanding through a porous plug from a chamber at pressure P_1 into a chamber held at lower pressure, P_2, as sketched in Fig. 4.4.2. As the gas flows from left to right, the pistons must move, changing the volumes to keep the pressures constant. The role of the plug is merely to support the pressure difference. Suppose that at the beginning of the experiment all the gas is in chamber 1, with volume V_1 at pressure P_1, and at the end it is all in chamber 2, with volume V_2 at pressure P_2. The piston on the left does work, $-\int P\,dV = P_1 V_1$ on the gas, but the gas does work $P_2 V_2$ on the piston on the right. If the whole system is isolated so that no heat leaks in or out, the net change in energy in the process is

$$E_2 - E_1 = P_1 V_1 - P_2 V_2 \qquad (4.4.24)$$

where E_2 and E_1 are, respectively, the energies of the gas when it is all in chamber 2 and all in chamber 1. We thus have

$$E_2 + P_2 V_2 = E_1 + P_1 V_1 \qquad (4.4.25)$$

or
$$W_2 = W_1 \qquad (4.4.26)$$

where W is the enthalpy; the process is isenthalpic.

We wish to know whether the temperature of the gas changes in this process—that is, under an isenthalpic change in pressure. The quantity of

4.4 Interacting Gases

interest, therefore, is $(\partial T/\partial P)_W$, called the Joule-Thomson coefficient. From Eq. (1.2.17),

$$dW = T\,dS + V\,dP \qquad (4.4.27)$$

so that

$$\left(\frac{\partial P}{\partial T}\right)_W = -\frac{T}{V}\left(\frac{\partial S}{\partial T}\right)_W \qquad (4.4.28)$$

Writing the entropy as a function of P and T, we have

$$dS = \left(\frac{\partial S}{\partial P}\right)_T dP + \left(\frac{\partial S}{\partial T}\right)_P dT \qquad (4.4.29)$$

or

$$\left(\frac{\partial S}{\partial T}\right)_W = \left(\frac{\partial S}{\partial P}\right)_T \left(\frac{\partial P}{\partial T}\right)_W + \left(\frac{\partial S}{\partial T}\right)_P \qquad (4.4.30)$$

Substitute Eq. (4.4.30) into (4.4.28) and solve for $(\partial P/\partial T)_W$, giving

$$\left(\frac{\partial P}{\partial T}\right)_W = -\frac{C_P/V}{1 + (T/V)(\partial S/\partial P)_T} \qquad (4.4.31)$$

where we have used $C_P = T(\partial S/\partial T)_P$. We now use the Maxwell relation, Eq. (1.2.24),

$$-\left(\frac{\partial S}{\partial P}\right)_T = \left(\frac{\partial V}{\partial T}\right)_P \qquad (4.4.32)$$

in Eq. (4.4.31) and invert the result to get

$$\left(\frac{\partial T}{\partial P}\right)_W = \frac{1}{C_P}\left[T\left(\frac{\partial V}{\partial T}\right)_P - V\right] \qquad (4.4.33)$$

This is what we were looking for.

Suppose that the gas is ideal. Then

$$T\left(\frac{\partial V}{\partial T}\right)_P = T\frac{\partial}{\partial T}\left(\frac{NkT}{P}\right) = \frac{NkT}{P} = V \qquad (4.4.34)$$

which means that $(\partial T/\partial P)_W = 0$; there is no change of temperature in the process. Any change that does occur must be a consequence of the interactions that cause the gas to depart from ideality. For the dilute gas of the previous subsection, the equation of state, Eq. (4.4.2), may be written

$$V = \frac{NkT}{P}[1 + \rho B(T)] \qquad (4.4.35)$$

In the correction term, $\rho B(T)$, we substitute the ideal gas equation, $\rho = P/kT$. Then

$$V = \frac{NkT}{P} + NB \qquad (4.4.36)$$

For this case, we find from Eq. (4.4.33) that

$$\left(\frac{\partial T}{\partial P}\right)_W = \frac{N}{C_P}\left[T\frac{dB}{dT} - B\right] \qquad (4.4.37)$$

Recall that at high temperature, $dB/dT \to 0$, and B is positive. The coefficient is therefore negative, meaning that a reduction in pressure leads to an increase in temperature; the gas warms in the process. However, at lower T, dB/dT is positive, and B itself eventually becomes negative, so the same process must produce refrigeration. If the approximate Eq. (4.4.17) is valid, then

$$\left(\frac{\partial T}{\partial P}\right)_W = \frac{N}{C_P}\left[2\frac{a}{kT} - b\right] \qquad (4.4.38)$$

and the temperature at which cooling begins, called the Joule-Thomson inversion point, is given by

$$kT_I = 2\frac{a}{b} \qquad (4.4.39)$$

obtained by setting $(\partial T/\partial P)_W = 0$. This is twice the temperature at which $B(T)$ changes sign in the same approximation or, once again anticipating Eq. (6.3.12), about seven times the critical temperature.

The physical reason that the process works is not hard to see. We have already argued that, at low enough temperature, the net effect of the interactions between gas atoms is attractive. When the gas expands, the atoms are slowed down in pulling out of each other's attractive potentials. Slower atoms, less kinetic energy, means a cooler gas. This process is not different, in principle, from what happens when a liquid evaporates.

When a substance is below its critical temperature, it can be caused to liquefy isothermally under sheer compression. At high enough pressure it forms a high-density liquid phase, separated by an interface from a low-density gas phase. Once self-condensed, the stuff is cooled by allowing the high-density phase to expand back to low density—that is, to evaporate. Above the critical temperature, nature will no longer provide phase separation for us, but if we force the issue, separating high-density and low-density states by means of the porous plug, we still get refrigeration when the gas "evaporates" through our artificial interface. As we have just seen, refrigeration can be obtained at temperatures as high as seven times the critical temperature of the working substance.

At temperatures that are not much below the inversion point, expansion provides little cooling, since the magnitude of the Joule-Thomson coefficient is small. However, the cooled, expanded gas can be used to precool more compressed gas by counterflow heat exchange, so that the next expansion starts at lower temperature and so on. These techniques combined were

4.4 Interacting Gases

eventually used in the conquest of hydrogen, and, later still, helium, the history of which we shall return to in Sec. 6.3.

d. Higher Densities

For densities at which the dilute gas approximation, Eq. (4.4.3), is no longer valid, we must consider higher-order terms in the virial equation of state, Eq. (4.4.1). These terms will depend, as we mentioned earlier, on collisions of successively larger numbers of particles or, in the language of Eq. (4.4.8), on successively higher-order clusters. In this section we shall derive a general formula for the virial coefficients in all orders in terms of cluster integrals. There are a number of ways to do this job. We shall do it here by means of an argument chosen because it introduces both the language and the methods we use in the next section in deriving a rather similar formula for an expansion involving $g(r)$ in powers of the density.

In the course of our arguments we shall want to make use of a quantity that relates Q_N, the configurational integral, to Q_{N-1}, the value it would have at the same V and T, but with one fewer particle in the system. Let us define the dimensionless ratio

$$\gamma = \frac{VQ_{N-1}}{Q_N} \qquad (4.4.40)$$

sometimes called the concentration activity coefficient. For fixed N, γ depends on N and we may write it γ_N, but in the limit $N \to \infty$, γ becomes independent of N. We shall also find it convenient to use

$$\xi = \rho\gamma \qquad (4.4.41)$$

which has the dimensions of reciprocal volume.

To see how these variables are related to the more familiar thermodynamic variables, consider Eq. (1.2.30) for the chemical potential, particularly,

$$\mu = \left(\frac{\partial F}{\partial N}\right)_{T,V} \qquad (4.4.42)$$

N is not really a continuous variable, since it must always be an integer, but for macroscopic systems Eq. (4.4.42) is quite the same as

$$\mu = F(T, V, N) - F(T, V, N - 1) \qquad (4.4.43)$$

or in terms of the partition functions, Z_N and Z_{N-1},

$$\mu = -kT \log \frac{Z_N}{Z_{N-1}}$$

$$= kT \log N\Lambda^3 \frac{Q_{N-1}}{Q_N} \qquad (4.4.44)$$

where we have used Eqs. (4.2.24) and (4.2.25). Substituting in Eqs. (4.4.40) and (4.4.41), we have

$$\mu = kT \log \rho \Lambda^3 \gamma \qquad (4.4.45)$$

and
$$\xi = \frac{1}{\Lambda^3} e^{\mu/kT} \qquad (4.4.46)$$

γ and ξ, like μ, may be thought of as functions of ρ and T. Using Eq. (1.2.35) for μ,

$$d\mu = -\frac{S}{N} dT + \frac{V}{N} dP \qquad (4.4.47)$$

we have that

$$\left(\frac{\partial \mu}{\partial \rho}\right)_T = \frac{1}{\rho} \left(\frac{\partial P}{\partial \rho}\right)_T \qquad (4.4.48)$$

If μ is known as a function of ρ at fixed T, Eq. (4.4.48) may be integrated to give an equation of state:

$$P = \int_0^\rho \rho' \frac{\partial \mu}{\partial \rho'} d\rho' \qquad (4.4.49)$$

where ρ' is a dummy variable and we have used the fact that $P \to 0$ if $\rho \to 0$. Remembering that T is fixed in the integration, we substitute Eq. (4.4.45) into this result to get

$$P = \rho kT + kT \int_0^\rho \rho' \frac{\partial \log \gamma}{\partial \rho'} d\rho' \qquad (4.4.50)$$

In the arguments to follow, we shall find an expansion of $\log \gamma$ in powers of the density

$$\log \gamma = -\sum_{m=1}^\infty \beta_m \rho^m \qquad (4.4.51)$$

This result will then reduce to the virial expansion by way of Eq. (4.4.50).

As we shall see, the arguments to be used here, basically ways of rearranging and collecting all the terms in Eq. (4.4.8), become rather involved. It could be worse, however. We will consistently make use of an assumption that greatly simplifies the mathematics. We assume that series such as Eq. (4.4.1) or (4.4.51) converge after a finite number of terms, even if $N \to \infty$. This is really a limit on the density, although somewhat ill-defined. Operationally, it has the advantage of allowing us to ignore clusters in which the number of particles approaches N itself. That will prove helpful. It will not be necessary to specify how many terms we may use in power series such as (4.4.1) and (4.4.51), only that the number be finite.

Let us get down to work. We need a way of categorizing the terms that

4.4 Interacting Gases

appear in Eq. (4.4.7), integrals over the terms in (4.4.8), in order to keep track of them while we manipulate and rearrange them. Imagine the following scheme: on each sheet of a stack of paper, we draw N dots, one for each atom in the system. We now use these sheets to form diagrams, by drawing lines that connect pairs of dots to each other. These lines represent the f_{ij}, where i and j are the atoms connected by the lines. Each sheet represents one term in the expression for Q_N formed from Eqs. (4.4.7) and (4.4.8):

$$Q_N = \int \cdots \int \left[1 + \sum_{i<j} f_{ij} + \sum_{\substack{i<j, k<\ell \\ i,j \neq k,\ell}} f_{ij} f_{k\ell} + \cdots \right] d\{N\} \quad (4.4.52)$$

For example, on one sheet all the dots will be unconnected except for the cluster shown in Fig. 4.4.3.

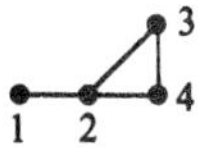

Fig. 4.4.3

That sheet is a diagram representing the integral

$$\int \cdots \int f_{12} f_{24} f_{23} f_{34} \, d\{N\} = V^{N-4} \int \cdots \int f_{12} f_{24} f_{23} f_{34} \, d\{4\}$$

All unconnected points integrate out as factors V. If a sheet has separate unconnected clusters, they appear as products of integrals; thus (Fig. 4.4.4)

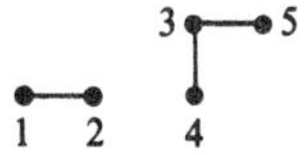

Fig. 4.4.4

with all other points unconnected is the term

$$V^{N-5} \int \cdots \int f_{12} f_{34} f_{35} \, d\{5\} = V^{N-3} \int f_{12} \, d^3 r_{12} \iint f_{34} f_{35} \, d^3 r_{34} \, d^3 r_{35}$$

Notice that one volume integral, for the origin, comes out of each separate cluster once we change the variables of integration—for example, to $\mathbf{r}_{12} = \mathbf{r}_2 - \mathbf{r}_1$ and so on. In order to represent all the terms in Q_N, we must draw all possible N point diagrams, representing $3N$-dimensional integrals, then add them up. Since each pair in each diagram can be either connected or not connected, and there are $N(N - 1)/2$ pairs, we shall need $2^{N(N-1)/2}$ sheets of paper, each large enough for N dots (this job becomes impossible, to be sure, if we let $N \to \infty$, but it should generally be adequate, say, to take $N \approx 10^{23}$). We will continue our instructions, assuming that the reader has prepared the necessary diagrams and has them at hand.

In view of the worldwide paper shortage we can reduce the number of sheets necessary if, on each sheet, we choose a certain particle to be number 1 but do not yet number the other points. Then, for example, the diagram represented in Fig. 4.4.3 with all other points unconnected actually represents $(N - 1)!/(N - 4)!$ diagrams, since that is the number of ways of choosing the three other points in the figure out of the remaining $N - 1$ particles. Let us proceed along this line and try to organize all the terms that appear in Q_N.

Pick a particular particle, to which we assign the number 1, and a diagram in which 1 joins exactly ℓ other particles in a cluster of $\ell + 1$ particles. This diagram contributes to Q_N the term

$$\int \cdots \int d\{\ell + 1\} \prod^{\ell+1} f_{ij} \int \cdots \int d\{N - \ell - 1\} \prod^{N-\ell-1} f_{ij} \quad (4.4.53)$$

where the first product is over all of the f_{ij}'s in the cluster containing 1, and the second is over all the remaining f_{ij}'s in the N-particle diagram. Now gather together all diagrams (i.e., all sheets of paper) having the same cluster of $\ell + 1$, regardless of what the other particles are doing, and add them up. Every term has the same first factor, and in the total this factor multiplies a sum over all possible diagrams involving the $N - \ell - 1$ remaining points. That sum is just $Q_{N-\ell-1}$, the configurational integral for the remaining atoms if they occupied the same volume without the $\ell + 1$ particles in the original cluster. Thus, the set of diagrams we have so far chosen contribute to Q_N the term

$$\int \cdots \int d\{\ell + 1\} \prod^{\ell+1} f_{ij} Q_{N-\ell-1} \quad (4.4.54)$$

The same contribution will be made regardless of which ℓ particles join 1 in the cluster, and there are

$$\frac{(N - 1)!}{\ell!(N - 1 - \ell)!} \quad (4.4.55)$$

ways of choosing the ℓ particles out of $N - 1$. It should be remembered throughout these arguments that the N particles are to be regarded as distinguishable, each tagged with a number. When Q_N is used to form, say, Z_N, we will divide by $N!$ to make them indistinguishable again. We now wish to gather together all clusters of $\ell + 1$ particles, each arising $(N - 1)!/$

4.4 Interacting Gases

$\ell!\,(N - 1 - \ell)!$ times. For example, all clusters of particle 1 plus two others are as shown in Fig. 4.4.5,

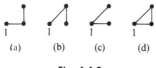

Fig. 4.4.5

each occurring $(N - 1)(N - 2)$ times. Diagrams (a) and (b) here differ only in the order of the two other particles, so that we do not wish to count under (a) the sketch in Fig. 4.4.6

Fig. 4.4.6

and then under (b), the one in Fig. 4.4.7,

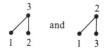

Fig. 4.4.7

since, for example, the first and last of these four represent the same integral, an integral that should appear only once in Q_N. It is to take care of this multiple counting that the $\ell!$ appears in the denominator of Eq. (4.4.55). In order to have a symbol for all the clusters of $\ell + 1$ particles, we define

$$b_{\ell+1} = \frac{1}{(\ell + 1)!} \sum^{(\ell)''} \int \cdots \int d\{\ell + 1\} \prod^{\ell+1} f_{ij} \qquad (4.4.56)$$

where $\sum^{(\ell)''}$ simply means sum over all clusters of ℓ particles plus particle 1. For example, $\sum^{(2)''}$ would mean sum over the four diagrams shown in Fig. 4.4.5. We now have, from all the $\ell + 1$ particle clusters, the combined contribution

$$\frac{(N - 1)!}{(N - 1 - \ell)!}(\ell + 1)b_{\ell+1}Q_{N-\ell-1} \qquad (4.4.57)$$

Finally, all the $2^{N(N-1)/2}$ diagrams are included if we just sum this expression over ℓ:

$$Q_N = \sum_{\ell} \frac{(N-1)!}{(N-1-\ell)!} (\ell + 1) b_{\ell+1} Q_{N-\ell-1} \qquad (4.4.58)$$

We have now succeeded in regrouping the terms in Eq. (4.4.52) and collecting them back together again. A noteworthy achievement perhaps, but the result, Eq. (4.4.58), is a little cumbersome. Things can be improved, however, by writing for a system of fixed N [see Eq. (4.4.41)]

$$\xi_N = \frac{N}{V} \gamma_N = N \frac{Q_{N-1}}{Q_N} \qquad (4.4.59)$$

where we have used Eq. (4.4.40). It follows that

$$\xi_{N-n} = (N - n) \frac{Q_{N-n-1}}{Q_{N-n}} \qquad (4.4.60)$$

We can thus form the product

$$\prod_{n=1}^{\ell} \xi_{N-n} = (N - 1) \frac{Q_{N-2}}{Q_{N-1}} (N - 2) \frac{Q_{N-3}}{Q_{N-2}} \cdots (N - \ell) \frac{Q_{N-\ell-1}}{Q_{N-\ell}}$$

$$= \frac{(N-1)!}{(N-\ell-1)!} \frac{Q_{N-\ell-1}}{Q_{N-1}} \qquad (4.4.61)$$

But recall that ξ_N becomes independent of N in a large system, so we can write

$$\prod_{n=1}^{\ell} \xi_{N-n} = \xi^{\ell} \qquad (4.4.62)$$

Dividing both sides of Eq. (4.4.58) by Q_{N-1} and using (4.4.40), (4.4.61), and (4.4.62), we have

$$\frac{V}{\gamma} = \sum_{\ell=0}^{\infty} (\ell + 1) b_{\ell+1} \xi^{\ell} \qquad (4.4.63)$$

Multiplying both sides by ξ, we find

$$\rho = \frac{1}{V} \sum_{\ell=0}^{\infty} (\ell + 1) b_{\ell+1} \xi^{\ell+1} = \frac{1}{V} \sum_{j=1}^{\infty} j b_j \xi^j \qquad (4.4.64)$$

This result is actually an expansion for the density in powers of ξ, which is really almost the opposite of what we want, an expansion for γ in powers of ρ.

In order to obtain the series we want, it will be necessary to start all over

4.4 Interacting Gases

again, regrouping the terms in Q_N in a slightly different way (the exercise we just went through will prove necessary, however). Before we start, it will be useful to understand one additional property of certain types of clusters. Consider the cluster shown in Fig. 4.4.8,

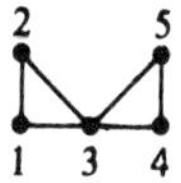

Fig. 4.4.8

which represents the factor

$$\int \cdots \int f_{12} f_{23} f_{13} f_{34} f_{45} f_{35} \, d\{5\}$$

We can hold r_3 fixed until the end, carrying out the other integrations by integrating over r_{13}, r_{33}, and so on. The term becomes

$$\int \cdots \int f_{12} f_{23} f_{13} f_{34} f_{45} f_{35} \, d^3 r_{13} \, d^3 r_{23} \, d^3 r_{34} \, d^3 r_{35} \, d^3 r_3$$

$$= V \iint f_{12} f_{23} f_{13} \, d^3 r_{13} \, d^3 r_{23} \iint f_{34} f_{45} f_{35} \, d^3 r_{34} \, d^3 r_{35}$$

It separates into independent factors. The general form here is of two clusters connected to each other at a single point. It is incidentally true that in the example given here the two factors are numerically equal, so that the term may be written

$$V \left(\iint f_{12} f_{23} f_{13} \, d\{2\} \right)^2$$

However, the more general point we are trying to illustrate is that whenever two (or more) clusters are joined at a single point, the resulting integral factors. Let us call the common point (point 3 in our example) a *node*. An alternative description of a node is that if every possible path from one point in a cluster to some other point must pass through the same intermediate point, the intermediate point is a node. Still another way of saying it is that if a cluster can be cut into separate clusters by snipping it at a single point, that point is a node. Whenever a node occurs, the subclusters joined at that point factor into independent integrals.

Let us introduce a new symbol for our diagrams. The symbol (Fig. 4.4.9)

Fig. 4.4.9

will represent any cluster with no nodes in it. Thus, two nodeless subclusters joined at a node become the symbol shown in Fig. 4.4.10.

Fig. 4.4.10

More generally, we can have the one given in Fig. 4.4.11

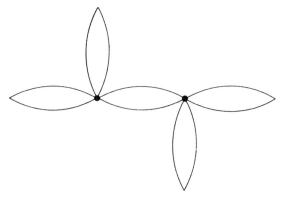

Fig. 4.4.11

and so on. A nodeless cluster or subcluster is said to be irreducible, since it cannot be reduced to simpler factors.

In the arguments leading to Eq. (4.4.63), we chose one central, core cluster, added up all the diagrams with that same core, and, finally, summed over possible core diagrams. The core cluster that we used there, which we called $b_{\ell+1}$, were of the most general kind, $\ell + 1$ connected particles, but with an undetermined number of nodes. We will now proceed by similar arguments except that, to start with, we shall choose as our core a subcluster that has no nodes, except possibly at particle 1. Such diagrams are still

4.4 Interacting Gases

partially reducible, and at the very end we shall go through one final step of reduction, eliminating the nodes at particle 1.

To begin, then, we choose some subcluster of particle 1 and k other particles connected to it, in which none of the other k particles is a node. We now build this subcluster up into the more general kind of cluster we dealt with before, by the following procedure. At particle 2, we attach any cluster of ℓ_2 further particles, making 2 a node unless ℓ_2 happens to be zero. We continue on, attaching a cluster of ℓ_3 points at particle 3 and so on, up to particle k. The new appended subclusters may have internal nodes—they do not need to be irreducible. When we finish, we wind up with a connected cluster of $L + 1$ points, where

$$L = k + \sum_{n=2}^{k+1} \ell_n \qquad (4.4.65)$$

All possible $L + 1$ point clusters can be built up this way.

If we now consider together all possible diagrams containing $L + 1$ particle clusters constructed in this way, we find that between them they contribute to Q_N a term with the following factors:

$$Q_{N-L-1} \qquad (a)$$

which arises from all possible combinations of the $N - L - 1$ particles not connected to our $L + 1$ clusters;

$$\frac{(N-1)!}{L!(N-L-1)!} \qquad (b)$$

which is the number of ways of choosing L-ordered particles out of $N - 1$ particles; and

$$L! \qquad (c)$$

which is the number of ways of permuting the L particles in the clusters. The scheme is to pick any L particles, then choose k out of these particles to form the core cluster, and, finally, ℓ_2, ℓ_3, etc. out of the remainder. k and all values of ℓ_n are fixed numbers so far, but we will consider all allowed diagrams of k points, dividing by $k!$ for the core cluster factor, and similarly for the appendages. For the core clusters we have

$$\frac{1}{k!} \sum^{(k)'} \int \cdots \int d\{k+1\} \prod f_{ij} \equiv V\beta'_k \qquad (d)$$

where $\sum^{(k)'}$ means the sum over all clusters of $k + 1$ that have no nodes except at 1; and

$$\sum_{n=2}^{k+1} \frac{1}{V\ell_n!} \sum^{(\ell_n)''} \int \cdots \int d\{\ell_n + 1\} \prod f_{ij}$$

where $\sum^{(\ell_n)''}$ once again [as in Eq. (4.4.56)] means sum over all connected clusters of $\ell_n + 1$ particles. The last term is the product of factors due to all of the appended subclusters, each with the appropriate combinatorial factor for internal ordering of the particles. These subclusters are of the same class as the ones defined in Eq. (4.4.56), and, in fact, we can rewrite this last term as

$$\frac{1}{V^k} \prod_{n=2}^{k+1} (\ell_n + 1) b_{\ell_n+1} \tag{e}$$

The factor V^{-k} comes out because there was a term V^{-1} in each of the k factors in the term originally written. That factor V^{-1} had been inserted there to change from an integral over $\{\ell_n\}$, the ℓ_n particles in the appended subcluster, to $\{\ell_n + 1\}$, in order to obtain the form used in Eq. (4.4.56) to define $b_{\ell+1}$.

The combined factors (a), (b), and (c) give us

$$Q_{N-L-1} \frac{(N-1)!}{(N-L-1)!} = Q_{N-1} \zeta^L \tag{4.4.66}$$

where we have used Eqs. (4.4.61) and (4.4.62). Thus the factors (a) to (e) together give us

$$V Q_{N-1} \zeta^L \frac{\beta'_k}{V^k} \prod_{n=2}^{k+1} (\ell_n + 1) b_{\ell_n+1} \tag{4.4.67}$$

We now redistribute the L factors ζ, keeping k of them, ζ^k, outside the product and the remaining $L - k$ inside the product by assigning ζ^{ℓ_n} to each term; we get

$$V Q_{N-1} \zeta^k \frac{\beta'_k}{V^k} \prod_{n=2}^{k+1} (\ell_n + 1) b_{\ell_n+1} \zeta^{\ell_n} \tag{4.4.68}$$

In order to get Q_N, we must now add up all contributions from clusters of all possible L (up to now we have considered all possible diagrams with the same L). To do so, we sum separately over k, as well as over all the ℓ_n inside the product, which sums over L by way of Eq. (4.4.65). The result is

$$Q_N = V Q_{N-1} \sum_k \left(\frac{\zeta}{V}\right)^k \beta'_k \prod_{n=2}^{k+1} \sum_{\ell_n} (\ell_n + 1) b_{\ell_n+1} \zeta^{\ell_n} \tag{4.4.69}$$

Strictly speaking, the sums over the ℓ_n are not independent of each other, but our approximation of limited density allows us to get away with pretending that they are. Each of the sums is of the form of Eq. (4.4.63), and hence of (4.4.64), which gives the density in powers of ζ. Since we expect expansions in powers of the density to converge in a finite number of terms when $N \to \infty$, we can invert that assumption and assume that expansions for the density will also converge. For this reason, the sums will be independent, and

4.4 Interacting Gases

we can take the upper limit of each to be infinity, and do the same in the sum over k, ignoring the restriction implied by $l_1 + 1 \leq N$. If we do so, then

$$\prod_{n=2}^{k+1} \sum_{\ell_n=0}^{\infty} (\ell_n + 1) b_{\ell_n+1} \zeta^{\ell_n}$$

$$= \left[\sum_{\ell=0}^{\infty} (\ell + 1) b_{\ell+1} \zeta^{\ell} \right]^k$$

$$= \left(\frac{V}{\gamma} \right)^k \tag{4.4.70}$$

where we have used Eq. (4.4.63). Finally, substituting Eq. (4.4.70) into (4.4.69), and using Eqs. (4.4.40) and (4.4.41), we have

$$\frac{1}{\gamma} = \sum_{k=0}^{\infty} \rho^k \beta_k' \tag{4.4.71}$$

We still have one additional stage of reduction to accomplish before we emerge from this combinatorial nightmare and arrive at the virial equation of state. The diagrams β_k' are factorable because particle 1 may still be a node. The general class of such clusters may be represented by a diagram that looks rather like a flower with particle 1 at the center (Fig. 4.4.12). The petals of the flower each represent any diagram that is fully irreducible—that is, in which all points are connected and there are no nodes. In analogy to b_ℓ and β_ℓ', let us define a symbol for these irreducible clusters:

$$\beta_m = \frac{1}{m!} \sum^{(m)} \int \cdots \int d\{m\} \prod f_{ij} \tag{4.4.72}$$

Fig. 4.4.12

where $\sum^{(m)}$ means sum over irreducible clusters of particle 1 plus m points. Consider a diagram, of the type Fig. 4.4.12, which happens to have exactly p petals and a total of $k + 1$ points,

$$k = \sum_{n=1}^{p} m_n \qquad (4.4.73)$$

where m_n is the number of points in the nth petal. If each petal can be any allowed diagram with m_n points, it will contribute to some term in Eq. (4.4.71) a factor

$$\beta_{m_n}\rho^{m_n} \qquad (4.4.74)$$

The p petals together give the factor

$$\prod_{n=1}^{p} \beta_{m_n}\rho^{m_n} \qquad (4.4.75)$$

If we now take together all diagrams with exactly p petals, but any total number of points, each factor in the product Eq. (4.4.75) becomes the same when we add them up, giving

$$\left(\sum_{m_n} \beta_{m_n}\rho^{m_n}\right)^p \qquad (4.4.76)$$

but we have counted this term $p!$ times too many, since the particles are ordered in the β'_k, but now we are counting separately all possible orderings of the petals. Thus, the contribution to Eq. (4.4.71) from all clusters with p petals is

$$\frac{1}{p!}\left(\sum_{m} \beta_{m}\rho^{m}\right)^p \qquad (4.4.77)$$

We now include all the terms in Eq. (4.4.71) by summing over p, including $p = 0$, to take into account the term $k = 0$—that is, the term in which particle 1 is isolated. We have then

$$\frac{1}{\gamma} = \sum_{p=0}^{\infty} \frac{1}{p!}\left(\sum_{m} \beta_{m}\rho^{m}\right)^p \qquad (4.4.78)$$

Now, recalling that

$$\sum_{p=0}^{\infty} \frac{x^p}{p!} = e^x \qquad (4.4.79)$$

we find at last

$$-\log \gamma = \sum_{m=1}^{\infty} \beta_{m}\rho^{m} \qquad (4.4.80)$$

which is just Eq. (4.4.51).

4.5 Liquids

The last few steps to obtain the virial equation of state should now be performed with stately ceremony. We use Eq. (4.4.80) to evaluate the last term in Eq. (4.4.50).

$$\int_0^\rho \rho' \frac{\partial \log \gamma}{\partial \rho'} d\rho' = -\sum_{m=1}^\infty m\beta_m \int_0^\rho \rho'^m d\rho'$$

$$= -\sum_{m=1}^\infty \frac{m}{m+1} \beta_m \rho^{m+1} \qquad (4.4.81)$$

which result we duly substitute into Eq. (4.4.50) to get

$$P = \rho kT \left(1 - \sum_{m=1}^\infty \frac{m}{m+1} \beta_m \rho^m \right) \qquad (4.4.82)$$

When we compare Eq. (4.4.82) to Eq. (4.4.1), we find that we now have an explicit expression for each of the virial coefficients, $B(T)$, $C(T)$, etc., in terms of integrals over all irreducible clusters of the corresponding number of particles, the β_m, defined in Eq. (4.4.72). You can see how this prescription works out in Prob. 4.6.

4.5 LIQUIDS

The last section was devoted to treating systems of interacting particles in a way that works best at low densities but becomes increasingly difficult to apply as the density increases. In this section we look into some ways that are used to study fluids, in which the approximations are not specifically low-density ones and hence the predictions are not necessarily restricted to that limit. It is perhaps too much to call the results a theory of liquids, but they are, at least, attempts to study dense gases, and we know that there is no difference, in principle, between a dense gas and a liquid.

Our approach will be to try to find closed-form expressions for $g(r)$ that fix up the dilute gas approximation, sketched in Fig. 4.4.1, making it look more like Fig. 4.2.4. Basically we want to include higher-order correlations between the particles in order to put more wiggles in $g(r)$ after the first big bump (of course, the first bump will be affected quantitatively as well). The three equations we shall arrive at are all of the form

$$g(r) + \frac{u(r)}{kT} = W(r) \qquad (4.5.1)$$

already anticipated in Eq. (4.4.23). The closed-form expressions for $W(r)$ will all involve $g(r)$ itself in rather complicated ways, but at least they do include contributions from all orders in the density. The first of these, the Yvon-Born-Green equation, Sec. 4.5a, has the virtue that the approximation used to arrive at it is one whose physical significance is clear. It turns out to be the least accurate of the three, however, in predicting the properties of reasonably dense fluids. In Sec. 4.5b we will find the Hypernetted Chain equation and the Percus-Yevick equation, by summing certain classes of cluster integrals. The approximations are to throw away those clusters that cannot be summed. In all three cases, the results give a way of predicting $g(r)$ given $u(r)$ and vice versa.

a. The Yvon-Born-Green Equation

In Sec. 4.3 we saw that the mean force between particles 1 and 2 was given by Eq. (4.3.4),

$$-\frac{\partial \varphi_N(r_{12})}{\partial \mathbf{r}_1} = -\frac{\int \cdots \int (\partial U\{N\}/\partial \mathbf{r}_1) \exp\left(-U\{N\}/kT\right) d\{N-2\}}{\int \cdots \int \exp\left(-U\{N\}/kT\right) d\{N-2\}} \quad (4.5.2)$$

where the potential for this force, φ_N, had been defined by

$$\varphi_N(r) = -kT \log g_N(r) \quad (4.5.3)$$

The force here results not only from the direct interaction between the two particles, $u(r_{12})$, but also from the influence that particles 1 and 2 have on the configurations of all the other particles, thermally averaged. This latter, indirect interaction between 1 and 2 is what puts the wiggles in $g(r)$; when they are absent, $\varphi_N(r)$ is just $u(r)$, and we are back to the dilute gas approximation.

Assuming pairwise additivity, we can write $U\{N\}$ as

$$U\{N\} = u(r_{12}) + u(r_{13}) + \cdots + u(r_{1N}) + \text{(terms not involving } \mathbf{r}_1\text{)} \quad (4.5.4)$$

so that

$$\frac{\partial U\{N\}}{\partial \mathbf{r}_1} = \frac{\partial u(r_{12})}{\partial \mathbf{r}_1} + \left[(N-2) \text{ terms of the form } \frac{\partial u(r_{1j})}{\partial \mathbf{r}_1}\right] \quad (4.5.5)$$

When this result is substituted into Eq. (4.5.2), the numerator on the right-hand side becomes

$$\frac{\partial u(r_{12})}{\partial \mathbf{r}_1} \int \cdots \int \exp\left(-\frac{U\{N\}}{kT}\right) d\{N-2\}$$
$$+ (N-2) \int d^3 r_3 \frac{\partial u(r_{13})}{\partial \mathbf{r}_1} \int \cdots \int \exp\left(-\frac{U\{N\}}{kT}\right) d\{N-3\} \quad (4.5.6)$$

4.5 Liquids

In the first of these terms, $\partial u(r_{12})/\partial \mathbf{r}_1$ has come out of the integration over particles 3 to N; the remaining $(N - 2)$ terms are identical, and $\partial u(r_{13})/\partial \mathbf{r}_1$ can then be taken out of the integration over particles 4 to N (by "integration over particles" we mean, of course, integration over the coordinates of the particles). Inside of the integral over particle 3, we now find isolated the quantity

$$\int \cdots \int \exp\left(-\frac{U\{N\}}{kT}\right) d\{N - 3\} \qquad (4.5.7)$$

which is related to the three-particle correlation function; from Eq. (4.2.46),

$$W_N^{(3)} = \frac{N(N - 1)(N - 2)}{Q_N} \int \cdots \int \exp\left(-\frac{U\{N\}}{kT}\right) d\{N - 3\} \qquad (4.5.8)$$

Using the same formula for two particles, we have

$$W_N^{(2)} = \frac{N(N - 1)}{Q_N} \int \cdots \int \exp\left(-\frac{U\{N\}}{kT}\right) d\{N - 2\} \qquad (4.5.9)$$

The integral in this expression is just the denominator in Eq. (4.5.2). Thus, when we divide Eq. (4.5.6) by the denominator, we find

$$\frac{\partial \varphi_N(r_{12})}{\partial \mathbf{r}_1} = \frac{\partial u(r_{12})}{\partial \mathbf{r}_1} + \frac{\int d^3 r_3 [\partial u(r_{13})/\partial \mathbf{r}_1] W_N^{(3)}(r_{12}, r_{23})}{W_N^{(2)}(r_{12})} \qquad (4.5.10)$$

Here we have a closed form for $\varphi_N(r_{12})$, based only on the pairwise additivity approximation. It is not really an advance, however, over, say, Eq. (4.2.49), connecting $g_N(r)$ to $g_N^{(3)}$. We know nothing about the three-particle correlation functions. To get any further, we must find some reasonable approximation for functions like $W_N^{(3)}$.

In order to have a physical basis for an approximation, it is useful to define a three-particle potential of mean force

$$\varphi_N^{(3)} \equiv -kT \log g_N^{(3)} \qquad (4.5.11)$$

In analogy to Eq. (4.5.2), it has the property that

$$-\frac{\partial \varphi_N^{(3)}}{\partial \mathbf{r}_1} = -\frac{\int \cdots \int (\partial U\{N\}/\partial \mathbf{r}_1) \exp(-U\{N\}/kT) \, d\{N - 3\}}{\int \cdots \int \exp(-U\{N\}/kT) \, d\{N - 3\}} \qquad (4.5.12)$$

is the mean force on one particle if we hold two others fixed. We now make what we might call the approximation of pairwise additivity of mean potentials

$$\varphi_N^{(3)} = \varphi_N(r_{12}) + \varphi_N(r_{23}) + \varphi_N(r_{13}) \qquad (4.5.13)$$

This is actually called the superposition approximation, and it was first suggested by Kirkwood.

Let us pause for a moment to try to get some feeling for what this approximation means. Consider, first, just two particles in the interacting

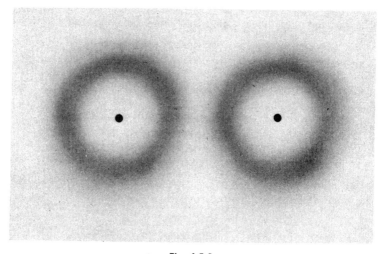

Fig. 4.5.1

medium. To a first approximation, there is a bump in the average density around each of the two particles, as in Fig. 4.4.1, which we represent in Fig. 4.5.1 by a smeared-out ring, actually a shell, of elevated density around each of the two. As the two particles are brought together, the forces between them, aside from the direct interaction $u(r_{12})$, arise because, for example, the shell around 1 causes the shell around 2 to become distorted, thereby exerting an additional force on 2. At some distances the additional force will be repulsive, at others, attractive. Consequently, there will be further, more diffuse rings in the density sketch, if we average over all second particles and start again. These bumps are the wiggles we are trying to introduce into $g(r)$. What we are seeing is why, even if the direct interactions are pairwise additive, they lead to complicated correlations among the positions of many particles.

Now suppose that we bring up a third particle with, in the first approximation, its own ring of mean density, like those in Fig. 4.5.1. The ring around particle 3 interacts with the one around particle 1, leading again to more complicated structure, and does the same with particle 2, just as we argued above. However, bringing up particle 3 has an additional consequence: by changing the density distributions around particles 1 and 2, it changes the forces acting between *them*. It is this final consequence that we have ignored in writing Eq. (4.5.13).

If we compare the preceding argument to the discussion of the pairwise additivity approximation itself, in Sec. 4.3b, we see that the two approximations are analogous but quite distinct; one certainly does not imply the other. The approximation that we have made here concerning the mutual effects of thermally averaged distributions of the density of other atoms is the same as the one we made earlier for the mutual effects of quantum mechanically averaged charge distributions within the atoms. However, the pairwise potentials of the earlier approximation do lead to correlations that are now neglected in the superposition approximation.

We can now use Eqs. (4.5.3) and (4.5.11) to write the superposition approximation in terms of distribution functions

$$g_N^{(3)} = g_N(r_{12})g_N(r_{23})g_N(r_{13}) \qquad (4.5.14)$$

and Eq. (4.2.45) to write it in terms of correlation functions

$$W_N^{(3)} = \frac{1}{\rho^3} W_N^{(2)}(r_{12})W_N^{(2)}(r_{13})W_N^{(2)}(r_{23}) \qquad (4.5.15)$$

[we have taken $N(N-1)(N-2)/V^3 = \rho^3$]. Substituting Eq. (4.5.15) into (4.5.10) and then using (4.2.45) again, we get

$$\frac{\partial \varphi_N(r_{12})}{\partial \mathbf{r}_1} = \frac{\partial u(r_{12})}{\partial \mathbf{r}_1} + \rho \int \frac{\partial u(r_{13})}{\partial \mathbf{r}_1} g_N(r_{13})g_N(r_{23}) \, d^3r_3 \qquad (4.5.16)$$

Replacing φ_N by $-kT \log g_N$ and taking the thermodynamic limit, we have

$$\frac{\partial}{\partial \mathbf{r}_1}\left[\log g(r_{12}) + \frac{u(r_{12})}{kT}\right] = -\frac{\rho}{kT} \int \frac{\partial u(r_{13})}{\partial \mathbf{r}_1} g(r_{13})g(r_{23}) \, d^3r_3 \qquad (4.5.17)$$

Equation (4.5.17) is our essential result, the Yvon-Born-Green equation. Let us, however, go through the few extra steps necessary to put it in the form of (4.5.1). Start by defining the function

$$E(x) = \int_\infty^x \frac{du(t)}{dt} g(t)\, dt \qquad (4.5.18)$$

Then
$$\frac{\partial E(r_{13})}{\partial \mathbf{r}_1} = \frac{\partial u(r_{13})}{\partial \mathbf{r}_1} g(r_{13}) \qquad (4.5.19)$$

Substitute Eq. (4.5.19) into (4.5.17), obtaining

$$\frac{\partial}{\partial \mathbf{r}_1}\left[\log g(r_{12}) + \frac{u(r_{12})}{kT}\right] = -\frac{\rho}{kT}\int \frac{\partial E(r_{13})}{\partial \mathbf{r}_1} g(r_{23})\, d^3r_3 \qquad (4.5.20)$$

Now integrate with respect to $\mathbf{r}_1$, giving

$$\log g(r_{12}) + \frac{u(r_{12})}{kT} = -\frac{\rho}{kT}\int E(r_{13}) g(r_{23})\, d^3r_3 + C \qquad (4.5.21)$$

where C is the constant of integration. We evaluate C where $r_{12} \to \infty$, keeping r_{13} fixed. The left-hand side is zero $[g(r_{12}) \to 1$ and $u(r_{12}) \to 0]$ and $g(r_{23}) \to 1$, so we have

$$0 = -\frac{\rho}{kT}\int E(s)\, d^3s + C \qquad (4.5.22)$$

Substituting this result back into Eq. (4.5.21), we can write

$$\log g(r_{12}) + \frac{u(r_{12})}{kT} = -\frac{\rho}{kT}\int E(r_{13})[g(r_{23}) - 1]\, d^3r_3 \qquad (4.5.23)$$

Equation (4.5.23), with E defined by Eq. (4.5.18), is the result we seek, the Yvon-Born-Green equation. An alternative way of writing it, using the same change of variables as in Eq. (4.5.70) below, is

$$\log g(r) + \frac{u(r)}{kT} = -\frac{\rho}{kT}\int E(|\mathbf{r} - \mathbf{y}|) h(y)\, d^3y \qquad (4.5.24)$$

b. The Hypernetted Chain and Percus-Yevick Equations

In this subsection we shall produce—derive would be too strong a word—we shall produce two widely used expressions, the Hypernetted Chain equation and the Percus-Yevick equation. The general method of attack is to use the topological properties of cluster diagrams to derive a power series in the density for $W(r)$ of Eq. (4.5.1), with cluster integrals as the coefficients. We then see how a certain subset of the cluster integrals, involving parts of the coefficients of all orders in the density, may be summed exactly, again by topological arguments. We then simply discard the diagrams that cannot be summed and arrive at the two equations in question.

4.5 Liquids

Our starting point is Eq. (4.2.39):

$$g_N(r) = \frac{V^2}{Q_N} \int \cdots \int \exp\left(-\frac{U\{N\}}{kT}\right) d\{N-2\} \quad (4.5.25)$$

Writing $U\{N\}$ by pairs, according to Eq. (4.5.4), we find that the factor $e^{-u(r_{12})/kT}$ is a constant in all the integrations, and it may be taken to the other side of the equation (recall that $r_{12} = r$, the argument of g_N). Using the f_{ij}'s introduced in Eq. (4.4.6), we write

$$g_N(r) \exp\left[\frac{u(r)}{kT}\right] = \frac{V^2}{Q_N} \int \cdots \int \prod_{i<j}{}' (1 + f_{ij})\, d\{N-2\} \quad (4.5.26)$$

where the symbol $\prod{}'$ means that the product excludes the term that we have just factored out, $(1 + f_{12})$.

The integrals appearing in Eq. (4.5.26) are to be represented once again by all possible N point diagrams, as in Sec. 4.4d. These new diagrams, however, differ from the old in that each has two points, 1 and 2, over which no integrations are to be performed. They also lack a direct line connecting 1 and 2, since f_{12} has been taken out. Let us call points that are not integrated over *fixed points* and represent them by open circles. All other points are *field points*, still represented by dots. A simple cluster is illustrated in Fig. 4.5.2, which represents the integral

$$\int f_{13} f_{23}\, d^3 r_3 \quad (4.5.27)$$

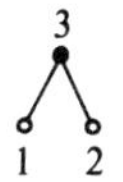

Fig. 4.5.2

As before, we wish to classify all possible clusters according to whether they factor and how they factor into independent integrals—that is, according to their reducibility. Here the rules are a bit different from those in Sec. 4.4d. For example, in Fig. 4.5.2, particle 3 is a node. In the earlier diagrams, where 2 would be a field point, the integral would be factorable, but here we see that Eq. (4.5.27) is not. Quite generally, a node on a path between 1 and 2 does not cause the cluster to be reducible. On the other hand, if we start with any cluster including both 1 and 2 and attach a further subcluster at any single point (including 1 or 2), that subcluster does represent an integral that factors out, just as it did in Sec. 4.4d.

Let us adopt the following strategem for describing clusters. We imagine the lines representing the f_{ij}'s to be electrical conductors, and we connect a battery between 1 and 2, so that a dc current flows through the paths connecting them. Then any point that does not carry current, either because it is entirely disconnected from 1 and 2 or because it belongs to a subcluster appended at a single point, will not appear in the reduced factor that includes 1 and 2. See Fig. 4.5.3. Here current flows only through the path 1–3–2, and the subclusters attached to 3 and 2 represent independent factors. We shall omit the battery in future diagrams.

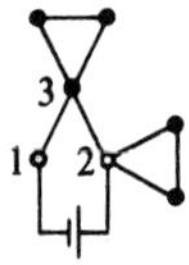

Fig. 4.5.3

Each of the N-point diagrams, then, breaks up into many factors, one of which represents the points that carry current. The important characteristic of the current-carrying cluster is that the integral it represents is a function of the distance between 1 and 2, r_{12}. None of the other factors depends on r_{12}; in fact, they are functions of T only. For example, the current-carrying part of Fig. 4.5.3 represents Eq. (4.5.27), which depends on r_{12}, whereas the subcluster attached to particle 2 represents the factor

$$\iint f_{24} f_{45} f_{25} \, d^3 r_4 \, d^3 r_5$$

This depends only on $\mathbf{r}_2$, which is arbitrary, since we are free to choose any origin.

Let us restrict our attention to clusters containing only current-carrying points. We can easily generalize later to all connected clusters by appending subclusters at each point, as we did in Sec. 4.4d. Even these current-carrying clusters are not completely irreducible. Consider, for example, the sketch of Fig. 4.5.4, which is equal to

$$\iint f_{13} f_{14} f_{23} f_{24} \, d^3 r_3 \, d^3 r_4 = \int f_{13} f_{23} \, d^3 r_3 \int f_{14} f_{24} \, d^3 r_4$$

$$= \left(\int f_{13} f_{23} \, d^3 r_3 \right)^2 \qquad (4.5.28)$$

Here the integral has factored because there are two (or more) independent paths between 1 and 2; that is, 1 and 2 are connected in parallel. We shall call such diagrams Parallel diagrams. A diagram belongs to this class if it contains points that are not connected except by paths that pass through

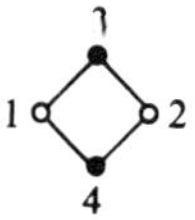

Fig. 4.5.4

either 1 or 2. Each branch of the parallel circuit between 1 and 2 can be arbitrarily complicated, so long as every point carries current. Parallel diagrams break up into factors, each of which is a function of r_{12}.

Conducting diagrams without parallel branches are completely irreducible. They may, nevertheless, be divided into two classes—those that have nodes and those that do not. A diagram with nodes can be thought of as a number of circuit elements connected in series, and we shall call them Series diagrams. All the current flowing from 1 to 2 passes through each node. Finally, irreducible diagrams without nodes are called Bridge diagrams. For example, Fig. 4.5.5 is a Bridge diagram, whereas Fig. 4.5.6 is a Series

Fig. 4.5.5

Fig. 4.5.6

Fig. 4.5.7

diagram, particle 4 being a node. The Series diagrams are of the form shown in Fig. 4.5.7, where 3 is the first node. The portion ○━● does not include any Series diagrams, but ●╌○ may have further nodes. Notice that the portion ○━● is not restricted only to Bridge diagrams; it could have parallel branches connecting 1 and 3, so long as there are no parallel branches connecting 1 and 2. It could also simply be f_{13}. Each branch of a Parallel diagram is made up of Series and Bridge diagrams.

Just as in Sec. 4.4d, we have extracted from all the possible N-point diagrams various types of central clusters with various degrees of reducibility. Most general are the connected diagrams, points connected by any path to 1 or 2 or both (although 1 and 2 may not be connected to each other). A more restricted, less reducible class are our fully conducting diagrams, in which 1 and 2 are connected by one or more parallel branches of Series and Bridge diagrams, all points carrying current. Then there are the irreducible diagrams, in which 1 and 2 are connected only by Series or Bridge diagrams. By arguments very similar to those used in Sec. 4.4d, it will be possible to write a power series for $g_N(r)e^{u(r)/kT}$ with conducting diagrams as coefficients, then rearrange that series into an exponential form, so that the series for $\log g_N(r) + u(r)/kT$ has the irreducible Series and Bridge diagrams for coefficients. In analogy to the β_k and β_k' of the earlier section, let us define

$$\delta_k' = \frac{1}{k!} \sum_{12}^{(k)\prime} \int \cdots \int d\{k\} \prod' f_{ij} \qquad (4.5.29)$$

where $\sum_{12}^{(k)\prime}$ means sum over all conducting diagrams of 1 and 2 plus k field points, and

$$\delta_k = \frac{1}{k!} \sum_{12}^{(k)} \int \cdots \int d\{k\} \prod' f_{ij} \qquad (4.5.30)$$

where $\sum_{12}^{(k)}$ means sum over all irreducible, Series plus Bridge, diagrams of 1, 2, and k field points. To complete the definitions, we should specify that $\delta_0 = \delta_0' = 1$.

Let us now try to organize the diagrams in Eq. (4.5.26), much as we did earlier ones in Eq. (4.4.52). Start with any diagram in which 1 and 2 are connected in a fully conducting cluster of $2 + k$ points (we can formally include $k = 0$, even though it does not conduct, to take account of those diagrams in which 1 and 2 are not connected to each other). Now attach any cluster of ℓ_1 points at particle 1, ℓ_2 at particle 2, ℓ_3 at particle 3, and so on, so that the complete central cluster has $L + 2$ points, where

$$L + 2 = \sum_{n=1}^{k+2} (\ell_n + 1) = k + 2 + \sum_{n=1}^{k+2} \ell_n \qquad (4.5.31)$$

Add together all diagrams in which 1 and 2 belong to an $L + 2$ point cluster constructed in this way. These diagrams together contribute to $g_N(r)e^{u(r)/kT}$ a term with the following factors:

$$\frac{V^2 Q_{N-L-2}}{Q_N} \qquad \text{(a)}$$

4.5 Liquids

where V^2/Q_N is the factor outside the integral in Eq. (4.5.26) and Q_{N-L-2} arises from all possible $N - L - 2$ point diagrams unconnected to the central cluster;

$$\frac{(N-2)!}{L!\,(N-L-2)!} \tag{b}$$

for the number of ways of choosing L-ordered particles out of $N - 2$;

$$L! \tag{c}$$

for the permutations of the L field points;

$$\delta'_k \tag{d}$$

for the fully conducting $k + 2$ point clusters we started with; and

$$\frac{1}{V^{k+2}} \prod_{n=1}^{k+2} (\ell_n + 1) b_{\ell_n + 1} \tag{e}$$

for the appended diagrams. This is the same term as factor (e) leading to Eq. (4.4.66), except that there were k appendages in that case and there are $k + 2$ in this case. The appended diagrams, and the symbol $b_{\ell+1}$ representing them, are the same in the two cases. Taking factors (a), (b), and (c) together, we have

$$\frac{V^2 Q_{N-L-2}}{Q_N} \frac{(N-2)!}{(N-L-2)!} = V^2 \frac{Q_{N-1}}{Q_N} \frac{Q_{N-2}}{Q_{N-1}} \frac{Q_{N-L-2}}{Q_{N-2}} \frac{(N-2)!}{(N-L-2)!}$$

$$= V^2 \left(\frac{\gamma}{V}\right)^2 \xi^L$$

$$= \gamma^2 \xi^L \tag{4.5.32}$$

where we have used Eq. (4.4.40) twice and Eqs. (4.4.61) and (4.4.62). Thus, terms (a) to (e) together give the factor

$$\gamma^2 \frac{\xi^k}{V^{k+2}} \delta'_k \prod_{n=1}^{k+2} (\ell_n + 1) b_{\ell_n + 1} \xi^{\ell_n} \tag{4.5.33}$$

We now sum this result over all L from 1 to infinity by summing over all k, and all ℓ_n separately, assuming as usual that the series converges in a finite number of terms when $N \to \infty$. Inside the product we have $k + 2$ factors

$$\sum_{\ell=0}^{\infty} (\ell + 1) b_{\ell+1} \xi^\ell = \frac{V}{\gamma} \tag{4.5.34}$$

according to Eq. (4.4.63). Substituting $(V/\gamma)^{k+2}$ into Eq. (4.5.33) in place of the product, and summing over k, we have now included all the possible diagrams and may write

$$g_N(r) \exp\left[\frac{u(r)}{kT}\right] = \sum_{k=0}^{\infty} \delta'_k \rho^k \qquad (4.5.35)$$

or in the thermodynamic limit

$$g(r) \exp\left[\frac{u(r)}{kT}\right] = \sum_{k=0}^{\infty} \delta'_k \rho^k \qquad (4.5.36)$$

Again, as in Sec. 4.4d, we can perform one further stage of reduction. This time, instead of factors consisting of the petals of a flower, we have somewhat less poetic factors corresponding to the branches of a parallel circuit. Take a conducting diagram with p branches and K field points,

$$K = \sum_{n=1}^{p} m_n \qquad (4.5.37)$$

where m_n is the number of field points in the nth branch (each branch, of course, is either a Series or a Bridge diagram). All such diagrams contribute to Eq. (4.5.36) a term consisting of p factors, each involving the irreducible diagrams of Eq. (4.5.30),

$$\prod_{n=1}^{p} \delta_{m_n} \rho^{m_n} \qquad (4.5.38)$$

To get all the terms in Eq. (4.5.36), we must sum over all values of the m_n inside the product, giving p identical factors equal to

$$\sum_{m=1}^{\infty} \delta_m \rho^m \qquad (4.5.39)$$

However, in summing over all possible diagrams in each branch independently, we have counted each diagram $p!$ times too many. Divide by $p!$ and, finally, sum over p [including $p = 0$ for the case $k = 0$ in Eq. (4.5.36)] to give

$$g(r) \exp\left[\frac{u(r)}{kT}\right] = \sum_{p=0}^{\infty} \frac{1}{p!} \left[\sum_{m=1}^{\infty} \delta_m \rho^m\right]^p$$

$$= \exp\left[\sum_{m=1}^{\infty} \delta_m \rho^m\right] \qquad (4.5.40)$$

or

$$\log g(r) + \frac{u(r)}{kT} = \sum_{m=1}^{\infty} \delta_m \rho^m \qquad (4.5.41)$$

This is the result we need.

Now that we have the power expansions we were looking for, let us

4.5 Liquids

pause for a moment to write the first few terms in Eqs. (4.5.36) and (4.5.41) explicitly. From the definitions we have

$$\delta'_0 = \delta_0 = 1 \tag{4.5.42}$$

[notice, though, that δ_0 does not appear in (4.5.41)]. In the next order,

$$\delta'_1 = \delta_1 = \begin{array}{c} 3 \\ \bigwedge \\ 1 \quad 2 \end{array} = \int f_{13} f_{23} d^3 r_3 \tag{4.5.43}$$

so that, to leading order in density, we have

$$g(r) \exp\left[\frac{u(r)}{kT}\right] = 1 + \rho \int f_{13} f_{23} \, d^3 r_3 \tag{4.5.44}$$

$$\log g(r) + \frac{u(r)}{kT} = \rho \int f_{13} f_{23} \, d^3 r_3 \tag{4.5.45}$$

This is the leading correction to the dilute gas approximation for $g(r)$. In the next order, we form all possible conducting diagrams with two field points (Fig. 4.5.8). All diagrams in Fig. 4.5.8 go into δ'_2. We do not include the one

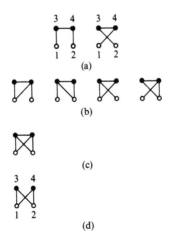

Fig. 4.5.8

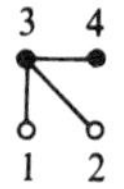

Fig. 4.5.9

in Fig. 4.5.9, since here no current passes through point 4 when a battery is connected across 1–2, which means that the integral over f_{34} comes out in a numerical factor that is not a function of $r = r_{12}$. The diagram on line (d) of Fig. 4.5.8, although it belongs to δ_2', does not belong to δ_2, since the paths 1–3–2 and 1–4–2 are independent parallel branches. This diagram is, in fact, just the one shown in Fig. 4.5.4, and its value is given in Eq. (4.5.28); it is the square of the diagram in Eq. (4.5.43) and may be written $(\delta_1)^2$. All diagrams on lines (a), (b), and (c) of Fig. 4.5.8 belong to δ_2; those on lines (a) and (b) are Series diagrams and that on line (c) is a Bridge diagram (it is the lowest order possible Bridge diagram). The two diagrams on line (a), although they form distinct diagrams in our counting, have the same value

$$\int\int f_{13}f_{34}f_{24}\, d^3r_3\, d^3r_4 = \int\int f_{14}f_{34}f_{23}\, d^3r_3\, d^3r_4 \qquad (4.5.46)$$

We regard 3 and 4 as distinguishable for counting purposes, but they make the same contributions to all integrals. Similarly, the four diagrams on line (b) have the same value. We can write

$$\delta_2' = \frac{1}{2!}\left\{2\,[\text{diagram}] + 4\,[\text{diagram}] + [\text{diagram}] + [\text{diagram}]\right\} \qquad (4.5.47)$$

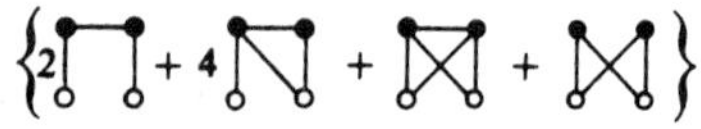

$$\delta_2 = \frac{1}{2!}\left\{2\,[\text{diagram}] + 4\,[\text{diagram}] + [\text{diagram}]\right\} \qquad (4.5.48)$$

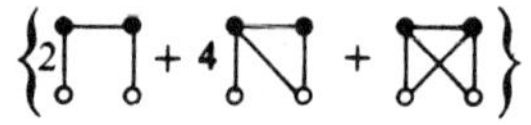

or in integral form

$$\delta_2' = \frac{1}{2!}\int\int f_{13}f_{34}f_{24}(2 + 4f_{23} + f_{14}f_{23})\, d^3r_3\, d^3r_4 + \frac{1}{2!}(\delta_1)^2 \qquad (4.5.49)$$

$$\delta_2 = \frac{1}{2!}\int\int f_{13}f_{34}f_{24}(2 + f_{23} + f_{14}f_{23})\, d^3r_3\, d^3r_4 \qquad (4.5.50)$$

Given these expressions, it is easy to write the order ρ^2 corrections to Eqs. (4.5.44) and (4.5.45). It should be obvious by now, however, that the problem of keeping track of all possible diagrams gets dramatically more difficult with increasing numbers of field points, so that generating successively

4.5 Liquids

higher-order terms in these series will not be the most fruitful course for us to follow.

On the other hand, we can, and shall, once again reclassify and regroup the diagrams in the series, a familiar activity by now. In the terms appearing in Eq. (4.5.41), points 1 and 2 are connected by Series and Bridge diagrams only. Let us define

$$S = \sum_{k} {}_{(S)} \delta_k \rho^k \tag{4.5.51}$$

and

$$B = \sum_{k} {}_{(B)} \delta_k \rho^k \tag{4.5.52}$$

where $\sum_{(S)}$ and $\sum_{(B)}$ mean, respectively, include only Series diagrams in the sum and include only Bridge diagrams in the sum. With these definitions we can write

$$\log g(r) + \frac{u(r)}{kT} = S + B \tag{4.5.53}$$

The sum in Eq. (4.5.36) involves not only S and B but also the single term 1, and diagrams with parallel branches each of the S or B types—that is, products of Series and Bridge diagrams. The easiest way to write Eq. (4.5.36) in terms of S and B is to exponentiate Eq. (4.5.53) and expand

$$g(r) \exp\left[\frac{u(r)}{kT}\right] = e^{S+B}$$

$$= 1 + S + B + \frac{1}{2!}(S + B)^2 + \cdots \tag{4.5.54}$$

With the definitions (4.5.51) and (4.5.52), this is exactly the same as the first member of Eq. (4.5.40). Let us formally define

$$P_1 = \frac{1}{2!}(S + B)^2 + \frac{1}{3!}(S + B)^3 + \cdots \tag{4.5.55}$$

P_1 is then a power series in ρ, one power of ρ for each field point, in which all the coefficients are Parallel diagrams. Notice that S and B themselves are not included in P_1—all diagrams in P_1 have at least two branches. Recall that the δ'_k included conducting diagrams with only one branch, so the power series in δ'_k, Eq. (4.5.36), is not simply P_1, but instead, putting Eq. (4.5.55) into (4.5.54),

$$g(r) \exp\left[\frac{u(r)}{kT}\right] = 1 + S + B + P_1 \tag{4.5.56}$$

Up to this point our diagrams have not included the direct line between 1 and 2, f_{12}, since we had factored out the term

$$1 + f_{12} = \exp\left[-\frac{u(r)}{kT}\right] \quad (4.5.57)$$

Let us now put this in our pictures, rewriting Eq. (4.5.56),

$$g(r) = (1 + f_{12})(1 + S + B + P_1)$$
$$= 1 + f_{12}(1 + S + B + P_1) + S + B + P_1 \quad (4.5.58)$$

When any diagram in S or B is multiplied by f_{12}, it becomes a sort of Parallel diagram. For example,

$$f_{12}\delta_1 = \underset{1 \quad 2}{\triangle} \quad (4.5.59)$$

has two parallel branches connecting 1 and 2. We can define the power series over these generalized parallel diagrams—the Augmented Parallel diagrams, P,

$$P = f_{12}(S + B + P_1) + P_1 \quad (4.5.60)$$

Then we can write

$$g(r) = 1 + f_{12} + S + B + P \quad (4.5.61)$$

or
$$h(r) = g(r) - 1 = f_{12} + S + B + P \quad (4.5.62)$$

The terms S, B, P, and P_1 are, of course, power series in ρ, one power of ρ for each field point in the coefficient diagram. Since it is easy to keep track of the powers of ρ by the number of field points, we shall say, somewhat loosely, that S is the sum of all Series diagrams, B the sum of all Bridge diagrams, and so on, leaving the powers of density (and the combinatorial factors) to be tacitly understood. In this sense, $h(r)$ can be seen from Eq. (4.5.62) to be given by the sum of all possible fully conducting diagrams connecting 1 and 2. It thus justifies its name, the total correlation function.

It turns out to be convenient to define an additional correlation function, $c(r)$, by

$$c(r) = f_{12} + B + P = h(r) - S \quad (4.5.63)$$

$c(r)$ differs from $h(r)$ in that Series diagrams are not included in its composition (although it does include $f_{12}S$, which is hidden in P). $c(r)$ is called the *direct correlation function*. It includes all contributions to the correlations between 1 and 2 except those that are mediated by single other particles, the nodes of the S diagrams; we can think of the nodes as the indirect correlations. The direct correlation function comes into its own in Chap. 6 when we study critical phenomena, since at the gas-liquid critical point, when the

4.5 Liquids

other correlation functions tend to blow up, $c(r)$ remains demurely well behaved. In this chapter, however, it serves principally as a mathematical convenience.

The basic reason why we have gone through all the manipulations in this subsection is because it turns out to be possible to sum the series S exactly. The machinery that we have developed so far will be useful in taking advantage of this trick. The argument is rather similar to ones that we have already used; we set out to regroup and add up the diagrams in S by means of their topological properties. Consider a typical diagram in S. We can represent it by Fig. 4.5.7, which we repeat here in Fig. 4.5.10. The balloon

○━━━●---○
1   3   2

Fig. 4.5.10

connecting 1 and 3 may consist of any subdiagram of the B or P class, or it could simply be f_{13}. The portion  includes subdiagrams from all classes.

Let us choose from all the diagrams belonging to S all those that have some particular diagram of ℓ points besides 3 and 2 in the part ●---○ When we add all these diagrams up, there will be a common factor for that part, which will be a function of r_{23}, and $\ell + 1$ powers of ρ for the ℓ field points plus point 3. Let us call the common factor

$$\rho^{\ell+1}\alpha_\ell(r_{23}) \tag{4.5.64}$$

This factor will multiply a sum over all diagrams between 1 and 3, excluding the Series diagrams—that is

$$f_{13} + P(r_{13}) + B(r_{13}) = c(r_{13}) \tag{4.5.65}$$

where we have emphasized that since the P and B diagrams connect points 1 and 3, they are functions of r_{13}, and we have used the first member of Eq. (4.5.63). All the diagrams in the class we are considering are then given by integrating over particle 3 the product of Eqs. (4.5.64) and (4.5.65),

$$\rho \int \rho^\ell \alpha_\ell(r_{23}) c(r_{13}) \, d^3 r_3 \tag{4.5.66}$$

To get all the Series diagrams, we must sum over all possible α_ℓ:

$$S = \rho \int \left[\sum_{\alpha_\ell} \rho^\ell \alpha_\ell(r_{23})\right] c(r_{13}) \, d^3 r_3 \tag{4.5.67}$$

But the α_ℓ can be any fully conducting diagram connecting 2 and 3, with any number, ℓ, of field points, so that if we sum over all possible α_ℓ, we get

$$\sum_{\alpha_\ell} \rho^\ell \alpha_\ell(r_{23}) = f_{23} + P(r_{23})$$
$$+ B(r_{23}) + S(r_{23})$$
$$= h(r_{23}) \tag{4.5.68}$$

We thus have for S

$$S = \rho \int h(r_{23}) c(r_{13}) \, d^3 r_3 \tag{4.5.69}$$

This is the result that we are seeking. Let us rewrite it after making the changes of variables

$$-\mathbf{y} = \mathbf{r}_{23}$$
$$\mathbf{r} = \mathbf{r}_{12} = \mathbf{r}_{13} - \mathbf{r}_{23}$$
$$= \mathbf{r}_{13} + \mathbf{y} \tag{4.5.70}$$

and since 2 is held fixed in the integration,

$$d^3 r_3 = d^3 y \tag{4.5.71}$$

Then
$$S = \rho \int h(y) c(|\mathbf{r} - \mathbf{y}|) \, d^3 y \tag{4.5.72}$$

or using Eq. (4.5.63),

$$h(r) = c(r) + \rho \int h(y) c(|\mathbf{r} - \mathbf{y}|) \, d^3 y \tag{4.5.73}$$

Equation (4.5.73) is called the *Ornstein-Zernike equation*, and it could have been used to define $c(r)$. We see that the total correlation function is a sum of the direct correlation function and a convolution of the total and direct correlation functions. Alternatively, we could say that the Series diagrams, S, is given by a convolution of c and h, integrated over a typical intermediate particle, whatever that means. In any case, if either $h(r)$ or $c(r)$ is known, the other may be obtained (numerically, if necessary) by means of Eq. (4.5.73). It turns out that the Fourier transforms of these quantities (which arise directly in scattering experiments; see Sec. 4.2) are easier to handle; we shall return to that point in the next section.

We are now ready to generate the two equations that give this subsection its title. From Eq. (4.5.53) we see that we would have a closed form for $\log g(r) + u(r)/kT$ if we could ignore the Bridge diagrams. No sooner said than done. Taking

$$B = 0 \tag{4.5.74}$$

4.5 Liquids

we have

$$\log g(r) + \frac{u(r)}{kT} = h(r) - c(r) \qquad (4.5.75)$$

Using Eq. (4.5.75), together with (4.5.73), and

$$h(r) = g(r) - 1 \qquad (4.5.76)$$

gives $g(r)$ from $u(r)$ alone, or vice versa. Equations (4.5.75) is called the *Hypernetted Chain equation*.

On the other hand, we can write

$$g(r) - c(r) = g(r) - [h(r) - S]$$
$$= g(r) - h(r) + S$$
$$= 1 + S \qquad (4.5.77)$$

But we also know, from Eq. (4.5.56), that

$$g(r) = \exp\left[-\frac{u(r)}{kT}\right](1 + S + B + P_1) \qquad (4.5.78)$$

whereupon, using Eq. (4.5.57),

$$g(r) = (1 + f_{12})(1 + S + B + P_1)$$
$$= (1 + f_{12})(1 + S) + (1 + f_{12})(B + P_1) \qquad (4.5.79)$$

Putting together Eqs. (4.5.77) and (4.5.79) gives

$$g(r) = (1 + f_{12})[g(r) - c(r)] + (1 + f_{12})(B + P_1) \qquad (4.5.80)$$

Once again we would have a closed form if we could ignore the last term on the right in Eq. (4.5.80) this time. Our wish is our command; setting

$$(1 + f_{12})(B + P_1) = 0 \qquad (4.5.81)$$

bringing $(1 + f_{12})$ back to the left as $e^{u(r)/kT}$ in the remaining terms of Eq. (4.5.80), and taking logs, we have

$$\log g(r) + \frac{u(r)}{kT} = \log\left[g(r) - c(r)\right] \qquad (4.5.82)$$

If we again use Eqs. (4.5.73) and (4.5.76), we have yet another way to find $g(r)$ given $u(r)$ and vice versa. Equation (4.5.82) is called the *Percus-Yevick equation*. Both the Hypernetted Chain and Percus-Yevick equations are closed forms involving at least some contributions from all orders in the density. The approximations used to arrive at them, however, are of somewhat uncertain significance. We shall analyze this situation a bit further in the next subsection.

c. Comments and Comparisons

According to our basic precepts of statistical mechanics, the equilibrium properties of a liquid at any temperature and density should be entirely determined by the energies of interaction between the constituent particles in various configurations. This chapter has been devoted to examining some of the ways in which people have tried to make that connection between microscopic properties and macroscopic behavior. Let us now look into some of the ways in which the results can be evaluated.

Let us suppose that we could start with a precise, quantum mechanical, first-principles calculation of the interatomic pair potential, $u(r)$, say, for argon. We could then calculate $g(r)$ from each of the equations just obtained: the Yvon-Born-Green, Hypernetted Chain, and Percus-Yevick—Eqs. (4.5.24), (4.5.75), and (4.5.82). (From now on we shall call them, respectively, the YBG, HNC, and PY equations.) The results could be compared to scattering experiments on liquid argon by means of Eq. (4.2.19). Moreover, $g(r)$, either experimental or theoretical, can be used to predict the equation of state of the liquid either by the fluctuation equation, Eq. (4.2.58), or by the pressure equation, Eq. (4.3.36), as well as to predict the heat capacity by Eq. (4.3.24). These, in turn, can be compared to experimental measurements. We thus have a network of cross-checks that can be used at various steps to ensure that we are always on the right track.

Unfortunately, all this manipulation does not work out too well. Not only can we not be sure that we are right, we cannot even be sure that we are wrong or, in any case, find out at what point we go wrong and by how much. We have already seen some of the causes of this dilemma in Secs. 4.2 and 4.3: the equations of state and the heat capacity are excessively sensitive to fine details of $g(r)$, whereas the overall shape of $g(r)$ is always pretty much the same. The result is that, on the one hand, we cannot expect accurate theoretical predictions of the equation of state and heat capacity, while on the other we cannot expect to rule out any theory on the basis of striking disagreement between predicted and measured $g(r)$. We must bear in mind that there are serious limitations to the experimental precision with which the quantities we need may be measured; moreover, there are even errors in calculating $g(r)$ from the theories, which are evaluated by trial solutions and iterated numerical integration. What is worse, however, is that we do not start with an exact form for $u(r)$; even that must be approximated in some way.

Take a specific example of how we might proceed. Suppose that we wanted to compare the results of the integral equations to the properties of some real liquids, say, argon and xenon. For a starting point in solving the integral equations, YBG, HNC, and PY, we must have an analytic form for $u(r)$. For this purpose, we choose the Lennard-Jones potential, Eq. (4.3.13),

$$u(r) = -4\varepsilon_0 \left[\left(\frac{\sigma}{r}\right)^{12} - \left(\frac{\sigma}{r}\right)^6 \right] \qquad (4.5.83)$$

4.5 Liquids

with the parameters ε_0 and σ fitted for argon and xenon.

$$\text{Ar:} \frac{\varepsilon_0}{k} = 119.8°\text{K} \qquad N\sigma^3 = 23.75 \text{ cm}^3/\text{mole}$$
$$\text{Xe:} \frac{\varepsilon_0}{k} = 225.3°\text{K} \qquad N\sigma^3 = 40.62 \text{ cm}^3/\text{mole} \qquad (4.5.84)$$

Once we obtain $g(r)$ in each case, we can compute the equation of state from Eq. (4.3.36)

$$\frac{P}{\rho k T} = 1 - \frac{\rho}{6kT} \int r \frac{du(r)}{dr} g(r) \, d^3r \qquad (4.5.85)$$

and compare the results to measurements of the equations of state of real liquid argon and xenon. It turns out, however, that we can simplify matters for ourselves, working things out for both argon and xenon at the same time, by rewriting the last term in Eq. (4.5.85) in terms of dimensionless parameters. We define the dimensionless quantities

$$\rho^* = \rho\sigma^3$$
$$r^* = \frac{r}{\sigma}$$
$$u^* = \frac{u}{\varepsilon_0}$$
$$kT^* = \frac{kT}{\varepsilon_0} \qquad (4.5.86)$$

With these definitions it is easy to see that Eq. (4.5.84) becomes

$$\frac{P}{\rho k T} = 1 - \frac{\rho^*}{kT^*} \int r^* \frac{du^*(r^*)}{dr^*} g(r^*) \, d^3r^* \qquad (4.5.87)$$

with the integral equations written entirely in terms of dimensionless quantities as well, for example,

$$\log g(r^*) + \frac{u^*(r^*)}{kT^*} = \log \left[g(r^*) - c(r^*) \right] \qquad (4.5.88)$$

for PY, and so on. Now we need not worry about whether we are dealing with argon or xenon; $P/\rho kT$ depends only on ρ^* and T^*. This is an example of the *Law of Corresponding States*, to which we shall return in Chap. 6. It is a consequence of the fact that, in the scheme we have set up, all the properties of the liquid depend on $u(r)$, which, in turn, has two parameters—a length σ and an energy ε_0. When we scale the temperature to ε_0/k and the density to σ^{-3}, all liquids become the same.

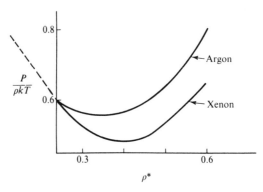

Fig. 4.5.11

In Fig. 4.5.11, we show a sketch of $P/\rho kT$ versus ρ^* for argon and xenon at $T^* = 1.4$. According to the Law of Corresponding States, the result should be one single curve for both substances. Clearly, the substances themselves do not obey the Law of Corresponding States for this value of T^*. There are two possible reasons: either the Lennard-Jones form for the pair potential is inadequate or the assumption of pairwise additivity is at fault. Either way, the integral equations used with the Lennard-Jones potential (or any other two-parameter form) cannot agree with both curves in Fig. 4.5.11; each integral equation will predict only one curve.

We cannot even conclude that the situation is hopeless, however. Real materials do come closer to obeying this law at higher temperatures—for example, at $T^* = 2.74$—as we shall see shortly. For orientation, the gas-liquid critical point in argon occurs at $T^* = 1.26$, $\rho^* = 0.32$.

Evidently direct comparison with experiment is too rigorous a test for the integral equation theories and pair potentials. There is an interesting and prolific class of work in this field that is neither true theory nor true experiment but something in between: the computer experiment. In these experiments a liquid is created in the imagination of a computer and its properties are computed. Two general techniques are used, one based on statistical mechanics and the other on particle dynamics. Typically, some number of particles (usually of the order of 10^2 to 10^3) are placed in a fixed volume with fixed total energy or temperature, and their law of interaction is specified (for example, a Lennard-Jones potential). Then in the statistical mechanics approach, configurations are generated by moving one particle at a time randomly and either accepting or rejecting the move according to a criterion designed to make all configurations appear with frequency proportional to $e^{-U(N)/kT}$ after an infinite number of moves. This is called the *Monte Carlo method*, the name and the technique both invented by von

Neumann. The other method is to start all the particles out with some set of initial positions and velocities, then let them bounce around obeying their equations of motion. The second technique is called *molecular dynamics* (in the case of argon or xenon, of course, *atomic dynamics*).

Computer experiments, like real experiments, are subject to experimental error. Because of the small number of particles in the molecular dynamics case, for example, any single "measurement" of a thermodynamic quantity is subject to substantial fluctuations (for 10^3 particles, $\sqrt{\overline{N}}/N \approx 0.03$; see Sec. 1.3f), so many measurements must be made in order to get accurate results. An elaborate program, let us say, to find the pressure at a number of temperatures and densities, or with different parameters in the interatomic potential, can be prohibitively expensive in computer time. Moreover, there may even be true many-body effects which are absent entirely in systems of 10^2 or 10^3 particles; remember, for example, that there is a real difference between $g(r)$ and $g_N(r)$, which goes away only in the limit $N \to \infty$. Such difficulties may be temporary, however, becoming less important as bigger and faster computers become available.

The first question that arises in connection with computer experiments is: Why bother trying to work out analytic theories at all if we can get everything we want to know from the computer? The second question is: Why bother doing computer experiments if we can measure whatever we want to know directly in the laboratory? The answer in both cases is that what we want is not merely accurate quantitative descriptions of nature but rather insight into what factors are physically important. The computer experiments serve as a useful intermediate step to help us sort things out.

The importance of the computer experiments is that it is we rather than nature who choose the laws to be obeyed. Basically YBG, HNC, and PY are competing theories of how liquids would behave if they were made up of particles that interacted pairwise. In the computer we can create a world in which interactions really are pairwise, with a potential that really is, say, Lennard-Jones, and thereby have the means to sort out the effects of pairwise additivity and the later approximations. The danger is that the theories become theories of computer experiments, the experiments test the theories, and a closed intellectual system is formed with nature left out entirely. One notices a tendency in this direction in the literature on liquids.

Nevertheless, the computer experiments play a useful role, of which the pairwise additivity problem is only one aspect. For example, in the molecular dynamics-type calculations, not only equilibrium properties but also transport properties, such as diffusion, may be studied.

A thorough evaluation of YBG, HNC, and PY as theories of the equilibrium liquid state would require comparing the results of these theories, for various forms of $u(r)$, at all temperatures and densities to computer results using the same potentials and to experimental measurements, both of scattering and thermodynamic quantities, for various substances. The intent of the study would be to isolate the errors introduced by specific approximations and estimate their magnitudes. We would then be in a position to decide where it seems worth trying to make improvements and whether all the essential physical features seem to be in hand. We do not have the results of such a grandiose scheme available to analyze, but many of the elements of the scheme have been worked out, and we can piece together part of the general picture by looking at them.

The first step is to find out what the integral equations actually predict. We will look at results using the Lennard-Jones potential only, since the other forms that have been tried do not seem to change things very much. The form of the predictions of the YBG equation is sketched in Fig. 4.5.12. We have plotted isotherms (constant T^*) in the P-V plane (reduced pressure, $P^* = P\sigma^3/\varepsilon_0$; reduced volume, $V^* = V/N\sigma^3 = 1/\rho^*$). In the cross-hatched region the equation is singular: there are no solutions.

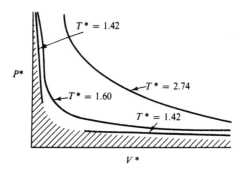

Fig. 4.5.12

When the equation goes singular, it is usually taken to mean that a phase transition occurs, and there is no longer a homogeneous fluid phase. At high P^* and low V^* the transition is presumably freezing, whereas at low P^* and high V^* it is condensation from the gas. The two transitions seem to merge smoothly, without the kink we see in real systems, as in Fig. 4.1.1. For $T^* \gtrsim 1.5$ there are no forbidden zones, while for $T^* \lesssim 1.4$ there are no solutions at all except at very low density. Between T^* of about 1.4 and 1.5, however, there are isotherms that enter the forbidden zone and then emerge again, indicating the existence of two distinct fluid phases at different densities. An example is the isotherm $T^* = 1.42$ sketched in Fig. 4.5.12. Those portions of isotherms in the upper left-hand corner of the diagram, which emerge from the forbidden zone, can be taken to represent a true condensed liquid phase. Unbroken isotherms are above the gas-liquid critical point, whereas those that enter the forbidden zone and come back out are subcritical. The critical point itself, in the YBG-Lennard-Jones system, is around $T^* \approx 1.5$

The PY and HNC equations also have critical behavior and therefore show what may be taken to be true liquid regions. However, in all three cases, there is a serious difficulty with the liquid region that is sufficiently demonstrated by the $T^* = 1.42$ isotherm in the YBG plot: the pressure of the liquid is much too high. The points at which the isotherm enters and leaves the forbidden zone cannot represent phase equilibrium between liquid and vapor because the pressures are not equal. When a liquid condenses, the attractive forces that come into play should bring the pressure so far below what it would be in a gas at the same density that coexistence between high- and low-density phases at the same P and T is possible. The integral equations tend to give pressures that rise monotonically with increasing density, so that true condensation never occurs (in Chap. 6 we shall study an equation of state, the van der Waals equation, which does predict condensation).

Of the three equations, only YBG predicts the freezing transition along the left-hand axis in the P^*-V^* plane.

From the observations we have just made it is clear that the integral equations will not produce good equations of state for the real liquid region. This expectation is confirmed by comparing what they predict to computer results, as in Table 4.5.1. We see that HNC and PY give pressures close to the ideal gas value (in which, of course, there are neither attractive nor repulsive forces), showing very little tendency to self-condense. In other words, the approximate form, Eq. (4.3.40), is never produced by the integral equations. The problem here cannot be the fault of the Lennard-Jones form, nor of the pairwise additivity approximation, since Monte Carlo calculations, using both of them, give much lower pressure, in agreement with real liquids. The fault lies in the correlations that have been neglected in arriving at the integral equations.

Well above the critical temperature, real substances, and computer simulations as well, show pressures that rise monotonically with density at

Table 4.5.1

System	$P/\rho kT$	
Ideal gas	1.00	$\begin{cases} T^* = 1.0 \\ \rho^* = 0.66 \\ \text{Lennard-Jones potential} \end{cases}$
HNC	0.8	
PY	0.7	
Monte Carlo	0.0 ± 0.05	

least up to the freezing curve. Here the inability of the integral equations to show self-condensation is less important, and we might hope they would do better. The predictions of the integral equations at $T^* = 2.74$ are shown in Fig. 4.5.13. Here we have shown not only the results computed from the pressure equation, Eq. (4.5.87), the one that we have been using up to now, but also the isothermal compressibility, K_T, from the fluctuation equation, Eq. (4.2.58). The two equations do not give equivalent results, being based on different assumptions (the fluctuation equation does not depend on pairwise additivity). Thus, the PY results fall between the other two in the left-hand graph, but PY gives the highest reciprocal compressibility on the right. It has seemed useless to point out up to now that the three theories agree with each other and nature as well at low densities—they are, after all, theories of departure from dilute gas behavior—but the agreement may be seen in these plots. Notice also the tendency of the YBG compressibility to level off as its freezing transition is approached.

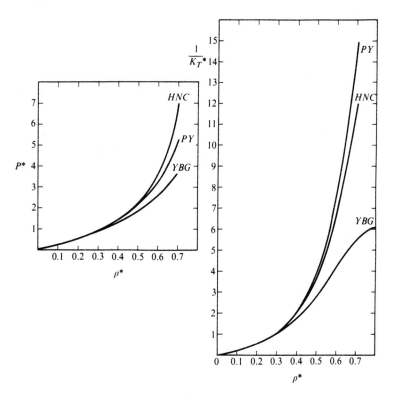

Fig. 4.5.13

4.5 Liquids

If we were to plot the Monte Carlo results for $T^* = 2.74$ in Fig. 4.5.13, the Monte Carlo curve in both graphs would fall essentially on top of the PY curves. It is this fact, the great accuracy of the Percus-Yevick equation up to quite high density in the high-temperature region, that gives impetus to the further study of the integral equations. It is also the reason that PY is now the most widely used of the three forms. It gives us a starting point, something right in the theories, from which we can hope to begin drawing conclusions.

One way to exploit the accuracy of the PY equation is to turn around somewhat the procedure that we have been discussing here. The PY equation presumably gives the $g(r)$ that real fluids at these temperatures would have if they interacted pairwise with a particular potential, $u(r)$. The game, then, is to take scattering data for real liquids and use them to work backward with the PY equation, deducing $u(r)$. The results can then be compared to independent measurements of $u(r)$ (from, say, molecular beam experiments) and the discrepancy credited to the failure of the pairwise additivity approximation. Operationally this is a bit easier than it sounds. The scattering experiments actually give us the quantity, Eq. (4.2.19),

$$S(Q) = 1 + \rho \tilde{h}(Q) \tag{4.5.89}$$

where we have written $\tilde{h}(Q)$ for the Fourier transform of $h(r)$

$$\tilde{h}(Q) = \int \exp(i\mathbf{Q} \cdot \mathbf{r}) h(r) \, d^3r \tag{4.5.90}$$

according to the PY equation, Eq. (4.5.82),

$$u(r) = kT \log \left[\frac{g(r) - c(r)}{g(r)} \right]$$

$$= kT \log \left[\frac{h(r) - c(r) + 1}{h(r) + 1} \right] \tag{4.5.91}$$

$h(r)$ is found from the reverse transform of $\tilde{h}(Q)$, and so comes easily out of the data, but we need $c(r)$ as well. That variable can be obtained from the Ornstein-Zernike equation, Eq. (4.5.73),

$$h(r) = c(r) + \rho \int h(y) c(|\mathbf{r} - \mathbf{y}|) \, d^3y \tag{4.5.92}$$

Multiply both sides by $e^{i\mathbf{Q} \cdot \mathbf{r}}$, and integrate over d^3r:

$$\tilde{h}(Q) = \tilde{c}(Q) + \rho \iint h(y) c(|\mathbf{r} - \mathbf{y}|) \exp(i\mathbf{Q} \cdot \mathbf{r}) \, d^3y \, d^3r \tag{4.5.93}$$

where $\tilde{c}(Q)$ is defined as in Eq. (4.5.90). Changing variables to $\xi = r - y$ and inverting the order of integration give

$$\tilde{h}(Q) = \tilde{c}(Q) + \rho \iint h(y)c(\xi) \exp[i\mathbf{Q} \cdot (\xi + y)] d^3\xi d^3y$$

$$= c(Q) + \rho \int h(y) \exp(i\mathbf{Q} \cdot y) d^3y \int c(\xi) \exp(i\mathbf{Q} \cdot \xi) d^3\xi$$

$$= \tilde{c}(Q) + \rho\tilde{c}(Q)\tilde{h}(Q) \qquad (4.5.94)$$

This convenient property of the Fourier transform of a convoluted integral allows us to get $\tilde{c}(Q)$ directly from the data, since

$$1 - \rho\tilde{c}(Q) = \frac{1}{1 + \rho\tilde{h}(Q)} \qquad (4.5.95)$$

Notice, incidentally, that by evaluating Eq. (4.5.95) at $Q = 0$ and noticing that, from Eq. (4.5.90),

$$\tilde{h}(0) = \int h(r) d^3r \qquad (4.5.96)$$

and

$$\tilde{c}(0) = \int c(r) d^3r \qquad (4.5.97)$$

we find a direct connection between integrals over $h(r)$ and $c(r)$:

$$1 + \rho \int h(r) dr = \frac{1}{1 - \rho \int c(r) dr} \qquad (4.5.98)$$

The fluctuation equation, Eq. (4.2.58), may thus be written in two ways:

$$1 + \rho \int h(r) d^3r = \rho k T K_T \qquad (4.5.99)$$

$$1 - \rho \int c(r) d^3r = \frac{1}{\rho k T K_T} \qquad (4.5.100)$$

These equations show that at the critical point, where $K_T \to \infty$, the integral over $h(r)$ blows up; that is, $h(r)$ becomes long ranged but $c(r)$ remains well behaved, a fact we shall make use of in Chap. 6. On the other hand, when a condensed fluid becomes incompressible, $K_T \to 0$, $h(r)$ is well behaved but $c(r)$ behaves strangely.

In any case, the procedure we have outlined, finding $u(r)$ from scattering data by way of the PY equation, makes possible a quantitative estimate of the extent of the failure of the pairwise additivity approximation. It must be

4.5 Liquids

remembered, however, that the results will be valid only over range of T^* and ρ^* where Percus-Yevick gives an accurate account of the behavior of pairwise interacting fluids.

There is another tack we can take, a less quantitative one, in which we assemble the information we have and try to deduce from it what the important physical principles are in governing the behavior of fluids. For example, the Monte Carlo calculations give results sufficiently close to the behavior of real liquids and gases to allow us to conclude that the errors due to the approximations of pairwise additivity and Lennard-Jones potential are not important *in principle*. To be sure, there are quantitative difficulties, but the essential qualitative features, including self-condensation, are present in pairwise, Lennard-Jones systems. We can thus take a new look at the approximations that went into the integral equations, in light of their successes and failures, in order to get some idea of what is important for various purposes.

Consider first the YBG equation. In a way this one is the easiest to deal with because the superposition approximation on which it is based has direct physical significance. Any failure of YBG (compared, e.g., to Monte Carlo) is due to the importance of the failure of *mean* forces to add pairwise.

Under all circumstances of T^* and ρ^*, YBG gives the worst quantitative results of the three equations. The averaged effect of a third particle on mean force between two others is thus always quantitatively important. Moreover, these forces are obviously significant as well in producing self-condensation. On the other hand, it is argued that the superposition approximation ought to be accurate not only at low density (where it becomes exact) but at very high density as well, because when the atoms are tightly squeezed together, only pairwise repulsions between them can really be important. Perhaps it is this feature that leads to an apparent freezing transition in the YBG system.

The HNC equation differs from Monte Carlo only in the absence of the Bridge diagrams. Since Monte Carlo gives condensation at zero pressure, and HNC does not (see Table 4.5.1), we are led to the remarkable conclusion that self-condensation is essentially produced by the Bridge diagrams (more accurately, by the correlated configurations of many particles represented by the Bridge diagrams). That fact ought to be important somehow.

PY differs from HNC in that even more diagrams are discarded. Yet PY is nearly always more accurate than HNC. It follows that whatever the full effects of the Bridge diagrams, part of their job must be to cancel at least partially the contributions of the other diagrams that are left out of PY. To get some idea of what this may mean, let us look explicitly at some diagrams in the first few orders of density.

The easiest way to make the comparison is to write out $c(r)$ explicitly up to order ρ^2. The total correlation function, $h(r)$, differs from $c(r)$ only in that it includes the pure Series diagrams, and it does so in both theories: from Eq. (4.5.63) and (4.5.60),

$$c(r) = h(r) - S$$
$$= f_{12} + B + P$$
$$= f_{12} + B + f_{12}(S + B + P_1) + P_1 \qquad (4.5.101)$$

The exact expression for $c(r)$ up to ρ^2 is thus

$$c(r) = \circ\!\!-\!\!\circ + \rho\, \triangle + \frac{\rho^2}{2}\left[2\,\square + 4\,\boxtimes + \boxtimes + \boxtimes + \boxtimes \right]$$

$$f_{12} \qquad f_{12}S \qquad f_{12}S \qquad f_{12}S \quad f_{12}B \quad f_{12}P_1 \quad B \quad P_1$$
$$\phantom{f_{12} \qquad f_{12}S \qquad f_{12}S \qquad f_{12}S \quad}XHNC \qquad XHNC$$

(4.5.102)

Under each term we have indicated the set of diagrams from which it comes. In the HNC approximation

$$B = 0 \qquad (4.5.103)$$

or
$$c(r)_{\text{HNC}} = f_{12} + f_{12}(S + P_1) + P_1 \qquad (4.5.104)$$

we have

$$c(r)_{\text{HNC}} = \circ\!\!-\!\!\circ + \rho\, \triangle + \frac{\rho^2}{2}\left[2\,\square + 4\,\boxtimes + \boxtimes + \boxtimes \right]$$

$$f_{12} \qquad f_{12}S \qquad f_{12}S \qquad f_{12}S \quad f_{12}S^2 \quad S^2$$
$$\phantom{f_{12} \qquad f_{12}S \qquad f_{12}S \qquad f_{12}S \quad\;\;}XPY \quad\;\; XPY$$

(4.5.105)

Equation (4.5.105) is formed from Eq. (4.5.102) by dropping the diagrams with $XHNC$ below them in (4.5.102).

A certain care must be taken in deciding which diagrams to omit. The approximation, Eq. (4.5.103), does not mean that each individual Bridge diagram is equal to zero but rather that the sum over Bridge diagrams, Eq. (4.5.52), totals zero. It does follow that the Bridge diagrams must sum to zero in each order of ρ. $c(r)_{\text{HNC}}$ is really derived from Eq. (4.5.54), with $B = 0$ and $e^{-u(r)/kT} = 1 + f_{12}$.

$$g(r)_{\text{HNC}} = (1 + f_{12})e^S$$
$$= (1 + f_{12})\left(1 + S + \frac{1}{2!}S^2 + \cdots\right) \qquad (4.5.106)$$

4.5 Liquids

$$h(r)_{\text{HNC}} = g(r)_{\text{HNC}} - 1$$
$$= f_{12}\left(1 + S + \frac{1}{2!}S^2 + \cdots\right) + S + \frac{1}{2!} + S^2 + \cdots$$
(4.5.107)

$$c(r)_{\text{HNC}} = f_{12}\left(1 + S + \frac{1}{2!}S^2 + \cdots\right) + \frac{1}{2!}S^2 + \frac{1}{3!}S^3 + \cdots$$
(4.5.108)

In Eq. (4.5.105) each term is labeled according to the term in (Eq. 4.5.108) from which it arose.

In the PY approximation, Eq. (4.5.81),

$$(1 + f_{12})(B + P_1) = 0 \quad (4.5.109)$$

we have, according to Eq. (4.5.79),

$$g(r)_{\text{PY}} = (1 + f_{12})(1 + S) \quad (4.5.110)$$

or
$$h(r)_{\text{PY}} = f_{12} + S + Sf_{12} \quad (4.5.111)$$

$$c(r)_{\text{PY}} = f_{12}(1 + S) \quad (4.5.112)$$

In terms of diagrams, then,

$$c(r)_{PY} = \underset{f_{12}}{\circ\!\!-\!\!\circ} + \rho \underset{f_{12}S}{\triangle} + \frac{\rho^2}{2}\left[2\underset{f_{12}S}{\square} + 4\underset{f_{12}S}{\boxtimes}\right] \quad (4.5.113)$$

to order ρ^2. This equation excludes those diagrams in Eq. (4.5.105) that are marked XPY.

The theories are the same to leading order in density, and both theories include contributions from all orders of density—that is, from clusters of all numbers of particles. However, in each theory certain terms are dropped in each order (after the first). Consider a moment what this means.

In the HNC approximation we ignore, in second order, the diagrams below.

$$\underset{f_{12}B}{\boxtimes} \quad \text{and} \quad \underset{B}{\boxtimes}$$

In higher orders we drop all $f_{12}B$ and pure B diagrams, as well as all Parallel diagrams that have Bridge diagrams among their parallel branches. Thus, in second order, we retain, for example,

$$\boxtimes = \int f_{12}f_{13}f_{14}f_{23}f_{24}\,d^3r_3\,d^3r_4 \quad (4.5.114)$$

while dropping

$$\vcenter{\hbox{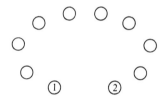}}\!\!\!\!\!\!\!\!\!\!\!\!\!\!\!\!\!\! = \int f_{12} f_{13} f_{14} f_{23} f_{24} f_{34}\, d^3 r_3\, d^3 r_4 \qquad (4.5.115)$$

The first gets a nonzero contribution from configurations in which atoms 3 and 4 are reasonably close to 1 and 2; for the second to be nonzero, 3 and 4 must also be close to each other. For a four-particle cluster this seems rather a poor approximation, because when 3 and 4 are close to both 1 and 2, they will necessarily be close to each other, so that the second integral above cannot be very much smaller than the first. However, in higher order, the approximation is more promising. Consider, for example, the particles with a physical configuration like that sketched in Fig. 4.5.14. This configuration and various distortions of it will contribute to Series terms in eighth order. Bridge terms, however, are nonzero in a much smaller region of configuration space, something like Fig. 4.5.15.

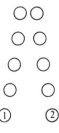

Fig. 4.5.14

Both HNC and PY take account of configurations like that in Fig. 4.5.14 but ignore those like Fig. 4.5.15. HNC includes but PY does not include configurations like Fig. 4.5.16—that is, products of Series diagrams with themselves. Excluding these as well as all Bridge diagrams somehow improves the accuracy of the result at high temperature.

The objective of all this is to see how we can go about devising a theory that will do a decent job of producing self-condensed liquid behavior. We have two insights to work with.

Fig. 4.5.15

4.5 Liquids

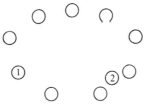

Fig. 4.5.16

1. The Bridge diagrams are of central importance in the self-condensed region.
2. Bridge diagrams cannot be very important in large clusters.

This suggests that all we need do is restore the low-order Bridge diagrams to HNC. Up to what order? If the cluster is small enough so that whenever

$$f_{34}f_{45}\cdots f_{jk}$$

is nonzero, 3 and k cannot be very far apart, so that f_{3k} will not be negligible, the Bridge diagrams ought to be included. We can estimate how far apart is far enough. In the true liquid range, typically,

$$\frac{\varepsilon_0}{kT} \approx 1$$

so that
$$f_{ij} = \exp\left[-\frac{u(r_{ij})}{kT}\right] - 1$$
$$\approx -\frac{u(r_{ij})}{kT}$$
$$\approx -4\left(\frac{\sigma}{r_{ij}}\right)^6 \qquad (4.5.116)$$

where we have assumed r_{ij} large enough to make $u(r_{ij})/kT \ll 1$. Thus, for $r_{ij} = 3\sigma$ we have

$$f_{ij} \approx -\frac{4}{(3)^6}$$
$$= -\frac{4}{729}$$
$$\approx -0.006 \qquad (4.5.117)$$

At $r_{ij} = 2\sigma$, the same sophisticated analysis gives

$$f_{ij} \approx -0.06 \qquad (4.5.118)$$

It would seem that we might stop worrying about Bridge diagrams beyond about 2σ separations—that is, beyond about second-neighbor spacing (very crudely). Each atom has roughly 12 first and second nearest neighbors in a liquid, so we are lead to propose the following scheme: let us modify HNC by adding on all contributions from Bridge diagrams up to about tenth order in density (12 particle clusters).

Here we run into difficulty. The number of possible cluster diagrams of 12 particles is $2^{(12)(11)/2} = 2^{66}$, a rather substantial number. To be sure, by no means all of them will be irreducible, and not all the irreducible ones are Bridge diagrams; nevertheless, the task of enumerating and computing all the proposed corrections, in each order, for every value of r promises to be a formidable one. Once it is done, we must still solve an integral equation to get $g(r)$ and find the thermodynamic predictions.

And if we were to carry out this program, and even if we were to discover that a condensed liquid resulted, what would we have gained? Perhaps an accurate equation of state, but that can easily be obtained by fitting experimental data, say, to a polynomial. Besides, if we used Lennard-Jones, the result would obey the Law of Corresponding States, which real liquids do not. The program might be worthwhile only if it gave some promise of revealing *why* Bridge diagrams are so important in condensed liquids.

The world is made of nucleons and electrons, which combine to form atoms, which combine in various ways to form successively more complex structures, including even you and me. All of this we understand very well, and yet there is no way to deduce, from elementary principles, the existence of you and me. The problem that we are dealing with in this section is one of the most elementary steps in that chain of deduction: how simple, identical atoms get together to form liquids. What we have seen is that the combinatorial problem—the correlations in the behavior of many, or even not so many, simple atoms—quickly eludes the grasp of our intuition. The essential job of the theory of liquids is to refine and enrich our intuition, until the central features of the liquid state seem as clear to us as, say, the vibrations of a harmonic crystal or the way in which cells synthesize DNA. The fact is that if one inquires closely enough, even our elegant and eminently successful theories of solids have their quantitative shortcomings. Their success lies in that we believe that we have a firm grasp of the basic principles of the problem. It is that kind of grasp that we lack in the liquid problem. All the formalism we have gone through is no substitute for the intuition we need. The hope, rather, is that it may lead us to that intuition.

BIBLIOGRAPHY

Discussion of much of the material found in this chapter may be found in a reasonably simple but seriously flawed monograph, P. A. Egelstaff, *An Introduction to the Liquid State* (New York: Academic Press, 1967). The treatment here, we are told, "...has been chosen for its physical content rather than mathematical rigour." That is a trying job, and it is not always done well in this book. There is, for example, the "derivation" presented for the Feynman theory of liquid helium, p. 219, arrived at by combining mutually exclusive approximations (we will discuss the Feynman theory in Chap. 5). One further example: the author argues that we can tell that liquids are less ordered than solids by noticing that they have larger entropy fluctuations (p. 9). He uses Eq. (1.3.194),

$$\overline{(\Delta S)^2} = kC_P$$

pointing out that, for example, in argon near the triple point, the liquid has $C_P/N = 5.06k$ and the solid $C_P/N = 3.89k$. Following his logic, we conclude that the ideal gas, $C_P/N = 2.5k$, is more ordered than either. There are other lapses of this sort. Also, following his dictum above, many derivations lack sufficient detail to be followed without outside help. Nevertheless, it does seem to be the only monograph written at a reasonably comprehensible level, and it presents a fair outline of the kind of work done in the field.

S. A. Rice and P. Gray, *The Statistical Mechanics of Simple Liquids* (New York: Interscience, 1965), is a formidable, authoritative tome, about which nothing is as simple as the title seems to imply. The authors say, of the notion that there is no acceptable theory of the liquid state, "It is the purpose of this book to show that this view is obsolescent." They then proceed, in mind-shattering detail, to come very close to proving the opposite.

It is particularly amusing to compare Rice and Gray to Landau and Lifshitz (see bibliography, Chap. 1) on the delicate issue of the pairwise additivity approximation. Early on, Rice and Gray (p. 15) assume pairwise additivity, "...as we do throughout the rest of the book...." Landau and Lifshitz, on the other hand, comment on the interaction energy of three atoms, "...(which in general is not equal to the sum [of pair potentials])...." So much for pairwise additivity. They agree to dismiss the matter in a parenthetical comment, drawing, however, opposite conclusions.

The idea of cluster expansions was first introduced by Joseph Mayer in the late 1930s, and a discussion of their use may be found in the book on statistical mechanics by Mayer and Mayer (bibliography, Chap. 1). Some useful discussion of scattering in fluids, for the determination of both structure and dynamic properties, appears in the monograph on critical phenomena by Stanley (bibliography, Chap. 6). The derivations we have used leading to Eqs. (4.4.82) and (4.5.41) were adapted from an article, E. E. Salpeter, "On Mayer's Theory of Cluster Expansions," *Annals of Physics*, **5**, 183(1958). The analysis and discussion of Sec. 4.5c are based in some measure on numerical solutions of the integral equations found in two articles: D. Levesque, "Etude des équations de Percus et Yevick, d'hyperchaîne et de Born et Green dans le cas des Fluides classiques," *Physica*, **32**, 1985(1966), in easy French, and G. J. Throop and R. J. Bearman, "The Pair Correlation Function and Thermodynamic Properties for the Lennard-Jones 6-12 Potential and the Percus-Yevick Equation," *Physics*, **32**, 1298(1966), in obscure English.

Finally, the remark by Galileo mentioned on page 246 comes from *Il Saggiatore* (Roma: Mascardi, MDCXXIII) p. 25. What he really wrote was: "*La Filosfia è scritta in questo grandissimo libro, che continuamente ci sta aperto innanzi à gli occhi (io dico l'universo) ma non si puo intendere se prima non s'impara à intendere la lingua, e conoscer i caratteri, ne quaili è scritto. Egli è scritto in lingua matematica....*" "(Natural) philosophy is written in this enormous book, which is continuously open before our eyes (I mean the universe) but it cannot be understood without first learning to understand the language, and recognize the characters in which it is written. It is written in mathematical language...."

PROBLEMS

4.1 **a.** Using a model of $g(\mathbf{r})$ that realistically represents a solid at finite temperature, find the temperature dependence of the scattered X-ray intensity. Make a sketch of $S(Q)$. You will need to know the mean square displacements of atoms about their equilibrium positions, a quantity you investigated in Prob. 3.2.

b. What happens if the crystal is two dimensional instead of three dimensional?

4.2 **a.** Using a simple model of a solid to give $g(\mathbf{r})$, show that at high temperature

$$\frac{3}{2}kT = \frac{\rho}{2}\int g(\mathbf{r})u(r)\,d^3r$$

b. For the same model, investigate the relations

$$P = \frac{N}{V}kT\left[1 - \frac{\rho}{6kT}\int r\frac{du}{dr}g(\mathbf{r})\,d^3r\right]$$

$$kT\left(\frac{\partial \rho}{\partial P}\right)_{V,T} = \rho\int [g(\mathbf{r}) - 1]\,d^3r + 1$$

What would these quantities be, using conventional solid formalism?

c. What does your model give for the heat capacity of the solid at low temperature?

4.3 **a.** Working entirely in the variable N formalism, show that

$$P = \rho kT\left[1 - \frac{\rho}{6kT}\int r\frac{du}{dr}g(r)\,d^3r\right]$$

b. Compare the compressibility computed from the preceding equation to the compressibility in the fluctuation equation of state. Look for rigorous conditions that, for example, might be applied to test the assumption of pairwise additivity.

4.4 Using the thermodynamic condition that the compressibility can never be negative, find a rigorous condition on $\int [g(r) - g_N(r)]\,d^3r$ [where $g_N(r)$ comes from the fixed N formalism]. Is the result testable? Does it depend on pairwise additivity? Under what conditions would you expect $g(r)$ to differ appreciably from $g_N(r)$?

4.5 **a.** Find the expression for the entropy in the dilute gas approximation.

b. Compare the dilute gas departure from ideality to the leading-order quantum corrections to the perfect gas (see Sec. 2.4). What is the criterion for ignoring the quantum correction relative to the dilute gas correction? Might there be any other quantum corrections to worry about?

c. Suppose that you wished to make use of experimental measurements of the equation of state of a gas in the dilute gas region to measure the parameters ε_0 and σ in the Lennard-Jones potential. Discuss how you would design the experiment and how you would analyze the data.

Problems

4.6 Write out explicitly the third and fourth virial coefficients in terms of $u(r)$, showing which three- and four-particle diagrams you use and which ones you do not use. What has happened to the problem of more than one collision at a time? For example, in our derivation of the coefficients of terms in the virial expansion, what did we do with configurations in which there were simultaneous two-particle collisions?

As mentioned in the preface, the problems in this book have all been consumer tested in the form of examination and homework problems in the course upon which the book is based. The section on liquids presents a special problem in that it is difficult to dream up nontrivial problems that can be done without computer techniques and whose solutions would not be publishable. Consequently, in this part of the course, the device of a term paper was always used. The following is an example:

Aph 105b   **Term Paper**   Due Friday, March 15, 1974

Topic: The Equilibrium Statistical Mechanics of Liquids: What one would like to know, and how to find out

Discuss the problem critically, evaluating the success or lack of it of the enterprise, and give at least one example of what you think ought to be done. Be specific: if you suggest an experiment, for example, be sure to estimate the precision and range of variables necessary and how that compares to what is ordinarily possible. The experiment must be designed to help answer whatever you think the most important question is. In reading your paper I will look for three things:

1. Your understanding of what we did in class
2. Your scientific judgment—that is, your ability to see what is important and where the difficulties lie
3. Your ability to plan, in detail, what to do next

You may use any source you can lay your hands on, but be sure to list all sources you use, published or otherwise, aside from the class notes and textbook.

FIVE

SOME SPECIAL STATES

5.1 INTRODUCTION

In the last three chapters we have surveyed the grand, pervasive states of matter: solids, liquids, and gases. In this chapter we look at some particular special states: superfluidity, superconductivity, and magnetism. We could equally well have chosen other states to study: plasmas (sometimes called the Fourth State of Matter), liquid crystals, two-dimensional matter in the form of adsorbed films. The choices that we have made serve to illustrate and reinforce certain general principles, but they do not do so uniquely.

There are certain general, sweeping tendencies in the properties of matter. At high temperatures and low densities, things tend to be random and disorganized, dominated by the behavior of single particles acting alone. This is the gas limit. At low temperature and higher density, matter goes into highly ordered states; moreover, the excitations out of the ordered state are collective, organized responses, which we can treat as quasiparticles. Phonons in solids are an obvious example. In between these limits we have behavior that is correlated and disorganized at the same time and hence much more difficult to understand than at either limit: the liquid state.

Within the liquid and solid states, there are, so to speak, internal subsystems that exhibit the same general tendencies. Superfluid helium is a perfectly ordered liquid whose excitations consist of phonons as well as other types that are possible only in a fluid. The high-temperature disordered state

5.2 Superfluidity

is the ordinary liquid. Superconductivity is a subtle, collectively ordered state of the electron gas in metals, the (relatively) disordered state being the degenerate Fermi gas. Ferromagnetism is an ordered state of magnetic moments, and paramagnetism is the corresponding disordered, single-particle state. In all three instances, the intermediate situation is a critical phase transition—that is, a phase transition without latent heat. Critical phase transitions are the subject of Chap. 6.

In our discussions in this chapter and the next, we shall make much use of the machinery and language that we have already developed: collective modes and quasiparticles, correlation functions and variational principles. It would be unfair to cite this usage as a proof of the underlying unity of nature, however. For one thing, these topics were chosen for study partly because they do make use of the ideas that we have already developed. For another, the recurrence of these themes reflects, to some extent, the paucity of genuinely new ideas in physics. We tend to think by analogy and therefore to impose analogies on our descriptions of nature. Nevertheless, much of our success in applying these analogies can be traced back to a handful of very deep principles in nature, particularly the ideas underlying the Second and Third Laws of Thermodynamics. All permissible states of a system are equally likely. Any system has only one possible ground state and must thus become perfectly ordered as absolute zero is approached. All our descriptions of equilibrium states of matter are, in effect, particular applications of these ideas. Electromagnetism and quantum mechanics govern the behavior of particles and dictate the individual possibilities when many particles get together. But it is the two ideas above that form the organizing principles of matter.

5.2 SUPERFLUIDITY

a. Properties of Liquid Helium

There is a triple point on the helium vapor pressure curve just about where we would expect to find solidification in any reasonable material. But helium, instead of freezing, undergoes a phase transition known as the lambda point because a plot of the heat capacity, which has a logarithmic infinity at that point, looks a bit like the Greek letter lambda. In spite of the infinite heat capacity, there is no latent heat—the transition is of the critical variety discussed in Chap. 6. Figure 5.2.1 is a sketch of the helium phase diagram and the heat capacity.

In some ways, the most remarkable of all the properties of liquid helium is the simple fact that it remains a liquid all the way down to absolute zero. The problem here is the Third Law of Thermodynamics, which tells us that the entropy must be zero; we must somehow account for a liquid without

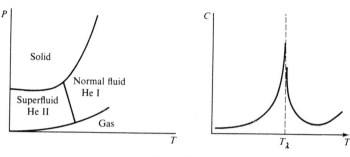

Fig. 5.2.1

entropy. Being a liquid, it can be stirred—atoms can move about throughout the occupied volume. For this reason, liquids have what we called communal entropy in Chap. 4. Somehow, at absolute zero, the superfluid manages to be a liquid without communal entropy.

When helium goes through the lambda point and becomes superfluid, it undergoes a change of state from a colorless transparent liquid into another colorless transparent liquid at the same density. Offhand, one would not expect the change to be visually dramatic, but it is quite a dramatic change and easily seen. When normal liquid helium is cooled by pumping away its vapor, it boils vigorously, swarms of tiny bubbles rushing to burst at the surface. But at the lambda point it suddenly stops boiling and becomes deadly calm—in the superfluid state it looks much like a very dry martini. The cessation of boiling is a consequence of its super heat conductivity. In an ordinary liquid, like water or normal helium, boiling occurs because hot spots develop, causing the vapor pressure to rise and overcome the local hydrostatic pressure. A bubble then forms and rises through the liquid. The superfluid, however, is too good a conductor of heat to allow hot spots to develop. The liquid continues to evaporate into gas (so that it can be further cooled by pumping away its vapor), but the evaporation takes place entirely at the liquid-vapor interface, leaving the bulk of the liquid quiet and undisturbed.

The most spectacular property of helium, the one that really gives the superfluid its name, is its ability to flow through small holes without resistance. If caused to flow slowly through a fine capillary, shown schematically in Fig. 5.2.2, then there is no pressure drop across the capillary. In any ordinary fluid there would have to be a pressure gradient through the tube, since some force would be necessary to overcome the effect of shear viscosity. In this sense, helium behaves as if it had no viscosity, and we are tempted to guess that zero viscosity is a property of the superfluid. The situation, however, turns out to be more complicated.

5.2 Superfluidity

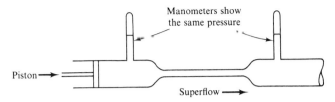

Fig. 5.2.2

Imagine a fluid flowing in the positive x direction with velocity v_x past a wall in the x-y plane. Atoms adjacent to the wall collide with it, lose momentum, and are brought to rest. The next layer of atoms collides with the first layer and is slowed down, and so on; there is a velocity gradient in the fluid, with the wall effectively exerting a drag force per unit area, F_D, given by

$$F_D = \eta \frac{\partial v_x}{\partial z} \tag{5.2.1}$$

where η is the coefficient of shear viscosity of the fluid. The situation is shown in Fig. 5.2.3. For an ordinary fluid flowing through the capillary shown in Fig. 5.2.2, a driving force would be necessary to overcome the drag force, and we would observe a lower pressure in the right-hand manometer than in the left. The pressure difference—or lack of one in the case of helium—is a measure of the viscosity of the liquid, which is basically the tendency of a fluid to come gradually to rest locally with respect to any surface it tries to flow past. According to this kind of measurement, the superfluid is evidently entirely inviscid.

There is, however, another way of measuring viscosity, and it gives a different result. It should not, in principle, matter whether we move the fluid past the wall or the wall past the fluid. If we hold the fluid far away at rest and try to move an object through it, the same viscous forces should exert a drag force on the object. Specifically, suppose that we pluck a violin

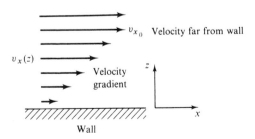

Fig. 5.2.3

string in a viscous fluid. The vibrations die away more rapidly than they would if the string were in a vacuum, and the difference in relaxation time is a measure of the viscosity. If we measure the viscosity of the superfluid in this way, we get a finite result, small, to be sure, but unquestionably inconsistent with the result of the previous experiment.

The only way to understand this peculiar behavior is to imagine that helium is composed of two different fluids that completely interpenetrate each other. One is normal, in the sense that it has an ordinary viscosity, and we shall call its density ρ_n and its velocity $\mathbf{u}_n$. The other fluid is super—it has no viscosity—and we assign to it a density ρ_s and a velocity $\mathbf{u}_s$. Now our picture of the capillary experiment is that the viscous forces bring the normal fluid entirely to rest, so that no forces are exerted, according to Eq. (5.2.1), whereas the superfluid part flows freely, giving the observed result. The vibrations of the violin string, on the other hand, are damped out by viscosity in the normal fluid, since the normal part cannot be simultaneously at rest with respect to the vibrating string and the walls of the container.

With this two-fluid model as a context, we can discuss some further properties of helium. The density at any point, ρ, is the sum of the two partial densities (in this discussion, ρ is the mass per unit volume)

$$\rho = \rho_s + \rho_n \tag{5.2.2}$$

It is approximately independent of temperature. The partial densities do depend on temperature, however, with the normal fraction ρ_n/ρ going from 1 at the lambda point to zero at zero temperature. The overall mass flow, $\rho\mathbf{u}$, where $\mathbf{u}$ is the center of mass velocity, can be described as the sum of two separate currents

$$\mathbf{j} = \rho\mathbf{u} = \rho_s\mathbf{u}_s + \rho_n\mathbf{u}_n \tag{5.2.3}$$

[we shall see later that Eq. (5.2.3) follows from a precise definition of ρ_n]. The normal part then obeys (more or less) normal hydrodynamics, but the super part has the important restriction,

$$\nabla \times \mathbf{u}_s = 0 \tag{5.2.4}$$

It is easy to see that Eq. (5.2.4) precludes the viscous flow situation depicted in Fig. 5.2.2, since, according to Eq. (5.2.4),

$$\frac{\partial u_{sx}}{\partial z} = \frac{\partial u_{sz}}{\partial x}$$

but there is no z component to the velocity, so u_{sx} can have no gradient in the z direction. Flow that obeys Eq. (5.2.4) is said to be irrotational.

Before continuing, we should mention that there is a catch in all of this. Superflow of the kind that we have been describing occurs only at low velocities. Just how low depends in a complicated (and not perfectly under-

5.2 Superfluidity

stood) way on the particular experiment, but if a certain critical velocity is exceeded in each experiment, resistance to flow sets in quite sharply. We shall return to this point at the end of our discussion of helium.

There is a certain confusion of nomenclature that we must learn to contend with here. Liquid helium above the lambda transition is call normal, and below, super; on the other hand, the superfluid itself has now been divided into super and normal parts. Although it does not help much, the normal liquid above the transition temperature T_λ is sometimes called He I and the state below T_λ, He II. There are thus He I, He II, ^{3}He, ^{4}He, and ^{6}He, this last being an unstable isotope with a radioactive decay half-life of less than 1 second.

Another interesting property of He II has come to be known as the *fountain effect*. This effect is the ability of the superfluid to balance a pressure difference with a temperature difference, provided that the normal fluid is unable to move.

Imagine two beakers connected by a capillary packed with fine powder. Normal fluid cannot move through the powder-packed capillary, but superfluid can, so it serves as a superleak connecting the two containers (Fig. 5.2.4). Now suppose that we heat the left-hand side a bit. Superfluid is drawn through the superleak toward the heater, and the left-hand level rises as in Fig. 5.2.4. If the two sides are reasonably well isolated thermally, we can turn off the heater and a sort of equilibrium will establish itself, in which the left side is at a higher temperature than the right, and there is also a higher pressure at the left-hand end of the superleak, applied by the extra hydrostatic head, $\rho g \, \Delta z = \Delta P$, g being the acceleration of gravity. It is found, for small differences, that ΔP and the excess temperature ΔT are connected by

$$\frac{\Delta P}{\Delta T} = \frac{S}{V} \qquad (5.2.5)$$

where S/V is the entropy per unit volume of the fluid. Notice that although the superfluid in bulk is a superconductor of heat, it becomes a nonconductor when the normal fluid is unable to move.

One final property, of great importance in understanding the nature of the superfluid, is its ability to sustain a persistent circular current at a definite

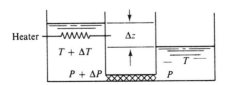

Fig. 5.2.4

velocity. To demonstrate, a doughnut-shaped container is packed with powder and filled with liquid helium. The apparatus is set into rotation about its central axis, as in Fig. 5.2.5, with the temperature above T_λ, so that the entire fluid is normal and goes into solid-body rotation. Still rotating, the temperature is reduced below T_λ, so that we now have rotating superfluid. At this point the container is brought to rest. If the superfluid is still rotating, angular momentum is stored inside, and thus it is a kind of gyroscope. If the axis is tilted, it precesses about its original direction with a rate proportional to the amount of stored angular momentum. It is found that at low velocities, if the temperature is well below T_λ, the stored current will persist for an indefinite amount of time.

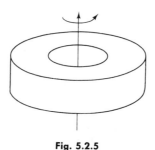

Fig. 5.2.5

So far we have only a very sensitive test of a property that we already asserted—if the normal fluid is clamped (by the fine powder), the superfluid flows without resistance and thus has no reason not to persist. However, there is a deeper point to this behavior. The stored angular momentum is $\ell \propto \rho_s <u_s>$, where $<u_s>$ is the average superfluid velocity inside. If the temperature is raised (but not too close to T_λ), ρ_s decreases and ℓ is observed to decrease. If we now lower the temperature again, ℓ increases again, reversibly: we have applied a reversible torque to the fluid, thereby changing its angular momentum, simply by changing the temperature. The temperature dependence of ℓ is the same as that of ρ_s and is, in fact, an excellent way of measuring how ρ_s depends on T. It is the velocity rather than the angular momentum or current that has gone into a fixed, persistent state, an observation that cannot be explained merely by zero viscosity.

Let us summarize the properties that we have enumerated for the superfluid:

1. There is a phase transition of the critical type into the superfluid state.
2. Helium is able to remain fluid down to absolute zero.
3. The superfluid is a superconductor of heat.
4. There is a two-fluid behavior, with the super part capable of inviscid flow.

5. Pressure differences can be balanced by temperature differences when the normal fluid part is clamped.
6. The superfluid can go into persistent current states of fixed velocity.

There is another peculiar aspect of superfluid behavior that one often hears about. He II drains itself out of any container it is held in by flowing through a film it forms on the walls. We have not listed this point separately because it is a consequence of property 4 above, together with another property that is not directly related to superfluidity: helium is the universal wetting agent. The helium-helium force is the weakest atomic force in nature, and it follows that the helium-anything else force is stronger; a helium atom would rather be next to anything other than another helium atom. Presented with the wall of a beaker, liquid helium quickly forms a film whose thickness is limited only because at some point the advantage of being reasonably near the wall is overcome by gravity. In this film, typically a few hundred angstroms thick, the normal fluid is clamped viscously to the wall, but the superfluid flows freely. Consequently, the films acts as a siphon, the helium in any beaker flowing out to seek the lowest gravitational level available. We could list as another property whatever the phenomena are that cause breakdown of superflow at some critical velocity, mentioned earlier. We shall return to it at the end of this section. Overall, however, the six properties listed above are the central ones that we must explain in order to understand superfluidity.

b. The Bose Gas as a Model for Superfluidity

Superfluidity in ^{4}He has long been regarded as being related to the Bose condensation of a perfect gas, examined in Sec. 2.6. In order to orient ourselves and understand better just what an adequate theory of liquid helium would require, let us take seriously the degenerate Bose gas as a model of the superfluid. We shall consider, in turn, each of the six superfluid properties listed earlier and see to what extent they are related to Bose gas behavior.

1. The phase transition

The Bose gas also undergoes a phase transition, and it is often pointed out that if its density were that of liquid helium (0.14 gram/cm^3), the transition temperature would not be very different from the lambda temperature, thereby helping to confirm the similarity between the two phenomena. The argument is specious, however, and the agreement mostly accidental. The Bose transition temperature increases with increasing density, whereas the lambda temperature has the opposite behavior. There is some critical density for Bose condensation to occur at any temperature, whereas the lambda transition runs only from the melting curve at 1.76°K to the vapor pressure curve at 2.17°K. The Bose condensation is a first-order phase transition, the lambda line a critical transition. Altogether, there is actually little similarity between the two.

To some extent the differences can be understood qualitatively as resulting from the fact that the Bose gas is taken as composed of point particles, while helium consists of atoms with finite volume. Thus, higher density in the Bose gas increases the statistical degeneracy by increasing the probability of multiple occupation of single-particle states, but in the real liquid higher density impedes the passage of atoms by each other, effectively decreasing the ability of the wave functions to overlap and decreasing the degeneracy. The greater the degeneracy, the higher the transition temperature. Furthermore, helium atoms cannot fall into a zero-volume condensate as the Bose particles do, which perhaps explains why the transition is not first order. Nevertheless, the Bose condensation does little to help us understand the lambda transition, and the transition remains the least well understood aspect of superfluidity. We shall discuss the problem further, together with other critical phase transitions, in Chap. 6.

2. Fluid at zero temperature

The Bose condensate, which occupies zero volume, does not seem a very helpful model for resolving the dilemma of how a fluid can have zero entropy, but, in fact, it is just the indistinguishability of the helium atoms that is needed to explain what happens. If we think of the atoms as being identical, then stirring them around without changing the density replaces atoms with other indistinguishable ones and thus gives us back the same state of the system. If other states are to be possible, so that there is some entropy, energy is required as well, and we are no longer at zero temperature. This picture forms the basis of the Feynman theory, to which we shall return later. The excited states of the system are longitudinal density fluctuations, or phonons, as in a solid, and another kind, called *rotons*, which can occur only in a liquid.

We have not explained why helium does not freeze, only why it does not have to. Like any other substance, it would freeze at finite T if by doing so it could reduce its free energy

$$F = E - TS$$

Except at $T = 0$, freezing will generally reduce S (by changing the kinds of excited states available), thereby increasing F. A substance will only freeze, then, if it can reduce its energy by doing so enough to compensate for the increase in the term $-TS$. This turns out to be difficult for helium to do, partly because of the weak attractive potential between helium atoms and partly because of the large zero-point energy that an atom as light as helium has when it is localized. Helium can be solidified, but only under applied pressure (see Fig. 5.2.1).

3. Superconductivity of heat

On this point we do get some guidance from the Bose gas. Below the condensation point the pressure of the Bose gas depends only on the temperature [see Eq. (2.6.26)]. A temperature gradient would cause a pressure gradient in the gas—an unbalanced force that would have to even itself out. Gases do not support pressure gradients; instead they relax themselves at the speed of sound. Thus, the Bose gas is, in fact, a superconductor of heat. As we shall see, something very much like this happens in liquid helium. The superconductivity of heat is a quite direct consequence of the two-fluid behavior of He II.

4. Two-fluid behavior

The two-fluid behavior of helium has an obvious counterpart in the Bose gas: for the super and normal fluids, we have the condensate and the excited particles. As in helium, the total density may remain constant, but the density of each component depends on temperature, going from all condensate at $T = 0$ to all excited at the transition temperature. The Bose gas differs from helium, however, in that the condensate does not have superfluid properties; it will not flow past a body without resistance. Of course, those particles in the condensate, being in a single quantum state, cannot be slowed down near a wall. But energy and momentum can be exchanged with a wall if particles are excited out of the condensate, and loss of energy and momentum will show up as resistance to flow. In order to be a superfluid, there must be some mechanism that prevents excitation out of the condensate when it flows past a wall. If no excitation can take place, then the flow will be inviscid.

Consider a process in which a body of mass M, momentum $\mathbf{P}$, and energy $E = P^2/2M$ creates an excitation of momentum and energy $\mathbf{p}$ and ε. The energy of the body afterward is

$$E' = E - \varepsilon \qquad (5.2.6)$$

and the momentum

$$\mathbf{P}' = \mathbf{P} - \mathbf{p} \qquad (5.2.7)$$

Square both sides of Eq. (5.2.7) and divide by $2M$:

$$\frac{P'^2}{2M} - \frac{P^2}{2M} = -\frac{\mathbf{P} \cdot \mathbf{p}}{M} + \frac{p^2}{2M} \qquad (5.2.8)$$

According to Eq. (5.2.6), the left-hand side of (5.2.8) is $-\varepsilon$, so we may write

$$\varepsilon = \frac{pP \cos \theta}{M} - \frac{p^2}{2M} \qquad (5.2.9)$$

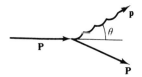

Fig. 5.2.6

where θ is the angle between **P** and **p** (see Fig. 5.2.6). Since the velocity of the object is simply

$$v = \frac{P}{M}$$

we have

$$v \cos \theta = \frac{\varepsilon}{p} + \frac{p}{2M} \qquad (5.2.10)$$

Now let us take $M \to \infty$ (the object could be, for example, the walls of a capillary through which the helium is flowing). Then since $\cos \theta \leq 1$, we have the condition

$$v \geq \varepsilon/p \qquad (5.2.11)$$

An excitation of energy ε and momentum p cannot be created in flow past an object unless Eq. (5.2.11) is satisfied. If ε/p has some minimum value, then at velocities lower than that value, the condensate is a superfluid. We thus appear to have accounted, in one stroke, both for the phenomenon of superfluidity and for its breakdown at some critical flow velocity. The observation that superflow can exist when Eq. (5.2.11) is not satisfied is known as the *Principle of Superfluidity*.

If our model of liquid helium is the perfect Bose gas, however, then the spectrum of the excited states is

$$\varepsilon = \frac{p^2}{2m} \qquad (5.2.12)$$

where m is the mass of a helium atom, and we therefore have resistance to flow at velocity

$$v \geq \frac{\varepsilon}{p} = \frac{p}{2m}$$

Excited states are available at any velocity down to zero, since p can have any value. The perfect Bose gas is not a superfluid.

We already know, of course, that the perfect Bose gas must be modified in some way before it can serve as an adequate model for superfluid helium. Clearly, we shall, among other things, have to abandon Eq. (5.2.12) as the energy spectrum of the excited states. Let us do so provisionally at this point. Unlike the perfect gas, the real atoms have both attractive and repulsive interactions with each other. Just as in the case of crystalline solids (see Chap. 3), the lowest energy state (the condensate) will consist of a uniform-density medium, with each atom as nearly as possible in a position of minimum potential energy in the force fields of its neighbors. Excitation of individual atoms—the picture that gives a spectrum of the type of Eq. (5.2.12)—will be very expensive energetically. If Δx is the mean distance between the left- and right-hand neighbors of an atom, the lowest quantum-excited state of the type of Eq. (5.2.12) will have momentum of the order $p \sim \hbar/\Delta x$ instead of the ideal gas value, $p \sim \hbar/L$, where L is the size of the box. In other words, the atoms are in boxes made by their neighbors, and so single-particle excited states have high energies. The lowest excited energy states of the system will be collective, and from our discussion of crystals we know what they will be. The lowest states will be long wavelength phonons, with spectrum

$$\varepsilon = c_1 p \qquad (5.2.13)$$

where c_1 is the speed of sound. According to Eq. (5.2.11), these cannot be excited at flow velocities less than c_1, so that, for this part of the spectrum at least, we can understand superflow. Let us now continue to examine the applicability of the Bose model, keeping this modification in mind.

5. The fountain effect

Consider two sealed containers of fixed volume, connected by a superleak, with N total particles, as in Fig. 5.2.7. The two sides (referred to by subscripts 1 and 2) are thermally isolated, but particles can flow from one side to the other by means of superflow—that is, in the form of condensate. The general condition for equilibrium of the system is to be deduced from Eq. (1.2.78) in the form

$$\delta E \leq T\,\delta S - P\,\delta V + \mu\,\delta N \qquad (5.2.14)$$

where $E = E_1 + E_2$, $S = S_1 + S_2$, $N = N_1 + N_2$, and $V = V_1 + V_2$. N and V are fixed, although N_1 and N_2 may change. Such changes, however,

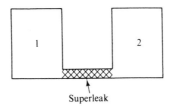

Superleak

Fig. 5.2.7

take place by means of flow of the condensate, which, being a single quantum state, carries no entropy; in other words, changes in N_1 and N_2 take place without changing S_1 and S_2. We thus have that $\delta S = \delta N = \delta V = 0$, or

$$\delta E \leq 0 \tag{5.2.15}$$

The system is in equilibrium when its energy is minimized at constant S, N, V; if N_1 tries to change by superflow, we require only that

$$0 = \left(\frac{\partial E}{\partial N_1}\right)_{S,N,V} = \frac{\partial E_1}{\partial N_1} + \frac{\partial E_2}{N_1}$$

$$= \left(\frac{\partial E_1}{\partial N_1}\right)_{S_1,V_1} - \left(\frac{\partial E_2}{\partial N_2}\right)_{S_2,V_2}$$

$$= \mu_1 - \mu_2 \tag{5.2.16}$$

Equilibrium requires only that the chemical potentials be the same on the two sides, not the pressures and temperatures, which means, for small changes, using Eq. (1.2.35)

$$0 = d\mu = -\frac{S}{N}dT + \frac{V}{N}dP$$

or

$$\frac{dP}{dT} = \frac{S}{V} \tag{5.2.17}$$

which is the observed result. The fountain effect is basically a result of the ability of mass to flow without entropy. Consequently, it strongly confirms the idea that the superfluid part is a single quantum state, without entropy, and so also confirms the wisdom of taking the Bose gas as our central model.

On the other hand, this treatment raises a new dilemma by focusing our attention on the chemical potential. The hallmark of the degenerate Bose gas is that its chemical potential is always equal to zero [Eq. (5.2.16) then follows trivially, since μ_1 and μ_2 are both always zero]. Real liquid helium, however, cannot have a zero chemical potential. Along the vapor pressure curve, it has the same chemical potential as its own vapor, which, in turn, is quite accurately a classical ideal gas (owing to its low density) and thus has a chemical potential that is large and negative (see Sec. 2.2). Yet the particles that are Bose condensed must have zero chemical potential—this is the only way that the number of excited particles can be variable and temperature dependent. Actually, we have already seen that it is not particles but rather collective motions that are excited out of the condensate. The fact that our strange new degenerate Bose state has a nonzero chemical potential is quite consistent with the fact that its excitation energy spectrum is collective rather than single-particle-like. The excited "particles" are quasiparticles, phonons for example, and *their* chemical potential is, in fact, zero, but that of the atoms is not.

5.2 Superfluidity

6. Persistent currents

Think of a Bose condensate, which consists of many particles, all in the same quantum state. The wave function of a single particle in the state may, in general, be complex, so we write it as

$$\psi(\mathbf{r}) = a(\mathbf{r}) \exp\left[i\gamma(\mathbf{r})\right] \tag{5.2.18}$$

where a is the amplitude and γ the phase. For the simplest case of a gas of particles in a box with periodic boundary conditions, a is constant, and $\gamma = \mathbf{q} \cdot \mathbf{r}$ for a state with momentum $\hbar\mathbf{q}$. The ground state is the state with $\mathbf{q} = 0$. We wish to generalize now, in order to be able to write the wave function for a ground state that is not necessarily at rest (external conditions, such as rotating walls, may force it to be in motion). Equation (5.2.18) is the most general wave function that we can write for one particle. The quantum mechanical current of particles in this state is

$$\mathbf{j} = \tfrac{1}{2}(\psi \hat{p} \psi^* + \text{complex conjugate}) \tag{5.2.19}$$

where the momentum operator is

$$\hat{p} = -i\hbar \, \nabla \tag{5.2.20}$$

Substituting this and Eq. (5.2.18) into (5.2.19), we find

$$\mathbf{j} = -\hbar a^2 \, \nabla \gamma(\mathbf{r}) \tag{5.2.21}$$

But the contribution of each particle to the current is

$$\mathbf{j} = m a^2 \mathbf{u}_s \tag{5.2.22}$$

where $\mathbf{u}_s(\mathbf{r})$ is the velocity field of the particles in this single quantum state. We have then

$$\mathbf{u}_s = -\frac{\hbar}{m} \nabla \gamma \tag{5.2.23}$$

Since $\mathbf{u}_s$ is the gradient of a scalar, we have immediately that

$$\nabla \times \mathbf{u}_s = 0 \tag{5.2.4}$$

so that the restriction on superflow, Eq. (5.2.4), drops out immediately as a bonus; it is a consequence of the single quantum state. The quantum mechanical phase, which is the same for all the particles, operates as a velocity potential for the superflow. Now think of a current flowing in a circular path, perhaps around some object in the fluid. We have around the path

$$\oint \mathbf{u}_s \cdot d\mathbf{l} = -\frac{\hbar}{m} \oint \nabla \gamma \cdot d\mathbf{l} \tag{5.2.24}$$

The integral on the right-hand side is the change in phase in going around a complete circuit and returning to the same point. The wave function, ψ, Eq. (5.2.18), must be single valued—it can have only one value at each point—but the phase, γ, at a given point can change by any integral multiple of 2π without changing ψ:

$$\oint \nabla\gamma \cdot d\mathbf{l} = 2\pi n \qquad (n = 0, \pm 1, \pm 2, \ldots) \qquad (5.2.25)$$

or

$$\oint \mathbf{u}_s \cdot d\mathbf{l} = nk_0 \qquad (5.2.26)$$

$$k_0 = \frac{h}{m} \approx 10^{-3} \text{ cgs} \qquad (\text{for } m = {}^4\text{He mass}) \qquad (5.2.27)$$

The quantity $\oint \mathbf{u} \cdot d\mathbf{l}$ in hydrodynamics is called the circulation; we have learned that, in a picture of superfluidity based on the Bose condensate, circulation is quantized in units of h/m.

If we imagine helium flowing around through a narrow tube closed on itself into a circle, the problem of the persistent current is easily resolved. The system is in some circulation quantum state, trapped at a fixed velocity. In order to change at all, the velocity must change by the same finite amount everywhere; stated differently, the phase γ must change by 2π at some point along the path. There are ways that this can happen, as we shall discuss later, but it is difficult, and it becomes increasingly unlikely at low velocities and low temperatures. Thus, the flow tends to persist at a fixed velocity.

Let us try briefly to summarize what we have learned up to now. The Bose degenerate gas—suitably modified to take account of the finite volume and interaction of real helium atoms—seems a promising model for the superfluid. We can expect it to form a zero-entropy fluid at zero temperature. There will be some phase transition where a finite amount of condensate starts to accumulate, although the details of the transition are not easy to obtain from the model. Super heat conductivity, superflow, the fountain effect, and persistent currents all find their analogs in Bose degenerate behavior. The modifications that we make mean that we must alter our language somewhat from what was suitable for discussing a perfect gas. We cannot think of a condensate of particles in the same single-particle ground state, and of excited single particles. Instead there is a ground state of the system as a whole, plus collective excitations out of that ground state. The excitations—quasiparticles called phonons and rotons—may, however, be thought of as a gas, quite closely analogous to the gas of single excited particles in the perfect Bose case. With that picture in mind, we can now turn to a phenomenological description of real liquid helium.

c. Two-Fluid Hydrodynamics

We shall not work out the two-fluid hydrodynamics of superfluidity in detail, anymore than we did the one-fluid hydrodynamics of ordinary liquids in Chap. 4. However, we shall discuss the problem a little, with two objectives: first, to try to capture some of the spirit in which it is done, so that our understanding at this level can be compared to the way in which superfluidity is described in more microscopic terms; and, second, we shall advance far enough to work out the two modes of sound propogation in helium that are associated with the existence of two independent velocity fields.

The hydrodynamics of the two-fluid model was worked out principally by L. D. Landau. To start with, said Landau, hydrodynamics is a pretty tough problem even when there is only one possible velocity at each point. With two velocities at each point, the problem looks close to impossible. Let us start by transforming ourselves into a frame of reference at rest with respect to the superfluid part, so that only the normal fluid moves, and there is only one velocity

$$\mathbf{w} = \mathbf{u}_n - \mathbf{u}_s \qquad (5.2.28)$$

An observer in this frame sees the quasiparticle gas drift by with velocity $\mathbf{w}$. If he measured the momentum per unit volume in the gas, $\mathbf{j}_0$, he would find it proportional to $\mathbf{w}$, with some constant of proportionality

$$\mathbf{j}_0 = \rho_n \mathbf{w} \qquad (5.2.29)$$

Equation (5.2.29) defines ρ_n. We can now define ρ_s by

$$\rho_s = \rho - \rho_n \qquad (5.2.30)$$

Notice that ρ_n is defined in a fundamental way, while ρ_s is merely what is left over. This situation corresponds to the fact that it is much easier to tell what is excited in liquid helium than what is condensed.

In order to find the momentum density in the laboratory frame, $\mathbf{j}$, we resort to a Galilean transformation. Since the super frame moves at velocity $\mathbf{u}_s$ with respect to the laboratory, the entire fluid (density ρ) must be transformed with this velocity:

$$\begin{aligned}\mathbf{j} &= \mathbf{j}_0 + \rho \mathbf{u}_s \\ &= \rho_n(\mathbf{u}_n - \mathbf{u}_s) + (\rho_n + \rho_s)\mathbf{u}_s \\ &= \rho_n \mathbf{u}_n + \rho_s \mathbf{u}_s \end{aligned} \qquad (5.2.31)$$

which is the same as Eq. (5.2.3).

In the laboratory frame, we must have conservation of mass

$$\frac{\partial \rho}{\partial t} + \nabla \cdot \mathbf{j} = 0 \qquad (5.2.32)$$

of momentum

$$\frac{\partial \mathbf{j}}{\partial t} + \nabla \cdot \boldsymbol{\pi} = 0 \quad (5.2.33)$$

and of energy

$$\frac{\partial e}{\partial t} + \nabla \cdot \mathbf{q} = 0 \quad (5.2.34)$$

$\mathbf{q}$ and the tensor $\boldsymbol{\pi}$, like $\mathbf{j}$, depend on both $\mathbf{u}_n$ and $\mathbf{u}_s$ and are undetermined at this point. e is the energy per unit volume of the fluid, $\boldsymbol{\pi}$ is the momentum flux tensor, and $\mathbf{q}$ is the energy flux density. There is also an equation which expresses the fact that only the normal fluid carries entropy

$$\frac{\partial}{\partial t}(\rho s) + \nabla \cdot (\rho s \mathbf{u}_n) = 0 \quad (5.2.35)$$

where s is the entropy per unit mass of the fluid.

In the superfluid frame, we write an identity for the energy density e_0

$$de_0 = \mu \, d\rho + T \, d(\rho s) + \mathbf{w} \cdot d\mathbf{j}_0 \quad (5.2.36)$$

Here we have used the fact that there is only one velocity, and it must be connected to the energy by

$$\frac{\partial e_0}{\partial \mathbf{j}_0} = \mathbf{w} \quad (5.2.37)$$

The equality sign in Eq. (5.2.36) means that we are assuming that flow takes place reversibly, with $\mathbf{w}$ and $\mathbf{j}_0$ as thermodynamic variables of state. $\partial e_0/\partial t$ is given in terms of $\partial \rho/\partial t$, $\partial(\rho s)/\partial t$, and $\partial \mathbf{j}_0/\partial t$ by the time derivative of Eq. (5.2.36). Furthermore, the quantities e, $\boldsymbol{\pi}$, and $\mathbf{q}$ are related to the corresponding quantities in the superfluid frame by Galilean transformations. For example, suppose that some element of fluid—without reference to whether it is normal or super—has a velocity v in the laboratory frame. Then its energy is, per unit volume,

$$e = e_i + \tfrac{1}{2}\rho v^2$$

where e_i is its internal energy. Its velocity in the superfluid frame is

$$\mathbf{v}_0 = \mathbf{v} - \mathbf{u}_s$$

so we have

$$e = e_i + \tfrac{1}{2}\rho(\mathbf{v}_0 + \mathbf{u}_s)^2$$
$$= (e_i + \tfrac{1}{2}\rho v_0^2) + \tfrac{1}{2}\rho u_s^2 + (\rho \mathbf{v}_0) \cdot \mathbf{u}_s$$
$$= e_0 + \tfrac{1}{2}\rho u_s^2 + \mathbf{j}_0 \cdot \mathbf{u}_s \quad (5.2.38)$$

where we have identified $e_0 = e_i + \frac{1}{2}\rho v_0^2$ and $\mathbf{j}_0 = \rho \mathbf{v}_0$. Equation (5.2.38) is the necessary transformation for energy density. Notice that the separation into normal and super parts is a model, whereas the argument leading to Eq. (5.2.38) is quite general.

A further equation is needed, for the superfluid acceleration. Since $\mathbf{u}_s$ is irrotational (Eq. 5.2.4), both $\mathbf{u}_s$ and its time derivative are given by the gradients of some scalar quantity

$$\frac{d\mathbf{u}_s}{dt} + \nabla \varphi = 0 \qquad (5.2.39)$$

where φ, like $\mathbf{q}$ and π, remains to be determined. The time derivative in Eq. (5.2.39) is the co-moving derivative, including both changes of $\mathbf{u}_s$ in time and steady-state changes of u_s with position:

$$\frac{d\mathbf{u}_s}{dt} = \frac{\partial \mathbf{u}_s}{\partial t} + \nabla(\tfrac{1}{2}u_s^2) - \mathbf{u}_s \times (\nabla \times \mathbf{u}_s) \qquad (5.2.40)$$

Since the last term in Eq. (5.2.40) is always zero, we may rewrite (5.2.39) as

$$\frac{\partial \mathbf{u}_s}{\partial t} + \nabla(\varphi + \tfrac{1}{2}u_s^2) = 0 \qquad (5.2.41)$$

There are now enough conditions to specify φ, π, and $\mathbf{q}$ and determine a complete set of equations, although to do so is hard work, which we shall not attempt. For the record, the results are

$$\varphi = \mu \quad \text{(the chemical potential)} \qquad (5.2.42)$$

$$\pi_{ik} = \rho_n u_{ni} u_{nk} + \rho_s u_{si} u_{sk} + P\,\delta_{ik} \qquad (5.2.43)$$

$$\mathbf{q} = (\mu + \tfrac{1}{2}u_s^2)\mathbf{j} + T\rho \mathbf{u}_n + \rho_n \mathbf{u}_n(\mathbf{u}_n \cdot \mathbf{w}) \qquad (5.2.44)$$

where
$$e_0 = T s \rho + \mu \rho + \rho_n w^2 - P \qquad (5.2.45)$$

Equation (5.2.45) defines P, the pressure. In ordinary fluid mechanics, the pressure is the force per unit area in a frame at rest with respect to the fluid, but that definition does not work here, where there are two velocities in the fluid. Notice that μ is here taken to be the chemical potential per unit mass and so has the dimensions of a velocity squared. If we put Eq. (5.2.42) into (5.2.41) and consider the steady state, $\partial \mathbf{u}_s/\partial t = 0$, with the superfluid at rest, $\mathbf{u}_s = 0$, we have

$$\nabla \mu = 0$$

The condition that the superfluid be at rest, in steady state, is that the chemical potential be uniform (not necessarily the temperature and pressure individually). What we have here is the fountain effect, Eq. (5.2.16). Although Eqs. (5.2.42) to (5.2.45) are difficult to derive, they are rather easier to verify (see Prob. 5.1).

We wish to find the modes of propagation of sound in this system. To do so, we begin by linearizing the equations, taking the velocities to be fluctuations about zero, as we did when studying ordinary sound in Sec. 3.2b. The equations we need are

$$\frac{\partial \rho}{\partial t} + \nabla \cdot \mathbf{j} = 0 \tag{5.2.46}$$

$$\frac{\partial \mathbf{j}}{\partial t} + \nabla P = 0 \tag{5.2.47}$$

$$\frac{\partial (\rho s)}{\partial t} + \rho s \nabla \cdot \mathbf{u}_n = 0 \tag{5.2.48}$$

$$\frac{\partial \mathbf{u}_s}{\partial t} + \nabla \mu = 0 \tag{5.2.49}$$

Equation (5.2.46) is the same as (5.2.32). Equation (5.2.47) comes from substituting (5.2.43) into (5.2.33) and dropping second-order terms in the velocities. Equation (5.2.48) is (5.2.35) linearized, and Eq. (5.2.49) is (5.2.39) linearized with (5.2.42) substituted in.

Equations (5.2.46) to (5.2.49), which are all the results of the two-fluid hydrodynamics we shall actually make use of, are relatively simple, and we can understand directly what they mean. The first two would be true under similar circumstances (small fluctuations about equilibrium) for any fluid; in fact, we used them in Sec. 3.2, Eqs. (3.2.35) and (3.2.36). Equation (5.2.48) merely tells us that the entropy density depends only on the flow of the normal fluid. Finally, Eq. (5.2.49) says that μ drives the superfluid in much the way that P drives the fluid as a whole [Eq. (5.2.47)]. As noted above, Eq. (5.2.49) is actually a dynamic generalization of the fountain effect.

If we now take the time derivative of Eq. (5.2.46) and the divergence of (5.2.47) and subtract to eliminate $\mathbf{j}$, we get

$$\frac{\partial^2 \rho}{\partial t^2} - c_1^2 \nabla^2 \rho = 0 \tag{5.2.50}$$

which is the equation for ordinary sound, with

$$c_1^2 = \frac{\partial P}{\partial \rho} \tag{5.2.51}$$

[See Eq. (3.2.41).] Rather than specify whether the derivative in Eq. (5.2.51) is to be taken at constant T or s, let us make an assumption that will greatly simplify our work. We shall assume, just as we did in Eq. (3.2.41), that P depends only on ρ, not on s or T, and, conversely, that s depends only on T, not on P or ρ. This is equivalent to assuming that $C_P = C_V$—the kind of approximation that one commonly makes for a condensed medium (we are,

5.2 Superfluidity

after all, dealing with a liquid). The approximation is good except very close to the lambda transition.

The approximation that we have made anticipates our result: we have found P-ρ waves (at constant s and T), ordinary sound, and we shall now find s-T waves (at constant P and ρ). We seek an equation of the form

$$\frac{\partial^2 s}{\partial t^2} - c_2^2 \nabla^2 s = 0 \tag{5.2.52}$$

where c_2 will be the velocity of the new kind of wave.

In Eq. (5.2.49) we have $\nabla \mu$, which we write as

$$\nabla \mu = -s \nabla T + \frac{1}{\rho} \nabla P \tag{5.2.53}$$

[This form can be deduced by writing Eq. (5.2.45) in differential form and subtracting Eq. (5.2.36); it is really just the familiar (1.2.35).] Substituting Eq. (5.2.47) into (5.2.53) gives

$$\nabla \mu = -s \nabla T - \frac{1}{\rho} \frac{\partial \mathbf{j}}{\partial t} \tag{5.2.54}$$

and substituting

$$\frac{\partial \mathbf{j}}{\partial t} = \rho_s \frac{\partial \mathbf{u}_s}{\partial t} + \rho_n \frac{\partial \mathbf{u}_n}{\partial t}$$

and (5.2.54) into (5.2.49), we get

$$\frac{\partial \mathbf{u}_s}{\partial t} - \frac{\rho_s}{\rho} \frac{\partial \mathbf{u}_s}{\partial t} - \frac{\rho_n}{\rho} \frac{\partial \mathbf{u}_n}{\partial t} - s \nabla T = 0$$

or

$$\frac{\rho_n}{\rho} \frac{\partial}{\partial t} (\mathbf{u}_s - \mathbf{u}_n) - s \nabla T = 0$$

which we may write as

$$\frac{\rho_n}{\rho s} \frac{\partial \mathbf{w}}{\partial t} + \nabla T = 0 \tag{5.2.55}$$

In order to put this result in a form that looks like (5.2.52), we take the divergence of Eq. (5.2.55) and substitute in

$$\nabla^2 T = \frac{\partial T}{\partial s} \nabla^2 s = \frac{T}{C} \nabla^2 s$$

where C is the specific heat, giving us

$$\frac{\partial}{\partial t} \nabla \cdot \mathbf{w} + \frac{Ts\rho}{C\rho_n} \nabla^2 s = 0 \tag{5.2.56}$$

We must now show that $(\partial/\partial t)\,\nabla\cdot\mathbf{w}$ is proportional to $\partial^2 s/\partial t^2$. Equation (5.2.48) may be written

$$s\frac{\partial\rho}{\partial t} + \rho\frac{\partial s}{\partial t} + \rho s\,\nabla\cdot\mathbf{u}_n = 0 \tag{5.2.57}$$

but from (5.2.46)

$$s\frac{\partial\rho}{\partial t} = -s\,\nabla\cdot\mathbf{j} = -s(\rho_s\,\nabla\cdot\mathbf{u}_s + \rho_n\,\nabla\cdot\mathbf{u}_n) \tag{5.2.58}$$

Substituting this into (5.2.57) gives

$$(\rho s - s\rho_n)\,\nabla\cdot\mathbf{u}_n + \rho\frac{\partial s}{\partial t} - s\rho_s\,\nabla\cdot\mathbf{u}_s = 0$$

or since $\rho s - s\rho_n = s\rho_s$,

$$s\rho_s\,\nabla\cdot\mathbf{w} + \rho\frac{\partial s}{\partial t} = 0 \tag{5.2.59}$$

The time derivative of (5.2.59) is the result we seek:

$$\frac{\partial}{\partial t}\nabla\cdot\mathbf{w} = -\frac{\rho}{s\rho_s}\frac{\partial^2 s}{\partial t^2} \tag{5.2.60}$$

Equation (5.2.60) combines with (5.2.56), finally, to give

$$\frac{\partial^2 s}{\partial t^2} - \frac{T\rho_s s^2}{C\rho_n}\nabla^2 s = 0 \tag{5.2.61}$$

The new mode thus propagates with a velocity whose square is

$$c_2^2 = \frac{T\rho_s s^2}{C\rho_n} \tag{5.2.62}$$

A sketch of c_2 versus temperature is given in Fig. 5.2.8. Over a wide temperature range, $c_2 \approx 20$ m/sec, compared to $c_1 \approx 240$ m/sec for ordinary sound.

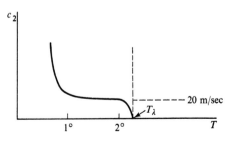

Fig. 5.2.8

5.2 Superfluidity

The new mode is called *second sound*. By default, ordinary sound in helium is referred to as *first sound*. This technique of nomenclature has been extended in both directions. There is something called zeroth sound in liquid ^{3}He that occurs when the degenerate Fermi behavior of ^{3}He (see Sec. 2.5) makes collisions so rare that first sound can no longer be propagated —that is, at very low temperature and very high frequency (for electrons in metals, zeroth sound corresponds to plasma waves). In superfluid ^{4}He there is also third sound, a kind of sloshing superfluid wave in the adsorbed film, and fourth sound, which is left as a problem for the reader (Prob. 5.2).

First sound is a density (and pressure) wave. The entire fluid oscillates back and forth, super together with normal. The two velocities have the same oscillations

$$\mathbf{u}_n = \mathbf{u}_s \quad \text{(first sound)} \tag{5.2.63}$$

Second sound, on the other hand, is an entropy (and temperature) wave at constant density. Since only the normal fluid carries entropy, the normal fluid must be oscillating back and forth, but there is no mass transport, so

$$\mathbf{j} = \rho_n \mathbf{u}_n + \rho_s \mathbf{u}_s = 0 \quad \text{(second sound)} \tag{5.2.64}$$

Thus, second sound is a kind of counterflow wave. As discussed above, if we wish to think of helium as a degenerate Bose system, the normal fluid must be identified with collective excitations out of the ground state. First sound (i.e., phonons) is one such excitation. Conceptually, the most useful way of thinking about second sound is that it is a sound wave in the gas of excitations.

The two-fluid model is very successful in describing the hydrodynamic behavior of liquid helium. We, however, are really interested only in how it relates to the general description of the properties of helium discussed earlier; let us see where the two-fluid model leaves us.

The two-fluid behavior is, of course, an assumption of the model; we can learn about its effects but not its causes. The thermodynamic behavior, such as the temperature dependence of the entropy, of ρ_n and ρ_s, and so on, is not treated here and must await the next section, which deals with the statistical mechanics of the system. On the other hand, we are now ready to explain the property of super heat conductivity. Heat, in liquid helium II, is not conducted diffusively as in most other media but coherently, in the form of second sound. A heat pulse thus travels at a speed of about 20 m/sec. We said earlier that the pressure in a Bose gas depends only on the temperature, so that heat travels at the speed of sound; that is just what is happening here, but the pressure is called the fountain pressure and the sound is second sound. In the Bose gas, a temperature fluctuation is a fluctuation in the number density of excited particles; here it is a fluctuation in ρ_n—that is, in the density of excitations or quasiparticles.

Let us now consider the statistical mechanics of those quasiparticles.

d. Phonons and Rotons

The picture of He II that has emerged so far from our discussions is one of a condensate, the superfluid part, the excitations from which are not single excited particles but some kind of collective excitations of the system. In order to work out the thermodynamic properties of the system by applying statistical mechanics, we need, as usual, to know the dispersion relation, $\varepsilon(p)$, of the modes that may be excited. This relation, like the two-fluid model, was first worked out by Landau, who guessed the form of $\varepsilon(p)$, guided by the behavior of the heat capacity. Feynman, as we shall see below, arrived at the same form from quantum mechanical arguments, and the curve has also been traced out by means of neutron scattering (just as the phonon spectrum of a solid may be deduced from inelastic scattering of neutrons). The curve is sketched in Fig. 5.2.9.

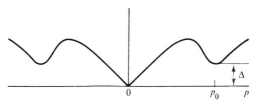

Fig. 5.2.9

As we have already guessed on general grounds, the linear part of the curve, which represents the lowest energy excitations, is simply a phonon spectrum

$$\varepsilon = c_1 p \quad \text{(phonon part)} \tag{5.2.65}$$

Excitations around the minimum of the curve are called rotons. Since we shall need an analytic expression for them, we merely expand $\varepsilon(p)$ around the minimum (where $\varepsilon = \Delta$ and $p = p_0$)

$$\varepsilon = \Delta + \frac{(p - p_0)^2}{2\mu_0} \tag{5.2.66}$$

The parameters are $\Delta/k \approx 8.5°\text{K}$, $\mu_0 \approx 0.2m$, where m is the mass of a helium atom and $p_0/\hbar \approx 2\text{Å}^{-1}$. Although μ_0 has the dimensions of a mass, it is merely a parameter and should not be thought of as the mass of a roton nor of anything else.

Excitations can exist at any point on the curve, but only those portions of the curve near the origin [Eq. (5.2.65)] and near the minimum [Eq. (5.2.66)] will contribute substantially to the thermal behavior of the liquid. At very low temperature, of course, only the low-energy, long wavelength phonons are excited. At higher temperature (approaching 1°K) the rotons

5.2 Superfluidity

become important because although a lot of energy (at least Δ) is needed to excite a roton, the density of states, proportional to $dp/d\varepsilon$, becomes infinite at the minimum. There is an infinite density of states at the maximum in the curve as well, but the energy there is twice as high as at the minimum, and so those states are unimportant (unlikely to be excited) relative to the rotons. Whatever these excitations are physically, they are certainly strange kinds of beasts, to judge from their properties. Around the minimum, they have momentum p_0 but group velocity ($d\varepsilon/dp$) equal to zero. To the left of the minimum, they go in the wrong direction (they have momentum and velocity in opposite directions). That part of the curve, between the maximum and the minimum, is called the region of anomalous dispersion. To the left of the maximum, the curve looks just like a branch of the phonon spectrum of a crystalline solid (see Chap. 3, Fig. 3.3.3). For a solid, the maximum would occur at the Brillouin zone boundary, and no new states would be found at higher values of q. Here, however, we are dealing with a liquid, in which there is no periodic structure capable of forming Brillouin zones. The part of the curve to the right of the maximum represents some new kind of motion, which is specifically characteristic of the fluid nature of the system. The name roton for the excitations around the minimum implies some kind of rotational state, and it may well be so, but no clear model of rotons has ever been agreed upon.

According to Eq. (5.2.11), superflow should be possible only up to a velocity given by the smallest value of ε/p on this curve. ε/p is the slope of a line from the origin to any point on the curve; its minimum value occurs at the roton minimum, where

$$\frac{\varepsilon}{p} = \frac{\Delta}{p_0} \simeq 60 \text{ m/sec}$$

Experimentally, one finds that superflow breaks down at much lower velocities, a point we shall return to later.

Phonons and rotons may be treated just as we have treated other quasiparticles. Having found their properties, we shall be able easily to deduce the behavior of a perfect gas of them and thus account for the thermodynamic behavior of He II, at least at reasonably low temperatures. Just as we did for phonons in solids and for electrons in metals, we have once again reduced our problem to that of the perfect gas, one that we know how to solve.

In fact, we have already solved the phonon part, which differs from Sec. 3.2 mainly in that there are only longitudinal modes, not transverse ones. At very low temperatures ($T \lesssim 0.6°K$), only the phonons are excited, and we know immediately that the heat capacity will be given by

$$C \propto T^3$$

with a Debye temperature that may be deduced from c_1, the speed of sound. Measurements confirm this, with $\Theta_D \simeq 30°K$ for the liquid under its own vapor pressure. At higher temperature the rotons become important, and their contribution must be calculated separately.

Consider a gas of perfect rotons in states whose energies are given by Eq. (5.2.66). The roton energies are more than about 8.5°K, whereas we wish to apply the theory we are working out only below about 2°K. Consequently, the probability of occupation of any single roton state is small; we can treat rotons as classical particles! The roton free energy is given by

$$F_r = -kT \log Z_r \qquad (5.2.67)$$

where the roton partition function, Z_r, is

$$Z_r = \frac{1}{N_r!} \left[\int \exp\left(-\frac{\varepsilon}{kT}\right) \frac{d^3p\, d^3r}{(2\pi\hbar)^3} \right]^{N_r} \qquad (5.2.68)$$

This is just Eq. (1.3.116) with a different form for the energies of the perfect gas particles. N_r is the number of rotons, whose equilibrium value is yet to be determined (unlike an ordinary classical gas, the number of rotons does not need to be conserved). Using Eq. (2.2.26) for $N_r!$, we can write the free energy for arbitrary N_r

$$F_r = -N_r kT \log \left[\frac{eV}{(2\pi\hbar)^3 N_r} \int \exp\left(-\frac{\varepsilon}{kT}\right) d^3p \right] \qquad (5.2.69)$$

The equilibrium value of N_r is found by minimizing F_r with respect to N_r:

$$0 = \left(\frac{\partial F_r}{\partial N_r}\right)_T = kT \left\{1 - \log\left[\frac{eV}{(2\pi\hbar)^3 N_r} \int \exp\left(-\frac{\varepsilon}{kT}\right) d^3p\right]\right\} \qquad (5.2.70)$$

$(\partial F_r/\partial N_r)_T$ is the chemical potential of the roton gas, which (like that of phonons) is equal to zero. Equation (5.2.70) requires that the argument of the logarithm be equal to e, so that

$$N_r = \frac{V}{(2\pi\hbar)^3} \int \exp\left(-\frac{\varepsilon}{kT}\right) d^3p \qquad (5.2.71)$$

The same result substituted into Eq. (5.2.69) gives, for the equilibrium free energy,

$$F_r = -N_r kT \qquad (5.2.72)$$

We must now compute N_r as a function of T and V. Substituting Eq. (5.2.66) into (5.2.71), we have

$$N_r = \frac{V}{(2\pi\hbar)^3} \exp\left(-\frac{\Delta}{kT}\right) \int_0^\infty \exp\left(-\frac{(p-p_0)^2}{2\mu_0 kT}\right) 4\pi p^2\, dp$$

$$= \frac{4\pi p_0^2 \exp(-\Delta/kT)}{(2\pi\hbar)^3} \int_{-\infty}^\infty \exp\left(-\frac{(p-p_0)^2}{2\mu_0 kT}\right) dp \qquad (5.2.73)$$

5.2 Superfluidity

To make the last step, we note that the exponential in the integrand goes rapidly to zero except when $p \approx p_0$, so the p^2 inside can be taken out as p_0^2, and the limits of integration can be taken from $-\infty$ to $+\infty$. We now change variables to $x = (p - p_0)^2/2\mu_0 kT$, bringing a factor $(\mu_0 kT)^{1/2}$ outside, and evaluate the remaining definite integral to give

$$N_r = \frac{2(\mu_0 kT)^{1/2} p_0^2 V}{(2\pi)^{3/2} \hbar^3} \exp\left(-\frac{\Delta}{kT}\right) \quad (5.2.74)$$

Since the free energy is $F_r = -N_r kT$ [Eq. (5.2.72)], we see that its basic dependences are

$$F_r(T, V) \propto VT^{3/2} \exp\left(-\frac{\Delta}{kT}\right) \quad (5.2.75)$$

F_r and all its derivatives will be dominated by the exponential temperature dependence, $e^{-\Delta/kT}$. In particular, the roton contribution to the entropy will be

$$S_r = -\frac{\partial F_r}{\partial T} = N_r k \left(\frac{3}{2} + \frac{\Delta}{kT}\right) \quad (5.2.76)$$

and to the heat capacity

$$C_r = T\frac{\partial S_r}{\partial T} = N_r k \left[\frac{3}{4} + \frac{\Delta}{kT} + \left(\frac{\Delta}{kT}\right)^2\right] \quad (5.2.77)$$

These quantities have the same basic exponential dependence as does N_r. C_r, added to the phonon contribution, can be compared directly to the measured heat capacity of helium (see Prob. 5.3).

We now have the number of rotons and could easily calculate the number of phonons—but how are these related to the two-fluid model of the previous subsection? What we need there is ρ_n, the normal fluid mass per unit volume. Are we to assign a mass to each roton and phonon? That is a procedure not to be contemplated. We return instead to the fundamental definition of ρ_n, Eq. (5.2.29). In the superfluid frame, we watch the excitation gas drift by with mean velocity **w**. Then ρ_n is the ratio of the observed momentum density, $\mathbf{j}_0$, to **w**. Suppose that the superfluid is actually flowing slowly through a capillary with velocity $\mathbf{u}_s$, so that the excitations are at rest in the laboratory and $\mathbf{w} = -\mathbf{u}_s$. Now, the energy that must be extracted from the wall of the capillary in order to create an excitation of a given momentum is no longer given by the curve in Fig. 5.2.9. That curve gives the energy of an excitation from the superfluid—that is, relative to the superfluid—measured in the superfluid frame. To find the energy relative to the wall, we must use the Galilean transformation, Eq. (5.2.38), with $\mathbf{w} = -\mathbf{u}_s$:

$$e = e_0 + \tfrac{1}{2}\rho w^2 - \mathbf{j}_0 \cdot \mathbf{w} \quad (5.2.78)$$

where subscript 0, as usual, refers to the superfluid frame. The energy to be extracted from the wall in order to create an excitation (at constant ρ and $\mathbf{w}$) is therefore

$$\delta e = \delta e_0 - \delta \mathbf{j}_0 \cdot \mathbf{w} \qquad (5.2.79)$$

Equation (5.2.79) means that when there is superflow, less energy is required to create an excitation with momentum in the direction of $\mathbf{w}$ than, say, in the opposite direction. It follows that there will be more of these excitations, which is the reason a net momentum density will be observed in the superfluid frame. The energy in the laboratory frame governs the number of excitations in each state, since the excitation gas must always be in thermal equilibrium with the wall. Of course, $\mathbf{w}$ is in the opposite direction from $\mathbf{u}_s$, so from the point of view of the laboratory, one sees a net polarization of the excitations, with momenta pointing upstream.

Equation (5.2.79) gives the changes in energy per unit volume. If one excitation arises in a unit volume, it will have energy $\varepsilon^* = \delta e$ in the laboratory frame, $\varepsilon = \delta e_0$ in the superfluid frame, and momentum $\mathbf{p} = \delta \mathbf{j}_0$ relative to the superfluid. The number of quasiparticles in each state is given by

$$n(\varepsilon^*) = n(\varepsilon - \mathbf{p} \cdot \mathbf{w}) = \frac{1}{\exp\left[(\varepsilon - \mathbf{p} \cdot \mathbf{w})/kT\right] - 1} \qquad (5.2.80)$$

Notice that $n(\varepsilon^*)$ diverges unless w satisfies the criterion for the maximum velocity of superflow, Eq. (5.2.11); a divergence in $n(\varepsilon^*)$ would mean a phase transition into the normal state. This observation is an alternative derivation of the Principle of Superfluidity.

We may now calculate ρ_n. The momentum per unit volume in the superfluid frame is

$$\mathbf{j}_0 = \frac{1}{(2\pi\hbar)^3} \int \mathbf{p} n(\varepsilon - \mathbf{p} \cdot \mathbf{w}) \, d^3p \qquad (5.2.81)$$

Taking $\mathbf{p} \cdot \mathbf{w} \ll \varepsilon$, we expand n,

$$n(\varepsilon - \mathbf{p} \cdot \mathbf{w}) = n(\varepsilon) - \mathbf{p} \cdot \mathbf{w} \frac{\partial n}{\partial \varepsilon}$$

When this result is substituted back into Eq. (5.2.81), the leading-order term integrates to zero by symmetry

$$\int \mathbf{p} n(\varepsilon) \, d^3p = 0$$

and we are left with

$$(2\pi\hbar)^3 \mathbf{j}_0 = -\int \mathbf{p}(\mathbf{p} \cdot \mathbf{w}) \frac{\partial n}{\partial \varepsilon} \, d^3p \qquad (5.2.82)$$

5.2 Superfluidity

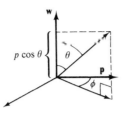

Fig. 5.2.10

If we take **w** along the $\theta = 0$ axis in polar coordinates, so $\mathbf{p} \cdot \mathbf{w} = pw \cos \theta$, then by symmetry with respect to φ (see Fig. 5.2.10), we can replace $\mathbf{p}(\mathbf{p} \cdot \mathbf{w})$ by $\mathbf{w} \, (p^2 \cos^2 \theta)$, or

$$(2\pi\hbar)^3 \mathbf{j}_0 = -\mathbf{w} \int p^2 \cos^2 \theta \frac{\partial n}{\partial \varepsilon} 2\pi p^2 \sin \theta \, d\theta \, dp$$

$$= -\mathbf{w} \frac{4\pi}{3} \int p^4 \frac{\partial n}{\partial \varepsilon} dp \quad (5.2.83)$$

Recalling the definition of ρ_n, we have

$$\rho_n = -\frac{4\pi}{3(2\pi\hbar)^3} \int p^4 \frac{\partial n}{\partial \varepsilon} dp \quad (5.2.84)$$

Let us finish the calculation explicitly for the phonon contribution (the reader will do the roton part in the course of solving Prob. 5.3). For the phonons, from Eq. (5.2.65), $d\varepsilon = c_1 \, dp$, or

$$(\rho_n)_{\text{ph}} = -\frac{4\pi}{3(2\pi\hbar)^3 c_1} \int p^4 \, dn \quad (5.2.85)$$

Integrate by parts:

$$\int p^4 \, dn = p^4 n \big|_0^\infty - 4 \int np^3 \, dp = -4 \int_0^\infty np^3 \, dp \quad (5.2.86)$$

(where we have used the fact that n goes rapidly to zero when $p \to \infty$). We have

$$(\rho_n)_{\text{ph}} = \frac{16\pi}{3(2\pi\hbar)^3 c_1} \int np^3 \, dp$$

$$= \frac{4}{3(2\pi\hbar)^3 c_1^2} \int (c_1 p) n(\varepsilon) 4\pi p^2 \, dp$$

$$= \frac{4}{3c_1^2} \cdot \frac{1}{(2\pi\hbar)^3} \int \varepsilon n(\varepsilon) \, d^3 p$$

$$= \frac{4 E_{\text{ph}}}{3 c_1^2} \quad (5.2.87)$$

where E_{ph} is the energy density of the phonons. Since we know (from Chap. 3) that $E_{ph} \propto T^4$, we have at low T, where only the phonons are excited,

$$\rho_n \propto T^4 \qquad (5.2.88)$$

Calculation of other aspects of the thermal behavior proceeds by means that are by now very familiar to us.

e. The Feynman Theory

Aside from tying up a few loose ends—critical velocities and the like—our principal remaining problem in understanding the essential nature of superfluid helium is to account for the shape of the excitation curve, Fig. 5.2.9. The theory of that curve, worked out by Feynman, is the matter now before us.

The essential question is why the helium liquid system has no available excited states below the curve in question; this absence of low-lying states, as we have seen [in Eq. (5.2.11)], is the key to the Principle of Superfluidity. The job, then, is to construct the lowest-lying excited states of the system. Before we can do so, we must have some idea of the nature of the ground state out of which the excitations are to be formed.

The ground-state wave function will depend on the positions of the N helium atoms in the system; let us call it $\psi_0\{N\}$, where the notation $\{N\}$ is the same one used in Chap. 4—it represents the position vectors of the N atoms, $\mathbf{r}_1, \mathbf{r}_2, \ldots, \mathbf{r}_n$. The probability of any particular configuration is proportional to $|\psi_0|^2$ and depends on $\{N\}$, with the normalization

$$\int \cdots \int |\psi_0|^2 \, d\{N\} = N \qquad (5.2.89)$$

$(1/N)|\psi|^2$ gives the quantum mechanical probability of a particular configuration in exactly the same sense that

$$\frac{\exp(-U\{N\}/kT)}{Q_N}$$

gave the thermal probability of a configuration of the N particles of a classical liquid in Chap. 4. Thus, for example, the quantity

$$\int \cdots \int |\psi_0|^2 \, d\{N - 2\}$$

is a two-particle correlation function. To simplify the writing, let us define

$$\rho_N = |\psi_0|^2 \qquad (5.2.90)$$

Then

$$\int \cdots \int \rho_N \, d\{N\} = N \qquad (5.2.91)$$

$$\int \cdots \int \rho_N \, d\{N - 1\} = \frac{N}{V} = \rho \qquad (5.2.92)$$

$$\int \cdots \int \rho_N \, d\{N - 2\} = \frac{N}{V^2} g(r) \qquad (5.2.93)$$

where $g(r)$ is the radial distribution function. Its meaning is the same as in Eq. (4.2.12) except that the average is quantum mechanical rather than thermal. Note how Eq. (5.2.93) compares to Eq. (4.2.39). The structure of the liquid, described by $g(r)$, will be much like that shown in Fig. 4.2.4; just as in the classical case, $g(r)$ can be measured by means of X-ray scattering. The probabilities of configurations of liquid helium in the ground state are not substantially different from those of any other liquid; the most probable configurations have the atoms evenly distributed in space.

To compute ψ_0, however, we must start with the Schrödinger equation

$$\mathcal{H}\psi_0 = E_0\psi_0 \qquad (5.2.94)$$

where E_0 is the energy of the ground state, and the Hamiltonian is

$$\mathcal{H} = -\frac{\hbar^2}{2m} \sum_{j=1}^{N} \nabla_j^2 + \sum_{i<j} u(r_{ij}) \qquad (5.2.95)$$

Here $u(r_{ij})$ is the same pairwise potential used in Chap. 4 (we will not worry here about inaccuracies due to assuming pairwise additivity). Obviously we are not going to solve Eq. (5.2.94) for 10^{23} or so particles, but we can make some general arguments about the form of the solution.

We have already described the general form of $|\psi_0|^2$. ψ_0 itself has a particularly useful property: it has no nodes; that is, it has the same sign for all configurations. This is a consequence, in part, of the Bose statistics obeyed by the helium atoms. If they obeyed Fermi-Dirac statistics, ψ_0 would have to be antisymmetric; it would change sign whenever two particles were interchanged. For the symmetric wave function $\psi_0\{N\}$, we can see why there are no nodes by assuming that there is a node, than showing that a lower energy state exists, thereby proving that the state with a node is not the ground state. Suppose that holding $N - 1$ particles fixed and moving the other causes ψ_0 to pass through zero, as in Fig. 5.2.11a. The energy of the state depends on $|\psi|^2$ (the potential energy term) and on $\psi_0^* \nabla^2 \psi_0$ (the

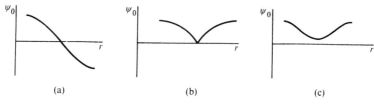

Fig. 5.2.11

kinetic energy term). This latter term, when it is to be integrated over the volume, may be written as $|\nabla\psi_0|^2$.† Since the energy depends only on $|\psi_0|^2$ and $|\nabla\psi_0|^2$, the energy of the state whose wave function is sketched in Fig. 5.2.11a must be the same as that in Fig. 5.2.11b, which has an energetically expensive cusp; the state in Fig. 5.2.11c has a lower energy. Quite generally, we can construct a state with smaller gradients (smaller kinetic energy) without larger $|\psi_0|^2$ (potential energy) if ψ_0 always has one sign. Accordingly, we arbitrarily choose the phase of the wave function such that ψ_0 is real and positive everywhere.

† This is a result that we shall use a number of times; let us work it out here. There is a theorem in vector analysis that

$$\int_V \nabla \cdot \mathbf{A} \, d^3r = \int_S \mathbf{A} \cdot d\mathbf{S}$$

where the integral on the right is over the surface of the volume over which we integrate on the left. $\mathbf{A}$ is some vector that depends on position in the volume. Now suppose that

$$\mathbf{A} = a\boldsymbol{\phi}$$

where a is a scalar function of $\mathbf{r}$ and $\boldsymbol{\phi}$ a vector that depends on $\mathbf{r}$. Then

$$\nabla \cdot \mathbf{A} = a \nabla \cdot \boldsymbol{\phi} + \boldsymbol{\phi} \cdot \nabla a$$

Substituting into the first equation, we have

$$\int a \nabla \cdot \boldsymbol{\phi} \, d^3r = \int a\boldsymbol{\phi} \cdot d\mathbf{S} - \int \boldsymbol{\phi} \cdot \nabla a \, d^3r$$

If $\boldsymbol{\phi} \cdot d\mathbf{S}$ is zero ($\boldsymbol{\phi}$ perpendicular to the normal at the surface) or if $a\boldsymbol{\phi}$ is zero at the surface, then

$$\int a \nabla \cdot \boldsymbol{\phi} \, d^3r = -\int \boldsymbol{\phi} \cdot \nabla a \, d^3r$$

For example, if $a = \psi^*$ and $\boldsymbol{\phi} = \nabla\psi$, where ψ is a wave function, then for either $\nabla\psi \cdot d\mathbf{S} = 0$ or for $\psi = 0$ at the boundary,

$$\int \psi^* \nabla^2 \psi \, d^3r = -\int |\nabla\psi|^2 \, d^3r$$

This is the result we have used above. We shall refer to this theorem in its various forms as the *divergence theorem*.

5.2 Superfluidity

So much for the ground state. We now get to our real business, which is to construct the lowest-lying excited states. We will call the excited state wave function ψ_1. It must be orthogonal to the ground state

$$\int \cdots \int \psi_1 \psi_0 \, d\{N\} = 0 \qquad (5.2.96)$$

Since ψ_0 is always positive, this integral can be zero only if ψ_1 changes sign; it must be positive for (roughly) half of all configurations and negative for the other half (this must be true of both the real and imaginary parts of ψ_1; for simplicity, we can discuss the real part, and the imaginary will do the same). Imagine that there is some particular configuration, $\{N_A\}$, in which ψ_1 has its largest positive value, and some other configuration, $\{N_B\}$, in which it has its largest negative value. The change in configuration, in going from $\{N_A\}$ to $\{N_B\}$, must involve the largest possible displacements of atoms if ψ_1 is to be a low-lying state. The reason is that if ψ_1 goes from positive to negative when, say, atom j moves a short distance, then $\mathbf{V}_j \psi_1$ is large and the energy is large. Let us therefore construct $\{N_B\}$ out of $\{N_A\}$ by taking atom j out of a position (say) on the left-hand side of the liquid and moving it a macroscopic distance to the right-hand side. Now the state that we have created this way has a rather high potential energy because there is a hole on the left and an extra atom jammed in on the right; we can obviously lower the energy by moving a few other atoms around a bit to smooth out the local densities. After this adjustment, let us look at the state that we have described. There are two possibilities:

1. After smoothing out locally, we find that $\{N_B\}$ has an extra atom on the right-hand side relative to $\{N_A\}$. This small-amplitude, large-wavelength change in density makes the real part of ψ_1 go from its maximum positive to its maximum negative value. We have just described a state with a long wavelength phonon. We already know that phonons exist, but they do not interfere with superfluidity.

2. After smoothing out, we find that the density is again uniform (i.e., we are looking now for other states, ones that cannot be constructed out of phonons and that must, therefore, have uniform density) and $\{N_B\}$ differs from $\{N_A\}$ principally in the fact that atom j has moved a macroscopic distance. But, of course, the atoms are indistinguishable—there is no way of telling which atom is atom j. It follows that whatever configuration $\{N_B\}$ we end up with, it can be constructed out of $\{N_A\}$ without moving any atom a long distance; any configuration of uniform density can be made out of any other without ever moving any atom more than about half an interatomic distance. ψ_1 therefore goes from maximum positive to maximum negative with only very small displacements of the atoms; it has large gradients and hence large energy. Here, then, is the crux of the matter; because of the

indistinguishability of the atoms, there can be no very low lying states aside from the phonons. Thus, the Principle of Superfluidity.

To make a more quantitative approach, we must guess about the nature of the lowest-lying nonphonon states that do exist and try to construct their wave functions. Feynman suggested a number of possibilities:

1. Perhaps the lowest-lying state is a ring of atoms that turn together in the liquid. If so, the lowest energy state of this type will involve the smallest ring that can rotate freely—say, a ring of six atoms. The change from $\{N_A\}$ to $\{N_B\}$ will then consist of rotating the ring through 30° as sketched in Fig. 5.2.12. Everything else is the same. This is the largest possible change;

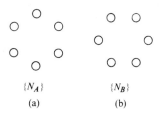

Fig. 5.2.12

rotating 60° brings $\{N_A\}$ back into itself. We can construct ψ_1 for this state as follows: define a function of position in the liquid $f(\mathbf{r})$ that is equal to $+\frac{1}{6}$ at each of the atomic positions in Fig. 5.2.12a and $-\frac{1}{6}$ at each position in Fig. 5.2.12b. It is equal to zero far away and varies smoothly between these values. We then write

$$\psi_1 = \left[\sum_{i=1}^{N} f(\mathbf{r}_i)\right] \psi_0\{N\} \qquad (5.2.97)$$

By summing over all atoms, we have made ψ_1 properly symmetric; it does not matter which atoms are at these sites, only whether there are atoms there. Multiplying by the ground-state wave function, ψ_0, puts into ψ_1 all the correlations built into ψ_0: the amplitude goes to zero when atoms get too close together, and so on. Furthermore, it is easy to see that ψ_1 will be orthogonal to ψ_0, since the sum will make equal positive and negative contributions when integrated over configurations. Different choices of $f(\mathbf{r})$ will give bigger rings and other variations, but Eq. (5.2.97) will be the general form.

2. We might imagine a state in which one single atom goes into an excited state in the box formed by its neighbors. Then ψ_1 changes from maximum positive to maximum negative as the atom moves across the little box it is trapped in. If we define $f(\mathbf{r})$ to go from $+1$ to -1 as $\mathbf{r}$ changes

5.2 Superfluidity

within the space normally occupied by an atom, then the state in which the ith atom is excited has the form

$$f(\mathbf{r}_i)\psi_0\{N\}$$

But, of course, the wave function must be symmetric, we must sum over all i, and we are back to the form Eq. (5.2.97), except that $f(\mathbf{r})$ has been redefined.

3. If the ith atom were alone in the volume containing the liquid, it would have excited state wave functions of the form

$$\exp(i\mathbf{q}\cdot\mathbf{r}_i) \qquad (5.2.98)$$

These are just the states of a free particle in a box (a particle of a perfect gas as in Chap. 2). A possible excited state of the liquid is for a single atom to move through the liquid almost as if it were free. The other atoms would have to move out of the way, of course—that is, there would be backflow—but let us ignore that refinement. Instead we multiply by ψ_0 once again to put in the correlations among all the other particles and sum over i as usual. The result still has the form Eq. (5.2.97) but $f(\mathbf{r}_i)$ now is given by Eq. (5.2.98).

All the excited states we can think of (actually, of course, all the excited states Feynman could think of) have the general form Eq. (5.2.97) (as we shall see later, even the phonons have this form). This suggests a procedure. Suppose that we assume that the excited state wave function is given by Eq. (5.2.97) with $f(\mathbf{r})$ undetermined (we do not, after all, know which of the states we have described has the lowest energy). We can then compute the energy of the state and minimize it with respect to $f(\mathbf{r})$. The result will give too high an energy unless the form we have guessed, Eq. (5.2.97), is exactly right, since nature knows how to choose the wave function that really does minimize the energy. Nevertheless, minimizing the energy with the approximate form that has been assumed should give us a fair picture of what is going on, provided that the form is shrewdly chosen.

Let us define

$$F_1 = \sum_i f(\mathbf{r}_i) \qquad (5.2.99)$$

Then we may write the Shrödinger equation for the excited state (subtracting off the ground state energy, E_0)

$$(\mathcal{H} - E_0)\psi_1 = \mathcal{H}F_1\psi_0 - E_0 F_1\psi_0$$

$$= -\frac{\hbar^2}{2m}\sum_j \nabla_j^2 F_1\psi_0 + \sum_{i<j} u(r_{ij})F_1\psi_0 - E_0 F_1\psi_0 \qquad (5.2.100)$$

where we have used Eq. (5.2.95). Now

$$\nabla_j^2 F_1 \psi_0 = \nabla_j \cdot (\psi_0 \nabla_j F_1 + F_1 \nabla_j \psi_0)$$

$$= 2\nabla_j\psi_0 \cdot \nabla_j F_1 + \psi_0 \nabla_j^2 F_1 + F_1 \nabla_j^2 \psi_0 \qquad (5.2.101)$$

so we can write

$$(\mathcal{H} - E_0)\psi_1 = -\frac{\hbar^2}{2m}\sum_j (\psi_0 \nabla_j^2 F_1 + 2\nabla_j\psi_0 \cdot \nabla_j F_1)$$
$$+ \left[-\frac{\hbar^2}{2m}\sum_j \nabla_j^2 \psi_0 + \sum_{i<j} u(r_{ij})\psi_0 - E_0\psi_0\right] F_1 \quad (5.2.102)$$

and we notice that the last term is just

$$\mathcal{H}(\psi_0 - E_0\psi_0)F_1 = 0 \quad (5.2.103)$$

Moreover,

$$\nabla_j \cdot (|\psi_0|^2 \nabla_j F_1) = |\psi_0|^2 \nabla_j^2 F_1 + 2\psi_0 \nabla_j\psi_0 \cdot \nabla_j F_1$$
$$= \psi_0(\psi_0 \nabla_j^2 F_1 + 2\nabla_j\psi_0 \cdot \nabla_j F_1) \quad (5.2.104)$$

Substituting Eqs. (5.2.103) and (5.2.104) into (5.2.102), we find

$$(\mathcal{H} - E_0)\psi_1 = -\frac{\hbar^2}{2m\psi_0}\sum_j \nabla_j \cdot \rho_N \nabla_j F_1 \quad (5.2.105)$$

where $\rho_N = \psi_0^2$, as we wrote in Eq. (5.2.90). The energy of the excited state (above the ground state), the quantity we wish to minimize, is

$$\mathcal{E} = \int \cdots \int \psi_1^*(\mathcal{H} - E_0)\psi_1 \, d\{N\}$$
$$= \int \cdots \int F_1^*\psi_0(\mathcal{H} - E_0)\psi_1 \, d\{N\}$$
$$= -\frac{\hbar^2}{2m}\int \cdots \int F_1^* \sum_j \nabla_j \cdot \rho_N \nabla_j F_1 \, d\{N\} \quad (5.2.106)$$

where we have made use of Eq. (5.2.105).

Equation (5.2.106) is a sum of integrals, one for each particle, each integral of the form

$$\int \cdots \int F_1^* \nabla_j \cdot (\rho_N \nabla_j F_1) \, d^3 r_j \, d\{N - 1\} \quad (5.2.107)$$

We can transform the integral over $\mathbf{r}_1$, using the footnote on page 350. Let $a = F_1^*$ and $\phi = \rho_N \nabla_1 F_1$. We find

$$\int F_1^* \nabla_1 \cdot (\rho_N \nabla F_1) \, d^3 r_1 = \int_S F_1^*\rho_N \nabla_1 F_1 \cdot d\mathbf{S} - \int \rho_N \nabla_1 F_1 \cdot \nabla_1 F_1^* \, d^3 r_1$$
$$(5.2.108)$$

We can take the boundary conditions on ψ_1 at the surface to be $\psi_1 = 0$. Then the surface integral in Eq. (5.2.108) politely drops out, since the same

5.2 Superfluidity

condition applies to ψ_0, hence ρ_N is equal to zero at the surface. Thus, Eq. (5.2.106) becomes

$$\mathscr{E} = \frac{\hbar^2}{2m} \sum_j \int \cdots \int (\nabla_j F_1^*) \cdot (\nabla_j F_1) \rho_N \, d\{N\} \quad (5.2.109)$$

Using Eq. (5.2.99), we see that

$$\nabla_j F_1 = \nabla_j f(\mathbf{r}_j)$$

since ∇_j operates only on $\mathbf{r}_j$, not on the coordinates of the other particles. $\mathscr{E}$ in Eq. (5.2.109) now consists of N identical integrals, each of them equal to

$$\frac{\hbar^2}{2m} \int \cdots \int \nabla_1 f^*(\mathbf{r}_1) \cdot \nabla_1 f(\mathbf{r}_1) \rho_N \, d\{N\}$$

$$= \frac{\hbar^2}{2m} \int \nabla_1 f^* \cdot \nabla_1 f \, d^3 r_1 \int \cdots \int \rho_N \, d\{N - 1\} \quad (5.2.110)$$

When we use Eq. (5.2.92), $\mathscr{E}$ becomes

$$\mathscr{E} = \frac{N\hbar^2}{2m} \rho \int \nabla f^* \cdot \nabla f \, d^3 r \quad (5.2.111)$$

This is the quantity that we must minimize with respect to the function $f(\mathbf{r})$. However, in doing so, we must be sure that the normalization stays fixed, since it simply counts the number of particles [see Eq. (5.2.89)]. Define

$$I = \int \cdots \int \psi_1^* \psi_1 \, d\{N\}$$

$$= \int \cdots \int F_1^* F_1 \rho_N \, d\{N\}$$

$$= \sum_{i,j} \int \cdots \int f^*(\mathbf{r}_j) f(\mathbf{r}_i) \rho_N \, d\{N\} \quad (5.2.112)$$

This double sum has in it N terms with $i = j$, all of which are the same:

$$\int \cdots \int |f(\mathbf{r}_1)|^2 \, d^3 r_1 \rho_N \, d\{N - 1\} = \rho \int |f(\mathbf{r})|^2 \, d^3 r \quad (5.2.113)$$

and $N(N - 1)$ terms with $i \ne j$, each of which is equal to

$$\int f^*(\mathbf{r}_1) f(\mathbf{r}_2) \left[\int \cdots \int \rho_N \, d\{N - 2\} \right] d^3 r_1 \, d^3 r_2$$

$$= \frac{N}{V^2} \int f^*(\mathbf{r}_1) f(\mathbf{r}_2) g(r_{12}) \, d^3 r_1 \, d^3 r_2 \quad (5.2.114)$$

where we have used Eq. (5.2.93). Altogether, then,

$$I = N\rho \int |f(\mathbf{r})|^2 \, d^3r + N\rho^2 \int f^*(\mathbf{r}_1) f(\mathbf{r}_2) g(r_{12}) \, d^3r_1 \, d^3r_2$$

$$= N\rho \int [\delta(r_{12}) + \rho g(r_{12})] f^*(\mathbf{r}_1) f(\mathbf{r}_2) \, d^3r_1 \, d^3r_2 \qquad (5.2.115)$$

We will have used again later for the quantity in brackets, which we define to be

$$G(r) = \rho g(r) + \delta(r) \qquad (5.2.116)$$

Then

$$I = N\rho \int f^*(\mathbf{r}_1) f(\mathbf{r}_2) G(r_{12}) \, d^3r_1 \, d^3r_2 \qquad (5.2.117)$$

We must now minimize the quantity $(\mathscr{E} - \varepsilon I)$, where ε is a Lagrange multiplier:

$$\delta\mathscr{E} - \varepsilon \, \delta I = 0 \qquad (5.2.118)$$

Being generally complex, $f(\mathbf{r})$ has two independent parts for each $\mathbf{r}$. We can vary f^*, holding f fixed, and look for the functional form that satisfies Eq. (5.2.117). For $\delta\mathscr{E}$ we have [Eq. (5.2.111)]

$$\delta\mathscr{E} = \frac{N\hbar^2 \rho}{2m} \int \delta(\nabla f^*) \cdot \nabla f \, d^3r \qquad (5.2.119)$$

Using the divergence theorem (page 350), we replace $(\delta \nabla f^*) \cdot \nabla f$ by $-\delta f^* \nabla^2 f$:

$$\delta\mathscr{E} = -\frac{N\hbar^2 \rho}{2m} \int \delta f^* \nabla^2 f \, d^3r \qquad (5.2.120)$$

In Eq. (5.2.117), replace $\mathbf{r}_1$ by $\mathbf{r}$ and $\mathbf{r}_2$ by $\mathbf{r} + \mathbf{r}_{12}$. Then

$$\delta I = N\rho \int \delta f^*(\mathbf{r}) f(\mathbf{r} + \mathbf{r}_{12}) G(r_{12}) \, d^3r_{12} \, d^3r \qquad (5.2.121)$$

Substituting Eqs. (5.2.120) and (5.2.121) into (5.2.118), we have

$$0 = \int d^3r \, \delta f^*(\mathbf{r}) \left[\frac{N\hbar^2 \rho}{2m} \nabla^2 f(\mathbf{r}) + \varepsilon N\rho \int f(\mathbf{r} + \mathbf{r}_{12}) G(r_{12}) \, d^3r_{12} \right]$$

$$(5.2.122)$$

This can be correct for arbitrary δf^* only if

$$\frac{\hbar^2}{2m} \nabla^2 f(\mathbf{r}) + \varepsilon \int f(\mathbf{r} + \mathbf{r}') G(r') \, d^3r' = 0 \qquad (5.2.123)$$

5.2 Superfluidity

The function we seek, $f(\mathbf{r})$, is the solution of this integrodifferential equation. Fortunately, Feynman knew the solution, which is

$$f(\mathbf{r}) = C_Q \exp(i\mathbf{Q} \cdot \mathbf{r}) \tag{5.2.124}$$

Substituted into Eq. (5.2.97), this will give us a different wave function, hence a different energy for each $\mathbf{Q}$, which is just what we want—a dispersion relation. When we put this solution into Eq. (5.2.123), we get

$$\frac{\hbar^2 Q^2}{2m} - \varepsilon \int e^{i\mathbf{Q}\cdot\mathbf{r}'} G(r') \, d^3r' = 0 \tag{5.2.125}$$

where we have divided through by $\exp(i\mathbf{Q} \cdot \mathbf{r})$. The integral in the second term is a familiar quantity. Using Eq. (5.2.116), we have

$$\int e^{i\mathbf{Q}\cdot\mathbf{r}'} G(r') \, d^3r' = 1 + \int e^{i\mathbf{Q}\cdot\mathbf{r}'} g(r') \, d^3r' \tag{5.2.126}$$

Aside from an irrelevant δ function at $Q = 0$, this is just the form of the scattering function, Eq. (4.2.13). Following the arguments given there to Eq. (4.2.19), we find that the structure factor is

$$S(Q) = \int e^{i\mathbf{Q}\cdot\mathbf{r}'} G(r') \, d^3r' \tag{5.2.127}$$

so that Eq. (5.2.125) becomes

$$\varepsilon = \frac{\hbar^2 Q^2}{2mS(Q)} \tag{5.2.128}$$

This ε is our Lagrange multiplier. It will turn out to be the energy of the quasiparticles, and Eq. (5.2.128) will be the required dispersion relation, but we must identify it formally. Putting Eq. (5.2.124) into (5.2.111) first, we have

$$\mathscr{E} = \frac{N\hbar^2}{2m} \rho C_Q^2 \int Q^2 \, d^3r = \frac{\hbar^2 Q^2}{2m} N\rho V C_Q^2 \tag{5.2.129}$$

and then into Eq. (5.2.117)

$$I = N\rho C_Q^2 \int e^{-i\mathbf{Q}\cdot\mathbf{r}_1} e^{i\mathbf{Q}\cdot\mathbf{r}_2} G(r_{12}) \, d^3r_1 \, d^3r_2$$

$$= N\rho C_Q^2 V \int e^{i\mathbf{Q}\cdot\mathbf{r}} G(r) \, d^3r$$

$$= N\rho V C_Q^2 S(Q) \tag{5.2.130}$$

Recall that $\mathscr{E}$ is the excitation energy and I the normalization. The quasiparticle energy is therefore just $\mathscr{E}/I$:

$$\frac{\mathscr{E}}{I} = \frac{\hbar^2 Q^2}{2mS(Q)} = \varepsilon_Q \tag{5.2.131}$$

This is the celebrated result of the Feynman theory. In Fig. 5.2.13 we have a sketch of $S(Q)$ and ε. At low Q, $S(Q)$ rises linearly, $S(Q) \propto Q$, so that ε is linear as well, and we have the phonon part of the spectrum, $\varepsilon = c_1 Q$, where c_1 is the speed of first sound. At higher Q, $S(Q)$ rises more sharply toward its peak; ε falls below its linear dependence, then turns over to form the roton minimum. Aside from the phonons, the Feynman theory gives energies that are too high, as we had anticipated; we have not chosen the best possible wave function. What we have done can be improved on with some difficulty, but it hardly matters; it is clear that we have in hand the essential outline of what is going on. We have accounted—mathematically at least—for the general shape of the excitation curve.

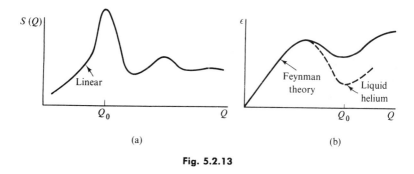

Fig. 5.2.13

Now that we have that general result, let us try to look back and see more clearly what we have done.

The Schrödinger equation for all the particles was never actually solved. Instead we have found how to modify the unknown ground-state wave function in order to form excited states. It is just as well not to know the ground state wave function; it is, after all, a function of $3N$ variables, where $N \sim 10^{23}$, and so cannot be expected to be a simple thing to write down. What is more, we do know what the ground state looks like: it looks like a liquid. The wave function will have its largest amplitude for configurations in which the atoms are randomly, but reasonably uniformly spaced. When we arrive at our result, Eq. (5.2.131), we have not yet put that fact in, and so it is still required. We put it in by using the experimental structure factor, $S(Q)$. The $S(Q)$ that is used, Fig. 5.2.13a, simply describes an ordinary liquid, not necessarily helium in its ground state. In fact, it is obviously not possible to measure $S(Q)$ for helium in its ground state; the $S(Q)$ actually used is measured in helium at its normal boiling point, 4.2°K, where it is not even a superfluid, just an ordinary liquid, as we said.

It may come as something of a surprise—it did come as a surprise to

5.2 *Superfluidity* 359

Feynman—that the result we obtained includes the phonons as well as the roton minimum we set out to get. Let us see how the wave function we have arrived at describes phonons. Although $\psi_0\{N\}$ has its largest amplitudes for configurations of reasonably uniform density, it also has finite amplitude for configurations in which there are density variations from place to place. These variations represent the zero-point oscillations of the phonon modes. The wave function for a state with one phonon in it is

$$\psi_{\text{ph}} = \sum_j C_Q e^{i\mathbf{Q}\cdot\mathbf{r}_j} \psi_0\{N\} \quad (5.2.132)$$

with small Q—that is, with $1/Q$ much larger than the average distance between atoms. The real part of ψ_{ph} is

$$\sum_j C_Q \cos(\mathbf{Q}\cdot\mathbf{r}_j)\psi_0\{N\} \quad (5.2.133)$$

Consider first a configuration, $\{N\}$, in which the density is uniform. $\psi_0\{N\}$ is just a number, the amplitude for that configuration in the ground state. This number is multiplied by a sum over the cosine terms, one for each atom. With uniform density there are the same number of positive and negative cosine terms of each magnitude, and so they sum to zero. ψ_{ph} has zero amplitude for uniform density (the argument obviously works equally well for the sine terms in the imaginary part). Now take a configuration in which there is a sinusoidal variation in density with wavelength equal to $2\pi/Q$ (along the direction $\mathbf{Q}$). $\psi_0\{N\}$ has a smaller, although finite, amplitude, but now we pick up more atoms in (say) the positive half-cycle of $\cos(\mathbf{Q}\cdot\mathbf{r})$ than in the negative, and ψ_{ph} has nonzero amplitude. ψ_{ph} just projects out of $\psi_0\{N\}$ those configurations with phonons of wave vector $\mathbf{Q}$. It is easy to verify that ψ_{ph}, in fact, all the excited state wave functions with $\mathbf{Q} \neq 0$, are orthogonal to ψ_0 (Prob. 5.4a).

The wave functions ψ_{ph} or, more generally, ψ_1

$$\psi_1 = \sum_j C_Q e^{i\mathbf{Q}\cdot\mathbf{r}_j} \psi_0\{N\} \quad (5.2.134)$$

describe excited states of the system that are $\mathbf{Q}$ eigenstates, with energy ε_Q, Eq. (5.2.131), representing a single excitation. Being a $\mathbf{Q}$ eigenstate, the excitation must be thought of as spread out uniformly through the system. A state with two excitations is

$$\psi_2 = \left(\sum_j C_{Q_1} e^{i\mathbf{Q}_1\cdot\mathbf{r}_j}\right)\left(\sum_k C_{Q_2} e^{i\mathbf{Q}_2\cdot\mathbf{r}_k}\right)\psi_0 \quad (5.2.135)$$

and has energy $\varepsilon_{Q_1} + \varepsilon_{Q_2}$ (Prob. 5.4b). In this way, we may build up the various states of the system or form localized excitations out of wave packets of excitations with different $\mathbf{Q}$'s. Each excitation may be considered an independent (noninteracting) quasiparticle as long as adding one to the system simply adds the energy we would deduce for that Q from Eq. (5.2.131). This is just the criterion for treating the quasiparticles as a perfect gas (see Sec. 1.1), and so we may apply the statistical mechanics of perfect gases to them. That is, in fact, exactly what we have done in the preceding subsection, Sec. 5.2d.

In a sense, we have completed our task. We have gone from the properties of superfluid helium to Landau's picture of the excitation spectrum needed to explain them and, finally, to Feynman's deduction of that excitation spectrum from the microscopic properties of helium atoms. There are, nevertheless, a few loose ends. We turn to them in the next subsection.

f. A Few Loose Ends

In Eq. (5.2.26) we reached the conclusion that circulation would be quantized in units of h/m in the condensate of a perfect Bose gas. Liquid helium is not a perfect gas, but now that we know its secrets, we can see how the argument works for the real stuff. We want now to modify the ground-state wave function, $\psi_0\{N\}$, to represent not an excited state but rather a flowing ground state, in order to see what kinds of flow are possible.

It is easy to set the entire system into uniform motion. The wave function

$$\psi = \exp\left(i\mathbf{q} \cdot \sum_j \mathbf{r}_j\right) \psi_0\{N\} \qquad (5.2.136)$$

represents the ground state with center of mass velocity

$$\mathbf{v} = \frac{\hbar \mathbf{q}}{m} \qquad (5.2.137)$$

In this case, we have obtained the total current from

$$\mathbf{j} = M\mathbf{v}$$

$$= \tfrac{1}{2}\left[\psi\left(\sum_i \hat{p}_i\right)\psi^* + \text{complex conjugate}\right] \qquad (5.2.138)$$

where M is the mass of the entire system

$$M = Nm \qquad (5.2.139)$$

Equation (5.2.138) is the many-body form of (5.2.19). Uniform motion of the center of mass is not exactly what we are interested in. We would like the velocity to change slowly, from place to place in the system, so that we can have macroscopic flow of some sort while still retaining the microscopic ground state. In a small region, then, where $\mathbf{v}$ is constant, the wave function must be essentially the one in Eq. (5.2.136). On the other hand, if the velocity

5.2 Superfluidity

is not to be uniform, **q** must change, gradually as we move about the system. We could try the wave function Eq. (5.2.136) with $\mathbf{q} = \mathbf{q}(\mathbf{R})$, **R** being simply a position in the fluid, or we can take this function inside the sum over atoms and write

$$\psi_{\text{flow}} = \exp\left[i \sum_j \gamma(\mathbf{r}_j)\right] \psi_0\{N\} \quad (5.2.140)$$

Now $\gamma(\mathbf{r})$ behaves just like the phase that we gave to each independent particle in the Bose gas condensate, Eq. (5.2.18), and we immediately reproduce the results, Eqs. (5.2.4) and (5.2.23) to (5.2.27). Real liquid helium, too, has quantized circulation.

The restriction on the flow, Eq. (5.2.4),

$$\nabla \times \mathbf{u}_s = 0$$

poses a dilemma. Suppose that we have a perfect, inviscid fluid in a bucket, and we set the bucket into rotation. If the fluid is inviscid, there is no mechanism to get it into rotation, but, nevertheless, in the thermodynamic equilibrium state of the system the fluid and bucket rotate all at the same angular velocity, say, ω, in solid-body rotation (Prob. 5.5a). The velocity field of solid-body rotation, **u**, obeys

$$|\nabla \times \mathbf{u}| = 2\omega \quad (5.2.141)$$

(Prob. 5.5b). Thus, superfluid helium cannot go into solid-body rotation because of its internal properties, although solid-body rotation is the equilibrium state of a perfect fluid.

The situation here bears a remarkably close analogy to a problem in superconductivity, which we discussed in Sec. 1.2e and to which we shall return in Sec. 5.3. Although a perfect conductor, by Lenz's law, will prevent an applied magnetic field from penetrating it, in the equilibrium state the field penetrates uniformly. A superconductor, however, for reasons of its own (Sec. 5.3), keeps the fields out even though it is costly of the appropriate free energy to do so (this is called the Meissner effect). Does a superfluid perform the analogous deed and keep the rotation field out?

The question can be answered experimentally by observing helium in a rotating bucket. The free surface of a fluid in rotation takes on the parabolic shape (Prob. 5.5c)

$$z = \frac{1}{2}\frac{\omega^2 r^2}{g} \quad (5.2.142)$$

where g is the acceleration of gravity; see Fig. 5.2.14. In liquid helium, we would expect the normal fluid part (the excitations) to rotate in any case, so we would predict

$$z = \frac{1}{2}\frac{\rho_n}{\rho}\frac{\omega^2 r^2}{g} \quad (5.2.143)$$

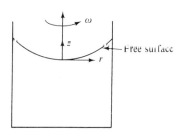

Fig. 5.2.14

if we expect the superfluid to remain stationary (the superfluid Meissner effect). Since ρ_n depends on T, the height of the meniscus should be temperature dependent. It turns out experimentally, however, that it is Eq. (5.2.142) that is obeyed—the entire fluid appears to go into solid-body rotation. Since solid-body rotation is forbidden by Eq. (5.2.4), we have a problem.

The solution to the problem—worked out independently by Onsager and by Feynman—is that the helium forms an array of vortex lines that imitate solid-body rotation. A familiar example of a vortex line is the whirlpool that forms over the drain when a bathtub empties. The flow is in circular paths about the line, as sketched in Fig. 5.2.15. If we integrate the velocity along one of the paths of constant velocity, we have, according to Eq. (5.2.26),

$$nk_0 = \oint \mathbf{u}_s \cdot d\mathbf{l} = 2\pi r u_s \qquad (5.2.144)$$

where r is the radius of the circle along which we have integrated. The velocity field of a vortex line is thus cylindrical, with the velocity varying with the distance from the line according to

$$u_s = \frac{nk_0}{2\pi r} \qquad (5.2.145)$$

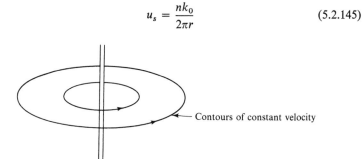

Fig. 5.2.15

5.2 Superfluidity

As r gets smaller, u_s gets bigger until, finally, centrifugal forces cause a hole to appear; classically, at least, the centrifugal forces are then balanced by the surface tension of the fluid. The hole is the vortex core. For quantized vortices in liquid helium, the core radius is of order 1Å; the core is basically a single line of missing atoms. Everywhere in the fluid, $\nabla \times \mathbf{u}_s = 0$, but in the core, where there is no fluid, there is vorticity ($\nabla \times \mathbf{u}$ is called the vorticity). This point can be seen by using the theorem

$$\int_S \nabla \times \mathbf{u} \cdot d\mathbf{S} = \oint \mathbf{u} \cdot d\mathbf{l} \qquad (5.2.146)$$

where the left-hand integral is over any surface terminating on the line of the right-hand integral. The vorticity in a fluid is the circulation per unit area; circulation in helium is quantized, in units of $k_0 = h/m$, and occurs in the form of quantized vortex lines. Fig. 5.2.16 indicates how an array of vortex lines can imitate solid-body rotation. The fluid is in counterclockwise rotation. The small circles are cross sections through vortex cores. The vortex cores are arrayed in concentric circles (the dotted lines). Between the two circles of cores, their velocity fields partially cancel. On the outside (to the right), they add together, so the fluid on the right has a larger velocity (in the counterclockwise direction) than that between the dotted circles. The vortex lines arrange themselves in order to do their best to cause the large-scale flow to resemble solid-body rotation, since that is the most favorable state (we shall see in Sec. 5.3 that something quite similar sometimes occurs in the analogous problem in superconductivity).

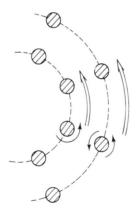

Fig. 5.2.16

The basic equations that describe the superfluid velocity field are

$$\nabla \times \mathbf{u}_r = 0 \qquad (5.2.4)$$

and
$$\nabla \cdot \mathbf{u}_s = 0 \qquad (5.2.147)$$

The second equation comes from Eq. (5.2.46) for steady-state ($\partial \rho / \partial t = 0$) flow at constant temperature and density. These equations are the same as Maxwell's equations for the behavior of the magnetic field in free space. The velocity field of a vortex line, Eq. (5.2.145), is mathematically the same as the magnetic field of a line current, with the circulation playing the role of the electric current. The force on a vortex core in a moving superfluid (Prob. 5.6b)

$$F = -\rho \mathbf{k} \times \mathbf{u}_s \qquad (5.2.148)$$

called the Magnus force in hydrodynamics, is formally analogous to the Lorentz force on a current-carrying wire in a magnetic field. In Eq. (5.2.148) **k** is a vector along the vortex core, with direction defined with respect to the sense of circulation by the right-hand rule and magnitude equal to the circulation, nk_0. Just as the energy density in a magnetic field is proportional to the square of the magnetic field, B^2, so the energy density in the superfluid field is proportional to the square of the velocity field; it is $\frac{1}{2}\rho_s u_s^2$. For a vortex line along the axis of a cylinder of radius R, the energy is

$$\begin{aligned}
\mathscr{E}_\ell &= \int \tfrac{1}{2}\rho_s u_s^2 \, d^3 r \\
&= \frac{l\rho_s}{2}\left(\frac{nk_0}{2\pi}\right)^2 \cdot 2\pi \int_{a_0}^{R} \frac{dr}{r} \\
&= \frac{ln^2 k_0^2}{4\pi} \log \frac{R}{a_0}
\end{aligned} \qquad (5.2.149)$$

where we have used Eq. (5.2.145) for u_s, a_0 is the radius of the core, and l is the length of the line. We see immediately that since the circulation per line is proportional to n and the energy is proportional to n^2, it is always preferable for vorticity to be single quantized. Henceforth we always take $n = 1$. Moreover, just as an electric current cannot end in free space, it follows from Eq. (5.2.147) that a vortex line must end on a wall, or a surface, or on itself, forming a vortex ring.

In Maxwell's day it was necessary to explain that electric currents and the magnetic fields they produce behave much like vortex lines and their velocity fields, since magnetic fields were a relatively new idea, but all physicists were acquainted with hydrodynamics. Times have changed. Now we can make use of our familiarity with magnetostatics to help visualize the behavior of vortex lines and velocity fields.

5.2 *Superfluidity* 365

There are two rather fundamental problems in superfluidity for whose solution quantized vorticity is commonly invoked. These are the critical velocity and the relaxation of metastable states of superflow. Let us finish this section with a discussion of those problems.

Equation (5.2.11) is a statement of the Principle of Superfluidity: there is some velocity below which a superfluid cannot interact with a foreign object. Below that velocity, it flows past the object without any resistance. Applying the criterion, the energies and momenta of phonons and rotons lead us to expect that this critical velocity will be roughly 60 m/sec, but in all experiments resistance-free superflow is actually found to break down at very much lower velocities, typically of the order of millimeters per second or, at most, centimeters per second. It would certainly help if some other excitation were available, with a ratio of energy to momentum more of this order of magnitude.

An object that seems to serve this purpose is a rather perverse beast called a quantized vortex ring. It is constructed by bending a vortex line around and having it end on itself. The result is a velocity dipole field, just as an electric current loop produces a magnetic dipole field. Calculation of the energy of a vortex ring is a rather formidable problem, but the answer turns out to be

$$\mathscr{E}_r = \frac{1}{2} \rho_s k_0^2 R_r \left(\eta - \frac{7}{4} \right) \tag{5.2.150}$$

where R_r is the radius of the ring, and

$$\eta = \log \frac{R_r}{a_0} \tag{5.2.151}$$

is a very slow function of R_r. We shall treat it more or less as a constant. The energy of the ring is not very different from what one might expect for a vortex line of length $l = 2\pi R$ [see Eq. (5.2.149)]. A vortex ring has no actual momentum (ignoring small-density variations near the vortex core), but the role of momentum is played by the impulse, $\mathscr{I}$. In order to create a vortex ring, or to change its size, a force must be applied for a finite amount of time. $\mathscr{I}$ is defined as the integral of the force over the time, $\int \mathscr{F} \, dt$, and it behaves just like a momentum, for purposes of conservation arguments like those leading to Eq. (5.2.11). The impulse of a circular vortex ring is

$$\mathscr{I} = \rho_s k_0 \pi R_r^2 \tag{5.2.152}$$

If we guess, then, that the critical velocity to be observed experimentally, $u_{s,c}$, will be given by Eq. (5.2.11) for the creation of a vortex ring, we get

$$u_{s,c} = \frac{\mathscr{E}_r}{\mathscr{I}} = \frac{k_0}{4\pi R_r} \left(\eta - \frac{7}{4} \right) \tag{5.2.153}$$

The smallest value of this velocity occurs for the largest radius, R_r, say, for a ring whose diameter is on the order of the size of the apparatus (for example, the width of the channel in Fig. 5.2.2). If we put $2R_r \approx d$, the channel width in various experiments, into Eq. (5.2.153), we actually obtain roughly the observed results.

As a complete explanation for the observed critical velocities, this argument is much too crude. For example, Eq. (5.2.150) gives the energy of a vortex ring in an infinite fluid. The energy of a vortex ring in a channel comparable to its own size would be quite different (much smaller). Nevertheless, we apparently have transformed a fundamental dilemma into a messy problem in hydrodynamics, and perhaps that is a step forward. At least it seems likely that most observed critical velocities are somehow associated with quantized vortex rings.

Ignoring the question of the precise critical velocity that one expects to find, as well as the (entirely unsolved) problem of how vorticity nucleates when able to do so, we still have an additional problem to consider: Exactly how does the production of vortex rings lead to the observed breakdown in superflow? The clearest example is the experiment sketched in Fig. 5.2.2. At the critical velocity, a difference in pressure is observed between the two manometers. We wish now to relate that pressure difference to the rate at which vortex rings are created in the channel.

Let us look at the channel again in Fig. 5.2.17. We have flow in the channel, at constant velocity, u_s, transporting helium from region 1 to region 2, at pressures P_1 and P_2. We imagine that the walls are excellent conductors of heat, and so the whole apparatus is kept at uniform temperature. It is easy to see that, according to the two-fluid model, we must have $P_1 = P_2$. If there were a difference between P_1 and P_2, then the pressure gradient in the channel would be related to the chemical potential gradient by

$$\nabla P = \rho \, \nabla \mu \qquad (5.2.154)$$

[see Eq. (5.2.53)]. With u_s uniform in space, we have from Eqs. (5.2.41) and (5.2.42)

$$\frac{\partial \mathbf{u}_s}{\partial t} + \nabla \mu = 0 \qquad (5.2.155)$$

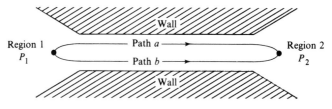

Fig. 5.2.17

5.2 Superfluidity

It follows that in steady state, when $\partial \mathbf{u}_s/\partial t = 0$, there could be no $\nabla \mu$, and so $P_1 = P_2$. Any effect leading to dissipation in this situation must be found outside the two-fluid model. That effect, of course, is the production of quantized vorticity at a critical velocity given roughly by Eq. (5.2.153). At that velocity, a pressure difference develops across the channel. Quantized vorticity was not included in the two-fluid model, so we must modify Eq. (5.2.155) when it is present. To see how, we consider the situation at a more microscopic level.

In Eq. (5.2.140) we have the wave function of the flowing helium. The superfluid velocity is related to the phase $\gamma(\mathbf{r})$ by Eq. (5.2.23)

$$\mathbf{u}_s = -\frac{\hbar}{m}\nabla\gamma \quad (5.2.23)$$

In regions 1 and 2 of Fig. 5.2.17, u_s is very small compared to its value in the channel, so we can take the phases in those regions, γ_1 and γ_2, to be uniform in space. The difference between γ_1 and γ_2 can be obtained by integrating $\nabla\gamma$:

$$\gamma_2 - \gamma_1 = \int_1^2 \nabla\gamma \cdot d\mathbf{l}$$

$$= -\frac{m}{\hbar}\int_1^2 \mathbf{u}_s \cdot d\mathbf{l} \quad (5.2.156)$$

The result of this integral, however, may depend on the path we take. In Fig. 5.2.17 we have sketched two paths from 1 to 2, path a and path b. The difference between the integrals along the two paths is

$$\int_{\text{path }a} \nabla\gamma \cdot d\mathbf{l} - \int_{\text{path }b} \nabla\gamma \cdot d\mathbf{l} = \oint \nabla\gamma \cdot d\mathbf{l}$$

$$= -\frac{m}{\hbar}\oint \mathbf{u}_s \cdot d\mathbf{l} \quad (5.2.157)$$

According to Eq. (5.2.25) or (5.2.26), this difference is not necessarily equal to zero. Should a quantized vortex core happen to be between the two paths, we would find a difference of 2π in $\int_1^2 \nabla\gamma \cdot d\mathbf{l}$, depending on which path we choose. In other words, when a vortex core crosses a path connecting 1 and 2, the phase difference measured along that path jumps by 2π. To take a simple picture, suppose that a vortex ring is created on the wall, encircling the channel (the vortex core is a ring in the plane perpendicular to the axis of the channel). Imagine that, staying always in the plane perpendicular to the axis, it pulls away from the wall, diminishes in size, and finally annihilates itself on the axis. At a certain moment, suppose that path a is outside the ring but path b passes through it. There is then a difference of 2π in $\int_1^2 \nabla\gamma \cdot d\mathbf{l}$

between path a and path b, since we then have $\oint \mathbf{u}_s \cdot d\mathbf{l} = h/m$. The core has passed through path a, changing $\gamma_2 - \gamma_1$ by 2π when measured along that path. At a later moment, both paths are outside the ring, there is no net vorticity inside, $\oint \mathbf{u}_s \cdot d\mathbf{l} = 0$, and the two paths agree on $\gamma_2 - \gamma_1$. The core has now passed through path b, causing its measurement of $\gamma_2 - \gamma_1$ to jump by 2π.

Now suppose that the process we have just described takes place at a steady average rate. Let $(1/\ell)(dn/dt)$ be the number of rings per second created in a length ℓ of wall and annihilated at the axis (or anywhere inside, for that matter). Then all paths between 1 and 2 will agree that $\gamma_2 - \gamma_1$ is changing at the average rate

$$\frac{1}{\ell}\frac{\partial}{\partial t}(\gamma_2 - \gamma_1) = \frac{2\pi}{\ell}\frac{dn}{dt} \qquad (5.2.158)$$

If this process is occurring uniformly along the length of the channel, then

$$-\nabla\gamma = \frac{\gamma_2 - \gamma_1}{\ell}$$

and

$$\frac{\partial}{\partial t}\frac{\gamma_2 - \gamma_1}{\ell} = -\frac{\partial}{\partial t}\nabla\gamma = \frac{m}{\hbar}\frac{\partial u_s}{\partial t} \qquad (5.2.159)$$

where we have used Eq. (5.2.23). If the superfluid is constrained by external forces to keep $\mathbf{u}_s$ constant, we can replace $\partial \mathbf{u}_s/\partial t$ in Eq. (5.2.159) by $\nabla\mu$; this is the modification that we were looking for in Eq. (5.2.155). Then, using Eq. (5.2.154), we see that the force needed to overcome the production of vorticity is

$$\frac{1}{\rho}\nabla P = -\frac{\hbar}{m}\frac{\partial}{\partial t}\nabla\gamma \qquad (5.2.160)$$

or integrating across the channel,

$$P_2 - P_1 = \frac{2\pi\hbar\rho}{m}\frac{dn}{dt} \qquad (5.2.161)$$

It is easy to see that the geometric details are not too important; the pressure difference depends only on the total rate of production of vorticity inside.

Before finishing our discussion of liquid helium, let us consider one final issue: when u_s is smaller than the critical velocity whose effects we have been discussing, can superflow really occur without any resistance whatever? In Sec. 1.2e we investigated the equilibrium state of a perfect conductor in an applied magnetic field. When the field is turned on, eddy currents are induced and the field is thereby prevented from penetrating into the sample. In equilibrium, there would be no eddy currents and the field would penetrate completely. The difficulty is that a perfect conductor, having no resistance

5.2 Superfluidity

whatever, has no way of reaching equilibrium. This difficulty is somewhat academic, there being no perfect conductors (superconductors, remember, are quite another matter), but we now have exactly the same dilemma in the not-so-academic case of superfluid helium.

Superflow in helium is not an equilibrium state, even if we make sure that $u_n = 0$ so that there is no ordinary dissipation to worry about. If there is superflow in some part of an isolated large system, the system could always raise its entropy if it could change the kinetic energy of the superflow into heat. We must ask, then, is there any way for this process to occur? Operationally, we can ask the question in two equivalent ways. We can either look for a very small pressure difference between the two manometers in the experiment sketched in Fig. 5.2.2, or we can ask whether the persistent superfluid current observed in the experiment sketched in Fig. 5.2.5 really persists undiminished forever. Let us continue to examine the channel experiment, leaving persistent currents to the reader to work out (Prob. 5.6e).

In our discussion of the critical velocity we assumed that quantized vorticity could somehow be produced mechanically, as a result of flow past the wall, since we had already assured ourselves that energy and momentum could be conserved were production to occur. We are now interested in conditions under which the production of a vortex ring would not conserve energy and momentum, at least not in the production process itself—that is, just when $u_s < u_{s,c}$. However, now something more subtle can happen.

If there is no pressure gradient across the channel, the helium flowing in the channel is in a metastable state. It would like to come to rest, dumping its kinetic energy into the walls as heat. If a vortex ring were somehow to arise, the phase across the channel would flip by 2π, producing a momentary deceleration of the superflow, and the whole system would end up in a thermodynamically more favorable state. The energy needed to do so is available in the form of heat, not, as we have seen, in the flow itself. It is, of course, very unlikely that the available heat will spontaneously organize itself into a vortex ring, but since the excitation of a vortex ring is, after all, a possible state of the system, it must occur from time to time in the course of ordinary thermodynamic fluctuations. Thus, there is a means by which superflow might decay, or, in the case of the channel, require a small pressure difference to maintain the current.

The heat energy available for creating the ring resides in the normal fluid and the wall, not the superfluid part, so we imagine the presence of a ring to be a possible state of the normal fluid and wall part of the system. We look for rings that are created at rest with respect to the normal fluid, hence moving at a velocity $\mathbf{v}_r = -\mathbf{u}_s$ with respect to the superfluid. The radius of the ring depends on its velocity, v_r, since the Magnus force, Eq. (5.2.148), on each part of the core produced by the velocity field arising from the rest of the ring must be balanced by motion of the ring through the fluid;

the smaller the ring, the faster it moves. Ignoring details of the core itself, we can estimate the group velocity of a vortex ring by using Eqs. (5.2.150) and (5.2.152):

$$v_r = \frac{\partial \mathscr{E}_r}{\partial \mathscr{I}}$$

$$\simeq \frac{k_0 \eta}{4\pi R_r} \qquad (5.2.162)$$

Roughly speaking, $v_r \sim 1/R_r$. Since $\mathscr{E}_r \sim R_r$, we have the odd result that the energy of a vortex ring goes inversely as its velocity! If we feed energy into a ring, somehow pushing on it, it slows down (but it grows bigger). In any case, the energy of a ring is given approximately by

$$\mathscr{E}_r \approx B \frac{\rho_s}{v_r} \qquad (5.2.163)$$

where we have gathered a number of constants into B. Using Eq. (5.2.38), the energy in the normal fluid-wall frame needed to create a ring is

$$\delta\mathscr{E} = \mathscr{E}_r + \mathscr{I} u_s = \frac{B' \rho_s}{u_s} \qquad (5.2.164)$$

where we have used Eq. (5.2.162) to eliminate R_r from Eq. (5.2.152) and have set $|v_r| = u_s$. B' gathers a new set of constants.

Now, the probability of a ring arising under random fluctuations is proportional to

$$\exp(-\delta\mathscr{E}/kT)$$

The constant of proportionality is the rate at which the helium tries to make rings out of heat. We do not know what that number is, but we might estimate it to be something like the interatomic spacing divided by the speed of sound—that is, some natural number associated with thermal fluctuations. In any case, if f_0 is that attempt frequency per unit volume of helium, the rate at which rings occur will be given by

$$\frac{dn}{dt} = f_0 V \exp\left(-\frac{B'\rho_s}{kTu_s}\right) \qquad (5.2.165)$$

where V is the volume of fluid in the channel. Substituting this result into Eq. (5.2.161), we have for the pressure difference across the channel

$$P_2 - P_1 = 2\pi \hbar N f_0 \exp\left(-\frac{B'\rho_s}{kTu_s}\right) \qquad (5.2.166)$$

where $N = \rho V/m$ is the number of helium atoms in the channel.

It is useless to try to make a numerical estimate of the expected pressure

5.3 *Superconductivity*

difference. We have no idea of what f_0 is, and we really do not know B' either, since the formulas used to derive Eq. (5.2.164) are (as we have noted earlier) valid only in an open bath, not in a narrow channel. The form of the result is interesting, however; $P_2 - P_1$ as a function of u_s has the form

$$y = e^{-1/x} \qquad (5.2.167)$$

As x goes to zero, y goes to zero faster than any power of x, which means that no matter how sensitive an apparatus we build to measure $P_2 - P_1$, there will always be some value of u_s below which $P_2 - P_1$ will appear to be zero. If we work very hard and make our apparatus ten times more sensitive, we will find that $P_2 - P_1$ goes to zero at a lower value of u_s, but only slightly, say, a few percent lower, and so on. There will always appear to be true superflow. We can check Eq. (5.2.165) with sufficiently sensitive apparatus, however, by testing the dependence of $P_2 - P_1$ on u_s and on temperature. We should have

$$\log (P_2 - P_1) = -\frac{B'\rho_s}{kTu_s} + \text{constant}$$

This dependence has been confirmed experimentally in cases where the velocity is lower than the ordinary critical velocity discussed earlier.

5.3 SUPERCONDUCTIVITY

a. Introduction

When Kammerlingh-Onnes succeeded in liquefying helium in 1908 (see Sec. 6.3 for more details), his laboratory at Leiden had as its own domain a whole new world to explore. Among the earliest experiments he did was a study of the electrical conductivity of metals. Mercury was particularly suitable for this purpose, since it could easily be purified by distillation. In 1911, using a sample of mercury, he discovered superconductivity.

The electrical resistivity of a pure metal is a smooth function of temperature, falling toward zero at zero temperature. Onnes discovered that the resistivity dropped abruptly to zero at a finite temperature; a voltmeter across the sample read zero while an ammeter in series with it still showed current flowing. The discovery was, of course, entirely unexpected.

Or was it? It is difficult at this distance in time to place ourselves in Onnes's mind as he chose what experiments to try in the newly available range of low temperatures, but it seems unlikely that he saw, as the culmination of his career, the adding of a few more points to a well-known curve. It is more probable that he expected something to happen. His working model of electrical conduction in metals would have been the classical picture presented by Drude and J. J. Thomson—a fluid of electrons flowing through a lattice of positive ions. When fluids are cooled, they tend to crystalize. A reasonable conjecture is that he expected the electron fluid to freeze and the metal therefore to become an insulator. The phase transition that he was looking for would show up by the resistivity abruptly becoming infinite. Instead it dropped to zero.

Onnes's reasoning, nevertheless, would have been sound and, in a fundamental sense, correct. The electron fluid does undergo a phase change, a kind of condensation, into an ordered state. It was only the details of the new state that Onnes could not have anticipated. Those details were not fully understood until Bardeen, Cooper, and Schrieffer (BCS) explained them in 1957.

The superconducting state has two central, unique characteristics. The first is the absence of electrical resistance, as Onnes discovered in 1911. The second, called the Meissner effect, is the fact that magnetic fields are expelled from the interior of a superconducting sample. We have already examined the relationship between these two properties in some detail in Sec. 1.2e. As we saw then, the Meissner effect does not follow as a result of the absence of electrical resistance. However, confusion over this point may help to explain the fact that the Meissner effect was not discovered for over two decades after the discovery of superconductivity.

Let us consider briefly how, in principle at least, the two properties—zero resistance and the Meissner effect—may each be established experimentally. Onnes's experiment is sketched schematically in Fig. 5.3.1a. The resistance of the sample is V/I, where V is the reading of the voltmeter and I the ammeter. There is no question that for a fixed I, V drops abruptly at a certain temperature, but the problem arises: Is V really zero, or is it some finite value below the sensitivity of our voltmeter? A better way to detect very small resistances is indicated in Fig. 5.3.1b. A circulating current is induced around a superconducting doughnut. The current loop produces a magnetic field, **B**, as shown, by which the presence and strength of the

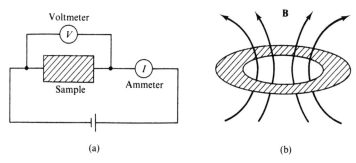

Fig. 5.3.1

5.3 Superconductivity

current may be detected. If there is any electrical resistance in the doughnut, the current will die out exponentially in time according to the law

$$I = I_0 e^{-t/\tau} \tag{5.3.1}$$

where I_0 is the current at time $t = 0$ and τ, the relaxation time, is given by

$$\tau = \frac{L}{R} \tag{5.3.2}$$

where L is the inductance of the loop and R the resistance. In this way, we can measure an arbitrarily small resistance merely by waiting long enough. In the purest of ordinary, normal metals at low temperature, such a current would die out in a small fraction of a second. In a superconducting material, a current of this kind has been observed, without measurable loss, over a period of years. Only a lower limit can be placed on the resistance of a superconductor. The number itself is of little interest, since it is probably related to something like the life expectancy of a graduate student rather than any intrinsic property of a superconductor.

The alert reader will have realized by now that the two experiments sketched in Fig. 5.3.1 are precisely analogous to the experiments on liquid helium sketched in Figs. 5.2.2 and 5.2.5. Other readers should draw out the details of the analogy (what plays the role of V in helium?, etc.).

In order to illustrate the existence of the Meissner effect, we must show that a superconducting sample will expel a stationary magnetic field. As we saw in Sec. 1.2e, a superconducting material will prevent a field from penetrating as the field is turned on, but that could be a consequence of its lack of resistivity; eddy currents induced by the changing field persist, but the final state is not necessarily one of thermodynamic equilibrium. The proper test is to place a piece of normal material, above its superconducting transition temperature, in a constant magnetic field. The field then penetrates the material thoroughly, as it would any ordinary, nonmagnetic material. The temperature is now reduced so that the sample becomes superconducting. The induction of eddy currents depends on the rate of change of the magnetic field, $\partial \mathbf{B}/\partial t$, but the field does not change during the cooling process. Nevertheless, it is found that the field is expelled when the superconducting transition occurs. At the end, the sample is magnetized, the supercurrents running in the surface producing a field that just cancels the applied one. The equilibrium magnetic equation of state is Eq. (1.2.87)

$$\mathbf{M} = -\frac{1}{4\pi} \mathbf{H} \tag{5.3.3}$$

where $\mathbf{M}$ is the magnetization and $\mathbf{H}$ the applied field. This is the Meissner effect.

As we know from our discussion in Sec. 2.5, electrons in metals do not form a classical fluid. The electron gas is highly degenerate in the Fermi-Dirac sense, and if no phase change occurs, it will go into a well-defined ground state, with zero entropy at zero temperature. Thus, from the point of view of the Third Law of Thermodynamics, there is no requirement that the electrons go into a more ordered state at low temperature. In fact, some metals evidently do not become superconducting. For example, there is no evidence at present that copper, or silver, or gold ever becomes a superconductor. Nevertheless, the superconducting state is a common one, found in the majority of pure elementary metals as well as in thousands of compounds and alloys. In addition to the two fundamental properties discussed above, the very fact that the state is so common is one of the factors that needs to be understood from a microscopic point of view.

One consequence of the widespread existence of superconductivity is that many of the details of behavior vary widely, depending on which material the state is studied in. At sufficiently low applied magnetic fields, all superconductors behave as we have described them. At very high applied fields, superconductivity is always destroyed; the state thus exists in a region of the H-T plane, as sketched in Fig. 1.2.3. Quantitative details of the curve vary from one material to another. T_c, the transition temperature in zero-applied field, varies from a fraction of $1°K$ up to tens of degrees; H_{c0}, the critical field at zero temperature from hundreds of gauss up to hundreds of thousands of gauss. Moreover, there are qualitative differences as well. Some materials always behave just as we have described, as long as they are superconducting. These are called type I superconductors. Others, called type II superconductors, change their nature somewhat if the applied field gets big enough, allowing some penetration of the applied field without becoming entirely normal. The reasons for these differences in the superconducting state in various materials will become apparent as we develop a deeper understanding of superconductivity.

The phenomena of superconductivity closely resemble those of superfluidity, in helium, studied in Sec. 5.2. As we saw, superfluidity is, in a deep sense, a consequence of the Bose nature of the helium atoms. Electrons, of course, are not bosons, but they form bosons in the superconducting state. A bound pair of electrons obeys Bose statistics. This point can easily be seen by forming an asymmetric wave function for a set of electrons as in Eq. (2.3.5) and then permuting the electrons by pairs only. Since the sign of the wave function changes twice (i.e., not at all) under permutations of pairs, the wave function for a system of pairs is symmetric; that is, the pairs obey Bose statistics.

Our explanation of superconductivity will ultimately involve bound electron pairs in a Bose-condensed state. Each pair has the same wave function

$$\psi_p = ae^{i\gamma} \quad (5.3.4)$$

5.3 Superconductivity

where the amplitude a and the phase γ may depend on position. We have already seen this form, of course, in, for example, Eq. (5.2.18). The reason that pairs form in this state, and the subtle way in which they do, is the subject of the microscopic theory of superconductivity. A system of Bose-condensed electron pairs does not suffice to explain the phenomena of superconductivity any more than the Bose-condensed perfect gas explains superfluidity. However, the basic properties of zero resistance and the Meissner effect do follow rather more easily from this picture than the properties of helium do from the Bose gas. For one thing, the electron pairs, unlike helium atoms, are, in fact, nearly a perfect gas. For another, the excitation spectrum, on which the Principle of Superfluidity is based, is a direct consequence of the pairing process and thus is already described once we have understood how pairing happens in the first place. Consequently, in the case of superconductivity (unlike superfluidity), most of the work involved in understanding the state will turn out to consist of understanding why the particles form bosons at all.

In Sec. 5.3b we will review and extend our discussion of superconductivity from the point of view of thermodynamics, in order to broaden our grasp of the basic phenomena involved. Sections 5.3c and 5.3d are devoted to the microscopic theory—that is, to the nature of the state in which electrons form pairs. Finally, in Sec. 5.3c we shall see how the paired electrons give rise to the macroscopic phenomena that we have been discussing.

b. Some Thermodynamic Arguments

In Sec. 1.2e the superconductor-perfect conductor problem was used as an example on which to exercise the techniques of thermodynamic variational principles. Let us start now by reviewing what we learned there about superconductivity.

A perfect conductor (a metal without any electrical resistance) would not expel a magnetic field in thermodynamic equilibrium. To do so requires work that may be turned into heat if the field is allowed to penetrate. A superconductor, however, does expel the field in equilibrium. The superconductor must gain some other energetic advantage by being superconducting that is great enough to overcome the disadvantage of having to expel the applied field.

According to our general variational principle, Eq. (1.2.78), a superconductor at a given T and $\mathbf{H}$ tries to minimize its Gibbs potential

$$\Phi = E - TS - \int \mathbf{H} \cdot \mathbf{M} \, d^3r \qquad (5.3.5)$$

In a perfect conductor or, for that matter, in a normal conductor, the field penetrates, there is no magnetization of the sample ($M = 0$), and the Gibbs potential is given by

$$\Phi_n = E_n - TS_n \qquad (5.3.6)$$

In a superconductor, the equation of state describing the Meissner effect is Eq. (5.3.3). Substituting (5.3.3) into (5.3.5), we find for the Gibbs potential of the superconductor

$$\Phi_{sc} = E_{sc} - TS_{sc} + \frac{1}{4\pi} \int H^2 \, d^3r$$

$$= E_{sc} - TS_{sc} + \frac{1}{4\pi} H^2 V \qquad (5.3.7)$$

where V is the volume of the sample (recall that we are always considering a long, cylindrical sample parallel to the applied field, so there are no demagnetizing effects). In each case, the energy E is given by Eq. (1.2.81)

$$E = E_0 + \frac{1}{8\pi} \int B^2 \, d^3r$$

$$= E_0 + \frac{1}{8\pi} B^2 V \qquad (5.3.8)$$

with $B = H$ for the perfect or normal case and $B = 0$ for the super case. E_0 is the energy in zero field. We thus have, respectively,

$$\Phi_n = E_{0,n} - TS_n + \frac{1}{8\pi} B^2 V$$

$$= F_{0,n} + \frac{1}{8\pi} H^2 V \qquad (5.3.9)$$

and

$$\Phi_{sc} = E_{0,sc} - TS_{0,sc} + \frac{1}{4\pi} H^2 V$$

$$= F_{0,sc} + \frac{1}{4\pi} H^2 V \qquad (5.3.10)$$

where $F_{0,n}$ and $F_{0,sc}$ are, respectively, the free energies of the normal and superconductors in zero field. Application of a field raises Φ_{sc} more than Φ_n. The difference is the extra work the superconductor must do in order to remain superconducting. It would not bother to do it were it not that $F_{0,sc}$ is lower than $F_{0,n}$ by more than the extra work required. At the critical field, $H_c(T)$, all advantage is lost and the superconductor goes normal. This is the field at which $\Phi_n = \Phi_{sc}$. Writing $\varphi = \Phi/V$ and $f = F/V$, we have

$$\varphi_n(T, H_c) = \varphi_{sc}(T, H_c) \qquad (5.3.11)$$

$$f_{0,n} - f_{0,sc} = \frac{H_c^2}{8\pi} \qquad (5.3.12)$$

5.3 Superconductivity

which is just Eq. (1.2.93). For reasons of its own, to be investigated later, it is inimical to the superconducting state to have a magnetic field in its interior. To expel a field, it is willing to do work up to $H_c^2/8\pi$ per unit volume. It follows that $H_c^2/8\pi$ per unit volume is the amount by which a metal is able to lower its free energy by the microscopic process of condensation into the superconducting state.

As we saw in Sec. 1.2g, a number of additional conclusions follow readily from Eq. (5.3.12). The super-normal phase transition in nonzero field is first order, with latent heat given by Eq. (1.2.133)

$$L = T(s_n - s_{sc}) = -\frac{TH_c}{4\pi}\frac{dH_c}{dT} \tag{5.3.13}$$

At T_c, H_c is equal to zero, and L goes to zero as well. However, there is a discontinuity in the heat capacity, c, which, like L, is related to the temperature dependence of H_c [Eq. (1.2.136)]:

$$c_n - c_s = -\frac{T}{4\pi}\left[\frac{dH_c}{dT}\right]^2 \tag{5.3.14}$$

The general forms of $H_c(T)$ and the heat capacity in zero field (Figs. 1.2.3 and 1.2.10) are repeated here in Fig. 5.3.2.

The order of magnitude of the difference in free energy density between a superconductor and a normal conductor at zero temperature is of some interest. In order to have an idea of what to expect, consider the energy involved in the condensation of a liquid or solid relative to the gas. The condensed medium is held together by the forces between the atoms, whose characteristic energy is the well depth of the interatomic potential (i.e., ε_0 in Fig. 4.3.1). Condensation first begins to occur at the gas-liquid critical temperature, T_c. As we might reasonably expect, this is just when kT is of order ε_0. By analogy, we can compare the superconducting critical temperature, again T_c, to the superconducting condensation energy per electron. At $T = 0$, all the electrons should be condensed, yielding an energy $H_{c0}^2/8\pi$ per unit volume relative to the uncondensed, normal state. For a crude estimate, take approximate values for tin, $T_c \approx 4°K$, $H_{c0} \approx 300$ gauss, and suppose

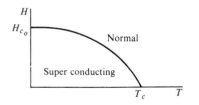
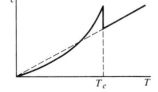

Fig. 5.3.2

that there are 10^{22} electrons per unit volume. Then the condensation energy per electron is $(9 \times 10^4)/(8\pi \times 10^{22}) \approx 4 \times 10^{-19}$ erg, while $kT_c \approx (1.4 \times 10^{-16}) \times 4 \approx 6 \times 10^{-16}$ erg. The condensation energy is too small to account for T_c in this way by some three orders of magnitude. This point alerts us to the fact that the condensation of electrons into superconducting pairs is a rather more subtle process than the condensation of atoms into a liquid.

A superconductor, as we have repeatedly seen, finds it necessary to run currents in its surface that cancel any applied magnetic field. Up to now we have taken the magnetic field **B** to be equal to the applied field **H** outside the sample and to be equal to zero inside. The change at the surface, however, is not discontinuous; there is some distribution of currents and a corresponding distribution of fields, **B** going from **H** outside to zero deep inside. Let us now investigate that distribution.

We are still dealing with a sample at constant T and **H**, and so the Gibbs potential, Eq. (5.3.5), is still to be minimized. However, where we previously neglected contributions from the surface of the superconductor, we must now take them into account, since that is what we are studying. In the surface region, the energy density has two terms that are lacking in the interior: one is the kinetic energy density of the electrons participating in the shielding currents; the other is the energy density in the field, $B^2/8\pi$, since the surface region is just where **B** has not yet reached zero. The energy, then, is

$$E = E_0 + \int \left(\frac{1}{2} n_s m^* v^2 + \frac{B^2}{8\pi} \right) d^3r \qquad (5.3.15)$$

where n_s is the density of superconducting charge carriers (they will turn out to be electron pairs), m^* is their effective mass, and v is their velocity. E_0, as before, is the energy in zero applied field. The field **B** is related to the supercurrent density, $\mathbf{j}_s$, by Maxwell's equation

$$\mathbf{j}_s = \frac{c}{4\pi} \nabla \times \mathbf{B} \qquad (5.3.16)$$

where c is the speed of light, and

$$\mathbf{j}_s = -n_s e^* \mathbf{v} \qquad (5.3.17)$$

Here $-e^*$ is the effective charge of the carriers (i.e., $e^* = 2e$, where $-e$ is the electron charge). Using Eqs. (5.3.16) and (5.3.17), we can write the kinetic energy of the electrons as

$$\frac{1}{2} n_s m^* v^2 = \frac{\lambda^2}{8\pi} |\nabla \times \mathbf{B}|^2 \qquad (5.3.18)$$

5.3 Superconductivity

where λ is a length, characteristic of the superconductor, given by

$$\lambda^2 = \frac{c^2}{e^{*2}} \frac{m^*}{4\pi n_s} \tag{5.3.19}$$

For reasons that will soon become obvious, λ is called the *penetration depth*. With the help of Eqs. (5.3.15) and (5.3.18), we can write the Gibbs potential, Eq. (5.3.5), as

$$\Phi = E_0 - TS + \int \left(\frac{\lambda^2}{8\pi}|\nabla \times \mathbf{B}|^2 + \frac{B^2}{8\pi}\right) d^3r - \mathbf{H} \cdot \int \mathbf{M}\, d^3r \tag{5.3.20}$$

We wish to find the distribution of fields, $\mathbf{B}(\mathbf{r})$. However, if we now proceed to minimize Φ with respect to $\mathbf{B}(\mathbf{r})$, we will find the solution $\mathbf{B} = \mathbf{H}$ everywhere; this is just the perfect conductor problem of Sec. 1.2e, reinforced by putting in the kinetic energy associated with any shielding currents (see Prob. 5.8a). Φ, as we have written it in Eq. (5.3.20), has no way of knowing that we are dealing with a superconductor, in which $E_0 - TS = F_0$ will jump to a higher value if we let the field in. In order to insert this fact, we minimize Φ subject to the condition that

$$\int \mathbf{M}\, d^3r = \text{constant} \tag{5.3.21}$$

under variations in $\mathbf{B}(\mathbf{r})$. In other words, we only consider variations, $\delta\mathbf{B}(\mathbf{r})$, that leave an arbitrary magnetization in the sample. Variations in $\mathbf{B}(\mathbf{r})$ will cause local variations in $\mathbf{M}(\mathbf{r})$, but we will conserve the integrated total magnetization and look for the best distribution under those circumstances.

The condition Eq. (5.3.21) is imposed by means of a Lagrange multiplier. Rather than minimize Φ, we minimize instead the quantity $\Phi + \mathbf{\varepsilon} \cdot \int \mathbf{M}\, d^3r$:

$$0 = \delta\left(\Phi + \mathbf{\varepsilon} \cdot \int \mathbf{M}\, d^3r\right)$$
$$= \int \left(\frac{\lambda^2}{4\pi} \nabla \times \mathbf{B} \cdot \nabla \times \delta\mathbf{B} + \frac{\mathbf{B}}{4\pi} \cdot \delta\mathbf{B}\right) d^3r - (\mathbf{H} - \mathbf{\varepsilon}) \cdot \int \delta\mathbf{M}\, d^3r \tag{5.3.22}$$

We now use the vector identity

$$\nabla \cdot (\mathbf{A} \times \mathbf{C}) = \mathbf{C} \cdot (\nabla \times \mathbf{A}) - \mathbf{A} \cdot (\nabla \times \mathbf{C})$$

with $\mathbf{A} = \nabla \times \mathbf{B}$ and $\mathbf{C} = \delta\mathbf{B}$ to write

$$(\nabla \times \mathbf{B}) \cdot \nabla \times \delta\mathbf{B} = -\nabla \cdot [(\nabla \times \mathbf{B}) \times \delta\mathbf{B}] + (\nabla \times \nabla \times \mathbf{B}) \cdot \delta\mathbf{B} \tag{5.3.23}$$

When the first term on the right in Eq. (5.3.23) goes into the volume integral in Eq. (5.3.22), we get

$$\int \nabla \cdot [(\nabla \times \mathbf{B}) \times \delta \mathbf{B}] \, d^3r = \int_{\text{surface}} (\nabla \times \mathbf{B}) \times \delta \mathbf{B} \cdot d\mathbf{S}$$

The last integral is over the surface. However, at the surface, $\mathbf{B} = \mathbf{H}$ or $\delta \mathbf{B} = 0$, so we may drop this term. Equation (5.3.22) now becomes

$$0 = \frac{1}{4\pi} \int \delta \mathbf{B} \cdot [\lambda^2 \nabla \times \nabla \times \mathbf{B} + \mathbf{B}] \, d^3r - \frac{(\mathbf{H} - \boldsymbol{\varepsilon})}{4\pi} \cdot \int \delta \mathbf{B} \, d^3r \quad (5.3.24)$$

where we have used $\mathbf{B} = \mathbf{H} + 4\pi \mathbf{M}$ and $\mathbf{H} = \text{constant}$, so that $\delta \mathbf{M} = \delta \mathbf{B}/4\pi$. To satisfy this equation for arbitrary $\delta \mathbf{B}$, we must have

$$\mathbf{B} - \mathbf{H} + \boldsymbol{\varepsilon} = -\lambda^2 \nabla \times \nabla \times \mathbf{B}$$

To choose the correct value of $\boldsymbol{\varepsilon}$, we note that deep inside the superconductor, we wish $\mathbf{B}$ to be constant ($\nabla \times \nabla \times \mathbf{B} = 0$) and equal to zero; thus, $\boldsymbol{\varepsilon} = \mathbf{H}$. Then

$$\mathbf{B} = -\lambda^2 \nabla \times \nabla \times \mathbf{B} \quad (5.3.25)$$

Equation (5.3.25) is called the London equation (London arrived at it by different arguments).

Equation (5.3.25) is easily solved for a semi-infinite slab of superconductor with an applied field parallel to its surface (Prob. 5.8b). The solution is sketched in Fig. 5.3.3. Outside the superconductor, $\mathbf{B} = \mathbf{H}$. Inside,

$$\mathbf{B} = \mathbf{H} e^{-x/\lambda} \quad (5.3.26)$$

and the shielding currents obey

$$\mathbf{j} = \mathbf{j}_0 e^{-x/\lambda} \quad (5.3.27)$$

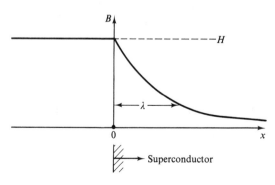

Fig. 5.3.3

5.3 Superconductivity

where j_0 is the current density at the surface. If **H** is applied in the y direction, **j** will run in the z direction (right-handed system). λ is thus the characteristic distance over which the field penetrates into the superconductor. In more complicated geometries the essential features of the solution will be the same, with the currents and fields falling off exponentially as we move in from the surface.

We can see now, in retrospect, just how this distribution of currents and fields occurred. In order for the field **B** to go from **H** outside to zero inside, there must be a current sheet at the interface. Because the superconductor does not like to let the field penetrate, it wishes the current sheet to be as narrow as possible so that the field is nearly discontinuous. On the other hand, since n_s has been taken to be constant, and $\mathbf{j}_s = -n_s e^* \mathbf{v}$, the narrower the current sheet, the higher **v** must be in order to produce enough total current to cancel the field. The kinetic energy of the electrons is proportional to v^2, so the narrower the current sheet, the higher this kinetic energy is. The kinetic energy favors a spreading out of the current sheet. It is in the competition between these two effects that Eq. (5.3.25) is forged.

In all this argument there has been a hidden assumption. Although the fields and currents change with a characteristic distance, λ, we have assumed that the superconducting state itself is quite constant on the scale of λ. In particular, n_s, the density of superconducting charge carriers, was taken to be independent of **r**. In effect, the amplitude of the wave function of the superconducting pairs, Eq. (5.3.4), must not have changed in the distance λ. Let us consider for a moment what that means.

Take a situation in which we can be sure that the amplitude must change with position. Suppose that a slab of superconductor is in intimate contact with a slab of normal conductor (say, a slab of tin and a slab of copper). Electrons flow freely across the interface, but deep inside the tin they are superconducting and inside the copper they are not. The amplitude of the superconducting state must somehow go from a finite value in the tin to zero in the copper. The amplitude cannot change too rapidly because to do so would be energetically expensive (the energy of the state has in it the term $\nabla^2 \psi_p$). As a result, we can be rather sure that the amplitude a will look something like the sketch in Fig. 5.3.4. The amplitude will change with some characteristic distance, which we call ξ, the *coherence length*.

In deriving the London equation, Eq. (5.3.25), we have assumed, in effect, that $\lambda \gg \xi$, so that starting at any interface, the material becomes fully superconducting much more quickly than the fields start to fall off. Superconductors in which this condition is satisfied are called type II superconductors, and only for these is Eq. (5.3.25) valid. In the opposite limit, $\xi \gg \lambda$, we have type I superconductors. A brief summary of the general characteristics of the two kinds of material follows.

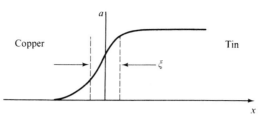

Fig. 5.3.4

Type I: $\xi \gg \lambda$ Generally soft, pure, poorly conducting metals, such as lead, indium, tin. Low critical fields (less than $\sim 10^3$ gauss) and low critical temperatures (less than $\sim 10°K$).

Type II: $\lambda \gg \xi$ Generally harder materials. Some pure elements (niobium) but more often alloys, compounds, and so on. High critical fields and temperatures.

The two lengths themselves are usually in the range of hundreds to thousands of angstroms. The relative magnitudes of the two characteristic lengths, ξ and λ, clearly have a great deal to do with the detailed properties of materials in the superconducting state. Since these lengths come into play at the surface of the superconductor, the different kinds of behavior are evidently due to some kind of surface effect. In order to understand the differences, we must study something that has a rather unlikely sound to it: the surface tension of a superconductor.

The part of the Gibbs potential that we have most recently been discussing is strictly associated with the surface and has a magnitude proportional to the surface area of the superconductor. If the surface area is free to change, this contribution to Φ must be positive, since if it is negative, Φ can be decreased merely by increasing the surface area, and the situation we are describing is therefore unstable. The surface area of the metal sample will not change—the energies we are dealing with are much too small for that to happen—but the interfacial area between a normal and a superconducting part of the same sample can increase. Even if no normal phase is present, the Gibbs potential per unit area of a hypothetical super-normal interface must be positive, or the sample could lower its Gibbs potential by creating bits of normal region at the surface. The Gibbs potential per unit interfacial area is the surface tension, and, as in any other kind of matter, the surface tension of a superconductor must be positive.

To compute the surface tension, we return to Eq. (5.3.20), writing it in the form

$$\Phi = \int \left[f_0 + \frac{\lambda^2}{8\pi} |\nabla \times \mathbf{B}|^2 + \frac{B^2}{8\pi} - \mathbf{H} \cdot \mathbf{M} \right] d^3r \qquad (5.3.28)$$

5.3 Superconductivity

where f_0 is the free energy density in zero field. We can eliminate **M** in favor of **B** and **H**, using Eq. (1.2.42),

$$\mathbf{M} = \frac{1}{4\pi}(\mathbf{B} - \mathbf{H}) \tag{5.3.29}$$

We also wish to gather together volume terms and surface terms in Eq. (5.3.28). **B** is nonzero only near the surface, so all the terms in the integral containing **B** will go into the surface part. Using Eq. (5.3.29), we find

$$\Phi = \int f_0 \, d^3r + \frac{1}{4\pi} \int H^2 \, d^3r + \frac{1}{8\pi} \int (\lambda^2 |\nabla \times \mathbf{B}|^2 + B^2 - 2HB) \, d^3r \tag{5.3.30}$$

Since **H** is everywhere constant, the second term on the right is just $(1/4\pi)H^2 V$, where V is the volume of the sample. The last term is a pure surface term, having zero contribution from the interior of the superconductor. The first term, however, still has both surface and volume contributions, since f_0 will generally differ near the surface from what it is inside. Without a microscopic description to work with, we cannot say exactly how f_0 will depend on the distance from the surface, but we can make a rough estimate from what little we already know about the microscopic behavior. Wherever the material is fully superconducting (i.e., far from the surface), its free energy density is

$$f_{0,\text{sc}} = f_{0,n} - \frac{H_c^2}{8\pi} \tag{5.3.31}$$

[Eq. (5.3.12)]. Near a super-normal interface, however, we argued that the material is not fully superconducting (see Fig. 5.3.4). The free energy density goes gradually from $f_{0,n}$ to $f_{0,\text{sc}}$, with a characteristic relaxation distance ξ. Not knowing the precise law governing the change, let us approximate the situation by saying that everywhere along the interface there is a strip of width ξ in which $f_0 = f_{0,n}$ instead of $f_{0,\text{sc}}$ and that $f_0 = f_{0,\text{sc}}$ everywhere inside the strip. If the surface area is S, there is a small volume, $S\xi$, in which the superconductivity is "damaged" by the presence of the interface, and the free energy is raised by $(H_c^2/8\pi)S\xi$. With this approximation,

$$\int f_0 \, d^3r = \int f_{0,\text{sc}} \, d^3r + \frac{H_c^2}{8\pi} \xi S$$

$$= f_{0,\text{sc}} V + \frac{H_c^2}{8\pi} \xi S \tag{5.3.32}$$

We thus have for Eq. (5.3.30)

$$\Phi = \left(f_{0,\text{sc}} + \frac{H^2}{4\pi}\right) V + \frac{H_c^2}{8\pi} \xi S$$

$$+ \frac{1}{8\pi} \int (\lambda^2 |\nabla \times \mathbf{B}|^2 + B^2 - 2HB) \, d^3r \tag{5.3.33}$$

For simplicity, let us evaluate the remaining integral, using the solution for the semi-infinite slab, Eq. (5.3.26). Since **B** is constant in the yz plane (i.e., the surface), we have

$$\int d^3r = \int dx \int dy\, dz = S \int dx$$

so that, substituting in Eq. (5.3.26), Φ becomes

$$\Phi = \left(f_{0,\text{sc}} + \frac{H^2}{4\pi}\right) V$$

$$+ S\left[\frac{H_c^2}{8\pi}\xi + \frac{1}{8\pi}\int (H^2 e^{-2x/\lambda} + H^2 e^{-2x/\lambda} - 2H^2 e^{-x/\lambda})\, dx\right] \quad (5.3.34)$$

The term in brackets multiplying S is the surface tension we seek. Let us call it σ. Integrating over x, we find for the surface tension

$$\sigma = \frac{1}{8\pi}(H_c^2 \xi - H^2 \lambda) \quad (5.3.35)$$

This equation cannot be taken too literally, since the first term is only approximate; nevertheless, it tells us what we want to know. For very small **H**, the interface of the superconducting state is always stable, since σ is positive. However, σ becomes equal to zero, and the interface becomes unstable approximately when

$$H^2 = H_c^2 \frac{\xi}{\lambda} \quad (5.3.36)$$

Now, if $\xi \gg \lambda$, the interface is stable for all fields up to H_c, whereupon the entire superconducting state is destroyed. That is the behavior of a type I superconductor.

On the other hand, if $\lambda \gg \xi$, then at some field considerably below H_c, the sample finds that it is desirable to create more super-normal interface, although the interior still finds it advantageous to be superconducting. It can effectively increase the super-normal interfacial area without substantially decreasing the superconducting volume by forming very small normal inclusions. Figure 5.3.5 indicates how the interface (cross-hatched) may be increased by forming an inclusion, even if no super-normal interface existed at the start. The plane of the sketch is perpendicular to the applied field. Each inclusion brings with it a small area of magnetic field—that is, a bit of flux. The smaller the inclusion, the more S is increased and the less V is decreased, so the efficiency of this process in reducing Φ is limited only by how small the smallest possible inclusion is.

The smallest inclusion of flux in a superconductor is determined in the same way as the smallest inclusion of vorticity in the superfluid. Flux is

5.3 Superconductivity

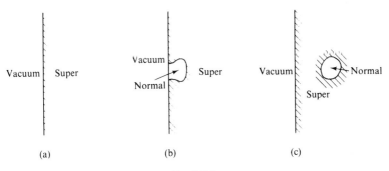

Fig. 5.3.5

quantized, into units called *fluxoids* or, by analogy to helium, *vortices* or *vortex lines*. The argument that leads to the quantization condition is the same one used in helium. Let us defer the details until Sec. 5.3e.

A material with $\lambda \gg \xi$ (i.e., a type II superconductor) thus does not have a complete Meissner effect up to the critical field, H_c. At a lower field, called H_{c_1}, the magnetic field begins to penetrate in the form of quantized fluxoids. At H_c, the bulk interior is ready to go normal, but the material can still remain superconducting if it has many vortices; it takes advantage of the negative surface tension by creating a lot of surface, and Φ remains lower than in the pure normal state. At some higher field, H_{c_2}, the bulk becomes normal, but even here there is a sheath of superconductivity on the surface to take advantage of the negative surface tension. Finally, at an even higher field, H_{c_3}, the material becomes completely normal. The phase diagram of a type II superconductor is sketched in Fig. 5.3.6. Equation (5.3.26) gives a crude estimate of H_{c_1}, the field at which the vortex state begins. We shall be able to do better, however, when we study the vortices themselves in Sec. 5.3e.

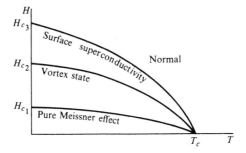

Fig. 5.3.6

c. Electron Pairs

The essential behavior of electrons in metals is that of the perfect Fermi gas, as we saw in Sec. 2.5. In that treatment, all interactions between electrons are neglected, a procedure that we justified then by arguing that they ought to be very weak. In this section we shall see that the ground state of the electrons cannot be the Fermi degenerate ground state if there is even a very weak net attractive interaction between them. The electrons instead form pairs, called Cooper pairs, which are the basic entities of the superconducting state. There are two important points to be understood from these arguments. The first is how the interaction between electrons comes to be attractive, and the second is how a very weak attraction between electrons can lead to the formation of pairs.

The direct interaction between electrons is, of course, the coulomb electrostatic force, which is repulsive. That force is largely screened by the positively charged medium in which each electron finds itself, but how does an attractive force between them arise? A great deal of complicated physics and chemistry is involved in the behavior of electrons in metals (see Sec. 3.6 for an indication). The theory of superconductivity succeeds because all of it is ignored except for the essential features that lead to the superconducting state. Let us undertake our explanation in that spirit.

Imagine two people on an old, sagging, nonlinear mattress. They tend to roll toward the middle, even if they don't like each other. That is, there is an attractive interaction. The cause of this interaction (remember, we are ignoring all features that are not pertinent to superconductivity) is that the people create distortions in the mattress, and the distortions are attracted to each other and try to merge. The attractive interaction between electrons occurs in somewhat the same way. The negatively charged electrons cause distortions of the lattice of positive ions, and the real attraction occurs between these distortions.

Pursuing the idea further, the electrons in the metal do not stand still but rather zip through the lattice at something like the Fermi velocity. The ions are attracted to the electrons but, owing to their large mass, move very slowly compared to the much lighter electrons. By the time the ions respond the electron is long gone, but it has, in effect, left behind a trail of positive charge, which is the lattice distortion we mentioned above. Another electron, traversing the same path, would find that its way had been prepared with the positive charge that it finds so attractive. We can imagine, if we wish, that the first electron created a phonon, which the second happily absorbs. This is the nature of the interaction between the two electrons. Notice that the interaction is strongest if the two electrons traverse exactly the same path—that is, if they have, say, equal and opposite momenta.

Descriptions of the kind just given cannot be taken too literally. For example, an electron cannot both follow a certain path (meaning it is

5.3 Superconductivity

localized in the crystal) and have a definite momentum (equal and opposite to its partner) without violating the uncertainty principle. However, with a bit of judicious compromise between localization in real and momentum space, the general picture stands and gives us some idea of where the attraction between electrons comes from. We shall try to do the quantum mechanics more correctly below.

The difficulty that arises now is that the interaction just described is, at best, a rather subtle one and very weak. Weak attractive interactions between particles do not, in general, lead to the formation of bound pairs. For example, an attraction between helium atoms leads to the condensation of many helium atoms into a liquid under appropriate conditions. However, helium atoms never form bound pairs, not even in the saturated vapor of liquid helium (this point is discussed further in Sec. 6.3). Why, then, do electrons form pairs under the very weak and subtle interaction we have been discussing?

To understand what happens, we shall perform a calculation that was first done by Cooper. We consider a perfect Fermi gas in its ground state, isolate two electrons from this gas, and examine the effect on them of an interaction of the type discussed. We will find that their state is changed regardless of how weak the interaction, and so the many-electron ground state we assumed to start with is unstable when there is any net attractive interaction.

In the absence of the interaction, the two electrons we isolate for examination are in momentum eigenstates in the Fermi sea. We choose electrons at the Fermi surface with momenta $\mathbf{q}$ and $-\mathbf{q}$ (the magnitude of each is the Fermi value q_F). With this choice, their center of mass is at rest, so that if they do form a Bose-like bound pair, it will be in the zero-momentum condensate of the Bose system. Moreover, as we argued above, we expect the interaction to be strongest for a pair with equal and opposite momenta. The spatial wave function of the pair, $\psi = \psi(|\mathbf{r}_2 - \mathbf{r}_1|)$, where $\mathbf{r}_2$ and $\mathbf{r}_1$ are the positions of the two electrons, obeys the Schrödinger equation

$$\frac{-\hbar^2}{2m}(\nabla_1^2 + \nabla_2^2)\psi + V(|\mathbf{r}_2 - \mathbf{r}_1|)\psi = (\varepsilon + 2\varepsilon_F)\psi \quad (5.3.37)$$

where V is the potential acting between them and $\varepsilon_F = \hbar^2 q_F^2/2m$ is the energy of each electron in the ground state when $V = 0$. ε, then, is the energy of the pair relative to the noninteracting state. If $\varepsilon < 0$, the pair is bound and the ordinary Fermi-degenerate ground state is unstable.

The wave function of the pair must be antisymmetric. The spatial part is symmetric, so in addition to specifying that the pair have opposite momenta, we must also specify that they have opposite spins, say, choosing the electrons $\mathbf{q}\uparrow$ and $-\mathbf{q}\downarrow$ ($+\mathbf{q}$, spin up; $-\mathbf{q}$, spin down).

We have not yet put into our picture the effect of all the other electrons present. Their effect is to keep the two we are interested in from occupying any states below the Fermi surface. In order to specify that condition, we must rewrite the Schrödinger equation, Eq. (5.3.37), in $\mathbf{q}$ space, where we know what the situation is. To do so, we first write the pair wave function as a linear combination of momentum eigenstates

$$\psi(\mathbf{r}) = \sum_{\mathbf{q}} g_{\mathbf{q}} e^{i\mathbf{q}\cdot\mathbf{r}} \tag{5.3.38}$$

where we have constructed each exponential from the two members of a pair in opposite momentum states

$$e^{i\mathbf{q}\cdot\mathbf{r}_2} e^{i(-\mathbf{q})\cdot\mathbf{r}_1} = e^{i\mathbf{q}\cdot(\mathbf{r}_2 - \mathbf{r}_1)} = e^{i\mathbf{q}\cdot\mathbf{r}} \tag{5.3.39}$$

If V were equal to zero, then in Eq. (5.3.38) only one of the coefficients $g_{\mathbf{q}}$ would be nonzero, and the electrons would be in the momentum eigenstates of the perfect Fermi gas. The effect of V is to scatter the pair, with some probability, into other states, say, from $\mathbf{q}$, $-\mathbf{q}$ into $\mathbf{q}'$, $-\mathbf{q}'$. This process mixes some amplitude for the state $(\mathbf{q}', -\mathbf{q}')$ into the wave function. In other words, the effect of V is that more than one of the coefficients $g_{\mathbf{q}}$ will be nonzero. On the other hand, no scattering can take place into states below the Fermi surface, for they are all fully occupied. We can ensure that condition by specifying

$$g_{\mathbf{q}} = 0 \quad \text{(for all } q < q_F\text{)} \tag{5.3.40}$$

It was for this purpose that we have switched into $\mathbf{q}$ space.

Substitute Eq. (5.3.38) into (5.3.37) to obtain

$$\frac{\hbar^2}{2m} \sum_{\mathbf{q}} g_{\mathbf{q}}(2q^2) e^{i\mathbf{q}\cdot\mathbf{r}} + V(r) \sum_{\mathbf{q}} g_{\mathbf{q}} e^{i\mathbf{q}\cdot\mathbf{r}} = (\varepsilon + 2\varepsilon_F) \sum_{\mathbf{q}} g_{\mathbf{q}} e^{i\mathbf{q}\cdot\mathbf{r}} \tag{5.3.41}$$

To get rid of the residual r dependence in this equation, divide by the volume, L^3, multiply by $\exp(-i\mathbf{q}'\cdot\mathbf{r})$, and integrate over volume, giving

$$\frac{\hbar^2 q^2}{m} g_{\mathbf{q}} + \sum_{\mathbf{q}} g_{\mathbf{q}'} V_{\mathbf{q}\mathbf{q}'} = (\varepsilon + 2\varepsilon_F) g_{\mathbf{q}} \tag{5.3.42}$$

where

$$V_{\mathbf{q}\mathbf{q}'} = \frac{1}{L^3} \int V(r) e^{i(\mathbf{q}-\mathbf{q}')\cdot\mathbf{r}} d^3r \tag{5.3.43}$$

and we have used

$$\frac{1}{L^3} \int e^{i(\mathbf{q}-\mathbf{q}')\cdot\mathbf{r}} d^3r = \delta_{\mathbf{q}\mathbf{q}'} \tag{5.3.44}$$

$V_{\mathbf{q}\mathbf{q}'}$, Eq. (5.3.43), is basically the matrix element for scattering the pair from pair state $\mathbf{q}$ (by which we mean $\mathbf{q}\uparrow$, $-\mathbf{q}\downarrow$) to pair state $\mathbf{q}'$. Equation (5.3.42) is called the Bethe-Goldstone equation.

We have now set up the machinery we need to investigate the fate of the

5.3 Superconductivity

pair that started out in state **q**, under the interaction $V_{qq'}$, provided that we can specify that interaction. We know the general nature of the interaction we want to insert from our arguments above. Somehow we need to place two people on a mattress in this calculation. The energy available for the interaction is the energy involved in lattice distortions or, in other words, the energy in the phonon spectrum of the solid. The largest interaction energy via the lattice is the energy of the highest-frequency phonon in the spectrum, roughly $\hbar\omega_D$, where ω_D is the Debye frequency. The potential V cannot scatter a pair from the state **q** to the state **q'** if these states are separated by more than $\hbar\omega_D$ in energy, because not enough energy is available in the lattice distortions to do so (we can imagine more complicated processes involving two or more phonons, but they are obviously higher order, less likely, and less important).

We expect, then, that $V_{qq'}$ will be nonzero only for states within $\hbar\omega_D$ of the Fermi surface. Recalling that $\varepsilon_F \gg \hbar\omega_D$ (see Fig. 3.5.6), we see that the states from which and into which electrons are scattered by the potential $V(r)$ are all of nearly the same energy; states **q** and **q'** really differ only in the directions of the **q** vectors. In order to progress in the calculation, let us simply assume that the potential is isotropic—independent of direction. Then $V_{qq'}$ is just a constant in a narrow shell above the Fermi surface:

$$V_{qq'} = -V$$

if

$$\frac{\hbar^2 q^2}{2m} < \varepsilon_F + \hbar\omega_D$$

and

$$\frac{\hbar q'^2}{2m} < \varepsilon_F + \hbar\omega_D$$

otherwise $V_{qq'} = 0$ \qquad (5.3.45)

where $V > 0$ so that $V_{qq'} = -V$ is an attractive potential. Using this simplification, Eq. (5.3.42) becomes

$$\frac{\hbar^2 q^2}{m} g_q - V \sum_{q'} g_{q'} = (\varepsilon + 2\varepsilon_F) g_q \qquad (5.3.46)$$

where the sum extends only over states in the interacting shell. Rearranging terms,

$$\left(\frac{\hbar^2 q^2}{m} - \varepsilon - 2\varepsilon_f\right) g_q = V \sum_{q'} g_{q'} = -D \qquad (5.3.47)$$

The middle term does not depend on **q** (**q'** is a dummy variable in the summation), so we have written it equal to a constant, $-D$, independent of **q**. Equation (5.3.49) is really two equations

$$D = -V \sum_{q'} g_{q'} \qquad (5.3.48)$$

and
$$g_q = \frac{D}{\varepsilon + 2\varepsilon_F - (\hbar^2 q^2/m)} \quad (5.3.49)$$

Substituting g_q from Eq. (5.3.49) into (5.3.48) gives

$$D = -V \sum_q \frac{D}{\varepsilon + 2\varepsilon_F - (\hbar^2 q^2/m)} \quad (5.3.50)$$

where we have changed the dummy variable from $\mathbf{q}'$ to $\mathbf{q}$. Defining the variable ξ,

$$\xi = \frac{\hbar^2 q^2}{2m} - \varepsilon_F \quad (5.3.51)$$

we can rewrite Eq. (5.3.50) (after canceling D from both sides) as

$$1 = V \sum_q \frac{1}{2\xi - \varepsilon} \quad (5.3.52)$$

ξ is just the energy of a single electron relative to the Fermi energy. The sum is to be carried out for values of ξ between zero and $\hbar\omega_D$. Changing the sum to an integral in the usual way, we have

$$1 = V \int_0^{\hbar\omega_D} \frac{\rho(\xi)\, d\xi}{2\xi - \varepsilon} \quad (5.3.53)$$

$\rho(\xi)$ is the usual single particle density of states [see, for example, Eq. (1.3.106)] except that ξ is the energy above the Fermi level, so that from Eq. (1.3.106)

$$\rho(0) = \frac{4\pi\sqrt{2}L^3 m^{3/2}}{(2\pi\hbar)^3} \varepsilon_F^{1/2} \quad (5.3.54)$$

In the narrow band between ε_F and $\varepsilon_F + \hbar\omega_D$, $\rho(\xi)$ hardly changes at all; we can take $\rho(\xi) \approx \rho(0)$ and write Eq. (5.3.53)

$$1 = V\rho(0) \int_0^{\hbar\omega_D} \frac{d\xi}{2\xi - \varepsilon}$$
$$= \frac{1}{2} V\rho(0) \int_{-\varepsilon}^{2\hbar\omega_D - \varepsilon} \frac{dx}{x}$$
$$= \frac{1}{2} V\rho(0) \log\left(\frac{\varepsilon - 2\hbar\omega_D}{\varepsilon}\right) \quad (5.3.55)$$

Solving for ε, we obtain

$$\varepsilon = \frac{2\hbar\omega_D}{1 - e^{2/\rho(0)V}} \quad (5.3.56)$$

5.3 Superconductivity

For weak potentials (i.e., small V), the exponential in the denominator dominates, and we find for ε

$$\varepsilon = -2\hbar\omega_n e^{-2/\rho(0)V} \qquad (5.3.57)$$

This is the result that we have been seeking. The energy of the pair is lower than it was in the pure Fermi ground state by this amount. The pair is not bound in the ordinary sense, since its overall energy, $2\varepsilon_F + \varepsilon$, is still quite positive, but it is quasi-bound, with energy lower than $2\varepsilon_F$.

Let us stop to see how this peculiar bound state has occurred. The reason we obtained a negative ε (in the limit of small V) is because we have taken a constant density of states, $\rho(0)$. This point can be seen by examining Eq. (5.3.53). As $V \to 0$, the integral must be very large to solve the equation. Suppose that the Fermi sea of other electrons were missing, and we were considering the interaction of two isolated particles through the weak potential, V. Without the other electrons, they would be able to populate states down to the origin in momentum space, and so the density of states would be $\rho(\xi) \propto \xi^{1/2}$. In this case, the integral is always small unless there is a zero in the denominator—that is, unless ε is positive. Thus, a weak potential does not lead to the formation of bound states between pairs of particles, as we said earlier. Instead some minimum strength is needed in the potential to form a bound state. Equation (5.3.57) gives us a bound state no matter how weak the potential, V, is.

In two dimensions, perfect gas particles would have a constant density of states rather than one that goes to zero at zero energy (see Prob. 2.6), and so a bound state of the kind we have seen here could occur. It is, in fact, a two-dimensional problem that we have solved, with the particles restricted to interact on (or near) the two-dimensional surface of the Fermi sea. The physical reason that a bound state can occur in our problem is that the importance of a weak interaction is greatly enhanced when the density of states is large. A large density of states means that many states are available that may easily be mixed into the wave function of the pair [see Eq. (5.3.38)], even by a weak potential. The result is that the wave function can be modified enough to result in a lower energy for the ground state.

The general form of Eq. (5.3.57) is the same as that of Eq. (5.2.167) describing the decay of superflow in helium. As V goes to zero, $-\varepsilon$ goes to zero faster than any power of V. In other words, there is no way to write ε as a power series, say, a Taylor expansion, in powers of V. That is one reason that the problem of superconductivity took so long to solve; the superconducting state cannot be constructed from the normal state by perturbation theory (even though the potential involved is small) because perturbation theory always leads to a power series, which is not the form characteristic of superconductivity.

The calculation we have just done does not describe the superconducting ground state. It only tells us that the normal ground state is unstable if there is even a weak attractive potential between the electrons. If there is an attractive potential acting between any one pair, it will act between all such pairs, changing their states as well as that of the pair we have isolated. But then the Fermi sea is no longer present in its unaltered form, and so the final state of the pair, which depended on the presence of the unperturbed Fermi sea, cannot be the one we have found. To find out what the superconducting ground state is really like, we must discover some way of treating all the electrons together, putting in the same interaction we have used here. We shall do so now.

d. The BCS Ground State

We know that the familiar Fermi ground state is unstable under an attractive interaction acting between pairs of electrons of opposite momentum and spin, $\mathbf{q}\uparrow$ and $-\mathbf{q}\downarrow$, but we do not know the form of the new ground state. That problem was first solved by BCS. We will work it out in a somewhat modified, simplified form.

Our general approach will be to use the same techniques developed earlier for examining the equilibrium states of thermodynamic systems in order to see how the perfect gas is changed by precisely the interaction we studied in the last section. Just as in Chap. 2, we will study the Landau potential

$$\Omega = E - TS - \mu N \qquad (5.3.58)$$

The idea is to put the Cooper interaction into the Fermi gas problem and find the distribution of particles among the single-particle states that minimizes Ω. According to the criterion, Eq. (1.2.78), Ω is the quantity to be minimized for a system at fixed T, V, and μ. We are to imagine our system to be in a particle bath, as we did in Chap. 2, so that N may change but μ remains fixed. The constant temperature in the problem is $T = 0$.

Just as in the Cooper calculation, we will not consider single electrons at all, only pairs of opposite $\mathbf{q}$ and spin. Instead of Eq. (5.3.38), in which $\psi(\mathbf{r})$ is constructed out of a linear combination of pair states in momentum space, we must now deal with the antisymmetrized wave function of the N electron system, all together at one time. Consequently, we shall need some useful notation.

For this purpose, we use the Dirac notation, but for pairs of electrons only, not for single electrons. Thus, the notation

$$|0\rangle \quad \text{or} \quad |1\rangle$$

will mean, respectively, that a pair state is either unoccupied or occupied. To reiterate: a state is available for occupation by a single electron in the perfect

5.3 Superconductivity

Fermi gas that has momentum $\hbar\mathbf{q}$, energy $\hbar^2 q^2/2m$, and spin up, and another state with momentum $-\mathbf{q}$, energy $\hbar^2 q^2/2m$, and spin down. The notation $|0\rangle$ means that both states are empty, while the notation $|1\rangle$ means that both are occupied. If necessary, we will use subscripts, $|0\rangle_\mathbf{q}$, to indicate that it is the pair $(\mathbf{q}\uparrow, -\mathbf{q}\downarrow)$ rather than $|0\rangle_{\mathbf{q}'}$, the pair $(\mathbf{q}'\uparrow, -\mathbf{q}'\downarrow)$, we are describing.

The wave function—in this notation it is more accurate to call it the wave vector—the wave vector of the N electron system is

$$|\psi\rangle = \prod_\mathbf{q} (a_\mathbf{q}|0\rangle + b_\mathbf{q}|1\rangle) \qquad (5.3.59)$$

where $a_\mathbf{q}$ and $b_\mathbf{q}$ are just numbers, giving the amplitudes for the pair state $\mathbf{q}$ to be empty or full. In the noninteracting ground state, $b_\mathbf{q} = 1$ and $a_\mathbf{q} = 0$ for all states up to the Fermi level (i.e., they are all occupied), and for all states above the Fermi level, $b_\mathbf{q} = 0$ and $a_\mathbf{q} = 1$. The manipulations of the pair state vectors are simple. For state $\mathbf{q}$,

$$\langle 0|0\rangle = 1$$
$$\langle 1|1\rangle = 1 \qquad (5.3.60)$$
$$\langle 0|1\rangle = 0$$

All such products commute when they refer to different states:

$$(a_\mathbf{q}|0\rangle_\mathbf{q} + b_\mathbf{q}|1\rangle_\mathbf{q})(a_{\mathbf{q}'}|0\rangle + b_{\mathbf{q}'}|1\rangle) = (a_{\mathbf{q}'}|0\rangle_{\mathbf{q}'} + b_{\mathbf{q}'}|1\rangle_{\mathbf{q}'})(a_\mathbf{q}|0\rangle_\mathbf{q} + b_\mathbf{q}|1\rangle_\mathbf{q})$$

The probability that a state is occupied or unoccupied is the square of the amplitude, $b_\mathbf{q}^2$ or $a_\mathbf{q}^2$. It follows that we must always have

$$a_\mathbf{q}^2 + b_\mathbf{q}^2 = 1 \qquad (5.3.61)$$

Furthermore, since $b_\mathbf{q}^2$ is the probability of finding a pair in state $\mathbf{q}$, it follows that

$$N = 2\sum_\mathbf{q} b_\mathbf{q}^2 \qquad (5.3.62)$$

As an example of how computations may be done with this notation, let us see how the overall state is normalized. We wish to compute

$$\langle\psi|\psi\rangle = \prod_\mathbf{q}(\langle 0|a_\mathbf{q} + \langle 1|b_\mathbf{q})\prod_{\mathbf{q}'}(a_{\mathbf{q}'}|0\rangle + b_{\mathbf{q}'}|1\rangle)$$

For each pair state, we have in the left-hand product a term of the form $(\langle 0|a + \langle 1|b)$ and in the right-hand product one of the form $(a|0\rangle + b|1\rangle)$. As we have just seen, all these terms may be written in any order. Let us bring together at the center of the double product the two terms for any state, say the kth. We have

$$\begin{aligned}\langle \psi|\psi\rangle &= \cdots(\langle 0|a_k + \langle 1|b_k)(a_k|0\rangle + b_k|1\rangle)\cdots \\ &= \cdots(a_k^2\langle 0|0\rangle + b_k^2\langle 1|1\rangle + a_kb_k\langle 0|1\rangle + a_kb_k\langle 1|0\rangle)\cdots \\ &= \cdots(a_k^2 + b_k^2)\cdots \\ &= \cdots(1)\cdots\end{aligned}$$

where we have used Eq. (5.3.60). In the same way, every other pair of terms from the left and from the right collapses together to give a factor 1. The result is

$$\langle\psi|\psi\rangle = 1$$

We must now write a Hamiltonian that will give us the many-body equivalent of Eq. (5.3.41) in this notation. For the kinetic energy part, we define an operator, $\mathscr{H}_{0\mathbf{q}}$, which has the properties

$$\begin{aligned}\mathscr{H}_{0\mathbf{q}}|1\rangle_\mathbf{q} &= \frac{\hbar^2 q^2}{m}|1\rangle_\mathbf{q} \\ \mathscr{H}_{0\mathbf{q}}|0\rangle_\mathbf{q} &= 0\end{aligned} \qquad (5.3.63)$$

and $\mathscr{H}_{0\mathbf{q}}$ does not act on any state other than $\mathbf{q}$. The operator for the kinetic energy of the system is then

$$\mathscr{H}_0 = \sum_\mathbf{q} \mathscr{H}_{0\mathbf{q}} \qquad (5.3.64)$$

To compute the kinetic energy of the system, we write

$$\langle\psi|\mathscr{H}_0|\psi\rangle = \prod_\mathbf{q}(\langle 0|a_\mathbf{q} + \langle 1|b_\mathbf{q})\left(\sum_{\mathbf{q}'}\mathscr{H}_{0\mathbf{q}'}\right)\prod_{\mathbf{q}''}(a_{\mathbf{q}''}|0\rangle + b_{\mathbf{q}''}|1\rangle) \qquad (5.3.65)$$

We now use the rules, Eqs. (5.3.60), (5.3.61), and (5.3.63), to eliminate most of the terms. It is easy to see that we are left with

$$\langle\psi|\mathscr{H}_0|\psi\rangle = \sum_\mathbf{q} b_\mathbf{q}^2 \frac{\hbar^2 q^2}{m} \qquad (5.3.66)$$

for the kinetic energy of the system.

In writing an operator for the potential energy, we allow ourselves to be guided by the Cooper calculation of the last section once again. The interaction we are interested in is one that scatters pairs from one state, $\mathbf{q}$, to another, $\mathbf{q}'$. Let us define an operator, $\hat{V}_{\mathbf{qq}'}$, which does just that. It always acts on two pair states at a time, and it has the property

$$\hat{V}_{\mathbf{qq}'}|0\rangle_{\mathbf{q}'}|1\rangle_\mathbf{q} = V_{\mathbf{qq}'}|1\rangle_{\mathbf{q}'}|0\rangle_\mathbf{q} \qquad (5.3.67)$$

5.3 Superconductivity

Here it has found state $\mathbf{q}$ occupied and $\mathbf{q}'$ empty and has scattered the pair from $\mathbf{q}$ into $\mathbf{q}'$, with strength $V_{\mathbf{qq}'}$ (just a number). However, if it does not find both $\mathbf{q}$ occupied and $\mathbf{q}'$ empty, it gives zero:

$$\hat{V}_{\mathbf{qq}'}|0\rangle|0\rangle = 0$$
$$\hat{V}_{\mathbf{qq}'}|1\rangle|1\rangle = 0 \qquad (5.3.68)$$
$$\hat{V}_{\mathbf{qq}'}|1\rangle_{\mathbf{q}'}|0\rangle_{\mathbf{q}} = 0$$

The Hamiltonian for the potential energy of the system (owing to this interaction only, of course) is

$$\mathscr{H}_1 = \sum_{\mathbf{q},\mathbf{q}'} \hat{V}_{\mathbf{qq}'} \qquad (5.3.69)$$

It is not hard to see that the operator we have introduced here performs exactly the same function as the potential in the Cooper argument. Starting as a potential, $V(r)$, between two electrons, it wound up scattering between pair states $\mathbf{q}$ and $\mathbf{q}'$ in $\mathbf{q}$ space because we constructed $\psi(\mathbf{r})$ out of these pair states. Matters are more complicated now because we cannot simply exclude scattering into all states up to the Fermi level; we do not yet know which states are occupied. We thus define an operator that has a result only if the initial state is occupied and the final state is not. Equation (5.3.43) gave the matrix element for that event in the simpler case. Now we have the machinery to handle the more general case. The potential energy Hamiltonian, Eq. (5.3.69), is the interaction we want between all pairs of pair states (notice that the double sum does not count each pair twice, since the interaction only goes one way—from $\mathbf{q}$ to $\mathbf{q}'$, not the reverse).

We now must compute the potential energy:

$$\langle\psi|\mathscr{H}_1|\psi\rangle = \prod_{\mathbf{q}} (\langle 0|a_{\mathbf{q}} + \langle 1|b_{\mathbf{q}}) \left(\sum_{\mathbf{q}'\mathbf{q}''} \hat{V}_{\mathbf{q}'\mathbf{q}''}\right) \prod_{\mathbf{q}'''} (a_{\mathbf{q}'''}|0\rangle + b_{\mathbf{q}'''}|1\rangle) \qquad (5.3.70)$$

We have a sum of terms in the middle, with a product on each side. Consider one term in the sum, with its two products

$$\cdots (\langle 0|a_{\mathbf{q}} + \langle 1|b_{\mathbf{q}})(\langle 0|a_{\mathbf{q}'} + \langle 1|b_{\mathbf{q}'})\hat{V}_{\mathbf{qq}'}(a_{\mathbf{q}'}|0\rangle + b_{\mathbf{q}'}|1\rangle)$$
$$\times (a_{\mathbf{q}}|0\rangle + b_{\mathbf{q}}|1\rangle)\cdots \qquad (5.3.71)$$

All other terms in the two products, of the form

$$\cdots (\langle 0|a_{\mathbf{k}} + \langle 1|b_{\mathbf{k}})\cdots(a_{\mathbf{k}}|0\rangle + b_{\mathbf{k}}|1\rangle)\cdots$$

commute with everything but their opposite numbers as we have seen, coming together to give the factor one. When we use the rules, Eqs. (5.3.67) and (5.3.68), the term shown in Eq. (5.3.71) has in it factors

$$\hat{V}_{\mathbf{qq}'}a_{\mathbf{q}'}|0\rangle_{\mathbf{q}'}a_{\mathbf{q}}|0\rangle_{\mathbf{q}} = 0$$

and so on, but also

$$\hat{V}_{qq'}a_{q'}|0\rangle_{q'}b_q|1\rangle_q = V_{qq'}a_{q'}b_q|1\rangle_{q'}|0\rangle_q \qquad (5.3.72)$$

This still gives zero when combined with the terms on the left in (5.3.71) except for

$$_q\langle 0|a_q {}_{q'}\langle 1|b_{q'}V_{qq'}a_{q'}|1\rangle_{q'}b_q|0\rangle_q = V_{qq'}a_qb_qa_{q'}b_q \qquad (5.3.73)$$

We get just this contribution from every term in the double sum in Eq. (5.3.70). The potential energy of the system is therefore

$$\langle \psi|\mathcal{H}_1|\psi\rangle = \sum_{qq'} V_{qq'}a_qa_{q'}b_qb_{q'} \qquad (5.3.74)$$

Equations (5.3.66) and (5.3.74) together give us for the energy of the system

$$E = \sum_q b_q^2 \frac{\hbar^2 q^2}{m} + \sum_{q,q'} V_{qq'}a_{q'}a_qb_{q'}b_q \qquad (5.3.75)$$

Let us pause now for a moment to see where we are going. Although we are using exclusively particular pairs of single-particle states together, these states themselves are still the solutions of the problem of a single particle in an empty box—that is, they are the correct states only for a perfect gas. We are now considering an interacting gas. In quantum mechanics the exact state of the interacting gas can be formed from a linear combination of the noninteracting states, since such states constitute a complete set, and that is what we are seeking to do. In the noninteracting case, each state available for occupation by a single particle is either occupied or it is not, in the ground state of the gas as a whole. For the interacting case, linear combinations are formed in which each single-particle state has some amplitude, between zero and one, whose square is the probability that the state is occupied. Our b_q is the amplitude for a pair of single-particle states taken together. We have guessed, on the basis of the Cooper argument, that the ground state will not involve different amplitudes for the states $q\uparrow$ and $-q\downarrow$. Our job now is to find the correct values of all the b_q's. In the ground state of the noninteracting Fermi system, all the b_q's are equal to one below the Fermi level and zero above. The weak interaction we are putting in will favor a more subtle distribution. In the noninteracting case, the kinetic energy of the system is minimized subject to the constraints on the system (the Pauli exclusion principle). Any other distribution of b_q's will increase the kinetic energy. However, it may also decrease the potential energy by enhancing the importance of the attractive interaction between the members of the pairs. It is the most favorable set of b_q's that we seek.

Thus, in order to find the ground state of our system, we should minimize the energy E with respect to the b_q's. However, when the b_q's change, the number of electrons may change with them through Eq. (5.3.62). We must

5.3 Superconductivity

place an appropriate constraint on the system or we will simply find that the lowest energy corresponds to $N = 0$. We can minimize the quantity

$$E - \lambda N$$

where λ is a Lagrange multiplier, which will later turn out to be the chemical potential, μ. Instead let us use our general variational principle. Our system is in a particle bath at fixed μ, and $T = 0$. The quantity to minimize is Ω, which at zero temperature is just

$$\Omega = E - \mu N \tag{5.3.76}$$

A rather deep point here should be mentioned. The b_q^2's that emerge from this calculation will be the purely quantum mechanical probabilities that the pair states be occupied. At nonzero temperature, however, we can use exactly the same procedure, except that we minimize

$$\Omega = E - TS - \mu N$$

where the entropy also depends on the b_q's. The b_q^2's that emerge then will be combined quantum mechanical and thermodynamic probabilities. Formally, we should construct all the possible quantum mechanical excited states, write the appropriate partition function, and so on, but all that is done at once by minimizing Ω at finite T. This is a nice illustration of the deep similarity between probabilities and averages and fluctuations in quantum mechanics and statistical mechanics. We will not actually study the finite-temperature behavior of superconductors in the way suggested because, alas, we would get the wrong answer. It is not because the procedure is incorrect, but because our guess that single electron states $q\uparrow$ and $-q\downarrow$ must have the same amplitude is only correct in the ground state. It is the price we pay for simplifying the BCS theory. We shall return to this point later.

In any case, back to the ground state. Using Eqs. (5.3.62) and (5.3.75), we find

$$\begin{aligned}\Omega &= \sum_{\mathbf{q}} \left(\frac{\hbar^2 q^2}{m} - 2\mu \right) b_\mathbf{q}^2 + \sum_{\mathbf{q},\mathbf{q}'} V_{\mathbf{q}\mathbf{q}'} a_{\mathbf{q}'} a_\mathbf{q} b_{\mathbf{q}'} b_\mathbf{q} \\ &= \sum_{\mathbf{q}} \left[2\xi_\mathbf{q} b_\mathbf{q}^2 + \sum_{\mathbf{q}'} V_{\mathbf{q}\mathbf{q}'} a_{\mathbf{q}'} a_\mathbf{q} b_{\mathbf{q}'} b_\mathbf{q} \right] \end{aligned} \tag{5.3.77}$$

where we have defined

$$\xi_\mathbf{q} = \frac{\hbar^2 q^2}{2m} - \mu \tag{5.3.78}$$

$\xi_\mathbf{q}$ is the energy of one electron in the pair relative to the Fermi level [see Eq. (5.3.51)]. The $a_\mathbf{q}$'s and $b_\mathbf{q}$'s are not independent but are related by

(5.3.61). In order to have only one parameter to deal with, we define the angle θ_q by
$$b_q = \cos \theta_q$$
$$a_q = \sin \theta_q \qquad (5.3.79)$$
so that
$$\Omega = \sum_q \left[2\xi_q \cos^2 \theta_q + \frac{1}{4} \sum_{q'} V_{qq'} \sin 2\theta_q \sin 2\theta_{q'} \right] \qquad (5.3.80)$$

Here we have used the trigonometric identity
$$2 \cos \theta \sin \theta = \sin 2\theta \qquad (5.3.81)$$

We now minimize Ω with respect to any one of the θ_q, say, θ_k:

$$0 = \frac{\partial \Omega}{\partial \theta_k} = -4\xi_k \cos \theta_k \sin \theta_k + \frac{1}{2} \cos 2\theta_k \sum_{q'} V_{kq'} \sin 2\theta_{q'}$$
$$+ \frac{1}{2} \cos 2\theta_k \sum_q V_{qk} \sin 2\theta_q$$
$$= -2\xi_k \sin 2\theta_k + \cos 2\theta_k \sum_q V_{qk} \sin 2\theta_q \qquad (5.3.82)$$

or
$$\tan 2\theta_k = \frac{1}{2\xi_k} \sum_q V_{qk} \sin 2\theta_q \qquad (5.3.83)$$

(Notice that in these last two equations q has become a dummy variable.) We now define the quantity Δ_k,
$$\Delta_k = -\frac{1}{2} \sum_q V_{qk} \sin 2\theta_q \qquad (5.3.84)$$
so that
$$\tan 2\theta_k = -\frac{\Delta_k}{\xi_k} \qquad (5.3.85)$$

Since $\tan 2\theta_k$ is negative, either $\sin 2\theta_k$ or $\cos 2\theta_k$ must be negative [we will have V_{qk} negative, in Eq. (5.3.84), to represent an attractive interaction]. If we choose the cosine to be negative, then
$$\sin 2\theta_k = \frac{\Delta_k}{\sqrt{\xi_k^2 + \Delta_k^2}}$$
$$\cos 2\theta_k = \frac{-\xi_k}{\sqrt{\xi_k^2 + \Delta_k^2}} \qquad (5.3.86)$$

To see why this is the correct choice of signs, we use Eq. (5.3.79) together with
$$\cos 2\theta = \cos^2 \theta - \sin^2 \theta$$

5.3 Superconductivity

and we see that

$$b_k^2 - a_k^2 = -\frac{\xi_k}{\sqrt{\xi_k^2 + \Delta_k^2}} \quad (5.3.87)$$

For states very far above the Fermi level, the weak interaction we are putting into the electron problem should make no difference, and so we expect in the limit as $\xi_k \to \infty$, $b_k^2 \to 0$ and $a_k^2 \to 1$; that is, we expect very high energy states to be empty. A glance at Eq. (5.3.87) shows that our choice of signs in Eqs. (5.3.86) has just this effect.

At this point we substitute Eqs. (5.3.85) and (5.3.86) into (5.3.83) and get

$$\Delta_k = -\frac{1}{2} \sum_q V_{qk} \frac{\Delta_q}{\sqrt{\xi_q^2 + \Delta_q^2}} \quad (5.3.88)$$

We are close to being on familiar ground. The right-hand side of this equation depends on **k** only through the interaction, V_{qk}. We now make the same approximations for the interaction as in the Cooper argument of Sec. 5.3c, and for the same reasons. The interaction can connect pair states that differ in energy by no more than $\hbar\omega_D$, and it is angle independent. Thus,

$$V_{qk} = -V \quad \text{if } |\xi_q| \leq \hbar\omega_D \text{ and } |\xi_k| \leq \hbar\omega_D$$
$$= 0 \quad \text{otherwise} \quad (5.3.89)$$

When we imposed the analogous condition in Eq. (5.3.45), we restricted the interactions to a shell of states above the Fermi level. Now there will be some amplitude for states to be available below the Fermi level as well, so the shell of interactions goes both above and below the Fermi level. Using Eq. (5.3.89) in (5.3.88), we find

$$\Delta_k = \frac{1}{2} V \sum_q \frac{\Delta_q}{\sqrt{\xi_q^2 + \Delta_q^2}} \quad (5.3.90)$$

The term on the right no longer depends on **k**, so Δ_k does not depend on **k**, nor does Δ_q depend on **q**. We write

$$\Delta_0 = \frac{1}{2} V \sum_q \frac{\Delta_0}{\sqrt{\xi_q^2 + \Delta_0^2}} \quad (5.3.91)$$

or

$$1 = \frac{1}{2} V \sum_q \frac{1}{\sqrt{\xi_q^2 + \Delta_0^2}}$$

$$= \frac{1}{2} V \int_{-\hbar\omega_D}^{+\hbar\omega_D} \frac{\rho(\xi) \, d\xi}{\sqrt{\xi^2 + \Delta_0^2}} \quad (5.3.92)$$

In the Cooper argument, the analogous events occurred in Eqs. (5.3.46) to

(5.3.53). We once again put in the fact that $\rho(\xi) \approx \rho(0)$ in the narrow shell of states within $\hbar\omega_D$ of the Fermi level and find

$$1 = \frac{\rho(0)V}{2} \int_{-\hbar\omega_D}^{\hbar\omega_D} \frac{d\xi}{\sqrt{\xi^2 + \Delta_0^2}}$$

$$= \rho(0)V \sinh^{-1}\left(\frac{\hbar\omega_D}{\Delta_0}\right) \qquad (5.3.93)$$

Solving for Δ_0 gives

$$\Delta_0 = \frac{\hbar\omega_D}{\sinh\left[1/\rho(0)V\right]}$$

$$\approx 2\hbar\omega_D e^{-1/\rho(0)V} \qquad (5.3.94)$$

for small V. The quantity Δ_0 is often referred to as the binding energy per electron in the formation of electron pairs. The situation is actually a bit more subtle. Let us examine the nature of the ground state we have discovered.

Using Eqs. (5.3.61) and (5.3.87), together with our result that Δ_0 does not depend on $\mathbf{q}$, we find that

$$2b_\mathbf{q}^2 = 1 - \frac{\xi_\mathbf{q}}{\sqrt{\xi_\mathbf{q}^2 + \Delta_0^2}} \qquad (5.3.95)$$

The $b_\mathbf{q}^2$'s are the probabilities that the original single-particle states are occupied. Obviously, for $\xi_\mathbf{q} \ll 0$, $b_\mathbf{q}^2 = 1$ and for $\xi_\mathbf{q} \gg 0$, $b_\mathbf{q}^2 = 0$. Between these limits, over a region whose width is of order Δ_0, $b_\mathbf{q}^2$ changes smoothly from 1 to 0, as sketched in Fig. 5.3.7. The distribution shown is reminiscent of that in Fig. 2.5.2 for the thermal probability of the occupation of these same states for normal electrons at finite temperature. The $b_\mathbf{q}^2$'s, however, are quantum probabilities in the ground state, at $T = 0$. In retrospect, it is not hard to see why this form has occurred. The influence of the interaction, $-V$, depends intimately on the number of unoccupied states nearby into which a pair might be scattered. We saw in the Cooper argument that a weak

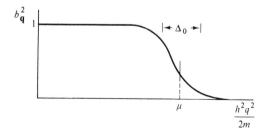

Fig. 5.3.7

5.3 Superconductivity

potential caused pairs to be bound simply because many states were available. The effect of the interaction is to cause some states that are unoccupied in the normal ground state (those above the Fermi level) to have some amplitude for being occupied and to leave states below the Fermi level with some amplitude for being unoccupied. This result, in turn, enhances the effect of $-V$ on other pairs, since there are now more electrons within $\hbar\omega_D$ of states with at least some amplitude for being unoccupied and therefore available for scattering into. The more the distribution spreads out, the more pairs there are available to take advantage of the attractive interaction. On the other hand, the kinetic energy of the system is increased by any change from the normal ground state. The more the distribution spreads out to lower its energy by taking advantage of $-V$ operating between pairs of electrons, the more it raises its kinetic energy. The calculation we have done has found the optimum compromise, the delicate balance.

If Δ_0 were the binding energy per electron in the pairing process, we would expect the energy of the superconducting ground state to be lower than the normal ground state by $N\,\Delta_0$. Instead the difference in energy is (Prob. 5.9a)

$$E_S - E_N = -\frac{\rho(0)\Delta_0^2}{2} \tag{5.3.96}$$

To compare this to $N\,\Delta_0$, use Eq. (5.3.54) for $\rho(0)$ and (2.5.7) for ε_F. Ignoring factors of order 1

$$\rho(0) \approx \frac{N}{\varepsilon_F} \tag{5.3.97}$$

so that

$$E_S - E_N \approx -N\,\Delta_0 \left(\frac{\Delta_0}{\varepsilon_F}\right) \tag{5.3.98}$$

From Eq. (5.3.94) we know that $\Delta_0 \ll \hbar\omega_D$, and we recall from Chap. 3 that $\hbar\omega_D \ll \varepsilon_F$; therefore, it is clear that the energy difference between super and normal ground states is quite small compared to $N\,\Delta_0$. In fact, typically, $\Delta_0/k \approx 10°$K, whereas $\varepsilon_F/k \approx 10^{4}$°K.

The difference $E_N - E_s$ per unit volume is simply what we have called the condensation energy at zero temperature. It is equal to $H_c^2/8\pi$ [see Eq. (5.3.12), for example]. We argued in Sec. 5.2b that this quantity is much smaller than kT_c per electron, and we now see that it is much smaller than $\Delta_0 \approx kT_c$. In a simple-minded picture, we might suppose that electrons form bound pairs with binding energy Δ_0 per electron. The critical temperature above which no pairs form ought to be of order Δ_0/k, and that is correct. The binding energy of the system at zero temperature ought to be $N\,\Delta_0$, however, and this expectation is in error by a factor Δ_0/ε_F or about 10^{-3}. We could fix up our picture by saying that only a small fraction, Δ_0/ε_F, of the electrons are close enough to the Fermi surface to participate in the binding process, but doing so would lead us to expect that only Δ_0/ε_F of the electrons are superconducting (i.e., carry supercurrent). In fact, as we shall see later, all the conduction electrons participate in superconductivity. The simple-minded picture will not work.

What does happen, as we already know, is nothing so gross as a direct binding of one electron to another. Instead there is a subtle and delicate balance in which each pair state is occupied with an amplitude that causes the pair state to be used most effectively in the binding of all other pairs, and in which the energy decrease in the whole binding process is nearly counterbalanced by the increase in kinetic energy that results from it. That is why $E_S - E_N$ is so small. Δ_0 is a collective property of the entire system, not to be associated with individual pairs. Moreover, it is true, as we can see in Fig. 5.3.7, that only a small fraction Δ_0/kT_F of the electrons are affected at all by the pairing process. To understand why all electrons are superconducting, even though only a few have been affected by the condensation, it helps if we change our point of view a little. The fact is that in the normal state all the electrons except those near the Fermi surface are, in some sense, already superconducting. Recall our discussion of normal electrical conduction in Sec. 3.6: electrons are freely accelerated from state to state through the Fermi sphere, to be scattered only at the surface. All electrons participate in conduction, but only those at the Fermi surface participate in resistance. Then, in the superconducting state, even those few get tied up collectively together and can no longer be excited with infinitesimal energy.

In order to understand the role of the quantity Δ_0, it is necessary to investigate the excited states of the system. The wave function that we have used, Eq. (5.3.59), like the Feynman wave function in superfluidity, represents the ground state accurately, but it is not the best possible choice for the excited states and so will not give the lowest-lying excited states. Nevertheless, like the Feynman wave function, it is good enough to give us the general idea of what is happening. The ground state is represented by Eq. (5.3.59) with the particular set of a_q's and b_q's we have computed. Any other choice of amplitudes will result in a higher energy and may thus represent an excited state, provided that it is normalized and is orthogonal to the ground state. For example, suppose that we select some particular pair state, $(\mathbf{k}\uparrow, -\mathbf{k}\downarrow)$, which has, in the ground state, amplitudes $a_\mathbf{k}$ and $b_\mathbf{k}$, and suppose that we construct an excited state in which all other amplitudes are unchanged, but $a_\mathbf{k}$ and $b_\mathbf{k}$ are replaced, respectively, by A and B. The excited state, $|\psi_{ex}\rangle$, will be normalized if

$$A^2 + B^2 = 1 \qquad (5.3.99)$$

and orthogonal to the ground state if

$$\langle \psi_{ex}|\psi \rangle = 0 \qquad (5.3.100)$$

5.3 Superconductivity

In this product, all the other pairs of states except for **k** commute toward the middle to find their opposite numbers giving contributions, as we saw earlier, of the form of factors

$$\cdots (\langle 0|a_{\mathbf{q}} + \langle 1|b_{\mathbf{q}})(a_{\mathbf{q}}|0\rangle + b_{\mathbf{q}}|1\rangle) \cdots = \cdots (a_{\mathbf{q}}^2 + b_{\mathbf{q}}^2) \cdots$$
$$= \cdots (1) \cdots \quad (5.3.101)$$

The net result is

$$\langle \psi_{ex}|\psi\rangle = (\langle 0|A + \langle 1|B)(a_{\mathbf{k}}|0\rangle + b_{\mathbf{k}}|1\rangle)$$
$$= Aa_{\mathbf{k}} + Bb_{\mathbf{k}}$$
$$= 0 \quad (5.3.102)$$

These conditions are enough to determine A and B, plus the energy of the new state, which is higher than the ground state energy by (Prob. 5.9b)

$$\varepsilon = 2\sqrt{\xi_{\mathbf{k}}^2 + \Delta_0^2} \quad (5.3.103)$$

In other words, the excited state is at least $2\Delta_0$ above the ground state in energy.

Equation (5.3.59), which we have used for the state of the superconducting system, suffers from the deficiency that the amplitudes for both members of the pair state, $\mathbf{q}\uparrow$ and $-\mathbf{q}\downarrow$, must be the same. In the more accurate state used by BCS, the amplitudes for the single members of the pair can change separately. The ground state that results is exactly the one we have worked out, but a lower-lying excited state may be formed by changing the amplitudes of only one member of a pair of states instead of both together. The resulting energy is just half that given in Eq. (5.3.103), so that the minimum excitation energy of the superconducting system is actually just Δ_0. For this reason, Δ_0 is, quite properly, known as the *energy gap*.

The excitation of a state of the system Δ_0 above the ground state is often referred to loosely as the "breaking of a pair." It is, however, really a collective excited state of the system, in which the amplitude for occupation of one member of a pair is altered and the energy Δ_0 occurs because of the effect of that alteration on the binding of all the other pairs. More highly excited states can, of course, be formed by altering more amplitudes. At finite temperature the system behaves as any other would; it fluctuates about among those of its possible excited states with approximately the right energy and number of electrons, T and μ being fixed.

In highly excited states, when many of the amplitudes have been altered from their ground state values, the mutual benefits to all pairs of the carefully chosen ground state distribution seriously deteriorates, and it takes less energy than Δ_0 to change the next set of amplitudes. At temperatures above zero the system fluctuates between states of this type, with energy gap less than Δ_0. In effect, the energy gap becomes a function of T, $\Delta(T)$, diminishing as T increases. The smaller $\Delta(T)$ is, the easier it is to excite the system, thus further decreasing $\Delta(T)$, and so on. At some point the process runs away, $\Delta(T)$ is driven to zero, and superconductivity is destroyed. That temperature is T_c, the superconducting transition temperature.

Treatment of the system at finite temperature requires the more accurate BCS representation of the superconducting state, which we do not wish to go into here. As we have already outlined, the procedure does not differ in principle from what we have already done. Ω, including the TS term, is minimized with respect to the probabilities, now both thermal and quantum mechanical, of occupation of the single-particle states. In the end Eq. (5.3.93) is replaced by

$$1 = \frac{\rho(0)V}{2} \int_{-\hbar\omega_D}^{\hbar\omega_D} \frac{d\xi}{\sqrt{\xi^2 + \Delta^2}} [1 - 2f(\mathscr{E})] \qquad (5.3.104)$$

where
$$f(\mathscr{E}) = \frac{1}{e^{\mathscr{E}/kT} + 1} \qquad (5.3.105)$$

and
$$\mathscr{E} = \sqrt{\xi^2 + \Delta^2} \qquad (5.3.106)$$

As we have said, Δ depends on T, and the transition to the normal state occurs where $\Delta(T) = 0$. Substituting $\Delta = 0$ into Eqs. (5.3.104) to (5.3.106), we find

$$kT_c = 1.14\hbar\omega_D e^{-1/\rho(0)V}$$

$$= \frac{1.14}{2} \Delta_0 \qquad (5.3.107)$$

We shall discuss this phase transition from a different point of view in Chap. 6.

We have seen in Secs. 5.3c and 5.3d that a weak attraction between pairs of electrons, acting through their effects on the lattice, leads to a serious alteration of the state of the electron system in a metal. What we have not seen is why this altered state behaves like a superconductor; that is, why it has the peculiar properties we described in Secs. 5.3a and 5.3b. Combining the macroscopic and microscopic descriptions of superconductivity is the purpose of the next section.

e. The Superconducting State

We now know that if the net, residual interaction between electrons is even weakly attractive, the ground state of the electron system is significantly altered from the noninteracting, perfect Fermi gas ground state. Our final job is to see why electrons in this new state behave the way superconductors are observed to behave. We must explain why current can flow

5.3 Superconductivity

without resistance and why superconductors have the Meissner effect. We will also see why magnetic flux, or vorticity, is quantized in superconductors and, as a result, just how the Meissner effect breaks down in type II superconductors.

In the new ground state electrons act together in pairs. Each pair has zero momentum and obeys Bose statistics. In short, the pairs form a Bose condensate. We have already discussed the properties of a Bose condensate extensively in Sec. 5.2b. It is not difficult to extend our arguments to a Bose condensate of electron pairs.

The question of whether the condensate is a superfluid depends, as we saw, not on the properties of the condensate but rather on those of the excitations. The condensate could flow without resistance if the energies and momenta of the possible excitations obeyed the Landau criterion, expressed in Eq. (5.2.11). If the smallest possible value of (ε/p) for the excitation is not zero, then the condensate flows without resistance up to velocity (ε/p). From the superconducting ground state, the lowest-lying excitation has energy Δ_0, which is associated with the appearance of an unpaired electron of net momentum $\hbar q_F$, the Fermi momentum. Thus, the critical velocity for the electron pairs is

$$v_c = \frac{\Delta_0}{\hbar q_F} \qquad (5.3.108)$$

and the critical current density is

$$j_c = -n_s e^* v_c = -\frac{n_s e^* \Delta_0}{\hbar q_F} \qquad (5.3.109)$$

In the case of superfluidity in liquid helium, the Landau criterion gave a critical velocity of two to three orders of magnitude too high; as we saw, other processes caused breakdown in the superflow long before the velocity for direct creation of rotons could be reached. For superconductors, the situation is simpler. Equation (5.3.109) gives a rather good estimate of the actual critical current density observed if for n_s we use half the total number of conduction electrons in the material. Since the observed critical current is given by the largest possible choice of both n_s and v_c, it follows that all the electrons must be participating in the conduction of supercurrent, not just those that are very close to the Fermi surface. This fact emphasizes a point that we made earlier: it would be a mistake to think that only a small fraction of the electrons participate in superconductivity.

The essential property of the Bose condensate itself is that all the particles—in this case, all the pairs—have the same wave function, Eq. (5.2.18),

$$\psi(\mathbf{r}) = a(\mathbf{r}) \exp[i\gamma(\mathbf{r})] \qquad (5.3.110)$$

No assumptions have been made here about what the single-particle wave function is; $\psi(\mathbf{r})$ is just the most general complex number we can write. However, there is an important restriction on this form if it is to represent the ground state of the system we are trying to describe. We wish to think of our pairs as noninteracting particles. The residual interaction between electrons was used up in forming the pairs, and so now the pairs themselves form a perfect gas. That picture makes sense only as long as the strong, direct coulomb interactions between one electron and another and between each electron and the lattice remain carefully balanced, as we have assumed them to be. The electron pairs are, of course, still charged, and if each pair has the wave function, Eq. (5.3.110), then the charge density in the electron gas is

$$-e^*n_s|a|^2$$

[where the normalization of Eq. (5.3.110) is $(1/V) \int |a|^2 \, d^3r = 1$]. The average charge density of the ions has the same magnitude and opposite sign. The crucial point here, however, is that the charges be neutralized, not on the average but in detail. The positive ion charge may be thought of as uniform (since we are interested in variations on the scale of the characteristic lengths of superconductivity, ξ or λ, both much larger than the distance between ions), and it follows that $|a|^2$ must be uniform as well. If $a(\mathbf{r})$ were to change from its constant value at some point, *all* the electrons would change their charge density at that point, thereby leading to a very large net charge locally, with very high energy. The result is that $a(\mathbf{r})$ is "stiff"; large forces are at work to keep it from varying spatially.

The argument we have just given is valid in the interior of a superconductor, far from any surfaces, and it is also valid at a free surface of a superconductor. In these circumstances, it means that the amplitude of the ground-state wave function is a constant, independent of position. It is not valid, however, at an interface between a superconductor and a normal conductor. As we move through the transition region between the two metals, the pairing interaction breaks down (it is present in the superconductor but not in the normal metal). The total electron charge density remains equal to the magnitude of the ion charge density (ignoring contact potentials), but near the Fermi surface the fraction of the electrons in condensed pairs changes from 1 in the superconductor to zero in the normal conductor. That is the situation depicted in Fig. 5.3.4, where the amplitude of the wave function is seen to fall to zero in a distance ζ.

Now that we understand the nature of the microscopic states of the normal and super-materials, the characteristic distance ξ is easy to estimate with the help of the uncertainty principle. On the normal side of the boundary, the electrons are in momentum eigenstates and thus cannot be localized at all. On the super side, however, pairs are formed by mixing together contributions from a range of momenta of the single-particle states. There is an indeterminacy in the momentum associated with an electron, $\hbar\,\delta q$, and hence a characteristic length ξ, given by

$$\xi\hbar\,\delta q \sim \hbar \qquad (5.3.111)$$

The single-particle eigenstates mixed into the pair are those for which b_q^2 makes its transition from 1 to zero in Fig. 5.3.7, covering an energy range Δ_0:

$$\Delta_0 \approx \delta\left(\frac{\hbar^2 q^2}{2m}\right)$$

$$\approx \frac{\hbar^2 q_F}{m}\delta q$$

$$= \hbar v_F\,\delta q \qquad (5.3.112)$$

where v_F is the Fermi velocity. From Eqs. (5.3.111) and (5.3.112),

$$\xi \approx \frac{\hbar v_F}{\Delta_0} \qquad (5.3.113)$$

This is the "size" of a pair and also the characteristic distance over which the wave function changes when it is forced to.

Let us consider now the same semi-infinite slab of superconductor sketched in Fig. 5.3.3. We wish to know the response of the superconducting electron pairs when a magnetic field is applied parallel to the surface in, say, the y direction. Will they exhibit the Meissner effect?

To begin with, we suppose that there are no fields and no currents. The phase of the wave function, Eq. (5.3.110), $\gamma(\mathbf{r})$, is a constant, which we can take equal to zero. The amplitude, $a(\mathbf{r})$, is either constant (if there is a vacuum or an insulator to the left of the interface) or a function of x only (if there is a normal metal there). Thus, $\psi = a(x)$ is the solution of the equation

$$\left[-\frac{\hbar^2 \hat{p}^2}{2m^*} + W(x)\right]a(x) = \varepsilon_0 a(x) \qquad (5.3.114)$$

where

$$\hat{p} = -i\hbar\nabla \qquad (5.3.115)$$

ε_0 is the ground state energy of the pair and $W(x)$ is a potential that merely describes conditions at the interface. We now wish to turn on a weak magnetic field. We know that a strong field will destroy the superconductivity, so we consider the response to a weak field, retaining contributions of leading order only in the field strength. When the field is turned on, the Hamiltonian, Eq. (5.3.114), is altered because the momentum operator, Eq. (5.3.115), becomes

$$\hat{p} = -i\hbar\nabla + \frac{e^*}{c}\mathbf{A} \qquad (5.3.116)$$

where **A** is the magnetic vector potential, defined by

$$\mathbf{B} = \nabla \times \mathbf{A} \qquad (5.3.117)$$

This is not enough to specify **A** uniquely; a gauge must be chosen. The convenient choice for this problem is the London gauge

$$\nabla \cdot \mathbf{A} = 0 \qquad (5.3.118)$$

(see Prob. 5.10 for the gauge invariance of the quantum mechanical equations under this choice). Then if **B** is to be in the y direction, $\mathbf{B} = B(x)\hat{\mathbf{y}}$, **A** is a vector in the z direction, $\mathbf{A} = A(x)\hat{\mathbf{z}}$, and

$$B(x) = \frac{\partial A(x)}{\partial x} \qquad (5.3.119)$$

When Eq. (5.3.116) is substituted into (5.3.114), the new terms generated by the presence of **A** are

$$\frac{i\hbar e^*}{c}\{\nabla \cdot [Aa(x)] + \mathbf{A} \cdot \nabla a(x)\} + \left(\frac{e^*}{c}\right)^2 |A|^2 a(x) \qquad (5.3.120)$$

The last term here is second order in the field strength, so we drop it. Inside the brackets, we have

$$2\mathbf{A} \cdot \nabla a(x) + a\nabla \cdot \mathbf{A} \qquad (5.3.121)$$

The second of these terms is zero by Eq. (5.3.118), and the first is zero because **A** is in the z direction, perpendicular to $\nabla a(x)$. The net result is that, to leading order, application of a magnetic field does not change the Schrödinger equation and hence does not alter the ground-state wave function.

We may thus calculate the currents that flow by using the ground-state wave function. The electric current density is obtained by multiplying Eq. (5.2.19), for the particle current, by $-n_s e^*/m^*$ and using Eq. (5.3.116) for the momentum operator. For later reference, let us use the more general wave function, Eq. (5.3.110), including a possible $\gamma(\mathbf{r})$,

$$\mathbf{j}_s = -\frac{n_s e^*}{m^*}\left[-\hbar|a|^2 \nabla\gamma(\mathbf{r}) + \frac{e^*}{c}|a|^2 \mathbf{A}\right] \qquad (5.3.122)$$

Let us restrict our attention now to type II superconductors, since it is for that case that we know the precise form of the Meissner effect, Eq. (5.3.25). We expect, then, that if Eq. (5.3.122) represents the Meissner effect, we will find currents and fields varying over a distance $\lambda \gg \xi$. The amplitude

5.3 Superconductivity

$a(x)$ is either constant or rises to its constant value over a distance ξ from the boundary; in either case, we can take it to be constant over the distances required by currents and fields to change. We thus replace $|a|^2$ by 1 and find

$$\nabla \times \mathbf{j}_s = -\frac{n_s e^{*2}}{m^* c} \nabla \times \mathbf{A} = -\frac{c\mathbf{B}}{4\pi\lambda^2} \quad (5.3.123)$$

where we have used Eqs. (5.3.19) and (5.3.117), as well as $\nabla \times \nabla\gamma(\mathbf{r}) = 0$. If we now use Eq. (5.3.16), we get

$$\nabla \times \nabla \times \mathbf{B} = -\frac{\mathbf{B}}{\lambda^2} \quad (5.3.124)$$

which is just the same as Eq. (5.3.25), the London equation for the Meissner effect. We have discovered that the Meissner effect is the property of a charged Bose condensate.

In Sec. 5.2b we investigated the behavior of an uncharged Bose condensate and found that the current was proportional to the gradient of a scalar, Eq. (5.2.21), and hence was irrotational. In the charged case, instead, the curl of the current is proportional to the magnetic field [Eq. (5.3.123)] and that is just the Meissner effect. The charged case also has an irrotational part in the current, provided that $\gamma(\mathbf{r})$ has a gradient, as we see in Eq. (5.3.122). In the uncharged Bose system, the curl-free current led to the quantization of circulation, Eq. (5.2.26). In superconductivity, it gives rise to the quantization of magnetic flux.

Consider the inclusion of magnetic flux shown as a normal region in Fig. 5.3.5c. The current $\mathbf{j}_s$ is nonzero only in the cross-hatched regions, falling off exponentially as we move into the pure superconductor. In this region then, we can set $\mathbf{j}_s = 0$ in Eq. (5.3.122), giving

$$\nabla\gamma = \frac{e^*}{\hbar c}\mathbf{A} \quad (5.3.125)$$

We can integrate this expression around any closed path that stays far from the cross-hatched regions:

$$\oint \nabla\gamma \cdot d\mathbf{l} = \frac{e^*}{\hbar c} \oint \mathbf{A} \cdot d\mathbf{l}$$

$$= \frac{e^*}{\hbar c} \int \mathbf{B} \cdot d\mathbf{S}$$

$$= \frac{e^*}{\hbar c} \varphi_m \quad (5.3.126)$$

where φ_m is the magnetic flux enclosed by our path of integration. The first member of Eq. (5.3.126) is the change in the phase of the wave function around a closed loop

$$\oint \nabla \gamma \cdot d\mathbf{l} = 2\pi n \qquad (5.3.127)$$

where n must be an integer if the wave function is to be single valued [see Eq. (5.2.25)]. Equation (5.3.126), together with (5.3.127), shows that the flux comes in quantized units

$$\varphi_m = n\varphi_{m0} \qquad (5.3.128)$$

where the quantum of magnetic flux is

$$\varphi_{m0} = \frac{(2\pi\hbar)c}{2e} \qquad (5.3.129)$$

Earlier, in Sec. 5.3b, we made a rough estimate of H_{c_1}, the field at which the complete Meissner effect breaks down in a type II superconductor [see Eq. (5.3.36)]. Now that we know more about the details of how the breakdown occurs, we can do a better job. The equilibrium state at fixed T and applied H is, as always, that with the lowest Gibbs potential. We must compare the Gibbs potential Φ of a state with complete Meissner effect to that with a distribution of flux quanta inside. Basically, we need to know whether the contribution to Φ of a fluxoid (alias flux quantum, vortex line) is positive or negative.

The fluxoid is essentially a cylinder of normal material, radius $\sim \xi$, with field H_c inside, and an exponentially decreasing field penetrating into the superconductor [notice that the magnetic fields and shielding currents of a fluxoid fall off much faster than the superfluid velocity field of a vortex line in helium, Eq. (5.2.145)]. The energy per unit length of a fluxoid is

$$\mathscr{E} = \frac{1}{8\pi} \int_\xi^\infty \left[B^2(\mathbf{r}) + \lambda^2 (\nabla \times \mathbf{B})^2 \right] d^3r + \text{core} \qquad (5.3.130)$$

where the core contribution, due to the cylinder of normal material, is $\sim \pi \xi^2 (H_c^2/8\pi)$. It is small and we shall ignore it. The result for $\mathscr{E}$ (see Prob. 5.11) is

$$\mathscr{E} = \left(\frac{n\varphi_{m0}}{4\pi\lambda} \right)^2 \log\left(\frac{\lambda}{\xi} \right) \qquad (5.3.131)$$

Unlike a vortex line in helium, the energy per unit length does not depend on the size of the sample, but it is proportional to n^2 as in helium. Thus, fluxoids in superconductivity, like vortex lines in helium, are always singly quantized.

In the vortex state (Fig. 5.3.6), where some distribution of fluxoids has penetrated the sample, the **B** field inside is no longer zero but has instead an average value

$$B = n_\varphi \varphi_{m0} \qquad (5.3.132)$$

where n_φ is the number of fluxoids per unit area in the plane perpendicular to the applied field. The Gibbs potential density of this array at zero temperature is

$$\Phi = n_\varphi \mathscr{E} - \frac{HB}{4\pi} = B\left(\frac{\mathscr{E}}{\varphi_{m0}} - \frac{H}{4\pi}\right) \qquad (5.3.133)$$

For sufficiently small H, this is positive and fluxoids are excluded. H_{c_1} is the field at which Φ changes sign

$$H_{c_1} = \frac{4\pi\mathscr{E}}{\varphi_{m0}} = \frac{\varphi_0}{4\pi\lambda^2} \log\left(\frac{\lambda}{\xi}\right) \qquad (5.3.134)$$

This equation replaces the crude estimate we made earlier, Eq. (5.3.36).

5.4 MAGNETISM

a. Introduction

When an external magnetic field **H** is applied to a piece of material, it usually responds in some way, so that the total field **B** inside the material does not remain equal to **H**. The difference is the magnetization of the sample **M** with the three fields related by Eq. (1.2.42)

$$\mathbf{B} = \mathbf{H} + 4\pi\mathbf{M} \qquad (5.4.1)$$

The response of the material can be described by its susceptibility, Eq. (1.2.109),

$$X(T, H) = \left(\frac{\partial M}{\partial H}\right)_T \qquad (5.4.2)$$

or by the magnetization itself, Eq. (1.2.63),

$$M = \frac{1}{V}\left(\frac{\partial \Phi}{\partial H}\right)_T \qquad (5.4.3)$$

Either $X = X(T, H)$ or $M = M(T, H)$ may be regarded as the magnetic equation of state of the material, to be predicted by a suitable understanding of its nature. We have already studied the thermodynamics of magnetism at some length in Chap. 1.

The magnetic susceptibility is a response function analogous to the compressibility, say, of a fluid. Unlike the compressibility, however, there are no stability conditions on its sign [see Eq. (1.2.120)]. The response of the material may be either to oppose the applied field or to intensify it. If the field is opposed, **M** is in the opposite direction to **H**, X is negative, and the material is said to be diamagnetic. If the field is intensified, **M** lines up with **H**, X is positive, and the material is said to be paramagnetic.

We have already dealt extensively with a special case of magnetic behavior, the Meissner effect in superconductivity, according to which a superconductor in low field is a perfect diamagnet with equation of state, Eq. (5.3.3). This case is unique because the response is purely macroscopic; large-scale currents flow in the surface of the material to shield out the applied field. In most other cases, the magnetic response of the material occurs basically at the atomic level.

At the microscopic level each molecule is either permanently magnetized or it is not, for reasons that we shall look into shortly. If the molecule has no permanent magnetization, or magnetic moment, its response to an applied field tends to be diamagnetic; we can think crudely of shielding currents being set up in its electron orbits. The same tendency occurs if the molecule has a magnetic moment, but the overall response is then always strongly dominated by the attempt of its magnetic moment to line up with the field. We shall study here only the latter kind of material, regarding the former, diamagnetic kind as essentially nonmagnetic. Our basic picture will be of a crystal in which we can think of each unit cell as having a net magnetic moment. As we shall see below, the magnetic moment is always associated with a net angular momentum in the electron system, often just with an unpaired electron spin. The set of magnetic moments in the material is often referred to briefly as the spin system.

Like other statistical systems, spin systems tend to be ordered at low temperature and disordered at high temperature. After seeing why atoms have magnetic moments in Sec. 5.4b, we will investigate the high-temperature thermal behavior of spin systems in Sec. 5.4c and the low-temperature thermal behavior in Sec. 5.4d. A more unified overall picture, including the phase transition between ordered and disordered states, is reserved for the next chapter, in Sec. 6.2.

b. Magnetic Moments

The magnetic properties of matter are a kind of afterthought of nature. A molecule, or an atom, or an ion in matter has a magnetic moment μ if any net angular momentum exists in its electron system. The state of the electron system, plus its consequent angular momentum, is a result of all the various interactions between all the particles, electrons, and nuclei in the material. Of all these interactions, the direct influence of the atomic magnetic field itself is among the weakest, and so the magnetic properties are, in a way, what is left over, being little influenced by magnetism itself. Even the interaction between separate magnetic moments on different ions in a

5.4 Magnetism

material is generally much stronger than the direct magnetic forces operating between them. Our principal interest in magnetic materials will be to study the macroscopic phenomena of paramagnetism (Sec. 5.4c), ferromagnetism (Sec. 5.4d), and the phase transition between those states (Sec. 6.2). However, it will be useful to have some idea of the kind of microscopic rules underlying the presence of magnetic behavior in materials, and that topic is what we turn to here. Ions in insulating salts will serve as our chief example.

The magnetic moment of a free ion is given by

$$\mu = -\mu_B(L + 2S) \tag{5.4.4}$$

where L is the net orbital and S the net spin angular momentum in the electron system, both in units of $\hbar$. The coefficient, called the Bohr magneton, is

$$\mu_B = \frac{e\hbar}{2mc} = 0.927 \times 10^{-20} \text{ erg/gauss}$$

The magnetic moments of nuclei are much smaller because nuclei have much greater masses than electrons, and the mass m appears in the denominator of the coefficient. The magnetic moment is usually written in terms of the total angular momentum J,

$$\mu = -g\mu_B J \tag{5.4.5}$$

where
$$J = L + S \tag{5.4.6}$$

The factor g varies between 1 and 2 to fix up the difference between the magnetic contributions of spin and orbital angular momentum, and it is given by

$$g = 1 + \frac{J(J + 1) + S(S + 1) - L(L + 1)}{2J(J + 1)} \tag{5.4.7}$$

We will try to get some general idea of what determines L, S, and J for a free ion and then see what happens when the ion is placed in a material.

The strongest force acting on the electrons is usually the electrostatic coulomb force due to the nucleus. If we first consider that interaction alone, we find possible states for the electron that are simply the familiar solutions of the Bohr hydrogen atom. The energy of the state depends only on the principal quantum number n and is proportional to $-1/n^2$. For any particular n, single electron states are available with orbital angular momentum ℓ equal to any integer from zero up to $n - 1$, all of them energetically degenerate. Moreover, the state with a particular n and ℓ still has $2\ell + 1$ possible projections of the orbital angular momentum—that is, an additional quantum number m_ℓ, which can have integer values from $+\ell$ to $-\ell$. Finally, there are two possible spin states, say, up and down, for each state with quantum numbers n, ℓ, and m_ℓ.

The other forces acting on the electrons, besides the electrostatic force from the nucleus, change the nature of the possible electron states. However, the changes are small enough so that the qualitative enumeration of states remains the same: we can still label all the possible states by the same set of quantum numbers, n, ℓ, m_ℓ, and up or down. What changes is the energy of a state with a given set of quantum numbers. For example, the energies are no longer simply proportional to $-1/n^2$, and, moreover, for a given n, not all allowed values of ℓ will have the same energy. It is these differences that govern the choice of quantum numbers of all the electrons in an ion and thereby determine the magnetic moment of the ion. Usually (not always) the magnetic moment of an ion is entirely determined by the set of quantum numbers that constitutes the ground state of its electron system.

In a subject replete with exquisite detail, we shall merely try to obtain some of the main outlines. Let us construct an ion from scratch, successively adding electrons to a nucleus that we imagine always to have abundant positive charge to balance all the electrons we use. The first electron goes into the $n = 1$ state, with spin, say, up. Only $\ell = 0$ is allowed. The state is designated $1s^1$, the first number being n, the letter representing ℓ according to the scheme

$$\begin{array}{c} s\ p\ d\ f\ g\ h \\ \ell = 0\ 1\ 2\ 3\ 4\ 5 \end{array} \ldots \text{etc.}$$

and the superscript saying that there is one electron in this state. For the second and all subsequent electrons, all states have their energies altered from the original energies by the interaction between electrons. In addition, the Pauli exclusion principle must be obeyed: no two electrons can have identical quantum numbers. We shall see how these conditions operate to influence the magnetic state of the ion.

The second electron can go into the $n = 1$, $\ell = 0$ state with spin down. That is the lowest energy state for two electrons; the configuration is $1s^2$, and no further electrons can have $n = 1$. The third electron will have $n = 2$, but now it can have ℓ equal either to zero or one. These states are degenerate in the absence of the $1s$ electrons, but in their presence the $\ell = 0$ state is preferred, because this state has more charge density near the nucleus; in the $\ell = 1$ state more of the charge density is far away and is thus shielded from the nucleus by the inner electrons. We therefore get $1s^2 2s^1$ for three electrons, $1s^2 2s^2$ for four electrons, and $1s^2 2s^2 p^1$ for five electrons. The $2p$ electron can choose to have $m_\ell = 1$ (i.e., orbital angular momentum parallel to its spin), or $m_\ell = 0$, or $m_\ell = -1$. The magnetic moment of the ion depends on this choice, but the electrostatic interaction with the nucleus and the other electrons now gives no basis for making it. It is, in fact, decided by a quite weak force, which is all that is left: the magnetic interaction between the electron's own orbital and spin magnetic moments, the spin-orbit interaction.

5.4 Magnetism

This interaction favors opposite orientations, so the ion has $L = 1$, $S = \frac{1}{2}$, and $|J| = |L - S| = \frac{1}{2}$. We also have

$$g = 1 + \frac{\frac{1}{2}(\frac{3}{2}) + \frac{1}{2}(\frac{3}{2}) - 1(2)}{2 \cdot \frac{1}{2}(\frac{3}{2})} = \frac{2}{3}, \text{ so } |\boldsymbol{\mu}| = \frac{2}{3}\mu_B J = \frac{1}{3}\mu_B$$

Let us consider one additional case in order to illustrate one more important and rather subtle way in which interactions decide the configuration that results in a magnetic moment. With six electrons we get the configuration $2s^2 p^2$ (we need not keep writing the closed $1s^2$ shell, since it is now always understood). If the first p electron has spin up, the second can have spin down and the same m_ℓ, or it can have spin up and a different m_ℓ. There is a difference in spin-orbit interaction between these possibilities, but it is small and the choice is dominated by far stronger considerations. If the two electrons have the same m_ℓ with opposite spins, their spatial wave functions are the same, and so their charge density distributions overlap completely. This is an unhappy choice, since both electrons are negatively charged. States with different m_ℓ have the same energy as far as the electron-nuclear potential is concerned, so the electron-electron interaction is the deciding factor, and the electrons choose to have different values of m_ℓ.

The two-electron system still has two choices to make. The spins are now free to be parallel or antiparallel, and the sum of the m_ℓ's (which gives L) can be $1 + (-1) = 0$ or $1 + 0 = 1$. Let us just look at how it decides the question of the spins.

Even though the two electrons are in different orbital states, their combined wave function must still be antisymmetric under exchange of particles. If the spins are antiparallel, the spin part of the wave function changes sign under exchange of particles, and so the spatial part must not change sign. The spatial part has the symmetric form, Eq. (2.3.3), for two electrons. If the spins are parallel, the spatial part must be antisymmetric, like Eq. (2.3.4). Ignoring normalization factors, the possible spatial wave functions are of the form

$$\psi \propto \phi_a(1)\phi_b(2) \pm \phi_a(2)\phi_b(1) \tag{5.4.8}$$

where the plus sign goes with ↑↓ and the minus sign goes with ↑↑. The difference in energy between these two possibilities will again be decided by the coulomb interaction between the two electrons. The potential for this interaction, which is repulsive and therefore positive, is

$$V = \frac{e^2}{|\mathbf{r}_1 - \mathbf{r}_2|} \tag{5.4.9}$$

where r_1 and r_2 are the positions of the two electrons. The energies of the two configurations have a term proportional to

$$\int \psi^* V \psi \, d^3r_1 \, d^3r_2 = K \pm J \qquad (5.4.10)$$

where
$$K = 2 \int V |\phi_a|^2 |\phi_b|^2 \, d^3r_1 \, d^3r_2 \qquad (5.4.11)$$

and
$$J = 2 \int \phi_a^*(1)\phi_b(1)\phi_b^*(2)\phi_a(2) \, d^3r_1 \, d^3r_2 \qquad (5.4.12)$$

Both K and J are positive. The minus sign in Eq. (5.4.10) belongs to the configuration ↑↑, which has the lower energy and is therefore favored.

What we have seen here is that the requirement that any system of electrons have an antisymmetric wave function has the effect of a kind of interaction between the spins. It is a quasiforce, which tends to align or disalign the spins, but it always operates by way of the other, real forces in the system. For example, starting with just a $1s^1$ electron, the next one can go into $1s^2$ with spin antiparallel to the first or into $2s^1$ with spin parallel. The $1s^2$ configuration is much favored because of the electron-nuclear coulomb force, and so the quasiforce results in the configuration ↑↓. On the other hand, the example we just worked out for the fifth and sixth electrons resulted in ↑↑. Regardless of which way it tends to line up the spins, this kind of quasiforce is known as the *exchange interaction*.

Suppose that we have some ion, and we wish to know the magnetic moment it will contribute when placed in a material. We must first know how the electrons are arranged among the possible values of n and ℓ quantum numbers and then, if there is an unfilled shell, what the values of m_ℓ and spin orientation are. The successive choices of n and ℓ as we add electrons to a nucleus are governed by the mechanisms just discussed, and when applied to complete neutral atoms, they give rise to the Periodic Table of the Elements. We shall take those choices as given. We are interested not in neutral atoms so much as in ions in the valence state in which they are commonly found in matter. Two series in the periodic table are of particular interest because of their magnetic behavior in matter: the rare earth series, and the iron group.

In the rare earth series (we are interested in elements 58, cerium, to 70, ytterbium) the 3+ ions have filled $5s^2p^6$ shells, but further in, the $4f$ shell is not filled. Since $\ell = 3$ in this shell, there are $2\ell + 1 = 7$ values of m_ℓ, each with two spin orientations, room for 14 electrons in all. If the shell is empty or full, the magnetic moment is zero, so we have 13 configurations to deal with. In the iron group ions, the outermost $3d$ shell, with room for 10 electrons, is the one that is not filled. The ions in this group can have valence

5.4 Magnetism

+2, +3, or +4, so the moment for a given element depends on the material it is in. The elements titanium (22) to copper (29) are in this group.

The problem is how to work out the magnetic moment once the number of electrons in the unfilled shell is given. For example, dysprosium, element 66, in the form Dy^{3+} has the configuration $4f^9$. What m_ℓ and spin values do those nine electrons take? The issue is to be decided by considering the contributions of the exchange and spin-orbit interactions in all possible configurations, a distastefully difficult task. Fortunately, a set of three simple rules, called *Hund's rules*, always gives the correct lowest-energy configuration. The rules are as follows:

1. $|S|$ is as big as possible.
2. $|L|$ is as big as possible once rule 1 has been satisfied.
3. $|J| = |L - S|$ if the shell is less than half-filled; $|J| = |L + S|$ if the shell is more than half-filled. If the shell is exactly half-filled, no rule 3 is needed, since rule 1, together with the Pauli principle, then gives $L = 0$.

For the example of Dy^{3+}, seven electrons with spin up, only two spin down give the largest possible S, 5/2. The sum of the m_ℓ's of the seven up electrons is zero, so the biggest L we can get comes from the other two with $m_\ell = 3$ and $m_\ell = 2$, or $L = 5$. The shell is more than half-filled, so $J = 5 + 5/2 = 15/2$. The spectroscopic notation for this arrangement is

$$^6H_{15/2}$$

The H means $L = 5$, just as h meant $\ell = 5$ for a single electron above. In the upper left, $6 = 2S + 1$; and at lower right, $15/2 = J$. Given J, L, and S, it is easy to work out the magnetic moment of the free ion using Eqs. (5.4.5) and (5.4.7).

The next step is to ask what happens when these ions are placed in a material. In a sense, we ought to start all over again because the material is not a collection of independent ions but actually more like a giant molecule. All the phenomena we have been discussing occur all over again, but now the system consists of many, many nuclei and many more electrons. Fortunately, however, a collection of independent ions is a reasonable approximation. In a single ion we saw that the basic states, enumerated by quantum numbers n and ℓ, due to the simple electron-nucleus interaction were perturbed, but not obliterated, by the more complicated considerations that arose when many electrons were present. In the same way, the basic ionic configurations we have just worked out may be perturbed, but they will not be obliterated by the new interactions that come into play when ions are placed in an environment created by other nearby objects. The reason is that the interactions within a single ion—the ones we have been discussing—are usually stronger than the interactions between electrons on different ions.

In the rare earth series, the magnetically active electrons are in an inner shell, protected from the environment by the outer $5s^2p^6$ electrons. The result is that the environment has little effect on the magnetic moment, which turns out just as it would in the free ion. In the iron group, on the other hand, it is the outermost electrons that contribute to the magnetic moment, and to these the environment can have an importance comparable to that of their own rather remote nucleus. It turns out that the seven spatial orbits of the $3d$ shell are generally distorted by the electric fields of other ions, so that they no longer all have the same energy for a single electron. The sevenfold degeneracy associated with $\ell = 3$ is lifted. There are still seven levels, and the first of Hund's rules still operates; the degeneracy is not lifted by enough to overcome the advantage of maximizing the total spin. However, no longer is any orbital angular momentum associated with the new states. The result is that the magnetic moments come out as if there were only spin, with $L = 0$ and $J = S$. This process is called quenching the orbital angular momentum.

As we shall see in Sec. 5.4c, it is possible under appropriate circumstances to measure the quantity

$$\mu^2 = g^2\mu_B^2 J(J + 1) \tag{5.4.13}$$

This information makes it possible to compare the actual magnetic moments to the ones that result from the arguments given above.

The next question in the natural order of things is, evidently, what is the effect of the interaction between the magnetic moments on separate ions in the same material? The exchange interaction will operate in one way or another, but will the result be to line the moments up parallel to each other, or antiparallel? This question turns out to be unanswerable in general; both types of behavior are actually observed in nature. But, in any case, we shall postpone not answering it until Sec. 5.4d. At this point another consideration arises.

As long as the interactions we are considering involve large energies, we can work things out simply by ranking them in quantitative order of importance and choosing the most favorable, or lowest energy, state as each new, less important effect is taken into consideration. The state of the system is just the quantum mechanical ground state, and the procedure we have been following is designed to give us a good approximation of that state. What we have been doing, in fact, is to describe in words the scheme known as perturbation theory in quantum mechanics. At some point along the way, however, the interaction energies we are considering must become so small that we can no longer assume that the system will be found in its ground state. That situation happens, of course, when the excited states we have been discarding are of order kT above the ground state. It is a very convenient fact that, for reasonable temperatures, we usually arrive at this point just when we finish considering the internal configuration of a single ion and start to deal with interactions between ions. Consequently, the internal state of the ion is a quantum mechanical problem, but the state of a many-ion system requires statistical mechanics as well. In the next two sections we deal with this latter problem, first in Sec. 5.4c, the case where the interaction between magnetic moments is negligible compared to kT, and then in Sec. 5.4d, the low-temperature problem, where the interaction again dominates the situation.

c. Paramagnetism

The magnetic moments in a material interact with each other with some characteristic energy, to which we shall return in Sec. 5.4d. If kT is large compared to the maximum value of this interaction energy, the moments may be taken to be independent or noninteracting. The system is then the magnetic analog of the perfect gas. The energy of a single moment $\boldsymbol{\mu}$ in an applied field is

$$\varepsilon = -\boldsymbol{\mu} \cdot \mathbf{H} = -\mu H \cos \theta \qquad (5.4.14)$$

where θ is the angle between $\boldsymbol{\mu}$ and $\mathbf{H}$. In the absence of a field, the spins (as we will loosely call them) are randomly oriented, so that there is no net magnetization; but when a field is applied, the magnetization of a unit volume of sample containing N spins is

$$M = \mu N \langle \cos \theta \rangle \qquad (5.4.15)$$

M points along the direction $\mathbf{H}$. The average ought to be taken over the $2J + 1$ orientations allowed quantum mechanically, but at high temperature it should suffice to work out the behavior classically. The average value of $\cos \theta$ may be computed, using the classical form of Eqs. (1.3.8) and (1.3.9):

$$\langle \cos \theta \rangle = \frac{\int e^{-\varepsilon/kT} \cos \theta \, [d^3r \, d^3q/(2\pi)^3]}{\int e^{-\varepsilon/kT} [d^3r \, d^3q/(2\pi)^3]}$$

$$= \frac{\int e^{\mu H \cos \theta / kT} \cos \theta \, do}{\int e^{\mu H \cos \theta / kT} \, do} \qquad (5.4.16)$$

All the integrations, top and bottom, cancel except for the angular part of the integral over position:

$$do = \sin \theta \, d\theta \, d\varphi = 2\pi \sin \theta \, d\theta \qquad (5.4.17)$$

If we write $x = \mu H / kT$ and $s = \cos \theta$ we get

$$\langle \cos \theta \rangle = \frac{\int_{-1}^{1} e^{sx} s \, ds}{\int_{-1}^{1} e^{sx} \, ds} = \frac{d}{dx}\left(\log \int_{-1}^{1} e^{sx} \, ds\right) \qquad (5.4.18)$$

but
$$\int_{-1}^{1} e^{sx}\, ds = \frac{e^x - e^{-x}}{x}$$

so
$$\langle \cos \theta \rangle = \frac{d}{dx}\left[\log(e^x - e^{-x}) - \log x\right]$$
$$= \frac{e^x + e^{-x}}{e^x - e^{-x}} - \frac{1}{x}$$
$$= \operatorname{ctnh} x - \frac{1}{x} \qquad (5.4.19)$$

We may thus write
$$M = N\mu L\left(\frac{\mu H}{kT}\right) \qquad (5.4.20)$$

where $L(x)$, called the *Langevin function* after the man who first did this calculation, is defined by
$$L(x) = \operatorname{ctnh} x - \frac{1}{x} \qquad (5.4.21)$$

In the high-temperature limit, $x \to 0$. We can expand
$$\operatorname{ctnh} = \frac{1}{x} + \frac{x}{3} - \frac{x^3}{45} + \cdots \qquad (5.4.22)$$

which diverges, but
$$L(x) \to \frac{x}{3} \qquad (5.4.23)$$

which does not. The magnetization becomes
$$M = \frac{N\mu^2 H}{3kT} \qquad (5.4.24)$$

and the susceptibility is
$$\mathrm{X}_T = \left(\frac{\partial M}{\partial H}\right)_T = \frac{N\mu^2}{3kT} \qquad (5.4.25)$$

The susceptibility is thus proportional to $1/T$. This result is called the *Curie law*. (It is probably irrelevant to note that Madame Curie was named correspondent in a celebrated divorce action taken against Paul Langevin.) If the average over orientations is performed quantum mechanically, the result is (Prob. 5.13)
$$M = Ng\mu_B J B_J\left(J\frac{\mu H}{kT}\right) \qquad (5.4.26)$$

where $B_J(x)$, the Brillouin function, is

$$B_J(x) = \frac{2J+1}{2J}\operatorname{ctnh}\left(\frac{2J+1}{2J}x\right) - \frac{1}{2J}\operatorname{ctnh}\left(\frac{x}{2J}\right) \qquad (5.4.27)$$

In the high-temperature limit, the resulting susceptibility is

$$X_T = Ng^2 J(J+1)\frac{\mu_B^2}{3kT} \qquad (5.4.28)$$

If we use Eq. (5.4.13) for the quantum mechanical square of the magnetic moment, this reduces exactly to the classical result, Eq. (5.4.25).

At temperatures with kT comparable to the interaction energy between spins, the magnetic system goes into an ordered state, as we shall see shortly. For temperatures high compared to the transition temperature, the Curie law is obeyed, and from measurements of the susceptibility it is possible to extract the values of μ^2, which we discussed in relation to the free ion values in Sec. 5.4b.

d. Ferromagnetism

In Sec. 5.4b we saw that there is a simple set of rules, Hund's rules, from which to deduce the magnetic ground state of an isolated ion. The outcome may be modified when the ion is placed in a crystal, but we can still think of a definite localized magnetic moment residing on a particular ion at a certain position in each unit cell. The result is a system or lattice of magnetic moments, which itself has some ground state determined by interactions of the same type already discussed. However, there is no simple set of rules for predicting the magnetic ground state of the many-body crystal system.

The principal influence on the choice of magnetic ground state is the exchange force discussed in Sec. 5.4b. It is by its nature a very short-range interaction, operating only if there is overlap between the spatial wave functions of the electrons involved. The range can be extended somewhat if the interaction operates indirectly, by way of electrons that do not contribute a magnetic moment of their own. In that case, the interaction is called superexchange. Just as exchange can lead to parallel or antiparallel alignment in the ground state of an ion, it can lead to parallel or antiparallel, or more complicated, ordered arrangements in the ground state of a crystal.

If, in the ground state, the magnetic moments all line up in the same direction, the material is said to be ferromagnetic. If neighboring moments face in opposite directions, the state is called antiferromagnetic. In a ferromagnetic material, the ground state has a macroscopic magnetization given by

$$\mathbf{M}_0 = N\mathbf{\mu} \qquad (5.4.29)$$

which follows directly from Eq. (5.4.15). This is a spontaneous magnetization, which occurs in the absence of any externally applied field. An antiferromagnetic material can be thought of as consisting of two interpenetrating sublattices magnetized in opposite directions. There is then no net spontaneous macroscopic magnetization. If there are two kinds of magnetic ion in the material with moments of different magnitude, these ions form sublattices that, in turn, may magnetize parallel or antiparallel to each other. The former possibility is called ferrimagnetism, the latter (which has a small net spontaneous magnetization) is called antiferrimagnetism.

In the rare earth and iron group salts, the exchange or superexchange mechanism usually (but not always) operates to produce an antiferromagnetic ground state. An example of how it works may be taken from manganese fluoride, MnF_2. The fluorine ions, F^-, do not themselves contribute any magnetic moment. Suppose that a fluorine ion has a pair of outermost p electrons (zero net L and S) that extend out toward an Mn^{2+} ion on either side. The manganese ion ($3d^5$ configuration, $^6S_{5/2}$ meaning spin 5/2 and no orbital contribution) tends by exchange to align its moment antiparallel to the nearby electron from the fluorine ion. That electron, in turn, is antiparallel to the other fluorine electron, which is antiparallel to the other Mn moment. The net result is that the Mn ions wind up antiparallel. This process would be referred to as antiferromagnetic coupling by means of superexchange. In real life the situation is more complicated, however. The Mn^{2+} ions form a bcc lattice with the corner ions magnetized one way and the body center ions the other. The F^- ions do not simply fall on the line between them, but the net result is still the same.

The elementary metals iron, nickel, and cobalt have ferromagnetic ground states. The ferromagnetic interaction presumably takes place by way of exchange with the conduction electrons in the metal: the ions each try to be antiparallel to the same conduction electrons and therefore wind up parallel to each other. It is, however, one of the deep embarrassments of solid-state physics that no workable theory of this interaction exists. Attempts to calculate the effective interaction between the magnetic moments in, say, iron, tend not even to give the right sign, much less the right magnitude. We are thus in the odd position of being able to explain the properties of a variety of exotic materials but unable to explain why a child's magnet attracts a scrap of iron.

Even if we cannot or do not wish to describe the interaction mechanism in detail, we know that the interaction exists, know its general form, and can go on now to study the statistical properties of magnetic systems at low temperatures. Given the ground state of the magnetic system, we wish to know what will happen to such thermodynamic properties as the magnetization and the heat capacity at finite temperature. In order to examine the problem, we, as usual, strip it to its essence. The energy of interaction

5.4 Magnetism

between the magnetic moments in a lattice can always be represented by the form

$$U = -\sum_{i<j} \alpha_{ij}\boldsymbol{\mu}_i \cdot \boldsymbol{\mu}_j \qquad (5.4.30)$$

where α_{ij} is a coefficient that gives the strength of the interaction between the ith and jth moments $\boldsymbol{\mu}_i$ and $\boldsymbol{\mu}_j$. If α_{ij} is positive, the interaction is ferromagnetic; if negative, antiferromagnetic. As in the instance of the vibrational states of a crystal lattice, the essential physics of the problem can be understood from the simplest imaginable example. We will consider a linear chain of moments, each with spin S and no orbital contribution, and with ferromagnetic nearest-neighbor-only interactions. Equation (5.4.30) can then be rewritten

$$U = -2J\sum_p \mathbf{S}_p \cdot \mathbf{S}_{p+1} \qquad (5.4.31)$$

where J is a positive constant called the exchange integral, $\mathbf{S}_p$ is the pth spin in the chain, and p runs from 1 to N, N being the number of spins in the chain (take $N \gg 1$ and assume periodic boundary conditions). Moreover, we can take $|\mathbf{S}| = \frac{1}{2}$ so that each spin can be imagined as having two possible orientations, up or down.

In the ground state, all the spins point in the same direction, let us say, in the $+z$ direction. The interaction energy is then $-2NJS^2$. The problem now is the same one we have faced a number of times: to construct the lowest-lying excited states. In this case, the answer seems obvious: since each spin can only be either up or down, and they are all up in the ground state, the first excited state must have one spin down. In the state with one spin down, say, the kth spin, only two terms in Eq. (5.4.31) differ from their values in the ground state: the $(k-1)$st term and the kth, each of which now contributes $+2JS^2$ to the energy. The state has energy $-2(N-2)JS^2 + 4JS^2$, or an energy $8JS^2$ above the ground state. The excitation spectrum of this system thus has an energy gap. From our experience with thermal systems with energy gaps, we can easily guess that the heat capacity of the spin system should be exponential in temperature, $C \sim \exp(-E/kT)$, where E is the energy gap. The departure of the magnetization from its zero-degree value will have this same form. In Chap. 6, in fact, we will work out exactly this model (in a three-dimensional material), and the magnetization result is given in Eq. (6.2.10).

The model that we have just discussed is somewhat analogous to the Einstein model of the vibrations of a solid, the essentially independent behavior of the spins giving rise to exponential behavior in the thermodynamic functions (see Sec. 3.2a). Like the Einstein model again, these predictions disagree with observation; both the heat capacity and the change in magnetization tend to go as $T^{3/2}$ in real materials. To fix up the Einstein model, we constructed collective rather than single-particle excitations—that is, phonons—with an energy spectrum that could go continuously to zero. Can we do the same here?

At first sight the answer seems to be no, since each spin must be either up or down, and therefore there is no way for the angle between neighboring spins to be anything but zero or 180°. That, however, is a misinterpretation of the quantization condition on the spins. The total projection of the spin angular momentum of any quantum mechanical system must change only in integral multiples of $\hbar$. If our system is a single spin, $S = \frac{1}{2}$, then it can change only from up to down, $\Delta S = 1$. In the measurement we wish to make (of the heat capacity or magnetization), it is not any single spin but rather the entire lattice that constitutes our system. Unless we are doing something that measures the spin projection at a single site, we cannot be sure it is either up or down; in fact, we cannot say anything about it. The excited states of the system may thus be more subtle than the ones we have already considered.

We will proceed in the following way: we will treat the spins classically, allowing them to point in any direction, and see what configurations are to be expected in such a system. We will then quantize the results by accepting only solutions with integral total spin projection. If that procedure seems strange, it may help to recall that we did precisely the same thing for the vibrations of a crystal lattice: we first did a classical normal mode analysis and then quantized the resulting collection of harmonic oscillators (see Sec. 3.3). We will now do a normal mode analysis of the spin system.

If the ith magnetic moment in a lattice found itself in an applied magnetic field **H**, its energy would be

$$u_i = -\boldsymbol{\mu}_i \cdot \mathbf{H} \qquad (5.4.32)$$

In the absence of an applied field, the ith moment nevertheless senses an interaction of a similar form trying to align it in some particular direction. From Eq. (5.4.30) the energy of the ith moment may be written

$$u_i = -\boldsymbol{\mu} \cdot \sum_j \alpha_{ij} \boldsymbol{\mu}_j \qquad (5.4.33)$$

We can formally define an effective magnetic field at the ith position

$$\mathbf{H}_{i,\text{eff}} = \sum \alpha_{ij} \boldsymbol{\mu}_j \qquad (5.4.34)$$

$\mathbf{H}_{i,\text{eff}}$ is not a real magnetic field; for example, it cannot be related to electric currents by means of Maxwell's equations. However, insofar as forces acting on $\boldsymbol{\mu}_i$ are concerned, it behaves like a magnetic field, and we shall use it only in that way. It is the same effective field that will emerge in Eq. (6.2.1) when we study the ferromagnetic phase transition.

5.4 Magnetism

For the case of the linear chain of spins, the effective field form of the energy may be constructed from Eq. (5.4.31), using (5.4.5) and (5.4.6),

$$U = \frac{1}{2}\sum_p \boldsymbol{\mu}_p \cdot \mathbf{H}_{p,\text{eff}} \qquad (5.4.35)$$

where
$$\boldsymbol{\mu}_p = -g\mu_B \mathbf{S}_p \qquad (5.4.36)$$

and
$$\mathbf{H}_{p,\text{eff}} = -\frac{2J}{g\mu_B}(\mathbf{S}_{p-1} + \mathbf{S}_{p+1}) \qquad (5.4.37)$$

What we are after here are the equations of motion of the spin system, something to play the same role in this analysis that Eq. (3.3.1) plays in the lattice vibrations analysis. There we wrote Newton's law for the acceleration of an atom when it is displaced from its equilibrium position. Here, instead, we take the torque on a magnetic moment displaced from its equilibrium direction parallel to a magnetic field

$$\boldsymbol{\tau} = \boldsymbol{\mu} \times \mathbf{H} \qquad (5.4.38)$$

and set it equal to the rate of change of the angular momentum

$$\hbar \frac{d\mathbf{S}_p}{dt} = \boldsymbol{\mu}_p \times \mathbf{H}_p \qquad (5.4.39)$$

This gives us three equations for each spin, $3N$ equations in all to solve simultaneously.

Before continuing with the mathematics, let us see what kind of solutions we should expect. By analogy to phonons, we should expect the lowest energy excitations to be some sort of long wavelength, low-frequency wave, in which each spin is only infinitesimally tilted with respect to its neighbor, so that the energy involved can be very small. Suppose, as shown in Fig. 5.4.1, that the spins in the ground state are spaced along the y axis, a distance a apart, pointing in the z direction. The effective fields are all in the z direction and there are no torques. In terms of components, the spins each have $|\mathbf{S}| = S_z$, with S_x and $S_y = 0$. Now suppose that we tilt one spin a bit, say, in the zy plane: $S_x = 0$ still, but $S_y \neq 0$. The effective field on the next

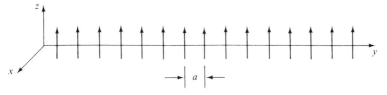

Fig. 5.4.1

spin in the chain is no longer in the pure z direction, but rather it is tilted a bit in the zy plane: by Eq. (5.4.38), the torque it applies is perpendicular to that plane—that is, along the x direction—so the next spin starts to move in that direction and so on. The point is that the wave takes place in the x and y components of the spins. The average values of S_x and S_y at each position remain equal to zero, but oscillations about zero take place. The average value of S_z is approximately $|S|$, relative to which any small changes in S_z may be ignored. With this point in mind, we write Eq. (5.4.39) in component form, making use of (5.4.36) and (5.4.37), and writing $S_{pz} = S =$ constant to get

$$\frac{dS_{px}}{dt} = \frac{2J}{\hbar} S[2S_{py} - S_{(p-1)y} - S_{(p+1)y}]$$

$$\frac{dS_{py}}{dt} = \frac{-2J}{\hbar} S[2S_{px} - S_{(p-1)x} - S_{(p+1)x}] \quad (5.4.40)$$

$$\frac{dS_{pz}}{dt} = 0$$

The rest of the game is familiar to us. We seek solutions of the form

$$S_{px} = u e^{i(pqa - \omega t)}$$
$$S_{py} = v e^{i(pqa - \omega t)} \quad (5.4.41)$$

which give, on substitution into Eq. (5.4.40),

$$-i\omega u = \frac{4JS}{\hbar}(1 - \cos qa)v$$

$$-i\omega v = -\frac{4JS}{\hbar}(1 - \cos qa)u \quad (5.4.42)$$

The determinant of the coefficients of u and v must be zero; that gives us the dispersion relation

$$\hbar\omega = 4JS(1 - \cos qa) \quad (5.4.43)$$

or, in the long wavelength limit we are interested in, $qa \to 0$, we have $\cos qa \to 1 - (qa)^2/2$, so that

$$\omega = \frac{2JS}{\hbar} a^2 q^2 \quad (5.4.44)$$

If we compare all these results to the phonon analysis, Sec. 3.3, we see that these equations of motion bear a certain similarity to the case of the vibrations of a linear chain (see Prob. 3.4). The principal difference is that we have a first time derivative in the spin case instead of the second derivative we dealt

5.4 Magnetism

with in Chap. 3. Thus, instead of $\omega^2 \propto q^2$ as for phonons, only one power of ω is brought down in Eq. (5.4.42) and we wind up with $\omega \propto q^2$ in the long wavelength limit.

It is easy to see from Eq. (5.4.42) that the x and y amplitudes are related by $u = iv$. The spin component in the xy plane thus executes circles with radius $u = v$, as sketched in Fig. 5.4.2a. In Fig. 5.4.2b we see the wave as it develops from one spin to the next, viewed in the xy plane.

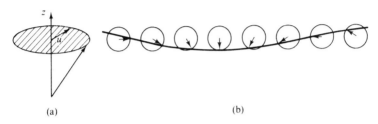

Fig. 5.4.2

The resulting z component of each spin (which really is time independent) is just

$$S_z = \sqrt{S^2 - u^2} = S\left(1 - \frac{u^2}{2S^2}\right) \quad (5.4.45)$$

the last step being the small amplitude limit, $u^2 \ll S^2$. The total z component of the spin of the N-particle system is NS_z. Our quantization condition is that NS_z must be an integer, and we are now prepared to apply it (it will be a constraint on the possible values of u). Since NS is the ground state value of NS_z it is sufficient to require that $NS - NS_z$ be an integer. Applying the condition to each mode, wave number q, we get

$$u_q^2 = \frac{2S}{N} n_q \quad (5.4.46)$$

where n_q may be any integer. Given this set of allowed amplitudes, it turns out (Prob. 5.15a) that the magnetic energy in a wave is related to the frequency ω_q and amplitude, represented by n_q, by the familiar equation

$$\varepsilon_q = \hbar \omega_q n_q \quad (5.4.47)$$

We thus have a new addition to our growing quasifamily of quasiparticles. These are called spin waves or, more in keeping with the tradition of family names, *magnons*. The form we have worked out here is applicable only to the simplest case of ferromagnetism; in antiferromagnets, for example, it turns out that $\omega^2 \propto q^2$, as for phonons, in the long wavelength limit (see Prob. 5.15b).

Each magnon changes the spin of the system by one unit. Magnons therefore have spin 1 and obey Bose statistics, and since their number is variable, the magnon gas has zero chemical potential [see Eqs. (3.2.72) and (3.2.73)]. In thermal equilibrium, the occupation number of each mode is given by

$$\bar{n}_q = \frac{1}{e^{\hbar\omega_q/kT} - 1} \qquad (5.4.48)$$

Each magnon reduces the magnetization of the system M by $g\mu_B$ from its saturation (zero temperature) value, $M_0 = g\mu_B NS$. Thus,

$$\frac{M - M_0}{M_0} = \frac{1}{NS} \sum_q \bar{n}_q \propto \int_0^\infty n(\omega)\rho(\omega)\, d\omega \qquad (5.4.49)$$

In the last step we have changed from a sum to an integral in the usual way, taking the upper limit to be infinity, since $n(\omega)$ goes rapidly to zero for $\hbar\omega \gg kT$, and we are seeking the low-temperature limit. We need to know $\rho(\omega)$ in order to obtain the temperature dependence we are looking for. Since $\omega \propto q^2$, just as for an ordinary free particle, we know that $\rho(\omega) \propto \omega^{1/2}$ [see Eq. (1.3.106)], so we have

$$M_0 \propto \int_0^\infty \frac{\omega^{1/2}\, d\omega}{e^{\hbar\omega/kT} - 1} = (kT)^{3/2} \int_0^\infty \frac{x^{1/2}\, dx}{e^x - 1} \qquad (5.4.50)$$

The change in magnetization goes as $T^{3/2}$. The energy of the magnon gas goes as $\int \omega n(\omega)\rho(\omega)\, d\omega \propto T^{5/2}$, and the heat capacity is proportional to $T^{3/2}$. These are the observed results we wished to explain.

All our discussion of magnetism up to this point has been concerned with phenomena that grow out of interactions between one electron and another or between one spin and another. Certain aspects of ferromagnetic behavior cannot be seen without stepping back and considering matters on a grander scale. Let us end this chapter by at least considering some of the sources of these phenomena.

Consider our linear chain of spins. We have neglected, in our discussion of the ferromagnetic ground state, to point out that if the spins are all lined up as shown in Fig. 5.4.1, they will themselves produce a magnetic field **B**, which extends through all of space and has an energy equal to $B^2/8\pi$ integrated over all of space. That energy should be included when deciding whether the state considered is the ground state. We can attempt to construct a lower energy state by, say, turning over half the spins, as in Fig. 5.4.3. In this scheme the exchange energy has been raised by $8JS^2$ [see the arguments following Eq. (5.4.31)], but the magnetic field energy is reduced, for, seen from far away, we notice only that half the spins are up and half down and

5.4 Magnetism

Fig. 5.4.3

so the fields nearly cancel. Thus, this configuration may well have a lower energy than what we previously called the ground state.

We can improve on this state by going gradually from the up regions to the down regions, somewhat as in the construction of spin waves; doing so will reduce the exchange energy compared to the state shown, where the change occurs abruptly. In fact, why have up regions and down regions? Perhaps the best arrangement is a kind of helix, the spin direction gradually shifting around in the xz plane as we move in the y direction. In fact, helical ground states are observed in nature (e.g., in MnO_2), but they are due to more complicated exchange interactions rather than to field energies. In any case, clearly we are still lacking some of the factors needed to determine the true ground state.

Crudely speaking, the principal points are as follows: It is true that a macroscopic sample uniformly magnetized in one direction is an energetically unfavorable state, because of magnetic field energy. On the other hand, in crystalline materials, there are axes along which the magnetization prefers to lie. This situation is a result of spin-orbit coupling and the fact that the spatial charge distributions associated with orbital states are pushed around by the anisotropic electric fields found in crystals. It is thus energetically costly for a spin to be in any direction other than, say, the $+$ or $-z$ direction. This is called the *anisotropy energy*. The third important point is that in changing from an up region to a down region, less exchange energy is added if the change is made slowly.

The net result is that macroscopic samples of ferromagnetic material tend to break up into *domains* of uniform magnetization, in different directions, separated by walls or transition regions of finite thickness. The size of the domains is large compared to atomic dimensions (a typical domain size is of order 0.1 mm), thereby justifying our previous discussion in which we ignored domains at the microscopic level, but they serve to cancel the large-scale magnetic fields. The size of the domain is a compromise (i.e., the result of minimizing the free energy) between the influence of the exchange and anisotropy energies, which tend to produce uniform magnetization, and the field energy, which desires the opposite. The thickness of the domain walls—typically a few hundred angstroms, in which the spins go gradually from up to down—is governed by competition between the exchange and anisotropy energies. The walls have positive free energy per unit interface area between domains. That free energy is the surface tension of the domain system, and it helps to ensure its stability. You can investigate the possibilities yourself in Prob. 5.16.

BIBLIOGRAPHY

Probably the best review ever written about liquid helium is K. R. Atkins's *Liquid Helium* (Cambridge, England: Cambridge University Press, 1959). It is brief, readable, and although considerably out of date by now, would still serve as a useful introduction to many aspects of the subject. Unfortunately, it has long been out of print, and it is almost impossible to get one's hands on a copy. Your library might have one, however. More recently, two vast compendia have come out: J. Wilks, *The Properties of Liquid and Solid Helium* (Oxford, England: Clarendon Press, 1967) and W. E. Keller, *Helium-3 and Helium-4* (New York: Plenum, 1969). Both are useful reference works if you have a problem in helium but are not very light bedtime reading. There is also an abridged version of Wilks' book, *An Introduction to Liquid Helium* (Oxford, England: Clarendon Press, 1970). A useful (and inexpensive) book is R. J. Donnelly's *Experimental Superfluidity* (Chicago: University of Chicago Press, 1967), based on notes from lectures given at the university. It includes a complete derivation of the two-fluid hydrodynamics, a job we delicately skipped over.

The backbone of our understanding of superfluidity comes from Landau as well as Feynman, and their own writings are a fine place to learn about the subject. L. D. Landau and E. M. Lifshitz, in *Fluid Mechanics* (Reading, Mass.: Addison-Wesley, 1959), pp. 507–522, discuss two-fluid hydrodynamics. The statistical mechanics of phonons and rotons is covered in Landau and Lifshitz, *Statistical Physics* (see bibliography Chap. 1), pp. 187–196. For Feynman's theory from the pen of the master himself, there is a well-known article in *Progress in Low Temperature Physics*, Vol. 1. Edited by C. J. Gorter (Amsterdam: North-Holland, 1964). However, more interesting sources are the original papers, where you can follow his reasoning in more detail, see him sometimes make mistakes (correcting them later), and generally see the ideas emerge. The references are R. P. Feynman, *Phys. Rev.*, **91**, 1301–1308(1953), and *Phys. Rev.*, **94**, 262–277(1954). He also discusses the theory in his book of lecture notes, *Statistical Mechanics* (Reading, Mass.: W. A. Benjamin, 1972). There is also a discussion of the Feynman theory in K. Huang, *Statistical Mechanics* (New York: Wiley, 1967).

Three books that are classics of the helium literature should be mentioned. W. H. Keesom, *Helium* (Amsterdam: Elsevier, 1942) is an early compendium of all sorts of information about helium the element, including helium the liquid. F. London, *Superfluids*, Vol. 2 (New York: Wiley, 1954) (Volume 1 of London's *Superfluids* is about superconductivity) and I. M. Khalatnikov, *An Introduction to the Theory of Superfluidity* (New York: W. A. Benjamin, 1965) are monographs.

There is a chapter on superfluidity in the book *An Introduction to the Liquid State*, by Egelstaff, discussed in the bibliography of Chap. 4. As pointed out there, it is not entirely dependable.

There are probably more people doing research on superconductivity today than there were scientists of any kind a century ago. At least a normal proportion of them have suffered the urge to write, and so monographs abound. One of the most elegant is P. G. De Gennes' *Superconductivity of Metals and Alloys* (New York: W. A. Benjamin, 1966). The book must be read with care. It is glib, even slippery in places (note, e.g., his derivation of the London equation, pp. 3 and 4). Nevertheless, a number of the arguments used here have been adapted from De Gennes (including the derivation of the London equation). A more painstaking treatment appears in *Theory of Superconductivity* by J. M. Blatt (New York: Academic Press, 1964). At very different levels, chapters on superconductivity are found in two of the books mentioned in the bibliography of Chap. 3: *Introduction to Solid State Physics* by Kittel and *Solid State Theory* by Harrison.

Chapters on magnetism may be found in virtually all books on solid-state physics, including Kittel's *Introduction* and Harrison's *Theory*. A detailed monograph on the subject is R. M. White's *Quantum Theory of Magnetism* (New York: McGraw-Hill, 1970). Discussions of the ground state configurations of atoms and ions are found in books on quantum mechanics and spectroscopy, but a particularly good source for an elementary but illuminating discussion is *The Feynman Lectures on Physics* (Reading, Mass.: Addison-Wesley, 1966), especially Vol. 3, Chap. 19.

PROBLEMS

Superfluidity

5.1 **a.** In ordinary fluid mechanics, the energy flux density (per unit mass) is given by

$$\mathbf{q} = \rho \mathbf{v} \left(\frac{1}{2} v^2 + \frac{e_i + P}{\rho} \right)$$

and the momentum flux density tensor by

$$\pi_{ik} = \rho v_i v_k + P \, \delta_{ik}$$

(subscripts i and k each refer to components x, y, and z). Show that the transformations of these quantities between superfluid and laboratory frames are

$$\mathbf{q} = \mathbf{u}_s(\tfrac{1}{2}\rho u_s^2 + \mathbf{j}_0 \cdot \mathbf{u}_s + e_0) + \tfrac{1}{2}u_s^2 \mathbf{j}_0 + (\pi_{0ik}) \cdot \mathbf{u}_s + \mathbf{q}_0$$

where the components of $(\pi_{0ik}) \cdot \mathbf{u}_s$ are $\pi_{0ik} u_{sk}$ and

$$\pi_{ik} = \rho u_{si} u_{sk} + u_{si} j_{0k} + u_{sk} j_{0i} + \pi_{0ik}$$

b. Verify that Eqs. (5.2.42) to (5.2.45) satisfy identically Eq. (5.2.36), together with (5.2.32) to (5.2.35), and (5.2.41).

c. Show that Eq. (5.2.36), together with (5.2.45), gives

$$\rho \, d\mu = dP - \rho s \, dT - \rho_n \mathbf{w} \cdot d\mathbf{w}$$

5.2 Consider a narrow channel in which the normal fluid is held at rest. Consider only one-dimensional flow, and assume, as we did in the text, that

$$dS = \frac{C}{T} dT \quad \text{and} \quad dP = c_1^2 \, d\rho$$

Momentum density is not conserved in this system, since forces can be transmitted to the walls. Using the other equations of the two-fluid model in linearized (small-amplitude) form, find the speed of the combined entropy and density waves in this system. The result is called fourth sound.

5.3 **a.** Compute the roton contribution to the normal fluid density.
  b. Suppose that you have just measured the heat capacity and normal fluid fraction of liquid helium in your laboratory and would like to use your data to test Landau's model of rotons and phonons. Your results are as shown in the table and $\rho = 0.15$ gram/cm^3. You learn that Landau's

T (°K)	C_V (j/gram/°K)	ρ_n/ρ
0.80	0.022	0.00082
0.85	0.034	0.0015
0.90	0.051	0.0027
0.95	0.074	0.0044
1.00	0.10	0.0069
1.05	0.14	0.010
1.10	0.19	0.015
1.15	0.25	0.020
1.20	0.32	0.028

paarmeters have been evaluated independently (by inelastic neutron scattering) to give $\Delta/k = 8.6°K$, $p_0/\hbar = 1.9 \text{Å}^{-1}$, and $\mu_0 = 0.16$ of the helium atomic mass, and the speed of sound is 237 m/sec. How would you use all this information to evaluate the theory?
  c. Show that
$$\lim_{T \to 0} c_2 = \frac{c_1}{\sqrt{3}}$$

5.4 **a.** Show that the excited-state wave functions in Feynman's theory are orthogonal to each other and orthogonal to the ground state.
  b. Show that if two excitations are present, the energy is the sum of the two independent excitation energies.
  c. According to the Feynman theory,
$$\lim_{Q \to 0} \varepsilon_Q = c_1 Q = \frac{\hbar^2 Q^2}{2mS(Q)}$$
so that
$$\lim_{Q \to 0} S(Q) = \frac{\hbar^2 Q}{2mc_1}$$
whereas according to Eq. (4.2.64),
$$\lim_{Q \to 0} S(Q) = \rho k T K_T$$
Is there any contradiction between these assertions?
  d. Show that a Feynman excitation of wave vector $\mathbf{Q}$ moving at velocity $\mathbf{u}$ relative to the ground state has energy $(\varepsilon_Q - \hbar \mathbf{Q} \cdot \mathbf{u})$.

5.5   **a.** Show that the equilibrium state of a fluid in a rotating bucket is solid-body rotation.

      b. Show that for a fluid in solid-body rotation at angular velocity ω, the fluid velocity field, $\mathbf{u}(\mathbf{R})$, obeys

$$|\nabla \times \mathbf{u}| = 2\omega$$

      c. Prove Eq. (5.2.142) for the free surface of a fluid in solid-body rotation.

5.6   **a.** Estimate the core diameter of a singly quantized vortex line, using the classical balance between centrifugal force and surface tension (the surface tension of liquid helium is 0.37 erg/cm^2).

      b. Show that the force on a vortex core of circulation $\mathbf{k}$ moving through a fluid at velocity $\mathbf{v}$ is

$$F = -\rho \mathbf{k} \times \mathbf{v}$$

(the Magnus force).

      c. Show that a vortex line cannot end in the midst of an incompressible fluid.

      d. Consider a quantized vortex line parallel to a plane wall. Find the force between the line and the wall as a function of distance from the wall. Find the equilibrium position if the superfluid is flowing with velocity u_s parallel to the wall and perpendicular to the vortex line. Finally, consider a circular tube big enough so that the wall can locally be considered plane. It is proposed that at some critical value of u_s, vortex lines can be created at the wall and pulled out into the fluid. Discuss the merits of this proposed mechanism for critical flow velocities.

      e. Consider a persistent superfluid current. According to the model of thermally activated vortex rings, how does the superfluid velocity depend on time, on temperature, and on the size of the channels through which it flows?

5.7   We have argued in this section that it is possible for a fluid to have zero entropy because moving indistinguishable atoms around does not create new states. Atoms other than helium are also indistinguishable from each other, however, and so the same kind of arguments may be applied to other liquids as well. This raises the question: What really is the origin of what we have called communal entropy in a liquid? If we consider liquid and solid phases of the same substance at the same density and temperature, why does the liquid normally have higher entropy?

Superconductivity

5.8   **a.** Show that if we follow the arguments, Eqs. (5.3.22) to (5.3.25), but without any restrictions on $M(\mathbf{r})$, the result is $\mathbf{B} = \mathbf{H}$ everywhere.

      b. Using Eq. (5.3.25), the London equation, together with Maxwell's equations, find the distributions of fields and currents when a uniform field $\mathbf{H}$ is applied parallel to a semi-infinite slab of superconductor [the solutions are Eqs. (5.3.26) and (5.3.27)].

      c. Even in a type I superconductor, the complete Meissner effect breaks down at $H < H_c$ if the sample is not a long cylinder as we have imagined it. The resulting situation is called the intermediate state. Find the value of

H/H_c for which the Meissner effect breaks down in a spherical sample (H is now the constant applied field far away from the sample).

5.9 **a.** Show that the difference in energy between the BCS ground state and the Fermi degenerate gas is

$$E_s - E_n = \frac{-\rho(0)\,\Delta_0^2}{2}$$

b. If each member of any pair of single-electron states, $\mathbf{q}\uparrow$, $-\mathbf{q}\downarrow$, always has the same amplitude for being occupied as the other member, show that the minimum excitation energy above the BCS ground state is $2\,\Delta_0$.

5.10 If we choose the gauge

$$\nabla \cdot \mathbf{A} = 0$$

for the magnetic vector potential, then any $\mathbf{A}$ can be replaced by

$$\mathbf{A}' = \mathbf{A} + \nabla X$$

where X is any scalar function of position.

Then $$\mathbf{B} = \nabla \times \mathbf{A}'$$

So $\mathbf{A}'$ is a perfectly good vector potential. Find a way of writing the phase of the ground-state wave function, γ, so that both the Schrödinger equation and Eq. (5.3.122) for the current are left unchanged by the transformation from $\mathbf{A}$ to $\mathbf{A}'$.

5.11 Find the energy per unit length of a superconducting fluxoid.

Magnetism

5.12 We wish to calculate the paramagnetic susceptibility of the Fermi degenerate perfect gas of electrons. It comes about this way: each electron has a magnetic moment $\boldsymbol{\mu}$ antiparallel to its spin. In an applied magnetic field $\mathbf{H}$ all the $\boldsymbol{\mu}$'s are either aligned parallel or antiparallel to $\mathbf{H}$. Those that are parallel have their energy lowered by $|\boldsymbol{\mu}\cdot\mathbf{H}|$; those antiparallel have their energy raised by $|\boldsymbol{\mu}\cdot\mathbf{H}|$. We can think of two gases, parallel and antiparallel, connected by the equilibrium condition that their chemical potentials be equal. We want the susceptibility in the zero field limit, so it is safe to assume that $|\boldsymbol{\mu}\cdot\mathbf{H}| \ll kT_F$. Give the answer in terms of T_F, the number of electrons per unit volume, N, and μ_B, the Bohr magneton.

5.13 Derive the quantum mechanical expressions, Eqs. (5.4.26) to (5.4.28), for the magnetization and susceptibility of a system of magnetic moments each with total angular momentum J.

5.14 **a.** Suppose that we measure the paramagnetic susceptibility X_T of a series of rare earth salts around room temperature. How can we test the data to see if the temperature is high enough for Eq. (5.4.28) to be a correct prediction?

b. In these measurements, it should be possible to predict the quantity

$$\frac{3kTX_T}{N\mu_B^2}$$

if we know the electronic configurations of the rare earth ions. The 13 ions have configuration $4f^x 5s^2 p^6$, where x runs from 1 to 13. The measured values of the preceding quantity are, respectively, 5.8, 12.3, 12.3, (unknown), 2.25, 11.6, 64, 90, 112, 108, 90, 53, and 20.2. Two ions are in serious disagreement with the predictions of Hund's rules because, as it turns out, the excited states of the ion are not at energies large compared to kT at room temperature. Find out which two ions disagree.

5.15 a. Show that for ferromagnetic spin waves of the type discussed in the text, $\varepsilon_q = n_q \hbar \omega_q$ when $q \to 0$.

b. Show that for the analogous spin waves in an antiferromagnetic material, $\omega^2 \propto q^2$. What is the form of the heat capacity and of the magnetization?

5.16 Try to formulate the problem of the overall ground state of a ferromagnetic body. You will need to guess reasonable forms for the exchange, anisotropy, and magnetic field energies, minimize the free energies, and wind up with domains and walls whose sizes depend on the parameters of your model. The idea is to get a feeling for the nature of the problem, so you are free to choose the simplest geometry that still includes all the important physical effects.

SIX

CRITICAL PHENOMENA AND PHASE TRANSITIONS

6.1 INTRODUCTION

The ways in which matter manages to change from one state to another may be grouped into two classes: those that involve a latent heat and those that do not. The former are first-order phase transitions. The latter, which we might call critical phase transitions, have been the subject of intense study in recent years and will be the principal subject of this chapter.

The reason for the intense interest in these critical phenomena is, as usual, not that they are any more important, say, than first-order transitions but rather because a good idea has come along and made progress possible. The idea, which gives rise to the scaling laws described in Sec. 6.7, is basically a way of applying dimensional analysis to the problem.

Dimensional analysis is a technique by means of which it is possible to learn a great deal about very complicated situations if you can put your finger on the essential features of the problem. An example is the well-known story of how G. I. Taylor was able to deduce the yield of the first nuclear explosion from a series of photographs of the expanding fireball in *Life* magazine. He realized that he was seeing a strong shock expanding into an undisturbed medium. The pictures gave him the radius as a function of time, $r(t)$. All that could be important in determining $r(t)$ was the initial energy release, E, and the density of the undisturbed medium, ρ. The radius, with the dimension of length, depended on E, ρ, and t, and he constructed a distance out of these

6.1 Introduction

quantities. E and ρ had to come in as E/ρ to cancel the mass. E/ρ has the dimensions $(\text{length})^5/(\text{time})^2$, so the only possible combination was

$$r(t) \propto \left(\frac{E}{\rho} t^2\right)^{1/5} \tag{6.1.1}$$

A log-log plot of r versus t (measured from the pictures) gave a slope of $\frac{2}{5}$, which checked the theory, and E/ρ could be obtained from extrapolation to the value of $\log r$ when $\log t = 0$. Since ρ, the density of undisturbed air, was known, E was determined to within a factor of order one. For the practitioner of the art of dimensional analysis, the nation's deepest secret had been published in *Life* magazine.

We shall apply a very similar kind of analysis to the problem of critical phase transitions. We shall not be interested in the coefficients, for example, E/ρ in Eq. (6.1.1), which can never be precisely determined by this kind of argument (there can always be, say, a factor of 2π lurking somewhere), but rather we shall concentrate on the exponents. A relation like Eq. (6.1.1) will be written in the form

$$r \sim t^{2/5}$$

There should be no ambiguity in the value of the exponent. We shall be able to answer questions such as: If the heat capacity of liquid helium near T_λ has the dependence

$$C \sim (T_\lambda - T)^{-\alpha'}$$

and the superfluid density vanishes according to

$$\rho_s \sim (T_\lambda - T)^\zeta$$

then how is α' related to ζ?

The energy release of the first nuclear device was not obvious to every reader of *Life* magazine, not even to those who were aware of the technique of dimensional analysis. The key to success is to identify the essential features of the problem, which can be hard to do until you have been shown how. You can convince yourself of that point by trying now to deduce the relation between α' and ζ mentioned above. If you have read Chap. 5, you know a good deal about superfluidity; no further properties peculiar to liquid helium will be introduced before we produce the promised result in Sec. 6.7. In order to develop some feeling for which properties are the essential ones governing behavior very close to the phase transition, we shall spend most of this chapter studying the nature of these transitions.

The common feature of all the transitions in question is the absence of a latent heat. This property may not seem like much to have in common, especially since the transitions appear to differ from each other in a number of central features. For example, the gas-liquid critical point is the endpoint of a first-order phase transition curve (in the P-T plane; see Fig. 1.2.5), where the latent heat goes to zero. At higher T and P, there is no distinction between the two phases. The superconducting-normal critical point is also the endpoint of a first-order phase transition curve (in the H-T plane; see Fig. 1.2.3), but one can never pass from the super to the normal conductor without suffering a phase transition. The phase transition between the ferromagnetic and paramagnetic states occurs at an isolated point (a certain temperature, in zero applied field, H), whereas the superfluid transition is actually a line (in the P-T plane; see Fig. 1.2.8) bounded at both ends by first-order phase transitions. Yet these four diverse transitions are the ones we shall study as examples, and we shall see that it is (almost) possible to study them together, as aspects of the same basic phenomenon.

We start with two venerable theories: the Weiss molecular field theory of the ferromagnetic transition and the van der Waals theory of the liquid-gas transition. It will turn out that although the theories seem to be of different types, concerning utterly different kinds of interactions and behavior, the results will be very similar. This will lead us to a discussion of how to draw analogies between the different phase transitions and then to a generalized theory that describes all four transitions together in the same way and with the same results, as the Weiss and van der Waals treatments do for their own special cases. These theories do not work—they give incorrect predictions—but it is from reasons for their failure that we are able to extract those crucial features that lead to the scaling laws. We will then make our dimensional arguments and, at the very end, try to evaluate whether they are, in fact, correct and where they leave us.

6.2 WEISS MOLECULAR FIELD THEORY

The kind of magnetic systems we discussed in Sec. 5.4 undergo a phase transition at some temperature T_c below which the system has a spontaneous magnetization. In trying to account for this transition, we begin by imagining that each magnetic moment finds itself not only in the externally applied magnetic field **H** but also in an internal field due to the effect of the other moments in the system. When the moments are randomly oriented, their effects on any one moment cancel, but if there is a net magnetization **M**, we assume that each moment senses an effective field $\mathbf{H}_{\text{eff}}$ proportional to **M**:

$$\mathbf{H}_{\text{eff}} = \lambda \mathbf{M} \quad (6.2.1)$$

Next we insert this field into the formula we developed for the high-temperature, noninteracting limit. This procedure is analogous to the one

6.2 Weiss Molecular Field Theory

we followed in Chap. 4, where we studied liquids by trying to insert interactions into a formalism best suited to the study of gases: We shall do much the same in the next section, when we study the liquid-gas phase transition. The individual moment finds itself in a field, $\mathbf{H} + \mathbf{H}_{\text{eff}}$, and reacts with a net alignment that produces a magnetization M, according to Eq. (5.4.24):

$$\frac{M}{H + H_{\text{eff}}} = \frac{C}{T} \tag{6.2.2}$$

where $C = N\mu^2/3k$. Putting in Eq. (6.2.1) and $X_T = M/H$, which is the susceptibility of the collective system, we have

$$\frac{C}{T} = \frac{X_T}{1 + \lambda X_T} \tag{6.2.3}$$

or
$$X_T = \frac{C}{T - \lambda C} \equiv \frac{C}{T - T_c} \tag{6.2.4}$$

At temperature $T_c = \lambda C$, the susceptibility becomes infinite. Below T_c, with no applied field, there is an instability in the system: as long as the spins are randomly oriented, each spin feels no effective field and there is no magnetization. However, fluctuations will always produce local magnetization, which aligns our spin, thereby further increasing the magnetization and so on. The only stable state is one with a permanent spontaneous magnetization. T_c is called the Curie temperature for a ferromagnetic transition.

If H_{eff} is actually a magnetic field, it must be due to the dipole fields of the other moments. Since these fall off as r^{-3}, only near neighbors will contribute, so $H_{\text{eff}} \sim \mu_B/a^3 \sim 10^3$ gauss, if $a \sim 2 \times 10^{-8}$ cm. We might then expect $T_c \sim \mu_B H_{\text{eff}}/k \sim 10^{-1}$ °K. Instead, these transitions are commonly observed at $T_c \sim 10^3$ °K, so there is a serious discrepancy. The magnetic field of the dipoles simply cannot cause ordering at 1000°K.

This problem brings to mind the case of superconductivity, where we found a quite similar discrepancy between the transition temperature and the apparent interaction energy involved in the transition. Once again the problem is that we have misinterpreted the interaction energy. The source of H_{eff} must be not the magnetic field of the dipoles but rather the exchange and superexchange mechanisms discussed in Sec. 5.4. As we said then, this is not actually a magnetic field but, for this purpose, behaves just as if it were one. In fact, T_c can be used to estimate the magnitude of the exchange coefficient J, for any particular system (see Prob. 6.1).

By simply putting the effective field into the Curie law, we have produced a good first picture of the phase transition, but it does not give us the magnetic equation of state—that is, the magnetization as a function of T. We can get that by using Eq. (5.4.26) before the high-temperature approximation is made

$$M = Ng\mu_B JB_J\left(\frac{\mu H}{kT}\right)$$

where J is, in this case, an angular momentum. Without any applied field, we take $H = H_{eff} = \lambda M$ and, for simplicity, just take spin $\frac{1}{2}$:

$$M = N\mu_B \tanh\left(\frac{\mu_B \lambda M}{kT}\right)$$

or defining reduced variables,

$$m = \frac{M}{N\mu_B} \tag{6.2.5}$$

$$t = \frac{kT}{N\mu_B^2 \lambda} \tag{6.2.6}$$

we have

$$m = \tanh\left(\frac{m}{t}\right) \tag{6.2.7}$$

This transcendental equation for $m(t)$ is solved graphically in Fig. 6.2.1 by plotting the left and right sides against m with t as a parameter. Solutions occur when the curves cross.

There is always a solution at $m = 0$, but for $t < 1$ a second solution appears at finite $m(t)$. For these cases, the $m = 0$ solution is physically unstable, as described above, and we have a spontaneous magnetization. The stable solutions have the form shown in Fig. 6.2.2. Figure 6.2.2 is the equivalent of a zero-field equation of state for the magnetic system.

The critical value $t_c = 1$ gives a transition temperature

$$T_c = \frac{N\mu_B^2 \lambda}{k} \tag{6.2.8}$$

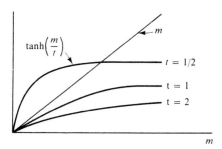

Fig. 6.2.1

6.2 Weiss Molecular Field Theory

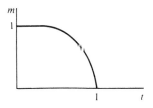

Fig. 6.2.2

which is to be compared to the Curie temperature from Eq. (6.2.4), plus (5.4.28),

$$T_c = \lambda C = \frac{Ng^2 S(S+1)\mu_B^2 \lambda}{3k} \quad (6.2.9)$$

For spin $\frac{1}{2}$, $g^2 S(S+1)/3 = 1$, so that these two results are the same. In fact, Eq. (6.2.9) gives the correct result for T_c for any spin.

In discussing the shortcomings of this model, we should not lose sight of the fact that it is remarkably successful. It is mathematically simple, physically transparent, and gives us a rather complete description of behavior that agrees with observation: Curie law at high temperature, a phase transition, and a spontaneous magnetization that varies continuously from zero to saturation as the temperature goes down. While interesting physics can be learned from its failures, it is essentially a matter of detail, to be understood as fine-scale improvements in what is a generally successful picture of the behavior of these magnetic systems.

One problem arises when we consider the way in which the magnetization approaches saturation at low temperature. When $m/t \to \infty$, $\tanh(m/t) \to 1 - 2e^{-2m/t}$, so that Eq. (6.2.7) gives us

$$\frac{M_0 - M}{M_0} = \exp\left(\frac{-2\lambda N\mu_B^2}{kT}\right) \quad (6.2.10)$$

where we have taken $M = M_0 = N\mu_B$ in the exponent.

This is the exponential dependence on T promised in Sec. 5.4c, and it contrasts with the observed $T^{3/2}$ behavior. We already know both the cause and the remedy for this problem. Just as in the Einstein model of crystals, we have considered the interaction of one particle (or spin) with the average effect of all the others. This assumption ignores the fact that the behavior of our one particle affects its neighbors and changes their average behavior; in short, we have ignored correlations in the statistical fluctuations of the particles (spins). In order to get the correct low-temperature thermal behavior, we constructed correlated fluctuations—that is, phonons in Sec. 3.2 and spin waves in Sec. 5.4. These correlated motions are important at low temperature because they require less energy than uncorrelated single-particle fluctuations, and therefore they are more easily excited.

What concerns us in this chapter is the detailed behavior near the phase transition. Here, too, the theory fails and essentially for the same reason. Correlated fluctuations, which have been ignored, become important. In testing the theory at the transition, one's first inclination is to look at the quantitative prediction of T_c but this involves what is essentially an adjustable parameter, λ, and thus is of little help. We do better by looking at the temperature dependence of X_T above T_c, as in Fig. 6.2.3. According to the theory, a plot of $1/X_T$ versus T should be a straight line with intercept at T_c. This prediction works rather well at high temperature, so that the high-temperature data "predict" T_c, but real systems tend to curve away from the straight line as the transition is approached, with the susceptibility actually becoming infinite at a lower temperature than expected.

In trying to understand why this situation occurs, it is useful to notice that the mean square thermodynamic fluctuations in the magnetization are given by

$$\overline{(\Delta M)^2} = \frac{kTX_T}{V} \qquad (6.2.11)$$

(Prob. 6.2). As the susceptibility blows up near the transition, the fluctuations in the magnetization become very large. There is no latent heat in this phase transition; the free energy is continuous in passing through the transition (we will formalize these ideas later); consequently, very close to the transition, it costs very little in free energy for the magnetization to flop around. As a result, the average magnetization acting on a single spin in Eq. (6.2.1) becomes meaningless, and the system is governed by its collective rather than its single-particle properties; that is, the fluctuations become more important than the average values. Unfortunately, we cannot deal with the problem here, as we did before, by constructing spin waves, since we can no longer call on the small-amplitude approximations that made life easy in Sec. 5.4.

Let us summarize what we have done in this section. We worked out the

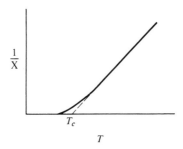

Fig. 6.2.3

6.3 The van der Waals Equation of State

behavior of a system of magnetic moments or spins by averaging over the states or fluctuations of a single spin under the average influence of all the others. Doing so produced a description that is quantitatively correct at high temperature (the Curie law) and has the qualitatively correct features of a phase transition into an ordered low-temperature state. At the phase transition, the susceptibility becomes infinite; and at lower temperatures, the random or disordered state is still a solution of the equation of state, but it is physically unstable.

When compared to real systems, we find that the transition temperature predicted on the basis of our model from high-temperature data turns out to be a bit too high. In addition, the temperature dependence of the susceptibility as the transition is approached and the low-temperature equation of state do not agree with real systems. These difficulties can be traced to the influence of collective or correlated fluctuations, which we ignored in our model.

In the next section we shall work out the van der Waals theory of liquid-gas systems. As we shall see, it will turn out that the preceding summary will serve, almost word for word, to describe that situation as well, and in subsequent sections we will adopt a point of view from which the two models will be regarded as aspects of the same theory.

6.3 THE VAN DER WAALS EQUATION OF STATE

The van der Waals theory, like the Weiss theory of the previous section, gives a complete qualitative description of the system it applies to—in this case, the gas-liquid system. The equation is

$$\left[P + a\left(\frac{N}{V}\right)^2\right](V - Nb) = NkT \tag{6.3.1}$$

where a and b are positive constants, fixed for each particular gas.

Many derivations, pseudoderivations, and plausibility arguments have been given for this equation. We shall work it out in a way that illustrates the points we wish to stress in comparing it to other theories.

In analogy to the last section, we start from the high-temperature limit, taking the gas to be basically ideal, and look for departures from ideality due to the average influence of all the other atoms on one single test atom. We consider a box of gas and the mean potential on the ith atom at position $\mathbf{R}$,

$$W_i(\mathbf{R}) = \sum_{j \neq i} u(r_{ij})$$

$$= \frac{N}{V} \int u(r)\, d^3r$$

$$= -\frac{N}{V} W_0 \tag{6.3.2}$$

In the second step we have found our average by assuming that the other atoms have a smoothed-out, continuous, fluidlike distribution in space. We temporarily neglect the repulsive part of $u(r_{ij})$, so $W_0 > 0$. The last step assumes that the test atom is far from the boundaries of the fluid.

Like the H field in the magnetic case, the pressure may have a local internal, effective value that differs from the external or applied value appearing in Eq. (6.3.1). There can, of course, be no unbalanced forces, but there can be changes in the local gas density and pressure due to spatial variations in potential, including W_i. In particular, if our test atom approaches the wall of the box, two things happen:

1. Our atom starts to sense the attractive potential of the atoms (of some other species) of which the wall is constructed. The result is to cause the local density to rise above the average value far away and the internal pressure to rise as well.

2. On the other hand, it finds that on its wall side there are fewer of its own kind of atoms, and W_i decreases below the value given in Eq. (6.3.2).

In many instances, the first effect is much larger than the second (in such cases, the fluid is said to wet the wall), but it plays no role at all in determining the applied pressure, which is the quantity of interest to us. The reason is that, although the wall exerts a force on the fluid, the fluid exerts an equal and opposite force on the wall, so there is no net effect on the applied pressure (which is really the force per unit area needed to keep the wall in place). On the other hand, the reduction in W_i near the wall is a measure of the tendency of the fluid to pull itself together, to self-condense, and this tendency does reduce the applied pressure. We thus ignore effect (1) and compute the applied pressure from effect (2) alone.

Evaluated at the wall, the integral in the second of Eqs. (6.3.2) extends only in the direction away from the wall, so that W_i is reduced by half from its value in the middle of the fluid:

$$W_i \text{ (wall)} = -\frac{N}{2V} W_0 \qquad (6.3.3)$$

This effect serves to reduce the local density

$$\left(\frac{N}{V}\right)_{\text{wall}} = \frac{N}{V} \exp\left[\frac{-[W_i(\text{wall}) - W_i(\text{middle})]}{kT}\right]$$

$$= \frac{N}{V} e^{-NW_0/2VkT}$$

6.3 The van der Waals Equation of State

We are looking for departures from the ideal gas limit, so we take the applied pressure to be that of an ideal gas with density $(N/V)_{\text{wall}}$:

$$P = kT \left(\frac{N}{V}\right)_{\text{wall}}$$

$$= kT \frac{N}{V} e^{-NW_0/2VkT}$$

$$\approx kT \frac{N}{V}\left(1 - \frac{NW_0}{2VkT}\right)$$

$$= kT \frac{N}{V} - \frac{1}{2} W_0 \left(\frac{N}{V}\right)^2 \tag{6.3.4}$$

We define

$$a = \frac{W_0}{2} = -\frac{1}{2}\int u(r)\, d^3r \tag{6.3.5}$$

In this way, we see that at low densities the attractions between atoms of our fluid produce a departure from ideality of the form

$$P = \frac{NkT}{V} - a\left(\frac{N}{V}\right)^2 \tag{6.3.6}$$

We have yet to put in the effects of the repulsive, hard-core part of $u(r_{ij})$, but let us stop to notice that Eq. (6.3.6) is reminiscent of the second-order virial equation

$$PV = NkT\left(1 + \frac{N}{V} B\right) \tag{6.3.7}$$

where we wrote, in Eq. (4.4.17),

$$B = b - \frac{a}{kT}$$

Comparing Eq. (6.3.5) to (4.4.17), we see a is basically the same quantity defined there, and b is due to the hard core, whose effect we now wish to include. Furthermore, the same correction term that appears in Eq. (6.3.6) may be found in (6.3.7). Let us proceed from Eq. (6.3.7). Rewrite it as

$$V = \frac{NkT}{P} + NB = \frac{NkT}{P} + N\left(b - \frac{a}{kT}\right)$$

where, as usual, we have substituted the ideal gas formula into the correction term. Continuing to do so, we have

$$P(V - Nb) = NkT - \frac{aPN}{kT} = NkT - a\frac{N^2}{V}$$

If $Nb \ll V$, we may write

$$P(V - Nb) + a\frac{N^2}{V} \approx P(V - Nb) + a\left(\frac{N}{V}\right)^2 (V - Nb)$$

or, finally,

$$\left[P + a\left(\frac{N}{V}\right)^2\right](V - Nb) = NkT \qquad (6.3.1)$$

Considerable sleight of hand is evident here, and what is more, we shall be most interested in the behavior of this equation just where the low-density approximation made in its derivation makes no sense. For example, at the critical point, we shall see that $V = 3Nb$, contrary to the assumption $Nb \ll V$. Obviously the value of the van der Waals equation rests principally on its empirical behavior rather than its theoretical foundation.

In Fig. 6.3.1 we have a sketch of a number of isotherms (P versus V at constant T and N) according to Eq. (6.3.1), one of them labeled T_c. The curves above T_c are monotonic, but those below develop a blip, which includes a region of positive slope, $(\partial P/\partial V)_T > 0$. As discussed in Sec. 1.2f, such behavior is forbidden. It is unstable and therefore nonphysical. The system must find some other way to get from the right side of the diagram to the left, instead of following the curve. It does so by means of a discontinuity in specific volume or in density; in short, there is a first-order phase transition.

This is the gas-liquid phase transition, condensation, or evaporation. As we ride from right to left along one of the $T < T_c$ curves—in the gas phase, but increasing the density and pressure—we eventually reach the vapor pressure of the gas. At that point the gas condenses at constant pressure; as we continue to decrease the volume, the system departs from the van der Waals curve and moves horizontally across to its new state—that is,

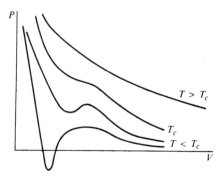

Fig. 6.3.1

6.3 The van der Waals Equation of State

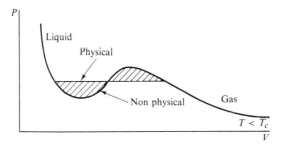

Fig. 6.3.2

at constant pressure equal to the vapor pressure as shown in Fig. 6.3.2. The system does not actually reach the instability itself but starts to condense instead at a lower pressure. The curve has no special feature to identify this pressure, but we can figure out where it is. We wish to identify the pressure at which the gas and liquid are in equilibrium, and we know that the condition for equilibrium is, in addition to equality of P and T, equality of the chemical potential μ in the two phases. At each point on the van der Waals curve (even the nonphysical part), the system has some chemical potential, so if we integrate along the curve from gas to liquid,

$$\mu_l - \mu_g = \int_g^l d\mu = \int_g^l \frac{V}{N} dP = 0$$

This means that the two shaded areas in Fig. 6.3.2 must be equal. This method of identifying the vapor pressure on any isotherm is called the Maxwell equal area construction. Once the vapor pressure is identified for each isotherm, a two-phase region on the P-V plot and a vapor pressure curve on a P-T plot can be drawn, and the result looks just like the gas-liquid portions of Figs. 1.2.5 and 1.2.6 of Sec. 1.2g.

Before going on to discuss the critical point itself, which is our objective, let us make some comments about the behavior at lower temperature. In this region the transition is not of the critical type but is first order instead.

It is clear from Eq. (6.3.1) that, for V/N large enough, the correction terms become unimportant and the van der Waals equation reduces to the ideal gas equation. At low temperatures the coexistence region spreads out, so that the gas may condense at large V/N, where it is very nearly ideal. This behavior does, in fact, occur; vapors in equilibrium with a condensed state well below T_c are often quite accurately ideal. Thus, it is not uncommon to have in equilibrium, at the same temperature and pressure, two states of the same matter—one self-condensed and the other virtually noninteracting. This peculiar situation reflects the fact that in spite of the attractive forces between them, atoms have little tendency to form small clusters: pairs, triplets, and so on. It is either all or nothing, condensed matter or ideal gas. We mentioned this point previously, in connection with the electron-pairing interaction in superconductivity, where it was pointed out that particles with weakly attractive interactions do not normally form bound pairs.

Also connected to this point is the existence of supercooled vapors—homogeneous phases that ought to condense but do not. The $T < T_c$ isotherms of Figs. 6.3.1 and 6.3.2 actually have two different kinds of instability: the part with positive slope, which is, so to speak, dynamically, explosively unstable and also that part inside the coexistence region but with negative slope. Figure 6.3.3 shows the various parts, with this one labeled metastable. The metastable part of the curve lacks stability in the thermodynamic sense; at the given temperature and volume there exists a state of lower free energy. As in the case of the perfect conductor shielding out an applied magnetic field, Sec. 1.2e, the existence of a thermodynamically preferable state does not ensure that there is a way to reach it. The main obstacle to condensation should be apparent from the last paragraph: two- or three-particle collisions do not lead to condensation; many particles must accidentally get together in order to discover that they are better off that way.

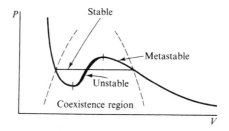

Fig. 6.3.3

In ordinary circumstances dust particles and other impurities that serve as nucleation centers are present, bringing together enough atoms to discover the advantages of the liquid state, so that condensation quickly ensues when the coexistence region is reached. Ionized atoms of the pure system also serve as nucleation centers, and the cloud chamber for the detection of cosmic rays and other high-energy particles is based on this principle. (All of this is also true if we enter the coexistence region from the other side, forming a superheated liquid. Thus, the principle of the bubble chamber.)

However, in a carefully controlled experiment it is possible to push a system along the metastable portion of the curve. An interesting question is: Can the true unstable part ever be reached? The answer is probably no; one cannot get very far into the metastable curve. In a pure, metastable system,

6.3 The van der Waals Equation of State

nucleation is a matter of time. If we wait long enough, random motions will lead to the formation of a cluster big enough to initiate condensation. As we move in along the metastable curve, the free energy advantage of condensation increases, the necessary cluster size decreases, and the waiting time decreases as well. Eventually it becomes too short, shorter than thermal equilibrium requires, for example, or shorter than the experimental time for changing external conditions, and the system can be pushed no further.

Let us now return to the critical point as it appears in the van der Waals equation. By examining the isotherms in Fig. 6.3.1, we can see that the isotherm at T_c is the one that is neither monotonic nor possessed of the blip mentioned; rather, it simply has a point of inflection. We thus have, at the critical point,

$$\left(\frac{\partial P}{\partial V}\right)_T = 0 \tag{6.3.8}$$

and
$$\left(\frac{\partial^2 P}{\partial V^2}\right)_T = 0 \tag{6.3.9}$$

These two conditions together with the equation itself, (6.3.1), give us three equations for the three unknowns, T_c, P_c, and V_c, the critical temperature, pressure, and volume, in terms of the parameters a and b.

Fortunately, there is an algebraically simpler way of obtaining these results. We start by rewriting the equation as a cubic in V. Taking $Nk = R$ and $N = N_0$, Avagadro's number, so that we are dealing with 1 mole, we get

$$V^3 - \left(N_0 b + \frac{RT}{P}\right)V^2 + \frac{aN_0^2}{P}V - \frac{abN_0^3}{P} = 0 \tag{6.3.10}$$

On each isotherm, at any P, there should be three solutions for V. Looking again at Fig. 6.3.1, we see that, for P close to what turns out to be the vapor pressure, the three solutions are real below T_c, merge together as we approach T_c, and two of the three become imaginary (for any pressure) above T_c. The critical point is just the point where all three solutions are the same:

$$(V - V_c)^3 = V^3 - 3V_c V^2 + 3V_c^2 V - V_c^3 = 0 \tag{6.3.11}$$

We now equate the coefficients of each power of V in Eqs. (6.3.10) and (6.3.11) and find

$$V_c = 3N_0 b, \quad P_c = \frac{a}{27b^2}, \quad RT_c = \frac{8aN_0}{27b} \tag{6.3.12}$$

It seems strange at first that numbers such as 8 and 27 show up in the coefficients of what is supposed to be a fundamental law, but these numbers are typically associated with a cubic equation: 2^3 and 3^3. In any case, if the values of a and b are known or measured (they depend on which particular substance we are studying), then we can predict the critical values of the variables through (6.3.12), as well as the coexistence curve and so on.

Later on we shall discuss in some detail the ways in which the van der Waals theory fails, but first let us discuss its successes; it was a very important theoretical advance. As we have already seen, when taken together with Maxwell's contribution of the equal area construction, it provides a complete qualitative description of the liquid-gas system. Quantitively, as we shall see, it has its shortcomings, but even then it is good enough to have played an important role in the first liquefaction of hydrogen and helium, the latter process being finally accomplished by Kammerlingh-Onnes at Leiden in 1908.

The theory first appeared in 1873 (it was van der Waals' doctoral thesis). Four years later, in 1877, fine sprays of liquefied nitrogen and oxygen were first produced, almost simultaneously, in Geneva and Paris, by Pictet and Cailetet, respectively. By 1883 Olzweski in Poland had been able to collect a few cubic centimeters of liquefied nitrogen and oxygen, enough to use as a refrigerant to study other substances, thus opening a vast new field of investigation. Of the permanent gases only hydrogen remained, and a race developed between Sir James Dewar, at the Royal Institution in London, and Onnes. As always, intense personal and nationalistic pride were involved in this competition. Dewar could claim as predecessor an earlier occupant of the Royal Institution professorship, Michael Faraday, who had started the enterprise of liquefaction of gases and production of low temperatures when he accidentally liquefied chlorine in 1823 (see Sec. 4.4c) and immediately grasped the possibilities presented by his discovery. Onnes, on the other hand, pointed out that it had been van Marum, a Dutchman, who had first knowingly, if accidentally, liquefied a gas in the course of testing Boyle's law with ammonia, still during the eighteenth century.

In any case, by the 1890s it was clear that Dewar was going to win, partly because of his invention of the double-walled vacuum flask that bears his name. Then in 1895 the rules suddenly changed when the British chemist, Ramsay, announced the discovery of terrestrial helium (it had been known from its spectral lines in the sun, from which it derives its name). Onnes dropped his stubborn refusal to use the Dewar flask and took out after helium. Liquid hydrogen was produced by Dewar in 1896.

The attempt to liquefy hydrogen and helium presented qualitatively new kinds of problems, quite aside from the more extreme requirements of thermal insulation and the like. Oxygen, for example, could be liquefied by a sort of cascade process: carbon dioxide liquefies under pressure at ordinary temperatures ($T_c \sim 304°K$); upon release of the pressure, it cools to a temperature where the next gas in the cascade may be liquefied under compression, and so on, until the vicinity of the oxygen critical temperature (155°K) is reached. Even so, it was important to know the critical parameters in advance, in order to decide which gases to use in the cascade, as

6.3 The van der Waals Equation of State

well as which pressures. In 1880 Amagat published P-V isotherms of a number of gases, including nitrogen, oxygen, ethylene, and hydrogen; and by 1882 Sarran had analyzed these data by means of the van der Waals equation, in order to find a and b for these gases and thus predict where liquefaction would occur. Olzweski, making use of these results, introduced ethylene into the cascade below carbon dioxide and proceeded to produce useful quantities of cryogenic liquids. He was able to reach about 125°K with the ethelyne. Expecting the critical parameters of oxygen to be about 160°K at 50 atm, he applied between 20 and 30 atm, then induced a sudden expansion, producing liquid oxygen at the boiling point.

However, there was a huge gap between the temperatures now accessible (70 to 80°K) and the expected critical point of the next gas, hydrogen (33° at 13 atm, which turned out to be quite accurate), and no intervening fluids to use in a cascade. The liquefaction would have to be done by a torturous process of Joule-Thomson expansion, heat exchange with newly compressed gas, reexpansion, and so on. Moreover, phrases such as "temperatures now accessible" cover a good deal of grief; Onnes set out to master Olzweski's cascade process for producing liquid nitrogen in 1884, and it took him ten years to do so.

Onnes had to measure his own P-V isotherms for the new gas, helium, at the temperature of liquid hydrogen, which he did once sufficient quantities of helium gas and of liquid hydrogen were accessible (in the sense discussed above). What he discovered was that a gap similar to that between nitrogen and hydrogen existed between hydrogen and helium ($T_c \sim 5°K$).

On July 10, 1908, at 5:45 A.M., Onnes and his students began liquefying the 20 liters of hydrogen they expected to need. It was not until 6:35 in the evening that the temperature in the circulating helium system first fell below that of the hydrogen. Soon afterward the last storage bottle of liquid hydrogen was attached to the system. It was noted that the expansion now produced warming instead of cooling, so they tried lowering the circulating helium pressure from 100 to 75 atm. At this point the thermometers in the reservoir intended for liquid helium became stable below 5°K, but there was no liquid to be seen, and the hydrogen was nearly gone.

It turned out, however, that there was no liquid to be seen only because the interface had managed to hide itself behind the thermometers. "The surface of the liquid," wrote Onnes, "was soon made clearly visible by the reflection of light from below, and that unmistakeably because it was clearly pierced by the two wires of the thermoelement. This was at 7:30 P.M. After the surface had once been seen, it was no more lost sight of. It stood out sharply defined like the edge of a knife against the glass wall."† Liquid helium was accessible.

---

† H. Kammerlingh-Onnes, "The Liquefaction of Helium" in Communications from the Physical Laboratory at the University of Leiden (Leiden, 1908) No. 108, p. 18.

All this labor required considerable faith in the belief that gas-liquid systems were all basically the same, even if no one had ever seen the liquid phase. This faith arose out of the repeated success of the van der Waals theory, which is essentially a universal equation of state, independent of the details of any particular substance once it has been properly scaled. If we define dimensionless quantities,

$$t = \frac{T}{T_c}, \quad p = \frac{P}{P_c}, \quad v = \frac{V}{V_c} \tag{6.3.13}$$

then Eq. (6.3.1) may be written

$$\left(p + \frac{3}{v^2}\right)(3v - 1) = 8t \tag{6.3.14}$$

In other words, when scaled to their own critical values, all fluids have the same equation of state. Even when the critical temperature, say, of hydrogen had never been obtained in a terrestrial laboratory, one could observe departures from ideal gas behavior that looked just like the departures observed in more condensable gases at suitably higher temperatures. As a result, not only was it possible to believe that hydrogen could be liquefied, but it was even possible to predict the necessary temperature and pressure. This basic similarity of all gas-liquid systems is known as the *Law of Corresponding States*.

Just as we did for the Weiss theory, we would like to make some comparison between the predictions of the van der Waals equation and the behavior of real systems for details in the immediate vicinity of the critical point. To begin, we notice that since the critical values of the three thermodynamic variables are given in terms of the parameters a and b in Eqs. (6.3.12), there must also be an equation relating them to each other if we eliminate a and b. The desired equation is

$$\frac{P_c V_c}{T_c} = \frac{(a/27b^2)(3N_0 b)}{8N_0 a/27b} = \frac{3}{8} = 0.375$$

It allows us to test the magnitude of the critical values, in contrast to the Weiss law, where T_c depended on the parameter λ.

If we form this combination for various substances, we find that it varies between about 0.23 (for water) to about 0.31 (for He^4). In general, the lighter, less-condensable gases have values closer to the van der Waals value, thus helping to explain its predictive utility in liquefying such gases. In any case, there is an unmistakable discrepancy between the van der Waals equation and real systems. We have already noted, in Sec. 4.5, that the Law of Corresponding States is not perfectly obeyed by real fluids.

If we now wish to continue, guided by our discussion in the last section,

6.4 Analogies Between Phase Transitions

we should note that the compressibility, $K_T = -(1/V)(\partial V/\partial P)_T$, becomes infinite at the critical point and that, according to Eq. (1.3.172), fluctuations in the density diverge as well. The result is that our picture of a single atom interacting with a smoothed-out average of all the other atoms becomes untenable, a fact that will help to account for the failure of the van der Waals equation in the critical region. Obviously K_T plays the same role here that X_T plays in the magnetic system, and the density is behaving like the magnetization. We could then investigate the temperature dependence with which K_T blows up in the van der Waals theory and find that it, like X_T in the Weiss theory, departs from the behavior of real systems.

We shall do so, but let us defer the details until later. At this point it is more interesting to pursue the analogy between the fluid and magnetic systems, so that we may compare the two theories to each other, as well as compare the systems to nature. The result of doing so will be to develop a generalized point of view, from which we shall be able to discuss many different kinds of phase transitions in the same language, at the same time.

6.4 ANALOGIES BETWEEN PHASE TRANSITIONS

At first the task of drawing analogies between magnetic and fluid phase transitions seems perplexing. We have already implied that the magnetization in the former case is somehow to be compared to the density in the latter, but the two quantities behave in very different ways. The spontaneous magnetization is simply zero above T_c, whereas it has a single magnitude that depends only on temperature below T_c. The density, on the other hand, is always finite, but, below T_c it breaks into two values that differ in magnitude. How are we to describe them in a way that makes them sound the same? We shall not be able to draw an exact analogy, and the differences will sometimes occur in annoying ways. However, we can do it well enough to be very useful.

Imagine, on the one hand, a magnetic sample, with no applied field ($H = 0$). For simplicity, let us suppose that there is one easy axis of magnetization—that is, some preferred axis along which the spontaneous magnetization forms (recall the anisotropy energy of Sec. 5.4). Then we can refer to the magnetization M as being either up or down, meaning in one direction or the other along the easy axis. For comparison, we consider a fluid system in a sealed container of constant volume, which we take equal to the critical value V_c, so that we could just as well say that the average density is equal to ρ_c, the critical density.

Above T_c, both systems are homogeneous, with $M = 0$ and $\rho = \rho_c$ everywhere. Let us compare, therefore, the behavior of the variables M and $(\rho - \rho_c)$, both of which are zero above T_c. Below T_c, the magnetic system develops a local, nonzero magnetization at each point; but as we discussed in Sec. 5.4, in equilibrium, it forms domains of opposing magnetization in order not to produce any large-scale magnetic fields, and so the average magnetization is still zero. The fluid system forms local regions of positive and negative $(\rho - \rho_c)$, but since V and N are fixed, the average value of $(\rho - \rho_c)$ is still zero.

Let us refer jointly to these quantities, M and $(\rho - \rho_c)$, as the *order parameter*. Anything proportional to them, such as m of Eq. (6.2.5) or $(1 - v)$ of Eq. (6.3.13), will do equally well $[(1 - v) = (\rho - \rho_c)/\rho \approx (\rho - \rho_c)/\rho_c$ near the critical point]. Then we can describe both transitions in the same words as follows: Above T_c, the order parameter is zero. Below T_c, the order parameter is locally nonzero and has more than one equilibrium value, up and down for the magnetic case, positive and negative for the fluid case. Since both are equilibrium values, the free energy density cannot depend on the sign or direction of the order parameter. Notice that with this description it is easy for us to assign order parameters to the superfluid and superconducting states as well. In both cases, we can choose ψ, the condensate wave function. For each state, the amplitude depends only on temperature, but the phase is continuously variable; in other words, it can point in any direction. In all cases, the amplitude of the order parameter goes continuously to zero as T_c is approached from below.

To continue with the analogy, we must consider the influence of H in the magnetic system and of P in the fluid. H and P are each thermodynamically conjugate to the order parameter and have the property of breaking up the symmetry between the multiple equilibrium values below T_c. That is, an increase in P at fixed T below T_c will drive $(\rho - \rho_c)$ to its liquid value, whereas a decrease will drive it to its gas value. An H field pointed up or down will cause the sample to have a net magnetization in the same direction. When a quantity has these properties, we shall call it the *conjugate field*.

At this point we can see why the analogies between different phase transitions are necessarily imperfect. For instance, the magnetic and fluid examples we described are not quite the same in terms of generalized quantities. In the magnetic case, we described the behavior of the order parameter (M) as the temperature changed at fixed conjugate field $(H = 0)$. In the fluid case, we observed instead the behavior of the order parameter $(\rho - \rho_c)$ as we changed T with the average value of the order parameter fixed $(V = V_c)$, and P left free to change. This discrepancy is less serious than it might be, since, as we have argued, in the magnetic case $H = 0$ implies that the average value of M is zero, so there may not be any important difference between the two systems. However, if we think of extending the analogy to the superfluid and superconducting cases, we run into trouble, for in these states (and certain other critical transitions as well), there is apparently no physical quantity that behaves like a conjugate field.

6.4 Analogies Between Phase Transitions

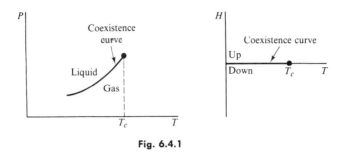

Fig. 6.4.1

Even if we restrict our attention to the magnetic and fluid cases, there are other differences. Speaking in terms of the generalized quantities, the different equilibrium values of the order parameter can coexist for any temperature below T_c at only one particular value of the conjugate field. In the magnetic case, this value is just $H = 0$; but for fluid systems, the appropriate value of P depends on temperature; it is just the vapor pressure, $P_0(T)$. Of course, we could fix this up, say, by defining the fluid conjugate field to be $P - P_0(T)$ below T_c and $P - P_c$ above. However, doing so would only cause difficulties elsewhere; for example, at a given temperature, below T_c, with zero conjugate field, the mean density would be indeterminate. Instead we shall make do with this imperfection in our analogy. It becomes significant in comparing the roles of certain thermodynamic derivatives. For example, C_H at $H = 0$ for the magnetic system is the heat capacity along the coexistence curve below T_c, and it is not at all comparable to C_P at $P = P_c$ in the fluid system. Some feeling for the behavior of the analog quantities may be obtained by studying Figs. 6.4.1 to 6.4.3.

Now that we know how to compare magnets and fluids, it is interesting to compare the two theories we have worked out: the Weiss theory and the van der Waals theory. What we are interested in is the detailed behavior

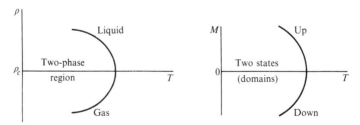

Fig. 6.4.2

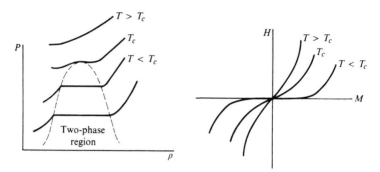

Fig. 6.4.3

of analogous quantities near the critical point. For example, in the Weiss theory, we saw that

$$X_T \propto (T - T_c)^{-1} \tag{6.4.1}$$

The susceptibility is the second derivative of the Gibbs potential with respect to the conjugate field at fixed T:

$$X_T = \left(\frac{\partial M}{\partial H}\right)_T = -\frac{1}{V}\left(\frac{\partial^2 \Phi}{\partial H^2}\right)_T \tag{6.4.2}$$

The analogous fluid quantity is

$$-\frac{1}{V}\left(\frac{\partial^2 \Phi}{\partial P^2}\right)_T = -\frac{1}{V}\left(\frac{\partial V}{\partial P}\right)_T = K_T \tag{6.4.3}$$

The isothermal susceptibility and compressibility are analogs.

In order to find the compressibility in the van der Waals equation, start from the reduced form, Eq. (6.3.14), solve for p, and expand in powers of the new variables,

$$\varepsilon = t - 1 = \frac{T - T_c}{T_c} \tag{6.4.4}$$

$$\theta = v - 1 = \frac{V - V_c}{V_c} \tag{6.4.5}$$

Notice that below T_c, θ will serve as the order parameter. We obtain (Prob. 6.7)

$$p = 1 + 4\varepsilon - 6\varepsilon\theta - \frac{3}{2}\theta^3 \tag{6.4.6}$$

Since $d\theta \propto dV$, we need

$$\left(\frac{\partial p}{\partial \theta}\right)_\varepsilon = -6\varepsilon - \frac{9}{2}\theta^2 \tag{6.4.7}$$

We evaluated X_T at $H = 0$; the proper (imperfect) analog is $\theta = 0$:

$$K_T \propto -\left(\frac{\partial V}{\partial P}\right)_{T,V_c} \propto -\left(\frac{\partial \theta}{\partial P}\right)_{\varepsilon, \theta=0} \propto \varepsilon^{-1} \tag{6.4.8}$$

Since Eq. (6.4.1) may be written $X_T \propto \varepsilon^{-1}$, X_T and K_T have the same dependence.

In the magnetic case, the reduced magnetization [Eq. (6.2.5)] can point in different directions below T_c, but it always has the same magnitude in both directions. Let us look at the magnitude of the two values of the order parameter for the van der Waals equation below T_c: using the Maxwell construction, we have

$$\int_1^2 v\, dp \propto \int_1^2 \theta\, dp = 0 \tag{6.4.9}$$

Substituting in $p(\theta)$ at constant ε from Eq. (6.4.6), we find

$$-\frac{9}{8}(\theta_2^4 - \theta_1^4) - 3\varepsilon(\theta_2^2 - \theta_1^2) = 0 \tag{6.4.10}$$

Since this is an identity for small ε, the physically acceptable solution is

$$\theta_1 = -\theta_2 \tag{6.4.11}$$

In other words, like the magnetic case, the two values of the order parameter have the same magnitude but different signs.

We can also find out how rapidly the order parameter vanishes when we approach T_c from below. From the van der Waals equation, with $\theta_1 = -\theta_2$ and $p_1 = p_2$ at any ε, Eq. (6.4.6) gives $\theta_2^2 = \theta_1^2 = -4\varepsilon$, or

$$\theta \propto (-\varepsilon)^{1/2} \tag{6.4.12}$$

The Weiss theory yields the same result (Prob. 6.10).

These basic similarities between the two theories suggest that we might be able to adopt a more general point of view, speaking of order parameters and conjugate fields, without specific reference to a particular system with a particular type of interaction between the constituents, and still obtain the same results. That is precisely what we shall do in the next section.

6.5 THE GENERALIZED THEORY

In this section we shall try to treat in a general way systems with an order parameter that goes continuously to zero at some critical temperature. We will develop a theory intended to be valid near T_c, which will turn out to include the Weiss and van der Waals theories in this region as special cases. This theory is principally due to Landau, although special cases of it have other names attached to them. The result will necessarily have the same shortcomings as the earlier theories, traceable primarily to the absence of correlated fluctuations. We will then try to include these fluctuations within the framework of the theory, and we will discuss the range of validity of the results.

Since we are going to try to work in general terms, let us define a generalized order parameter $m(\mathbf{r})$, which may vary from place to place in the system, and a generalized conjugate field h. The basic procedure is to expand $f(m, T)$, the free energy density, in powers of $m(\mathbf{r})$ with coefficients that depend on T. However, as we have seen, analogies between phase transitions are a bit dangerous, and, besides, it should be evident from earlier discussions that the free energy is not always the proper quantity to minimize. We shall therefore mention frequently how various details apply to particular systems, and we start by indicating, in Table 6.5.1, how we apply the theory in the four relevant systems described in this book. In each case, we are going to make a functional variation of the total free energy of the system, F, with respect to the internal variable, $m(\mathbf{r})$. For the magnetic and superconducting cases, Φ, the Gibbs potential, is the correct quantity to minimize at constant H, but since $H = 0$, $F = \Phi$ (see Sec. 1.2e for further discussion of this point with regard to superconductivity). The arguments of Sec. 1.2d show how to go from the third column to the fourth in the table.

Table 6.5.1

System	Order parameter, $m(\mathbf{r})$	Held constant	Minimize
Fluid	$\rho(\mathbf{r}) - \rho_c$	T, V	$F(T, V)$
Magnetic	$M(\mathbf{r})$	$T, H = 0$	$\Phi(T, H)$
Superconducting	$\psi(\mathbf{r})$	$T, H = 0$	$\Phi(T, H)$
Superfluid	$\psi(\mathbf{r})$	T, V	$F(T, V)$

The free energy of the system, in all cases, is given by

$$F = \int f(m, T) \, d^3r \tag{6.5.1}$$

where we are to understand that $m = m(\mathbf{r})$. The trick now is to identify all the lowest-order terms in the expansion of $f(m, T)$ in powers of $m(\mathbf{r})$ and $(T - T_c)$. We take account of the fact that the free energy density can depend not only on m itself but also on its derivatives. For example, in the fluid system below T_c, if we go from a liquid region to a gas region, we must pass through a region of changing density. This region is the interface, and

6.5 The Generalized Theory

it must have a surface tension, which is an increase in the free energy density above the value it would have in either of the homogeneous phases. Thus, a gradient in m, in this case—and, in fact, quite generally—will increase f. The leading-order terms in powers of m and its derivatives for $f(m, T)$ in Eq. (6.5.1) are

$$f(m, T) = f_0(T) + \alpha(T)m^2 + \tfrac{1}{2}\beta(T)m^4 + \gamma(T)|\nabla m|^2 \quad (6.5.2)$$

Here $f_0(T)$ is the free energy density when $m(\mathbf{r}) = 0$. The odd-order terms, m, m^3, etc. are left out because, as we argued in the previous section, the free energy density cannot depend on the sign (or direction) of the order parameter. Also, in the case of the fluid system, we can argue that since $\int_V (\rho - \rho_c) d^3r = 0$, any odd-order term in f would drop out of Eq. (6.5.1) in any case. f cannot depend on ∇m by isotropy. It could depend on $\nabla^2 m$, but any such term is integrated over the volume when put into Eq. (6.5.1), whereupon, by means of the divergence theorem, just as we have used it in Secs. 5.2 and 5.3, it becomes a surface term, plus a term in $|\nabla m|^2$, which is included in the last term of Eq. (6.5.2). For a sufficiently large system, any surface contribution to F may be neglected, so we drop the surface term. This leaves us with Eq. (6.5.2).

a. Equilibrium Behavior

Now let us consider a homogeneous system, $m(\mathbf{r}) =$ constant. Then there are no gradients, and if m is complex (superconductivity and superfluidity), we can choose the phase to make it real. When we minimize F

$$\frac{\partial F}{\partial m} = V(2\alpha \overline{m} + 2\beta \overline{m}^3) = 0$$

The solutions are

$$\overline{m}^2 = \left(-\frac{\alpha}{\beta}, 0\right) \quad (6.5.3)$$

We want m to have a real, nonzero solution below T_c but not above T_c. This requirement will help us choose the leading-order terms in the expansions of $\alpha(T)$ and $\beta(T)$:

$$\alpha(T) = \alpha_0 + a_0(T - T_c) + \cdots \quad (6.5.4)$$

$$\beta(T) = b_0 + \beta_0(T - T_c) + \cdots \quad (6.5.5)$$

and, incidentally,

$$\gamma(T) = \gamma_0 + g_0(T - T_c) + \cdots \quad (6.5.6)$$

Now, from our conditions on Eq. (6.5.3), we want $-\alpha/\beta$ to be positive and finite below T_c, go to zero at T_c, and become negative above T_c. To accomplish this, we must have $\alpha_0 = 0$, $a_0 \neq 0$, and $b_0 \neq 0$. We retain only the leading-order term in each series (since such terms are always multiplied by powers of the small quantity m), and we get

$$f = f_0(T) + a_0(T - T_c)m^2 + \frac{b_0}{2} m^4 \qquad (6.5.7)$$

(for a homogeneous system, it does not matter whether we speak of f or F). In Fig. 6.5.1 we plot f versus m from Eq. (6.5.7). For $T > T_c$, f has a minimum at $m = 0$; but for $T < T_c$, there is a minimum at a finite value of m and a maximum at $m = 0$. These properties are shared with both the Weiss and van der Waals theories, for which the order parameter had stable solutions that were zero above T_c and nonzero below, with an unstable solution at zero below T_c.

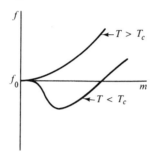

Fig. 6.5.1

Equation (6.5.3) gives us, for the nonzero solution below T_c,

$$\bar{m}^2 = -\frac{a_0(T - T_c)}{b_0} \qquad (6.5.8)$$

$\bar{m}$ will thus generally have two values of the same magnitude but opposite sign, and the magnitude has the temperature dependence

$$\bar{m} \propto (-\varepsilon)^{1/2} \qquad (6.5.9)$$

Once again these results are identical to those of the van der Waals and Weiss theories.

Although we have not formally proved it, from now on we shall take it as established that in the critical region this theory is a generalization of the Weiss and van der Waals theories, giving the same results. It has become conventional to refer to these theories together as the classical theories, not classical as distinct from quantum mechanical, but simply because classical is what we usually call the theory that came just before the one we believe now. This nomenclature helps to lend a certain legitimacy to the current theory, the scaling laws discussed later in Sec. 6.7.

6.5 The Generalized Theory

The van der Waals and Weiss theories produced equations of state, essentially $m(T, h)$, which made it easy to find the susceptibility, or the response of the order parameter to the conjugate field. The approach here gives us $f(T)$ instead, and that fact makes it easy to find the heat capacity. To do so, we substitute (6.5.8) into (6.5.7), obtaining

$$f = f_0(T) - \frac{a_0^2(T - T_c)^2}{2b_0} \qquad (6.5.10)$$

Then the specific entropy is

$$s = -\left(\frac{\partial f}{\partial T}\right) = s_0(T) + \frac{a_0^2(T - T_c)}{b_0} \qquad (6.5.11)$$

Above T_c, the order parameter is zero, and the system has entropy $s_0(T)$. Below T_c, the second term arises; it is negative, meaning that the order parameter reduces the entropy as it should. But it starts from zero at T_c so that the entropy is continuous; there is no latent heat. The specific heat is given by

$$c = T\left(\frac{\partial s}{\partial T}\right) = c_0 + \frac{a_0^2}{b_0} T \qquad (6.5.12)$$

The specific heat thus has a finite discontinuity, $\Delta c = (a_0^2/b_0)T_c$ at T_c. Notice that the relevant heat capacity is C_V for the fluid and superfluid systems, but C_H with $H = 0$ in the other two.

Of the four cases we are considering, the fluid, superfluid, and magnetic cases are all found, in practice, to have infinite heat capacities at the critical point. Consequently, this last result is apparently another instance of the failure of the classical theories. However, as we have already seen, there is a finite jump in the heat capacity at the superconducting transition, so it is possible that the classical theory is more nearly applicable for this one. Let us pause for a moment and consider some details of this application.

For superconductivity, the classical theory is called the Ginsburg-Landau theory, and it does, in fact, work exceedingly well. Although it was proposed long before there was any microscopic understanding of superconductivity, and in a fundamental sense has been superseded by the BCS theory, it is, in fact, still the principal working theory of superconductivity. The reason is that it is much simpler and easier to use than BCS, and it correctly predicts how superconductors will behave under a wide variety of circumstances. Its utility derives largely from applications to superconductors in strange geometries, including surfaces, carrying supercurrents, in magnetic fields, and so on. However, we shall be content to see how it works in regard to the phase transition, for a large, homogeneous sample, in zero field.

Ginsburg and Landau started as we have, writing

$$f = f_0 + \alpha|\psi|^2 + \frac{\beta}{2}|\psi|^4 + \gamma|\nabla\psi|^2 \qquad (6.5.13)$$

where we have reverted to the more suggestive notation, ψ, for the order parameter. α and β are chosen just as we have done it. γ, as we have argued, does not go through zero or change sign, since it is always energetically expensive to change the order parameter, so the constant term in Eq. (6.5.6) is to be chosen. In fact, Ginsburg and Landau made the inspired guess

$$\gamma = \frac{\hbar^2}{2m^*} \qquad (6.5.14)$$

where m^* was originally taken to be the electron mass, but we now take it to be the mass of an electron pair. It is then required that

$$\frac{\partial F}{\partial \psi^*} = \frac{\partial}{\partial \psi^*} \int f(\mathbf{r}) \, d^3r = 0$$

and the divergence theorem is again applied, to obtain the form

$$-\frac{\hbar^2}{2m^*}\nabla^2\psi + \beta|\psi|^2\psi = -\alpha\psi \qquad (6.5.15)$$

which, of course, looks beguilingly like the Schrödinger equation, with the order parameter playing the role of a wave function. Not quite, however. This is actually a nonlinear, second-order differential equation; the term playing the role of the potential, $\beta|\psi|^2$, depends on ψ, the solution.

Since α and β are the same as they were in the more general treatment, the heat capacity discontinuity is still given by Eq. (6.5.12). In addition, the density of superconducting pairs, n_s, is just the square of the order parameter, which has the temperature dependence given in Eq. (6.5.8),

$$n_s \propto |\psi|^2 = \frac{a_0}{b_0}(T_c - T) \qquad (6.5.16)$$

This dependence near T_c agrees both with the BCS theory and with the behavior of real superconductors. With appropriate normalization, a measurement of $n_s(T)$ and one of Δc_H will suffice to give a_0 and b_0.

There is one more quantity to consider, one that will become increasingly important to us. We can rewrite Eq. (6.5.15) as

$$\left[-\frac{\hbar^2}{2m^*\alpha}\nabla^2 + \frac{\beta}{\alpha}|\psi|^2 + 1\right]\psi = 0$$

Obviously $-\hbar^2/2m\alpha$ has the dimensions of a length squared, and this length

6.5 The Generalized Theory

is a measure of how quickly ψ can change in space. This quantity is just what we called ξ, the coherence length, in Sec. 5.3, for example, Eq. (5.3.113):

$$\xi = \left(\frac{\gamma}{\alpha}\right)^{1/2} = \frac{\hbar v_F}{\Delta} = \left(-\frac{\hbar^2}{2m^*\alpha}\right)^{1/2} \propto (T_c - T)^{-1/2} \qquad (6.5.17)$$

This relation tells us, also correctly, that Δ, the energy gap, has the same temperature dependence as ψ,

$$\Delta \propto (T_c - T)^{1/2} \qquad (6.5.18)$$

Notice that in the more general theory the coherence length is simply

$$\xi = \left(\frac{\gamma}{\alpha}\right)^{1/2} \qquad (6.5.19)$$

The meaning of ξ in the other systems will become more evident a little later.

All this discussion makes remarkably good sense for superconductivity, but as we have already indicated, it does not do so well for any of the other systems, not even for superconductivity's sister state, superfluidity. Our problem is perplexing; we want to know not only why this theory does not work, but also why it sometimes does.

b. Fluctuations

We have mentioned a number of times that the difficulties in the classical theories lie in their neglect of the effects of correlated fluctuations very close to the critical point. We shall now look into this problem, working out a way to incorporate fluctuations. The results will not be adequate to give agreement with experiment where there was previous disagreement, but they will help us to judge the validity of the classical theories, to see why those ideas work better for superconductors than other states, and to understand, at least qualitatively, the phenomenon of critical opalescence. Critical opalescence is the strong scattering on light observed in fluids close to the critical point.

In order to gain some insight into the situation, we shall now make a peculiar kind of argument. We shall argue that if the Landau generalized theory is correct for each reasonably small part of the system, it cannot be correct for the system as a whole. Reasonably small is a vague phrase, but here it can have only one possible meaning, since there is only one length in the problem, the coherence length, $\xi = (\gamma/\alpha)^{1/2}$. Just as in the superconducting case, and by the same argument, ξ is the characteristic distance required by the order parameter to change its value. Therefore, if we choose parts of the system that are smaller than ξ, the order parameter will always have a fairly definite value in each part. This is what we mean by reasonably small.

Let us consider a particular reasonably small subsystem. Given the temperature (and conjugate field), the free energy, both of the system and of the subsystem, will be a minimum if the order parameter of the subsystem (remember, it has a fairly definite value) is

$$\bar{m} = \left[\frac{a_0(T_c - T)}{b_0}\right]^{1/2} \quad (T < T_c) \tag{6.5.20}$$

$$\bar{m} = 0 \quad (T > T_c) \tag{6.5.21}$$

However, m is a thermodynamic variable and actually takes on various values, each with some definite probability; that is, it fluctuates. $\bar{m}$ is the most probable value, to be sure, but it can have other values, $m \neq \bar{m}$, with probability

$$w(m - \bar{m}) \propto e^{-[f(m)-f(\bar{m})]/kT} \tag{6.5.22}$$

We shall formalize this result in Eq. (6.5.28) below.

The local free energy density, f, depends on m, as shown in Fig. 6.5.1; in all cases,

$$f(m) \geq f(\bar{m}) \tag{6.5.23}$$

Remember, now, these are not fluctuations out of equilibrium but rather the equilibrium state is an average over these fluctuations, weighted by the probability distribution, Eq. (6.5.22). Thus, the subsystem, since it always fluctuates into states with free energy at least as big as $f(\bar{m})$, must have an average free energy density larger than $f(\bar{m})$. But the free energy of the whole system, according to the Landau theory, is basically a sum of all the $f(\bar{m})$'s of the subsystems, so it follows that the Landau theory must always underestimate the real free energy, with the discrepancy being due to fluctuations in the order parameter.

The real question is: How important is all this? After all, the effect is mitigated somewhat by the $\gamma|\nabla m|^2$ term in Eq. (6.5.2), which increases the free energy cost and therefore decreases the probability of nearby subsystems having different order parameters; in a sense, this term tries to lock the subsystems together and prevent independent fluctuations. In fact, the fluctuations are not very important unless we get very close to T_c.

Consider the free energy density, Eq. (6.5.2), very close to T_c. Since $\bar{m}$ is either zero (above T_c) or very small (below), and $\alpha(T)$ is close to zero, the cost of a fluctuation in m gets to be very small, and, according to Eq. (6.5.22), the probability of the fluctuation becomes correspondingly large. There is, of course, the expense of changing m to worry about, the $\gamma|\nabla m|^2$ term, but after our experience with spin waves and phonons in earlier chapters, we know exactly how to handle that: we construct slowly changing, long wavelength, highly correlated fluctuations. These fluctuations become increasingly probable (in quasiparticle language, increasingly easy to excite) as T_c is approached, and they always have the effect of raising the free energy

6.5 The Generalized Theory

above the Landau value. We thus conclude that if the Landau theory is correct on a scale smaller than ξ, it cannot give the correct dependences for the free energy of the system as a whole.

Nevertheless, this problem cannot always be very important; the theory does, after all, work for superconductivity. Furthermore, if the problem in the other cases is the one we have outlined, we should be able to fix things up by investigating the fluctuations. Let us try to do so.

The fluctuations are most easily discussed in terms of the correlation functions, with which we became familiar in Chap. 4. Of particular interest to us now is $G(r)$, which is related, in the fluid case, to the other functions we used then by

$$\rho G(r) = \rho\, \delta(r) + \rho^2 h(r)$$

$$= \rho \int S(q) e^{-i\mathbf{q}\cdot\mathbf{r}} \frac{d^3 q}{(2\pi)^3}$$

$$= \langle \rho(0)\rho(r) \rangle - \rho^2 \tag{6.5.24}$$

[To within an additive constant, ρ, this is the same quantity defined in Eq. (5.2.116).] We shall return shortly and work out the behavior of the correlated fluctuations near the critical point, making use of the language we developed in Chap. 4 to deal with liquids. However, let us first deduce the structure factor from the Landau formalism we are using here. The order parameter $m(\mathbf{r})$ replaced $\rho(r) - \rho$, and for the sake of our general arguments, let us replace $\rho G(r)$ by $m_0 G(r)$, where m_0 is an appropriate low-temperature order parameter—that is, the saturation magnetization, the critical density, ρ_c, and so on. We have, then,

$$m_0 G(r) = \langle m(r) m(0) \rangle - \overline{m}^2 = \langle [m(r) - \overline{m}][m(0) - \overline{m}] \rangle \tag{6.5.25}$$

Multiply both sides of (6.5.25) by $V e^{i\mathbf{q}\cdot\mathbf{r}}$ and integrate over $d^3 r$. The Fourier transform of the left-hand side of this equation is $m_0 S(q)$, where $S(q)$ is the same structure factor we used before. For the right-hand side we may write

$$\int \langle [m(\mathbf{r}_1) - \overline{m}][m(\mathbf{r}_2) - \overline{m}] \rangle e^{i\mathbf{q}\cdot(\mathbf{r}_2 - \mathbf{r}_1)}\, d^3 r_1\, d^3 r_2$$

$$= \left\langle \left| \int [m(\mathbf{r}) - \overline{m}] e^{i\mathbf{q}\cdot\mathbf{r}}\, d^3 r \right|^2 \right\rangle \equiv V^2 \langle |m_q|^2 \rangle$$

or

$$\langle |m_q|^2 \rangle = \frac{m_0 S(q)}{V} \tag{6.5.26}$$

The arguments used here should be familiar from Chap. 4. The only tricky point is that, since we are concerned here only with fluctuations in the amplitude of m, we have taken m to be real; that is, $m_q^* = m_{-q}$. m_q is defined in Eq. (6.5.26). We now calculate $\langle |m_q|^2 \rangle$ from the Landau theory.

To do so, we shall expect that the probability of having a given value of m_q is given by

$$w(m_q) \propto \exp\left(-\frac{|m_q|^2}{2\langle |m_q|^2 \rangle}\right) \tag{6.5.27}$$

which is a form of Eq. (1.3.55) of Chap. 1. This Gaussian distribution is to be deduced from the fluctuations in the free energy

$$w[m(\mathbf{r})] \propto \exp\left\{-\frac{(F[m(\mathbf{r})] - F(\overline{m}))}{kT}\right\} \tag{6.5.28}$$

Equation (6.5.28) differs from (6.5.22) in that since we now know that the fluctuations will be correlated over long distances, we are no longer dividing the system into subsystems; F is the total free energy. Equation (6.5.28) may be derived as follows.

For each type of system it may be seen that its fluctuations are statistically independent of fluctuations in T. Suppose that the sample is to be attached to a medium that acts purely as a temperature bath. When $m(\mathbf{r})$ fluctuates, the medium, as always, remains in equilibrium. Thus, for the medium,

$$\delta E' = T \, \delta S'$$

while for our sample

$$\delta E = T \, \delta S + \delta F[m(\mathbf{r})]$$

since T is constant and $E = F + TS$ quite generally defines the free energy. For the combined medium and sample,

$$\delta S_t = \delta S + \delta S' = -\frac{\delta F}{T}$$

where we have used $\delta E_t = \delta(E + E') = 0$. When this equation for δS_t is substituted into Eq. (1.3.52), we arrive at Eq. (6.5.28).

In writing the Gaussian distribution, Eq. (6.5.27), we have already assumed that the fluctuations will be small (see the arguments of Sec. 1.3b), so we may as well rewrite Eq. (6.5.2) as

$$f(m, T) = f(\overline{m}, T) + [a_0(T - T_c) + 3b_0\overline{m}^2](m - \overline{m})^2 + \gamma|\nabla m|^2$$

$$= f(\overline{m}, T) + a_1(m - \overline{m})^2 + \gamma|\nabla m|^2 \tag{6.5.29}$$

where we have expanded $f(m) - f(\overline{m})$ in powers of $(m - \overline{m})$. The coefficient a_1 will have different values above and below T_c, depending on whether $\overline{m}^2$ is zero or $a_0(T_c - T)/b_0$. We now take

$$m - \overline{m} = \sum_q m_q e^{-i\mathbf{q} \cdot \mathbf{r}} \tag{6.5.30}$$

6.5 The Generalized Theory

Substitute (6.5.30) into (6.5.29) and integrate over volume, to get

$$F(m) - F(\bar{m}) = V \sum_{q=0}^{\infty} (a_1 + \gamma q^2)|m_q|^2 \tag{6.5.31}$$

Comparing this result with Eqs. (6.5.28) and (6.5.27), we have

$$\langle |m_q|^2 \rangle = \frac{kT}{2V(a_1 + \gamma q^2)} \tag{6.5.32}$$

It is possible now to reinvert this relation and get the behavior of $G(r)$, but before we do so, it is worthwhile to consider the meaning of this result in the language we developed to study liquids in Chap. 4. We shall also rederive Eq. (6.5.32), using different arguments.

The quantity $\langle |m_q|^2 \rangle$ is basically the structure factor, according to Eq. (6.5.26). Recalling that we have defined a_1 by

$$a_1 = a_0(T - T_c) + 3b_0\bar{m}^2 \propto (T - T_c) \tag{6.5.33}$$

we see that at the critical point

$$S(q) \propto \frac{1}{q^2} \tag{6.5.34}$$

a result valid for small q, so that $S(0) \to \infty$. Let us consider what this means for the fluid system.

In Sec. 1.3f, where we discussed thermodynamic fluctuations, we found for the fluctuations of the volume of a fixed number of particles, Eq. (1.3.172)

$$\frac{\overline{(\Delta V)^2}}{V^2} = \frac{kT}{V}\left(-\frac{1}{V}\frac{\partial V}{\partial P}\right) = \frac{kT}{V} K_T \tag{6.5.35}$$

or if the volume is fixed and the number fluctuates,

$$\frac{\overline{(\Delta N)^2}}{N^2} = \frac{kT}{N}\frac{\partial \rho}{\partial P} = \frac{kT}{N} \rho K_T \tag{6.5.36}$$

the same result. Thus, at the critical point, where $K_T \to \infty$, these fluctuations diverge. As we pointed out in Sec. 6.2, the same is true of fluctuations in the magnetization near the ferromagnetic critical point. Notice that in the derivation of Eq. (1.3.172) we did not assume that the fluctuations were small. The same argument that gave Eq. (1.3.172) as an exact result also led in Chap. 4 to Eq. (4.2.58), which may be written

$$kT\left(\frac{\partial \rho}{\partial P}\right)_T = \rho \int h(r)\, d^3r + 1 \tag{6.5.37}$$

Here the integral on the right-hand side, although formally taken over an infinite volume, usually gives a finite result because $h(r)$, the total correlation function, quickly falls to zero with increasing r. Clearly, at the critical point, $h(r)$ becomes long ranged, which is just another way of saying that the fluctuations become highly correlated. The connection with what we have been doing becomes clearer when we note that

$$S(q) = 1 + \rho \int h(r) e^{i\mathbf{q}\cdot\mathbf{r}} d^3r \qquad (6.5.38)$$

or
$$S(0) = 1 + \rho \int h(r) d^3r = kT \left(\frac{\partial \rho}{\partial P}\right) \qquad (6.5.39)$$

which is merely Eq. (4.2.64). The divergence of $S(0)$, the long-range correlation of the fluctuations in density, and the divergence of the generalized susceptibility (in this case, K_T) are all the same phenomenon.

It is worth remembering that $S(q)$, besides being mathematically related to the density fluctuations, is also proportional to the intensity of scattered light (Sec. 4.2a). Light is strongly scattered near the critical point; essentially, when the range of the correlated density fluctuations becomes of order of the wavelength of the light, the medium turns opaque. This phenomenon is known as critical opalescence. The results being discussed here were actually obtained long before the Landau theory originated. The theory was worked out by Ornstein and Zernike, around 1917, in order to explain critical opalescence in the framework of the understanding of the liquid gas transition that had been obtained from the van der Waals theory. During the course of this work, they first introduced the idea of the direct correlation function $c(r)$, which we encountered in Chap. 4 in discussing the liquid state. It may help to understand the connection between the dense fluids discussed there and the critical fluids discussed here if we rederive Eq. (6.5.32) from the liquid-state formalism of the earlier chapter.

We defined $c(r)$ to be the sum of all nonseries diagrams of the cluster integrals and then found it was related to $h(r)$ by

$$h(r) = c(r) + \rho \int c(|\mathbf{r} - \mathbf{r}'|) h(r') dr' \qquad (6.5.40)$$

with Fourier transforms related by

$$\tilde{c}(q) = \frac{\tilde{h}(q)}{1 + \rho \tilde{h}(q)} \qquad (6.5.41)$$

[Equations (6.5.40) and (6.5.41) arise from (4.5.73) and (4.5.94).] Ornstein and Zernike originally defined $c(r)$ by means of Eq. (6.5.40), and we can now see why they did so. Close to T_c, we know that $\tilde{h}(q) \to \infty$ as $q \to 0$. But from Eq. (6.5.41) this simply means that $\tilde{c}(0) \to 1/\rho$ at T_c. Thus, at the critical point, when everything else is blowing up at low wave numbers, $\tilde{c}(q)$

6.5 The Generalized Theory

is placidly finite; we have at least one correlation function that minds its own business. It was for this property that $c(r)$ was first introduced.

Since $\tilde{c}(q)$ remains well behaved at T_c, we can try to expand it in a Taylor series about $q = 0$:

$$\tilde{c}(q) = \tilde{c}(0) + \eta q^2 + O(q^4) \quad (6.5.42)$$

We cut off the series because we are interested only in the long-range, low q behavior. The odd-order terms drop out for the same reason that terms like ∇m were omitted from the Landau theory, because the system is isotropic. For example, the coefficient of the linear term is

$$\frac{\partial}{\partial q} \tilde{c}(q) = \left[\frac{\partial}{\partial q} \int c(r) e^{i\mathbf{q}\cdot\mathbf{r}} d^3r \right]_{q=0}$$

$$= i \int (r \cos \theta) c(r) d^3r$$

$$\propto i \int_{-1}^{1} \mu \, d\mu = 0$$

where $\mu = \cos \theta$. The even-order terms stay because, in the next order for example, the μ integral is $\int_{-1}^{1} \mu^2 \, d\mu \neq 0$. Now we have Eq. (6.5.38),

$$S(q) = 1 + \rho \tilde{h}(q) \quad (6.5.43)$$

but

$$1 - \rho \tilde{c}(q) = 1 - \frac{\rho \tilde{h}(q)}{1 + \rho \tilde{h}(q)} = \frac{1}{1 + \rho \tilde{h}(q)} = \frac{1}{S(q)}$$

so that

$$S(q) = \frac{1}{1 - \rho[\tilde{c}(0) + \eta q^2]} \quad (6.5.44)$$

At T_c, $\tilde{c}(0) = 1/\rho$, and so $S(q) \propto 1/q^2$ just as before; Eq. (6.5.44) is basically the same as Eq. (6.5.32) except that, in this case, our arguments have not given us the temperature dependence of the coefficients, $\tilde{c}(0)$ and η.

We now revert back to the Landau notation. From Eqs. (6.5.26) and (6.5.32),

$$S(q) = \frac{kT}{2m_0} \frac{1}{a_1 + \gamma q^2} \quad (6.5.45)$$

We can now see how the correlation function, $G(r)$, behaves by taking an inverse Fourier transform

$$G(r) = \int S(q) e^{-i\mathbf{q}\cdot\mathbf{r}} \frac{d^3q}{(2\pi)^3}$$

$$= \frac{kT}{2m_0 \gamma} \int \frac{e^{-i\mathbf{q}\cdot\mathbf{r}}}{(a_1/\gamma) + q^2} \frac{d^3q}{(2\pi)^3}$$

The necessary integral is†

$$\int \frac{e^{-i\mathbf{q}\cdot\mathbf{r}}}{\xi^{-2}+q^2} \frac{d^3q}{(2\pi)^3} = \frac{1}{4\pi} \frac{e^{-r/\xi}}{r} \quad (6.5.46)$$

where

$$\xi = \left(\frac{\gamma}{a_1}\right)^{1/2} \quad (6.5.47)$$

Finally, for the correlation function, we obtain

$$G(r) = \frac{kT}{8\pi m_0 \gamma} \frac{e^{-r/\xi}}{r} \quad (6.5.48)$$

Equation (6.5.48) means that, near the critical point, the correlations fall off in a basically exponential way, with a characteristic length, which we can get by using Eq. (6.5.33).

$$\xi = \left(\frac{\gamma}{a_1}\right)^{1/2} = \begin{cases} \dfrac{\gamma^{1/2}}{[a_0(T-T_c)]^{1/2}} & (T > T_c) \\[2ex] \dfrac{\gamma^{1/2}}{[2a_0(T_c-T)]^{1/2}} & (T < T_c) \end{cases} \quad (6.5.49)$$

The correlation function $G(r)$ is sketched in Fig. 6.5.2. We see that the coherence length is playing basically the same role that it did in superconductivity; essentially, it measures how much space the order parameter needs in order to change substantially and, in particular, how far we must

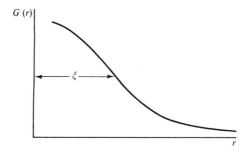

Fig. 6.5.2

† To derive this transform, note that $y = e^{-kr}/r$ is a solution of $\nabla^2 y - k^2 y = -4\pi\, \delta(r)$; multiply both sides of this by $\exp(-i\mathbf{q}\cdot\mathbf{r})$ and integrate by parts to give

$$\int y \exp(-i\mathbf{q}\cdot\mathbf{r})\, d^3r = \frac{4\pi}{k^2+q^2}$$

6.5 The Generalized Theory

go before a fluctuation dies out. We also see that at the critical point ξ becomes infinite. The fluctuations are very strongly correlated, and

$$G(r) \propto \frac{1}{r} \quad \text{at } T_c \tag{6.5.50}$$

Naturally $h(r)$ has this same dependence, and so the compressibility, following Eq. (6.5.37), depends on the volume of the system and becomes infinite in the thermodynamic limit. We shall refer to the ideas we have worked out here as the Ornstein-Zernike theory.

Let us summarize briefly what we have done. If a sample had a homogeneous order parameter, the Landau theory would give us its most probable value, $\bar{m}$. The theory can also be used to calculate the probability that the order parameter changes from place to place, provided that it does not change too rapidly. We have now computed the spectrum of the fluctuations that occur in this slow way and found that they are correlated over a range that grows to infinity as the critical point is approached. Thus, a fluid will have increasingly large patches of high and low density; a magnetic-system, patches of up and down spin.

We have yet to test the assumption that the order parameter does not change too rapidly. The function $G(r)$ tells us by how much the fluctuations of the order parameter differ, on the average, a distance r apart. Since we have assumed, in effect, that it does not change much over a distance ξ, what we want is that $G(\xi)$ be small compared to $\bar{m}$ itself. More precisely,

$$m_0 G(\xi) = \langle [m(\xi) - \bar{m}][m(0) - \bar{m}] \rangle \ll \bar{m}^2 \tag{6.5.51}$$

If this condition were not true, for example, we could not cut off the expansion of the free energy density, Eq. (6.5.2), at the $|\nabla m|^2$ derivative, nor, equivalently, could we cut off the expansion of $c(r)$ in Eq. (6.5.42) at the q^2 term. Large values of q would be necessary to tell us about behavior at shorter distances, and it is not clear that an expansion in powers of q (or derivatives of m) would be helpful at all. Substituting Eqs. (6.5.8) for $\bar{m}^2$ and (6.5.48) for $G(\xi)$ into (6.5.51), we find

$$\frac{kT}{8\pi\gamma} \frac{e^{-1}}{\xi} \ll \frac{a_0(T_c - T)}{b_0} \tag{6.5.52}$$

Taking $T = T_c$ on the left-hand side, and using $\xi = [\gamma/2a_0(T_c - T)]^{1/2}$, we can rearrange terms to get

$$\frac{\sqrt{2} b_0 k T_c^{1/2}}{8\pi e a_0^{1/2} \gamma^{3/2}} \ll \left(\frac{T_c - T}{T_c}\right)^{1/2} \ll 1 \tag{6.5.53}$$

The second inequality comes from our initial assumption—we started out by expanding the free energy about the critical point.

What Eq. (6.5.52) or (6.5.53) gives us, then, is at best a window, a range of temperatures not too close to T_c, but not too far away either, in which the classical theories are applicable (the shrewd reader will notice that we have quietly started including Ornstein-Zernike in the classical theories, an unmistakable clue that it will be superseded before the chapter concludes). In order to decide whether, for each type of transition, an applicable range of temperatures exists, it will be necessary each time to find ways to evaluate the constants a_0, b_0, and γ.

Evaluation of these quantities may take some ingenuity. What we need, basically, are the heat capacity discontinuity, Eq. (6.5.12), the temperature dependence of the order parameter, Eq. (6.5.8), and the coherence length, Eq. (6.5.49). But the theory is generally not obeyed very close to T_c, so we must find sensible ways to estimate these quantities a little away from T_c. What would the heat capacity discontinuity be if the heat capacity were finite? Actually, finite discontinuities are found even though the heat capacity itself becomes infinite. For liquid helium near the lambda transition, for example, it is found approximately that

$$C = \begin{cases} A \log (T - T_c) + B & (T > T_c) \\ A' \log (T_c - T) + B' & (T < T_c) \end{cases}$$

with $A \approx A'$. Then we can take $\Delta C = B' - B$. The coherence length may be estimated in a number of ways. There are scattering data: measurements of $S(q)$. According to Eq. (6.5.45), a plot of $1/S(q)$ versus q^2 (called an Ornstein-Zernike-Debye plot) will give a_1 and γ from its slope and intercept. Lacking that, there are various indirect estimates, such as the width of a magnetic domain wall or the core radius of a quantized vortex line in superfluidity.

We shall leave to the reader the pleasure of finding appropriate kinds of data and making the estimates for the various systems. One point, however, can be cleared up immediately—the reason the theory works so well for superconductivity. The lower limit of applicability of the theory goes as $\gamma^{-3/2}$. Using Eq. (6.5.49) at $T = 0$, we have, for the zero-temperature coherence length, $\xi_0 = \gamma^{1/2}/(2a_0 T_c)^{1/2}$. Roughly, then, the lower limit of Eq. (6.5.53) goes as ξ_0^{-3}. Of all the systems we have studied, ξ_0 is by far the largest in superconductivity; here it is nearly macroscopic—thousands of angstroms in some cases. By contrast, it is about 1 angstrom (the core radius of a vortex line) in liquid helium. In superconductivity, then, the range of applicability is very wide, the lower limit being closer to T_c than we will probably ever reach experimentally. In superfluidity, there may be no range of applicability at all. Magnetic systems tend to be intermediate, with the theories working, both according to Eq. (6.5.53) and in real life, over a few decades of $(T_c - T)/T$. This argument explains why the classical theory seems to work for superconductivity, but not for the other cases.

6.6 CRITICAL POINT EXPONENTS

We have now developed some very general ways of thinking about critical phase transitions and of describing them. As we have seen, the various systems are not quite identical to each other, and yet, from an appropriate point of view, we can, in fact, work on them together. The purpose of this section is to focus attention on questions sufficiently general so that the answers will allow us to compare the various phase transitions to each other and to the theories, without undue disturbance from the details of differences between them. We made a start in the last two sections, but there we were primarily interested in making the classical theories seem universal. Our job now is to include real phenomena in our discussion.

The universal property of critical phase transitions, theoretical or otherwise, is that various quantities become either zero or infinite. The question we shall find ourselves asking for the rest of this chapter is, essentially, how fast does this process occur? Actually, we have already begun asking this question. For example, in Sec. 6.4 we found that in both the Weiss and van der Waals theories, the order parameter

$$m \propto (T_c - T)^{1/2} \tag{6.6.1}$$

while the generalized susceptibility

$$X \propto (T - T_c)^{-1} \tag{6.6.2}$$

However, as we saw in Sec. 6.2, real magnetic systems cease to obey Eq. (6.6.2) near T_c (see Fig. 6.2.3), and what we really want now is a quantitative way to express that kind of disagreement.

As we did in Sec. 6.4, let us define

$$\varepsilon = \frac{T - T_c}{T_c} \tag{6.6.3}$$

Now we can write

$$X \sim \varepsilon^{-\gamma} \tag{6.6.4}$$

which, informally, defines γ and the notation $\sim$. We shall do this more formally a bit later. Equation (6.6.4) is to be read "The susceptibility goes as ε to the minus γ." γ is an example of a critical point exponent—that is, an indication of how fast something becomes singular (either zero or infinite). We can now say: for the classical theory, $\gamma = 1$, but, experimentally, one typically finds γ between about 1.1 and 1.4. The situation is thus expressed in a way that is independent of annoying details.

One of the quantities that diverges at the critical point is the number of critical point exponents it is possible to define. Miraculously, however, the notation for the principal ones is reasonably standard—the exponent we have defined to be γ is always given that name—and there are even rules about which ones are worth defining. For example, if we hold $T = T_c$ ($\varepsilon = 0$, the critical isotherm) and reduce the conjugate field, h, then the order parameter goes to zero. We define δ by

$$m \sim h^{1/\delta} \qquad (6.6.5)$$

Now, the susceptibility diverges as $h \to 0$, so we could define, say e, by

$$X \sim h^{-e}$$

But then we notice that

$$X = \frac{\partial m}{\partial h} \sim \frac{\partial}{\partial h}(h^{1/\delta}) \sim h^{1/\delta - 1} \qquad (6.6.6)$$

so that

$$e = 1 - \frac{1}{\delta}$$

The definition of e would thus be redundant and is not done. The critical point exponents, by design, should never be related directly by thermodynamics; they should be independent.

On the other hand, we can sometimes use thermodynamic arguments to put limits on the behavior of certain exponents. For example, we know that the susceptibility is to be infinite at $\varepsilon = h = 0$, which means that the exponent of h in Eq. (6.6.6) must be negative, so that $(1/\delta) - 1 \leq 0$. In other words,

$$\delta \geq 1 \qquad (6.6.7)$$

This constraint is not too restrictive (δ is usually found to be between 3 and 5), but it is an example of a thermodynamic inequality for critical point exponents. There are others, as we shall mention below.

Other conventions regarding the exponents exist. They are defined to be positive, which is why we write $\varepsilon^{-\gamma}$ in Eq. (6.6.4), so that $\gamma \geq 0$. Also, if a given quantity becomes singular when approaching from both sides of T_c, then two exponents are defined, with the one below T_c being primed:

$$\begin{aligned} X &\sim \varepsilon^{-\gamma} & (T > T_c) \\ X &\sim (-\varepsilon)^{-\gamma'} & (T < T_c) \end{aligned} \qquad (6.6.8)$$

Of course, certain exponents are meaningful only on one side of T_c. For the order parameter, we define

$$m \sim (-\varepsilon)^\beta \qquad (6.6.9)$$

and there is no corresponding definition to be made above T_c. It would be considered bad form to speak of β', even though it is a below-T_c exponent.

Before proceeding, we need a formal definition of exactly what we mean by a critical point exponent. If we write

$$f(\varepsilon) \sim \varepsilon^\lambda \qquad (6.6.10)$$

6.6 Critical Point Exponents

we mean

$$\lambda \equiv \lim_{\varepsilon \to 0} \frac{\log f(\varepsilon)}{\log (\varepsilon)} \qquad (6.6.11)$$

Thus, λ is the critical point exponent even if $f(\varepsilon)$ is really given by

$$f(\varepsilon) = \varepsilon^\lambda (1 + A\varepsilon^y + \cdots) \qquad (y \geq 0) \qquad (6.6.12)$$

More to the point, it is obvious from Fig. 6.2.3 of Sec. 6.2 that $X \propto \varepsilon^{-1}$ over some range of ε for real systems, but since only the limit $\varepsilon \to 0$ is involved in Eq. (6.6.11), we cannot conclude that $\gamma = 1$. Also, the formal definition helps us in certain special cases. For example, suppose that $f(\varepsilon) \sim \log \varepsilon$, as in the heat capacity of liquid helium, mentioned above. Then substitution into (6.6.11) gives $\lambda = 0$. A zero exponent can mean a finite discontinuity, or a cusplike singularity, or a logarithmic infinity.

We are now prepared to define some of the standard exponents. The ones we shall be most interested in appear in Table 6.6.1.

Certain of these exponents require special comment. Notice that α and α' are not defined in analogous ways for the fluid and magnetic cases. For the magnetic transition we take C_H—that is, we keep the conjugate field

Table 6.6.1

Exponent	Fluid	Magnetic	Classical value	Comment
α'	$C_V \sim (-\varepsilon)^{-\alpha'}$	$C_H \sim (-\varepsilon)^{-\alpha'}$	0	Below T_c
α	$C_V \sim \varepsilon^{-\alpha}$	$C_H \sim \varepsilon^{-\alpha}$	0	Above T_c
β	$\rho_L - \rho_g \sim (-\varepsilon)^\beta$	$M \sim (-\varepsilon)^\beta$	$\tfrac{1}{2}$	Below T_c only
γ'	$K_T \sim (-\varepsilon)^{-\gamma'}$	$X_T \sim (-\varepsilon)^{-\gamma'}$	1	Below T_c
γ	$K_T \sim \varepsilon^{-\gamma}$	$X_T \sim \varepsilon^{-\gamma}$	1	Above T_c
δ	$P - P_c \sim \|\rho - \rho_c\|^\delta \times \text{sgn}\,(\rho - \rho_c)$	$H \sim \|M\|^\delta \,\text{sgn}\, M$	3	Critical isotherm
ν'	$\xi \sim (-\varepsilon)^{-\nu'}$	Same	$\tfrac{1}{2}$	Below ⎫ (classical values
ν	$\xi \sim \varepsilon^{-\nu}$	Same	$\tfrac{1}{2}$	Above ⎭ from Ornstein–Zernike)
η	$G(r) \sim \|r\|^{-(d-2+\eta)}$	Same	0	d = dimensionality
		Special cases		
μ	$\sigma \sim (-\varepsilon)^\mu$			Surface tension
ζ	$\rho_s \sim (-\varepsilon)^\zeta$			Superfluid density

fixed—whereas in the fluid case, we consider C_V. This difference is an obvious consequence of the real difference between the two kinds of transition discussed in Sec. 6.4, and it should be kept in mind. Notice also that v and v' cannot be deduced from thermodynamic quantities, such as the free energy or the equation of state, since they depend directly on the fluctuations. Finally, η is of a different nature from the others and requires separate explanation.

According to the Ornstein-Zernike theory, we found that, at T_c, $G(r) \sim r^{-1}$. This computation was done, naturally, in three dimensions. Had we taken the integral over only two dimensions in Eq. (6.5.46), we would have found $G(r) \sim \log r$ at the critical point, a physically absurd result, for it would mean that the correlations grow rather than fall off with distance. Here is another instance of the special importance of fluctuations in two dimensions, related to the fact that there is no Bose condensation, and there can be no crystals in two dimensions (see Prob. 6.3). The nonphysical result suggests that in two dimensions, and perhaps in three as well, something other than Ornstein-Zernike is needed to describe the fluctuations. The two-dimensional Ornstein-Zernike result can be written $G \sim r^{-0}$, in three dimensions $G \sim r^{-1}$, and, quite generally, $G \sim r^{-(d-2)}$. We insert η as indicated in the table in order to generalize the result properly.

We are now in a position to ask, from the point of view of the critical point exponents: Is the Ornstein-Zernike theory simply an extension of the other classical theories, allowing us to obtain v, v', and η, for example, or does it actually give us different results for the other exponents? Using Eqs. (6.5.24), (6.5.39), and (6.5.48), and keeping only the singular parts, we find for the fluid case

$$K_T \sim \frac{\partial \rho}{\partial P} \sim \int G(r)\, d^3r$$

$$\sim \int \frac{e^{-r/\xi}}{r} r^2\, dr$$

$$\sim \xi^2 \int e^{-x} x\, dx$$

$$\sim \xi^2$$

But since

$$\xi \sim \varepsilon^{-1/2}$$

we have

$$K_T \sim \varepsilon^{-1}$$

or $\gamma = 1$, the same result as the other classical theories. However, it is possible for the Ornstein-Zernike theory to give rise to critical exponents different from those of the classical theories without fluctuations. (See Prob. 6.13.)

6.6 Critical Point Exponents

As we pointed out above, there are no direct thermodynamic relations between the critical point exponents, but there are relations in the form of inequalities, which are generally consequences of fundamental thermodynamic inequalities. To get the idea, you are asked in Prob. 6.14 to prove that

$$\alpha' + 2\beta + \gamma' \geq 2 \qquad (6.6.13)$$

This is a consequence of the thermodynamic relation, Eq. (1.2.111),

$$C_H - C_M = T \frac{[(\partial M/\partial T)_H]^2}{X_T} \qquad (6.6.14)$$

which, together with

$$C_M \geq 0 \qquad (6.6.15)$$

gives

$$C_H \geq \frac{T[(\partial M/\partial T)_H]^2}{X_T} \qquad (6.6.16)$$

Equation (6.6.13), which follows from (6.6.16), is called the Rushbrooke inequality. Notice that the comparable fluid relation

$$C_P \geq \frac{T}{V} \frac{[(\partial V/\partial T)_P]^2}{K_T} \qquad (6.6.17)$$

does not lead to the same inequality, although, of course, the inequality may still be obeyed.

Needless to say, we would not have gone to all this trouble to define critical point exponents if we were going to conclude that the classical theory was right after all or that it could be adequately fixed up by means of the Ornstein-Zernike theory. The fact is that, for each kind of phase transition, there is a critical region close to the critical point where the classical theories fail: specifically, the critical point exponents are found to differ from their classical values. We can expect the critical region to be found where the inequalities of Eqs. (6.5.52) and (6.5.53) of the last section fail. There are, of course, very grave experimental difficulties associated with measuring the critical point exponents, especially since one can never be sure that, say, ε is small enough to satisfy Eq. (6.6.11). We shall return to a discussion of how experiments are to be evaluated in Sec. 6.8. Nevertheless, overwhelming evidence that real systems depart from the classical predictions exists. For example, as we have noted before, the classical theories tend to give $\gamma = 1$, whereas γ is commonly found, in magnetic systems, to be about 1.3. We have also already mentioned discrepancies between classical predictions and observations for various systems in the heat capacity. For ν, $\frac{1}{2}$ is predicted, whereas ~ 0.7 is commonly found. For β, $\frac{1}{2}$ is again the prediction, but 0.3 is closer to the typical experimental result, and so on. In the next section we shall work out arguments that although they do not lead us to understand why the exponents are what they are, at least give us reason to believe that they may be related to each other.

6.7 SCALING LAWS: PHYSICS DISCOVERS DIMENSIONAL ANALYSIS

The failure of the classical theories in the immediate vicinity of the critical point still leaves us with the task of trying to organize and characterize the behavior of matter at these phase transitions. To review again what we know, these are all transitions at which the energy functions and entropy are continuous and at which one of the phases has an order parameter whose amplitude goes continuously to zero. It is also characteristic that, below T_c, the order parameter has more than one equilibrium value: up or down magnetization, gas or liquid density, continuously variable phase in superconductivity.

The way in which the classical theories fail may be helpful in guiding us to a new approach. We tried expanding the free energy density in powers of the order parameter and temperature around the critical point. There is only one way to do so, and the result is the Landau theory. The trouble with this procedure is that it ignores local fluctuations in the order parameter, and these fluctuations become increasingly likely near the critical point, since they cost very little in free energy. When we tried to include fluctuations, we had to assume that they did not change too rapidly, and the result, basically the Ornstein-Zernike theory, was still not satisfactory. All of this suggests (possibly out of desperation) that if we are to reach the core of the matter, we should ignore the mean value of the order parameter altogether and concentrate entirely on the fluctuations.

As the transition is approached, the range over which random fluctuations are correlated grows longer and longer, finally becoming infinite at the transition itself. Instead of trying to expand the free energy density in powers of the order parameter whose average value goes to zero, perhaps we can make progress by expanding in inverse powers of the average range of the correlations, the quantity we have called ξ. This quantity has the clear advantage of not varying from place to place in the system, and, of course, it focuses attention on the fluctuations, which we now know to be most important.

Before we start writing equations, let us try to state a clear hypothesis. We choose the simplest one possible under the circumstances:

Nothing matters except ξ

What we mean by this statement is that in the immediate vicinity of the phase transition, the system itself knows only what ξ is, not how it was arrived at. "I have," it says to itself, "a correlation length of 1.2 cm," and that is all it knows about its state so long as it is homogeneous. In Fig. 6.7.1 we have a generalized phase diagram. The abscissa to the left of the origin is the coexistence curve; phases 1 and 2 could be, for example, magnetization up and down or density liquidlike and gaslike. h could be the magnetic field, or $(P - P_c)$. The origin is the critical point, ε the temperature parameter. We

6.7 Scaling Laws: Physics Discovers Dimensional Analysis

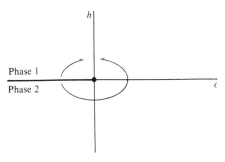

Fig. 6.7.1

have drawn a hypothetical contour of points with $\xi = 1.2$ cm. The idea is that the system has no way of knowing where on that contour it is. It follows immediately that it cannot tell whether ε is positive or negative; all critical point exponents defined separately above and below T_c must be the same: $\alpha = \alpha', \gamma = \gamma', \nu = \nu'$. The hypothesis makes no sense unless these relations are correct.

Now let us try to expand an appropriate energy function. We choose the Gibbs potential, Φ, since its independent variables, ε and h, are the most easily manipulated (it will turn out not to make any difference which energy function we choose). In order to have a dimensionless expansion parameter, we choose a subsystem with linear dimension L and expand in powers of (L/ξ):

$$\Phi - \Phi_0 = a \left(\frac{L}{\xi}\right)^n + \beta_1 \left(\frac{L}{\xi}\right)^{2n} + \cdots$$

Only the leading-order term will be of interest, since we want the limit when $\xi \to \infty$. The Φ_0 we have subtracted off is the uninteresting, nonsingular part.

What, then, is n, the power of the leading-order term? If we choose $n = 3$, then in leading order, the Gibbs potential density

$$\varphi = \frac{\Phi - \Phi_0}{\text{volume}} = \frac{\Phi - \Phi_0}{L^3}$$

will be independent of L, as it should be. We thus make this choice, or, more generally, $n = d$, where d is the dimensionality of the system, and write†

$$\varphi \sim \xi^{-d}$$

---

† For the benefit of the experts, this hypothesis is equivalent to what is known as "strong scaling." We could have made it weaker by not specifying the power of ξ.

Now we must know how fast ξ blows up as the critical point is approached from various directions—that is, the shape and scale of the contour in Fig. 6.7.1. Our hypothesis tells us nothing about them, so we assume that if we approach at $h = 0$,

$$\xi \sim |\varepsilon|^{-x} \qquad (h = 0)$$

or at $\varepsilon = 0$,

$$\xi \sim |h|^{-y} \qquad (\varepsilon = 0)$$

These are merely definitions of x and y. They will remain the fundamental unknowns of scaling theory. We are now prepared to derive the scaling laws, which will give us all the critical point exponents in terms of x and y, or of each other, but always with two unknowns.

Comparing to our earlier definitions, we see that

$$x = \begin{cases} v' & (T < T_c) \\ v & (T > T_c) \end{cases}$$

or, in other words,

$$v = v'$$

which is already an experimentally verifiable prediction.

To get α and α', we note that

$$C \sim \frac{\partial^2 \varphi}{\partial \varepsilon^2} \sim \varphi \varepsilon^{-2} \sim \xi^{-d} \varepsilon^{-2}$$

$$\sim (|\varepsilon|^{-x})^{-d} \varepsilon^{-2} \sim |\varepsilon|^{dx-2}$$

Since $C \sim \varepsilon^{-\alpha}$ and $C \sim (-\varepsilon)^{-\alpha'}$, we have

$$-\alpha = -\alpha' = dx - 2$$

or $\qquad dv = dv' = 2 - \alpha = 2 - \alpha'$

(three equations in four unknowns, v, v', α, α'; y has not yet entered).

Now let us get β, defined by $m \sim (-\varepsilon)^{\beta}$, where m is the order parameter. We have

$$m \sim \frac{\partial \varphi}{\partial h} \sim \varphi h^{-1} \sim \xi^{-d} h^{-1}$$

This gives us m as we vary h, with $\varepsilon = 0$, whereas what we want is m as we vary ε with $h = 0$. But the point is that the system does not know what we are manipulating. We really change ξ by changing h^{-y}, but that is exactly the same as changing it by varying ε^{-x}; the quantities $h \sim \xi^{-1/y} \sim |\varepsilon|^{x/y}$ are all the same:

$$m \sim \xi^{-d}(-\varepsilon)^{-x/y} \sim (|\varepsilon|^{-x})^{-d}(-\varepsilon)^{-x/y} \sim (-\varepsilon)^{dx-x/y}$$

6.7 Scaling Laws: Physics Discovers Dimensional Analysis

or
$$\beta = x\left(d - \frac{1}{y}\right)$$

We also have δ, defined by
$$h \sim m^\delta$$

and
$$m \sim \frac{\partial \varphi}{\partial h} \sim \varphi h^{-1} \sim \xi^{-d} h^{-1} \sim (h^{-y})^{-d} h^{-1} \sim h^{dy-1}$$

or
$$m \sim h^{1/\delta} \sim h^{dy-1}$$

$$\frac{1}{\delta} = dy - 1$$

And, furthermore, γ' defined by
$$X_T \sim (-\varepsilon)^{-\gamma'}$$

$$\sim \frac{\partial^2 \varphi}{\partial h^2} \sim \varphi h^{-2}$$

$$\sim h^{dy-2}$$

$$\sim [(-\varepsilon)^{x/y}]^{dy-2}$$

$$\sim (-\varepsilon)^{dx-2/y}$$

$$\gamma' = dv' - \frac{2x}{y} = \gamma$$

(the last without further ado).

Eliminating y and x from all equations, the results to this point may be summarized as

$$2 - \alpha = 2 - \alpha' = dv = dv' = \gamma + 2\beta = \gamma' + 2\beta = \beta(\delta + 1)$$

What we are doing here is a kind of dimensional analysis except that only the dimensions of those quantities that go to zero or infinity come into play. Thus, the heat capacity, $C = T(\partial S/\partial T)$, normally has the same units as entropy, but, for these purposes, the T out in front of $\partial S/\partial T$ does not count, since it remains finite, and $C \sim S\varepsilon^{-1}$. The singular part of any energy density function goes away like ξ^{-d}, and ξ is the only length in the problem. Everything may then be reduced to powers of, say, ξ: $\varepsilon \sim \xi^{-1/x}$, $h \sim \xi^{-1/y}$, $S \sim \varepsilon^{1-\alpha} \sim \xi^{(\alpha-1)/x}$, and so on. These rules, carefully applied, give us all the results of the scaling hypothesis.

Let us illustrate the point by deriving three more scaling laws, now using purely dimensional arguments.

We defined the exponent η by
$$G(r) \sim r^{-(d-2+\eta)}$$

It follows that
$$G(\xi) \sim \xi^{-(d-2+\eta)}$$

but $G(\xi)$ will go to zero like an order parameter squared [see Eq. (6.5.25)] and $m \sim (-\varepsilon)^\beta$:

$$\xi^{-(d-2+\eta)} \sim (-\varepsilon)^{2\beta} \sim [(\xi^{-1/x})]^{2\beta}$$

or
$$d - 2 + \eta = \frac{2\beta}{x} = \frac{2\beta}{v'}$$

This is the scaling law for η; to put it into the string of equations, we write

$$\frac{d\gamma'}{2 - \eta} = dv' = \cdots \text{ etc.}$$

One of the most striking visual features of the gas-liquid critical point is that the interface between the phases vanishes. This fact means that one of the quantities which vanishes at the critical point must be σ, the surface tension, or free energy per unit interface area. As discussed earlier, there must be a positive σ associated with any interface. Accordingly, we define a new critical point exponent, μ:

$$\sigma \sim (-\varepsilon)^\mu$$

But σ has the dimensions of an energy per unit area, or an energy density times length

$$\sigma \sim \varphi\xi \sim \xi^{-d+1} \sim [(-\varepsilon)^{-x}]^{-(d-1)}$$

$$\mu = v'(d - 1)$$

which enters into the chain of equations as

$$\frac{d\mu}{d - 1} = dv' = \cdots \text{ etc.}$$

Finally, we treat an interesting special case, the superfluid fraction in liquid He^4. From our discussion of superfluidity (Sec. 5.2), it is easy to see why one is tempted to identify $(\rho_s)^{1/2}$ as the amplitude of the order parameter and write $\rho_s \sim (-\varepsilon)^{2\beta}$. In this way, β could be deduced from measurements (e.g., the speed of second sound). However, except for purposes of comparing the exponents of superfluidity to those of other phase transitions (a point we shall return to in the next section), to do so would be an academic exercise. The reason is that there is no h field in superfluidity; that is, there is no variable, thermodynamically conjugate to the order parameter, which can be used to break up the symmetry of its multiple equilibrium values. Consequently, the exponents γ, γ', and δ are meaningless, and it follows that there is no way of testing any scaling law involving β.

Nevertheless, an appropriate (if ad hoc) scaling argument can be made. We start by defining a new exponent

$$\rho_s \sim (-\varepsilon)^\zeta$$

6.7 Scaling Laws: Physics Discovers Dimensional Analysis

Now we recall from our discussion of superfluidity that ρ_s is really defined only in superflow; ρ_n is the momentum density per unit velocity in the superfluid frame, and ρ_s is defined to be the part left over. We therefore imagine some very slow motion and take the singular part of φ to be

$$\varphi \sim \xi^{-d} \sim \tfrac{1}{2}\rho_s u_s^2$$

Now, u_s, the superfluid velocity, is given by the gradient of the phase of the order parameter, so we write

$$u_s^2 \sim \frac{|\nabla \psi|^2}{|\psi|^2}$$

This quantity has the dimensions of an inverse length squared, and there is only one length in the problem

$$u_s^2 \sim \xi^{-2}$$

We have

$$\xi^{-d} \sim (-\varepsilon)^\zeta \xi^{-2}$$

or

$$d = \frac{\zeta}{v'} + 2$$

Since $d = 3$, $\zeta = v'$. For the lambda transition in liquid helium, the heat capacity is logarithmically infinite; that is, $\alpha' = 0$. Using $dv' = 2 - \alpha' = 2$, we have $v' = \tfrac{2}{3}$, or $\zeta = \tfrac{2}{3}$. Measurements of ρ_s along the vapor pressure curve have given $\rho_s = 0.667 \pm 0.006$.

These, then, are the static scaling laws. For each argument we have introduced one new exponent and produced one new equation, always leaving two unknowns. Many other exponents can be (and therefore have been) defined for special purposes and may usually be related to those we have by the same sort of arguments.

These laws are not fundamental and, in fact, may not be correct. Whether correct or not, they serve an important purpose: they give focus and direction to research on critical phase transitions, for they raise well-defined questions that can be dealt with experimentally. In particular, the following questions arise:

1. Do the scaling laws work? Are the equations we have given verified experimentally?
2. If the laws are (more or less) correct, are all phase transitions (more or less) the same? That is, do they have the same values of x and y? Are there definite characteristics (e.g., continuously variable or discretely variable order parameters) that distinguish different classes of phase transitions?

The answers to these questions are not necessarily simple. In the next section we shall try to evaluate and discuss this situation.

6.8 EVALUATION AND SUMMARY: TESTING THE SCALING LAWS

In testing the scaling laws of the previous section, it is necessary to measure the various critical point exponents in various different systems and see whether the predictions are verified. These measurements, as well as the application of their results to test the laws, involve a number of particular difficulties; in order to evaluate the results intelligently, we must be aware of what these difficulties are and how they operate. To get a feeling for the problems involved, let us consider in some detail how we would go about measuring a particular, if typical, exponent. We shall suppose that we are setting out to measure the heat capacity exponents, α and α', for the gas-liquid critical point in, say, argon.

Our general plan will be to measure the heat capacity at the critical density, C_V, as a function of T, then plot log C_V versus log $|(T - T_c)/T_c|$. The results should tend to straight lines for $|T - T_c|$ small, and the slopes will be equal to α and α'. Figure 6.8.1 is an example of what we might expect to find. The two curves are for $T \gtrless T_c$, and the scaling law $\alpha = \alpha'$ is verified if they both tend to the same slope at small $|(T - T_c)|$.

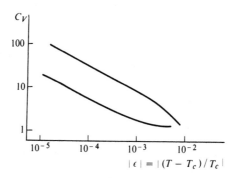

Fig. 6.8.1

The proper amount of argon is placed in a sealed calorimeter of fixed volume, cooled to an appropriate temperature, and isolated together with a heater (probably a coil of resistance wire) and the thermometer (say, a calibrated thermistor). The heat capacity is measured by putting a known amount of heat, ΔE, in via the heater and measuring the consequent temperature change, ΔT. Since the volume is fixed, $C_V = (\partial E/\partial T)_V \approx \Delta E/\Delta T$ if ΔE was small enough. These operations may involve difficulties but, so far, none that are peculiar to critical point measurements.

We can now begin to enumerate some of the special problems that will occur if we wish to put our data in the form of Fig. 6.8.1.

6.8 Evaluation and Summary: Testing the Scaling Laws 485

1. For an ordinary heat capacity measurement, a sufficient criterion that ΔE was small enough is that $\Delta T \ll T$. We actually measure

$$\frac{1}{\Delta T} \int_T^{T+\Delta T} C_V \, dT$$

that is, an average of C_V over the range ΔT. If $\Delta T \ll T$, C_V generally will not change very much in ΔT, and once we have (approximately) $C_V(T)$, corrections can be made for small changes of C_V in ΔT. However, in order to make the plot in Fig. 6.8.1, the region we average over must obviously be small compared to $|T - T_c|$; we have the much more restrictive condition

$$\frac{|\Delta T|}{T_c} \ll |\varepsilon| \ll 1$$

In practice, the sensitivity of our thermometer will be limited, say, by electronic noise in the readout circuit, so that the minimum error in T will be some δT, with the resulting condition—in order to know ΔT accurately—that $\Delta T \gg \delta T$. The net consequence is that we are restricted to values of ε given by

$$|\varepsilon| \gg \frac{|\Delta T|}{T} \gg \frac{\delta T}{T} \tag{6.8.1}$$

Remember that we are setting out to test a prediction valid in the limit $|\varepsilon| \to 0$.

2. We must somehow find out just what the critical temperature T_c is. It is always much easier to measure a temperature difference, ΔT, than to calibrate for an absolute temperature. For argon, $T_c \approx 10^2 \, °K$. Suppose that we can measure ΔT of $10^{-3} \, °K$ with an error $\sim 10\%$ (i.e., $\delta T \approx 10^{-4} \, °K$). By the criterion (6.8.1), we should then be able to measure down to values of $|\varepsilon| \approx 10^{-4}$ (since $|\Delta T|/T \approx 10^{-5}$ and $|\delta T|/T \approx 10^{-6}$). However, to do so, we must know T_c to much better than one part in 10^4. It is not enough that it be measured and tabulated somewhere with that accuracy; we must know the resistance of our thermistor at $T = T_c$ with that precision.

The numbers that we have used are hypothetical. Our purpose will always be to measure to values of $|\varepsilon|$ smaller than whatever data are already available, and an absolute calibration of our thermometer to the necessary precision will usually be a major research effort itself, even if a sufficiently accurate value for T_c exists. In general, we will instead try to measure T_c ourselves in the course of our own experiment. We can try to pick it out by using the heat capacity data themselves, thus introducing, in effect, an additional adjustable parameter for use in trying to make the results look like Fig. 6.8.1. Or if we are going to be more rigorously impartial about testing the laws, we might devise some independent means of finding T_c (measuring some other property that diverges). Suppose that we do the latter. Now, given our value of T_c, we analyze our data, making the plot indicated in Fig. 6.8.1. There are two possible results. One is that things work out just as they should, and our report verifying the scaling law is off for publication in the next mail. The other is that things work out badly; no straight lines, or the two lines are not quite parallel. In this case, we reexamine our techniques carefully, particularly our measurement of T_c, looking for errors; a small change in T_c will considerably alter the limiting behavior in Fig. 6.8.1. We may even redo the experiment, but in any case we search for some (honest) way of making matters come out right. In general, if they come out right to begin with, we will not search nearly so hard to find some way to make them come out wrong (it will be all too easy to find). Regardless of whether the scaling laws are correct or not, there is a strong bias in favor of verifying them, if that is what we wish to do. Finally, if we could not make things come out right in spite of all our efforts, it is always possible that our value of $|\varepsilon|$, no matter what it was, was too large for the laws to be applicable; there is no criterion except to take the limit as $|\varepsilon| \to 0$. There is thus a built-in tendency to prove them right, and, strictly speaking, it is impossible to prove them wrong.

A number of other special problems are presented by critical point measurements.

3. Owing to the effect of gravity, the pressure will not be uniform in the calorimeter. Instead, in equilibrium, we will have

$$\frac{\partial P}{\partial y} = \rho g \qquad (6.8.2)$$

where y is the height in the cell and g the acceleration of gravity. Thus, the critical pressure and density will occur at only one horizontal plane, and, to make matters worse, the density variations due to this effect will tend to be large, since $\delta \rho = (\partial \rho / \partial P)_T \, \delta P$ and $(\partial \rho / \partial P)_T \to \infty$ at the critical point. We can make corrections for this effect if we have an accurate equation of state to use very close to the critical point, and such an equation of state can be generated from the scaling laws themselves. So we are put in the position of having to assume that the laws are correct in order to test them. To be sure, this effect can be minimized by making the sample as small as possible. On the other hand, however—

4. Even if the scaling laws are strictly correct, we expect them to apply strictly only for an infinite sample. We certainly do not expect the predictions to hold when ξ, the correlation length, becomes of order of the size of the sample. Once again we can correct for this problem if necessary but only by using the scaling laws themselves.

5. There are other kinds of experimental difficulties as well. For example, the thermal relaxation time tends to infinity at the critical point. Consequently, as we approach the critical point, we must wait longer and longer

6.8 Evaluation and Summary: Testing the Scaling Laws

for thermal equilibrium in our heat capacity measurements. Errors are then introduced into the results, for there is always some loss (or gain) of heat, δE, due to imperfect isolation of the calorimeter.

The experiments, then, are extremely difficult and have built into them all sorts of factors that make the impartial tests not nearly so impartial as we might like to believe. Of course, none of these difficulties would make any difference if the predictions were far off; that fact would quickly be discovered. But even here there is no consolation; they cannot be far off. Many data were already available when they were first formulated; to some extent, they were proposed because they agreed with what was already known. Thus, they cannot be absurdly wrong; if they are to be found wrong, the disagreement must be rather subtle. It is in light of all this that we must try to make our evaluation of whether the predictions are verified.

Although the specific problems we have discussed are peculiar to the field of critical phenomena, the general tendency is not; there is a bias in scientific investigation in favor of verifying the scientist's prejudices. Yet the success of the scientific enterprise is largely due to the fact that even the most attractive ideas are eventually discarded if they turn out to be wrong. There must be countermechanisms at work to protect us from going too far astray, and it is worth pausing a moment to see what they are.

First, it should be remembered that scientists are generally intelligent, invariably honest, and usually aware of the pitfalls outlined. They are honest because they believe that dishonesty in science is always uncovered. This belief derives from every scientist's most deeply held article of faith: that what he is studying is an objective reality, that an experiment can be repeated by someone else and its results reproduced. With nature itself acting as the great impartial arbiter, it obviously does not pay to cheat. This honesty, dignified as the "scientific ethic" in order to imply that the scientist's personal integrity somehow has something to do with it, extends into the area of analysis and evaluation of data and would seem to give us some protection from being mislead. However, one can see that this protection is not entirely adequate by considering what would happen if it really worked —that is, if the scientist, as a matter of integrity, approached his work in a genuinely disinterested, impartial way. The result would be to remove the driving force from science. Scientists are willing to do prodigious, wholly irrational amounts of work in order to prove that they are right or to be able (politely) to point out that someone else is wrong. It is the convinced advocate that keeps the enterprise going. Few people are driven by a passion for the impartial collecting of data. Thus, the mechanisms that tend to bias the results of measurements, even when recognized and openly confronted, still tend to operate.

On the other hand, their operation does not necessarily always lead us astray. When a theory (like scaling) is first proposed, the exciting thing to do is to show that it is right. If immediately shown to be wrong, then it was not really such a good idea; it will soon be forgotten, and the experiment (no matter how clever) that showed it to be wrong will be forgotten with it. If it is verified instead, one has helped to produce an important advance, a new insight into nature. And so, unless the idea is a bad one, there is a tendency at first to confirm it. As evidence in its favor accumulates, the idea becomes better established (scaling theory, for instance, transforms into scaling law), and at this point, if its correctness is not yet proved beyond doubt, its importance certainly is. The idea becomes a part of the received wisdom. Now an important inversion occurs. It becomes more exciting to show that it is wrong. Further verification would just be adding points to a well-known curve, but now falsification amounts to combating a crusty prejudice that stands in the way of new insights. All those factors that at first tended to ensure verification of the idea now operate in reverse, allowing one to doubt the original interpretation, allowing scope for an attack on the entrenched idea. At this point, with clever, resourceful, passionate advocates operating on both sides of the issue, one has reason to hope that some approximation to reality will emerge. The issue will eventually be settled by the sheer weight of accumulated evidence, all of it scrutinized critically from both sides.

By now the reader will be aware that he is not going to be told in any simple way whether the scaling laws are correct or not. Instead what we can say is that they have become a part of the received wisdom. Their importance is established beyond question. In order to get some feeling for what the experiments give, we can consider some results that have been gathered together in Table 6.8.1. These results primarily come from an attempt made in 1967 to survey and evaluate the status of scaling theory, and they give a rather good idea of the situation as it appeared then.

The top row gives the results of classical theory, with v, v', and η calculated from the Ornstein-Zernike theory [those results marked by (OZ)]. The second and third rows, the two- and three-dimensional Ising models, introduce an active field of theoretical study that we have not discussed, the use of calculable models to test scaling. The Ising model is a grid of points in space, each of which has two possible states: spin up and spin down if we are thinking of it as a model of ferromagnetism; occupied and unoccupied if it is a model of the liquid-gas transition. The energy of a point in each state depends on the states of each of its neighbors: The model exhibits a phase transition of the critical type. A partition function can be formed for it, and its properties (e.g., critical point exponents) calculated, exactly in two dimensions, and numerically in three dimensions. The questions of whether and why it obeys scaling can thus be explored without many of the bothersome problems associated with experiments on real systems. These and

6.8 Evaluation and Summary: Testing the Scaling Laws

Table 6.8.1

System	$2-\alpha$	$2-\alpha'$	$d\nu$	$d\nu'$	$\dfrac{d\gamma}{2-\eta}$	$\gamma+2\beta$	$\gamma'+2\beta$	$\beta(1+\delta)$	$\dfrac{d\mu}{d-1}$
Classical	2	2	1.5 (OZ)	1.5 (OZ)	1.5 (OZ)	2	2	2	–
2D Ising	2	2	2	2	2	2	2	2	–
3D Ising	1.87 ±0.12	1.93 ±0.04 0.16	1.93 ±0.01	–	1.933 ±0.008	1.87 ±0.01	1.94 ±0.05	1.93 ±0.05	–
Ferromagnet	1.92 ±0.08	1.92 ±0.08	1.95 ±0.09	–	2.08 ±0.12	1.99 ±0.09	–	1.7 ±0.2	–
Gas-liquid	1.8 ±0.2	1.88 ±0.12	–	~1.7*	–	2.06 ±0.2	1.7 ±0.3	1.87 ±0.14	1.91 ±0.03
He^3 gas-liquid (quantum fluid)	1.7 to 2	1.8 to 2	–	–	–	1.81 ±0.1	1.8 ±0.2	1.6 ±0.2	–

Data mostly from L. Kadanoff et al., *Rev. Mod. Phys.*, **39**, 395–431, 1967.
* More recent, from E. H. Stanley, *Introduction to Phase Transitions and Critical Phenomena* (New York and Oxford: Oxford University Press, 1971).

other models have played an important role in suggesting theoretical hypotheses about the behavior of matter.

The results in the next three rows were compiled from many reports of experiments on various materials. The form of the table itself has a strong bias built into it. If scaling works, all the numbers in each row should be the same. Furthermore, if all transitions are basically the same, then all the numbers in the table should be the same as each other. Should either or both of these comparisons turn out to be favorable, we may be satisfied that scaling is confirmed. If not, we must go on to ask: What was $|\varepsilon|$ for each measurement? How were the data analyzed to choose a mean value among the measurements for different materials? And so on. A glance at the table seems to indicate that scaling has largely been verified, but a certain amount of discretion is needed in judging this, even aside from the points we have already raised. For example, the first column shows that all the values of $2 - \alpha$ may well be the same. How impressive is this (possible) agreement? α is defined to be positive by means of $C \sim \varepsilon^{-\alpha}$; it follows that the excess entropy goes as $S - S_0 \sim \varepsilon^{-\alpha+1}$. Since the entropy is always continuous in these transitions $-\alpha + 1 \geq 0$ or $0 \leq \alpha \leq 1$. Thus, the range of values in the table, $2 - \alpha = 1.7$ to 2, is really a good fraction of the maximum possible range. Then, again, look at the gas-liquid row. It does not conclusively choose scaling over the classical values given in the first row.

Nevertheless, the influence and significance of the scaling laws are established beyond doubt. They do work, at least approximately, possibly exactly. They have focused attention on the critical point exponents as the quantities of interest in critical phase transitions and have correlated an enormous amount of experimental information by means of a few simple rules. Finally, and of no little importance, simply because it is an attractive and testable idea, it has stimulated a good deal of increasingly careful and precise measurement.

In the early 1800s, Prout, misinterpreting the results of Dalton's measurements, suggested that all atomic weights might be integral multiples of that of hydrogen. This simple, testable idea, with its obviously revolutionary implication of subatomic structure, stimulated a century of increasingly sophisticated measurement of atomic weights by chemists, culminating in the resolution of the paradox when Aston discovered isotopes with a kind of mass spectrometer. Earlier, however, Prout's hypothesis simply could not be put to rest. There was at least one glaring exception—chlorine with mass 35.5—and a good number of cases where an integral multiple was just outside of experimental error. Yet it was so nearly right that there had to be something in it. The analogy to the present situation of scaling is inescapable. There may be one or more clear exceptions and perhaps some near misses in scaling as well; and even a thorough confirmation of scaling would not amount to a profound understanding of these phase transitions. It may even turn out that the scaling laws have served only to misdirect our attention from the really essential features of these phenomena. Even in that case, it seems likely that the large body of experimental work they stimulated will serve to ensure the historical importance of the scaling laws. Workers in the field of scaling laws can take heart in that Prout was not entirely forgotten; Rutherford named the proton after him.

Out of all the work that has been done to test scaling ideas in the past few years, it is possible to detect a consensus forming on certain points. One is that there are indeed differences between classes of phase transitions. For example, those whose order parameter can have two distinct values (as in the liquid-gas critical point) differ from those with a continuously variable phase (as in superfluidity). Moreover, some of the scaling laws seem better established than others (those involving the dimensionality, d, are somewhat suspect). Finally, an interesting idea has emerged, based on the observation that the classical theories would be correct if the world had more than four dimensions (see Prob. 6.12). It appears that it might be possible to estimate the absolute values of the critical exponents by starting from their correct values in a $4+$-dimensional world and seeing how they are changed by lowering the dimensionality. As these ideas make their presence felt, new questions arise and new problems are debated. And that we can identify as progress.

BIBLIOGRAPHY

Many of the results of the classical theories are worked out in Landau and Lifshitz, *Statistical Physics* (see bibliography in Chap. 1), scattered in various places throughout the book. The van der Waals equation is in Paragraph 74, its critical exponents in Paragraph 83, the Ornstein-Zernike theory in Paragraph 119, and a version of the generalized theory is applied in Paragraphs 137 and 138. In the last ten years, experimental and theoretical work inspired by scaling ideas has spawned an immense research literature but little of the more permanent interpretive kind of writing. The only book so far devoted to the subject seems to be H. E. Stanley's *Introduction to Phase Transitions and Critical Phenomena* (New York and Oxford: Oxford University Press, 1971). The book is generally lively and readable but perhaps excessively theoretical in orientation; there is remarkably little consideration of whether the scaling laws are supported by experiment. *Cooperative Phenomena Near Phase Transitions* (Cambridge, Mass.: MIT Press, 1973), edited by Stanley, has recently appeared as a kind of sequel. It contains a bibliography of the field, a section of reprints of experimental articles, including, very appropriately, Thomas Andrews' celebrated Bakerian Lecture of 1869 (Andrews, successor to Faraday at the Royal Institution, London, coined the term critical point), and, finally, a compilation of abstracts of theoretical articles that reminds one strikingly of an old Sears & Roebuck catalog.

The data in Table 6.8.1 were mainly taken from a fine review article by L. Kadanoff et al., *Rev. Mod. Phys.*, **39**, 395–431(1967). The article includes a discussion of classical theory as well as scaling theory. Also used in the preparation of this chapter was a set of lecture notes obtained from the University of Minnesota, *Lectures on the Theory of Phase Transitions*, by N. G. van Kampen (1968).

After writing this chapter, your author undertook, as a sort of homework assignment, to try an experiment in the field of critical phenomena. It was done in collaboration with Professor F. Scaramuzzi and Dr. A. Savoia of the C.N.E.N. Laboratory, Frascati, Italy. Frascati is justly famous for its fine white wine. Nevertheless, the results have appeared in *Physical Review* (May 1974). All the pitfalls discussed in Sec. 6.8 presented themselves as expected, and your faithful author had no more success in escaping them than anyone else has.

PROBLEMS

6.1 Using the Weiss theory, find the exchange coefficient, J, of a material whose Curie temperature is known to be T_c [J is defined in Eq. (5.4.31)].

$$\text{Answer:} \quad J = \frac{3kT_c}{2zS(S+1)}$$

where z is the coordination number and S the spin.

6.2 Show that the mean square fluctuations in magnetization are given by

$$\overline{(\Delta M)^2} = \frac{kT X_T}{V}$$

6.3 Show that the following things cannot exist in two dimensions:
  a. Bose condensation
  b. Crystalline solid
  c. Ferromagnet
  The first two were done in problems earlier in the book. You may get a clue from them about how to do the third.

6.4 Show that the preceding arguments for crystals and magnets do not work if you use a mean field theory (you *can* have two-dimensional solids and ferromagnets in mean field theory).

6.5 We fixed up the Landau theory of phase transitions by putting in fluctuations. Could we have done the same for our theory of the Bose condensation? Explain.

6.6 Many substances, when adsorbed on a clean surface, act like two-dimensional gases or liquids and even have a two-dimensional critical point. Using van der Waals theory and Lennard-Jones potentials, estimate the ratio of the two-dimensional critical temperature of a substance to the three-dimensional critical temperature.

$$Answer: \quad \frac{T_{c2}}{T_{c3}} = 0.45$$

6.7 Expand the reduced van der Waals equation about the critical point to obtain Eq. (6.4.6),

$$p = 1 + 4\varepsilon - 6\varepsilon\theta - \tfrac{3}{2}\theta^3$$

Why do we retain the term in θ^3?

6.8 Using van der Waals theory, show that if gas and liquid with reduced volume v_1 and v_2 are in equilibrium, then

$$6\left(\frac{1}{v_1} - \frac{1}{v_2}\right) + \frac{24t}{9}\left(\log\frac{3v_1 - 1}{3v_2 - 1} + \frac{1}{3v_2 - 1} - \frac{1}{3v_1 - 1}\right) = 0$$

How would you use this result to find the vapor pressure of the liquid?

6.9 Show that the generalized susceptibility above T_c has the same critical point exponent in the Weiss, van der Waals, and Landau theories.

6.10 Show that in Weiss theory, near the Curie point,

$$m \sim (-\varepsilon)^{1/2}$$

6.11 Find the necessary experimental data to evaluate the range of applicability of classical theory to the fluid, magnetic, superfluid, and superconducting cases, by means of Eq. (6.5.53).

6.12 Using the criterion expressed by Eq. (6.5.52), show that the classical theory would be valid at the critical point if the world had more than four dimensions.

6.13 Find α, the heat capacity exponent above T_c, for a fluid that obeys the Ornstein-Zernike theory and has pairwise additive potentials whose attractive long-range part $\sim r^{-n}$. What is the condition that C_V remain finite?

Problems

6.14 Show that in the magnetic case

$$\alpha' + 2\beta + \gamma' \geq 2$$

Explain why this relation cannot be proved in the fluid case

6.15 Show that according to the scaling hypothesis, if the singular part of the Gibbs potential density $\varphi \sim \xi^{-d}$, then the singular part of the free energy density, f, has the same dependence.

6.16 Define what is meant by the critical point exponents α, α', β, γ, γ', δ, ν, ν', η, and μ in terms of measurable quantities for the following phase transitions:

    a. Gas-liquid transition in xenon.
    b. Ferromagnetic transition in nickel.
    c. Superconducting transition in tin.
    d. Superfluid transition in He^4.

6.17 Consider the following experiment: at temperatures close to the gas-liquid critical temperature of xenon, the index of refraction of light, n, is measured as a function of height in a long vertical tube. Assuming that the density ρ may be deduced from n:

$$\rho \propto \frac{n^2 - 1}{n^2 + 2}$$

which critical point exponents may be measured in this experiment, and which scaling laws may be tested?

6.18 Suppose that some quantity χ is measured in the range $10^{-5} < \varepsilon < 10^{-2}$ and fitted to the law

$$\chi \sim \varepsilon^{-a}$$

However, $T_c \cong 100°K$ but can be determined in the experiment only within a range $2m°K$ wide. Discuss how much uncertainty is introduced into the exponent a.

INDEX

Acoustic branch, 169–174
Acoustic waves (*see* Sound waves)
Amagat, 451
Amorphous solid, 229, 247
Andrews, Thomas, 267, 491
Anharmonicity in crystals, 192
Anisotropy energy, 429, 435, 453
Antiferromagnetic state, 421–422, 427–428
Antisymmetric wave function, 106–107, 415–416
Argon:
  equation of state, 304
  pair potential, 303
Aston, 490
Atkins, K. R., 430

Boltzmann, Ludwig, 1
Boltzmann statistics (*see* Gas, ideal; Classical fluids; Classical statistics)
Boltzmann's constant, 4
Bose condensation, 99, 127–139 (*see also* Superfluidity, Bose gas model; Superconductivity, Bose condensate of electron pairs)
Bose-Einstein statistics, 99, 105–107, 108–109
Bosons, 107 (*see also* Bose-Einstein statistics)
Bravais lattices, 178–180
Brillouin function, 421
Brillouin zone, 170, 190, 211, 214–221
Bubble chamber, 448

Bardeen, Cooper and Schrieffer (BCS), 372, 392–404, 461–462
Basis, 175
Bethe-Goldstone equation, 388
Blackbody radiation, 157
Blatt, J. M., 431
Bloch's theorem, 196, 204–207
Bohr, Niels, 144
Bohr atom, 413
Bohr magneton, 413

Cailetet, 450
Cell theories of liquids, 229
Circulation, quantized (*see* Superfluidity, quantized circulation; Vortex lines; Vortex rings)
Classical fluids:
  statistical mechanics of, 68–70, 237–248
  validity of classical approximation, 247–248

Index

Classical statistics, 61–65, 87–90 (see also Classical fluids; Gas, ideal)
Clausius-Clapeyron equation, 39
applied to Bose condensation, 138
Cloud chamber, 448
Cluster integrals, 262, 271–283, 288–301
diagrams:
Augmented Parallel, 298
Bridge, 291–292, 311–316
irreducible, 278, 291–292
nodes of, 277
Parallel, 290–292
Series, 291–292, 312–313
sum of all Series, 298–300
Coexistence curve (see Phase diagram)
Coherence (or Correlation) length, 381–385, 406–407, 463, 470–472, 478, 486
Collective modes (see Normal modes)
Compressibility, 30
critical behavior, 245, 453, 456
equation of state (see Fluctuation equation of state)
Concentration activity coefficient, 271
Condensation (see Gas-liquid phase transition)
Conductivity:
electrical, 214
thermal, of insulators, 191–194
Configurational integral, 69, 238
Conjugate field, 454, 474, 478, 480–481
Cooper pairs, 386–392 (see also Superconductivity)
Correlated fluctuations, 230, 441–443, 463–472
Correlation function (see also Direct correlation function; Radial distribution function; Total correlation function):
n-particle, 241–243
Ornstein-Zernike theory, 469–472, 476
Correlation length (see Coherence length)
Corresponding states, law of, 303–304, 316, 452
Critical opalescence, 463, 468
Critical point, 34, 35, 267, 310, 438, 449–453
Critical point exponents, 473–477 (see also Scaling laws)
experimental values, 483, 489
inequalities, 477
measurement of, 484–487
Crystal structures, 175–183

Curie, Madame, 420
Curie law, 420–421, 439–441
Curie temperature, 439–442
Cyclotron resonance, 127

Dalton, John, 115, 490
Davy, Sir Humphrey, 115
de Broglie wavelength, 70, 102–103, 238, 248
Debye model of solid, 154–160
Debye temperature, 158, 195
Degenerate gases, 98–99 (see also Bose condensation; Fermi degenerate ground state)
slightly, 110–114
de Gennes, P. G., 431
Degrees of freedom, 62, 160–162, 164, 224
Dekker, A. J., 222
de la Tour, Caignard, 267
Density of states, 8, 51, 64
many body, 71–72
photon, 67
single particle, 66
vibrational modes in a solid, 174–175
Detailed balance, principle of, 87–89
Dewar, Sir James, 450
Dewar flask, 450
Diagrams (see Cluster integrals)
Diamagnetism (see Magnetic susceptibility, diamagnetic)
Dimensional analysis, 436–437, 478, 481–483
Dirac notation, 392–396
Direct correlation function, 298–299, 311–314, 468–469
Discrete states, criterion for integrating over, 110
Dispersion relation, 164
of phonons and rotons, 342–343
of various quasiparticles, 195
Distinguishability (see Indistinguishable particles)
Divergence theorem, 350
Domains, magnetic (see Magnetic domains)
Donnelly, R. J., 430
Drude, 371

Easy axis, 453 (see also Anisotropy energy)
Effective magnetic field, 424–425, 438–439
Effective mass, 120, 126–127, 220–221

Egelstaff, P. A., 316-317, 430
Ehrenfest, Paul, 1
Einstein, Albert, 53, 144, 148
Einstein model of solid, 147-152, 423-424
Einstein temperature, 148
Electron-phonon scattering, 213
Electrons in metals, 195-222 (*see also* Superconductivity)
 dispersion relation (energy versus momentum), 120, 198, 201
 energy gaps (and bands), 198-211
 perfect Fermi gas model, 114-127
 wave function, 196-202
Energy, potential, of a fluid, 248-260
Ensembles, 89-90
 canonical, 90
 grand canonical, 90
 microcanonical, 90
Enthalpy, 15
Entropy, communal, 247, 322, 433
Equal probabilities postulate, 3, 61, 87-89
Equilibrium, condition for, 3, 93 (*see also* Variational principles)
Equipartition, law of, 146, 151
Ergodic hypothesis, 87-88
Exchange energy, 428-429
Exchange integral (or coefficient), 423, 439, 491
Exchange interaction (or force), 416, 418, 421-423, 439
Extended zone scheme, 201, 219
Extremum principles (*see* Variational principles)

Faraday, Michael, 21, 266-267, 450, 491
Fermi degenerate ground state, 115-118
 instability of, 387, 392
Fermi-Dirac statistics, 99, 105-108, 114-127 (*see also* Electrons in metals; Superconductivity)
Fermi energy, 115
Fermi level in metals and semiconductors, 217-222
Fermi momentum, 115
Fermi sphere, 115, 212-215, 402
Fermi surface, 115, 212-222
Fermi temperature, 115, 195, 212-222
Fermions, 107 (*see also* Fermi-Dirac statistics)
Ferrimagnetism, 422

Ferromagnetism, 421-430
Feynman, R. P., 342, 348-360, 362, 402, 430, 431
Field points, 289 (*see* Cluster integrals)
First law of thermodynamics, 10
 for magnetic work, 22
Fitzgerald, Edward (*see* Khayyám, Omar)
Fixed points, 289 (*see* Cluster integrals)
Floquet's theorem (*see* Bloch's theorem)
Fluctuation equation of state, 244-245, 310, 317
Fluctuations, 9, 71-83 (*see also* Correlated fluctuations)
 quantum mechanical (*see* Uncertainty, quantum mechanical)
Fluid (*see also* Classical fluids; Liquids):
 critical point exponents, 475
 order parameter, 458 (*see also* Order parameter)
Free energy, 14
 for magnetic work, 22
 minimization of, 24
 statistical formula, 48, 51
Fugacity, 70

Galilean transformation, 335-337, 345-346
Galileo, 246, 317
Gamma functions, 67
Gas:
 dense (*see* Virial equation of state)
 dilute, 260-266
 radial distribution function, 264-266
 structure, 263-266
 ideal, 99-105
 criterion, 59, 100, 102, 104-105
 definition, 57
 entropy, 102
 equation of state, 60, 100-102
 heat capacity, 102
 quantum corrections, 111-114
 interacting, 260-283
Gas-liquid phase transition, 446-449 (*see also* Critical point)
Gaussian distribution, 53, 79-82, 96, 466
Gibbs, J. Willard, 89, 93
Gibbs distribution, 48
Gibbs free energy (*see* Gibbs potential)
Gibbs potential, 15
 for magnetic work, 22
 minimization of, 25
Ginsburg-Landau theory, 461-463

Index

Glass, 229
Gorter, C. J., 430
Grand partition function, 49, 53-55 (*see also* Partition function)
Grüneisen constant, 225

Harmonic approximation, 160, 224-225
Harmonic oscillator, 96, 147
Harrison, W. A., 223, 431
Helium, phase diagram, 37, 40 (*see also* Superfluidity)
Helmholz free energy (*see* Free energy)
Holes, 220-222
Holes, black, 255
Huang, K., 93, 430
Hund's rules, 417, 421, 435
Hypernetted chain equation, 284, 289-301, 302-316

Impurity mode,
Indeterminacy (*see* Uncertainty)
Indistinguishable particles, 55-57, 95, 96, 105-107, 274, 328, 351-352
Iron group elements, 416-418, 422
Ising model, 488-489

Joule-Thomson process, 140, 268-270, 451

Kadanoff, L., 489, 491
Kammerlingh-Onnes, 371-372, 450-451
Keesom, W. H., 430
Keller, W. E., 430
Khalatnikov, I. M., 430
Khayyám, Omar, 12
Kirkwood, 285
Kittel, C., 222, 223, 431
Kronig-Penney model, 202-211

Lambda point (*see* Superfluid phase transition)
Landau, L. D., 335, 342, 360, 430, 432

Landau, L. D. and Lifshitz, E. M., 23, 92-93, 139, 222, 317, 430, 491
Landau criterion (*see* Superconductivity, critical current density; Superfluidity, principle of)
Landau potential, 19
  minimization of, 25
  statistical formula, 47
Landau theory of phase transitions, 457-472
Langevin, Paul, 420
Langevin function, 420
Lattice, 175 (*see also* Crystal structures)
Lattice translation vector, 176
Lennard-Jones potential, 251
Levesque, D., 317
Life magazine, 436-437
Lindemann melting formula, 223
Linear chain (*see* Normal modes)
Liouville's theorem, 87-88
Liquefaction, 266-268, 450-451
Liquid, definition, 227-228
Liquid metals, 231
Liquid semiconductor, 231
Liquids, 282-316
London, F., 430
London equation, 380, 409, 431, 433
London gauge, 408, 434

Magnetic:
  critical point exponents, 475
  domains, 429-430, 472
  field:
    energy, 19, 428-429
    relation of B, H, and M, 19, 411
  moments, 412-419
  order parameter, 458 (*see also* Order parameter)
  susceptibility, 31, 411
    critical behavior, 442, 456
    diamagnetic, 412
    of electrons (*see* Pauli paramagnetism)
    paramagnetic, 411, 420, 434-435
  variables in thermodynamics, 19-23
Magnetism, 411-430
Magnetization, 19, 411
  critical behavior, 457
  ferromagnetic, 421
  mean square fluctuations, 442, 491
  paramagnetic, 419-420

Magnetization (*cont.*)
  of perfect conductor, 25–27
  spontaneous, 440–443
  of superconductor, 20, 27, 376–381 (*see also* Meissner effect)
Magnons (or Spin waves), 427–428, 435, 441
Magnus force, 364, 369
Martini, very dry, 322
Maxwell, James C., 61, 364
Maxwell equal area construction, 447, 457
Maxwell relations, 16
Mayer, Joseph and Mayer, M., 93, 317
Mays, Willie, 43
Mean field theory, 145, 149, 492 (*see also* Einstein model of solid; Weiss molecular field theory; van der Waals theory)
Meissner effect, 361–362, 372–373, 375, 405, 407–409, 433–434
Metalization, principle of, 114 (*see also* Electrons in metals)
Metals (*see* Electrons in metals)
Molecular dynamics, 304
Momentum, conservation of in a crystal, 183–194
Monte Carlo method, 304–309, 311
Mott, N. F., and Jones, H., 223
Mott transition, 210

Nernst's theorem (*see* Third law of thermodynamics)
Node, 277 (*see also* Cluster integrals)
Nonequilibrium, meaning of thermodynamic quantities, 9
Normal modes, 151, 160–175
  of linear chain, 164–175
  of linear triatomic molecule, 164

Olzweski, 450, 451
Onsager, 362
Optical branch, 169–174
Order parameter, 454, 457
  critical point exponent, 473–475
  Landau theory, 457–472
Ornstein-Zernike-Debye plot, 472
Ornstein-Zernike equation, 300, 309–310
Ornstein-Zernike theory, 468–472, 476, 477, 478, 488

Pair potential, 250–252
Pairwise additivity, 252–260, 287, 304–311, 317
  equation of state, 258
  heat capacity, 255
  of mean potentials (*see* Superposition approximation)
Paramagnetism, 419–421
Paris, Dr., 266
Partition function, 49, 53–55
  of ideal gas, 69–70
  of interacting fluid, 68–70, 238, 247 (*see also* Configurational integral)
Pauli exclusion principle, 107, 114 (*see also* Antisymmetric wave function; Fermi-Dirac statistics)
Pauli paramagnetism, 434
Percus-Yevick equation, 284, 289–301, 302–316
Perfect conductor (*see* Magnetization of perfect conductor)
Periodic Table of the Elements, 416
Periodic zone scheme, 201, 219
Persons, 156
Phase diagram, 34–41, 95
  generalized, 478–479 (*see also* 455–456)
  helium, 36–37, 40
  solid-liquid-gas, 35–36, 227–228
  superconductor, 28, 40
  water, 36, 39
Phase separation of He^3-He^4 mixture, 140–141
Phase space, 62, 83–90
  volume of fundamental cell, 63–65
Phase transitions, 436–490
  analogies between, 453–457 (*see also* Landau theory of phase transitions; Critical point exponents; Scaling laws)
  Landau theory, 457–472
  order of, 132–136
Phonons, 155–160, 183–195, 331, 342–348, 358–359, 441
Photons, 157, 194–195, 202
Pictet, 450
Potential of mean force, 249
  for three particles, 285
Prout, 490

Quantized circulation (or vorticity) (*see* Superfluidity, quantized circulation;

Index

Vortex lines; Vortex rings)
Quantized flux (see under Superconductivity: fluxoids, quantized magnetic flux, vortex state)
Quantum statistics (see Bose-Einstein statistics; Fermi-Dirac statistics)
Quasiparticles, 67, 194–195, 332 (see also Electrons in metals; Holes; Magnons; Persons; Phonons; Photons; Rotons)
Quenching of orbital angular momentum, 418

Radial distribution function, 234–236, 239–248, 349
theories of (see Gas, dilute; Hypernetted chain equation; Percus-Yevick equation; Yvon-Born-Green equation)
Ramsay, 450
Rare earth series, 416–418, 422, 434–435
Reciprocal lattice, 186–191
Reduced zone scheme, 201
Reif, F., 93
Relatation time, limits on, for applicability of thermodynamics, 87
Rice, S. A. and Gray, P., 317
Rotons, 328, 342–348, 358
Rushbrooke inequality, 477
Rutherford, 490

Salpeter, E. E., 317
Sarran, 451
Savoia, A., 491
Scaling laws, 478–490
Scaramuzzi, F., 491
Sears and Roebuck, 491
Second law of thermodynamics, 12
Second sound, 341
Semiconductors, 216–222
  amorphous and liquid, 231
  intrinsic, 222
Semimetals, 216
Sound waves, 151, 152–154
Spectroscopic notation, 417
Spin degeneracy, 110
Spin-orbit interaction, 414–415
Spin waves (see Magnons)
Stanley, E. H., 317, 489, 491

Statistically independent variables, 81–82
Stirling's formula, 103
Structure factor, 236, 245, 357–358, 465–469
Structure of fluids, 229, 231–248
  measurement of, 232–237
  statistical mechanics of, 237–246
Superconductivity, 371–411
  binding energy of electron pairs, 391, 400–402
  Bose condensate of electron pairs, 405–410
  coherence length, 381–385, 406–407, 463
  critical current density, 405
  critical (or transition) temperature, 377–378, 404, 438
  energy gap, 403
  fluxoids, 385, 410, 434
  order parameter, 458 (see also Order parameter)
  penetration depth, 379–385
  quantized magnetic flux, 409–411
  surface tension, 382–384
  thermodynamics of, 27–29, 40–41, 375–385
  type I and type II, 381–382
  vortex state, 385, 411
Supercooled vapors, 448–449
Superexchange, 421, 422
Superfluid, rotating, 361–362
Superfluid film, 327
Superfluid flow, 322–325, 329–331, 360–362, 433
Superfluid fraction (or density):
  critical point exponent, 437, 475
  scaling argument, 482–483
Superfluid persistent currents, 325–326, 333–334, 369, 433
Superfluid phase transition (or Lambda transition), 321–322, 327–328, 437–438, 458, 461, 472
Superfluidity, 37, 40, 321–371
  Bose gas model, 327–334
  critical velocity, 324–325, 365–366, 433
  dissipation due to fluctuations, 369–371
  Feynman theory, 348–360, 432
  fountain effect, 325, 331–332
  fourth sound, 341, 432
  order parameter, 458 (see also Order parameter)
  principle of, 330, 346, 348, 352, 365
  quantized circulation, 334, 360–363

Superfluidity (cont.)
quantized vorticity (see under Superfluidity: quantized circulation, vortex lines, vortex rings)
second sound, 341
super heat conductivity, 322, 325, 329
third sound, 341
two fluid model, 324, 329-331, 335-341, 345, 366-367
vortex lines, 362-364, 433, 472
vortex rings, 365, 367-370
Surface tension, 459
critical point exponent, 475, 482
of magnetic domains, 430
of superconductor, 382-384
Superposition approximation, 285-287, 311
Susceptibility, critical point exponent, 473-476 (see also Magnetic susceptibility)
Symmetric wave function, 106-107, 349-350, 415-416
Symmetries of crystals (see Crystal structures)

van Kampen, N. G., 491
van Marum, 266, 450
Variational principles in thermodynamics, 23-29
applied to:
compressibility, 31
heat capacity, 32
perfect conductor, 26
phase equilibrium, 38-41
phase transitions, 458-462
superconductors, 28, 375-384, 392-401, 410
superfluid, 328, 332, 344, 353-358
Virial equation of state, 260, 271-283, 445
Viscosity, 323-324
von Laue conditions, 186, 237
von Neumann, 305
Vortex lines (see under Superconductivity: fluxoids, quantized magnetic flux, vortex state; under Superfluidity: quantized circulation, vortex lines, vortex rings; Vortex rings)
Vortex rings, 61, 365

Taylor, G. I., 436
Thilorier, 267
Third law of thermodynamics, 13, 149
Thomson, J. J., 61, 118, 371
Thomson, William (Lord Kelvin), 61, 268
Throop, G. J. and Bearman, R. J., 317
Tolman, R. C., 93
Total correlation function, 235, 298
Transistor, 222
Triple point, 34-36, 38

Umklapp processes, 191-194
Uncertainty:
quantum mechanical, 7, 61-66, 83-87, 106
statistical, 7 (see also Fluctuations)
Unit cell, 143, 162-163, 175-183
primitive, 176-182
Wigner-Seitz, 176

van der Waals theory, 438, 443-453, 456-457, 492

Weiss molecular field theory, 438-443, 492
White, R. M., 431
Wilks, J., 430

Xenon:
equation of state, 304
pair potential, 303
X-ray diffraction:
fluids, 232-237
solids, 178, 185-186, 232-237

Yvon-Born-Green equation, 283, 284-288, 302-316

Zeroth law of thermodynamics, 10
Zeroth sound, 341
Ziman, J. M., 222
Zymansky, M., 93

A CATALOG OF SELECTED
DOVER BOOKS
IN SCIENCE AND MATHEMATICS

A CATALOG OF SELECTED
DOVER BOOKS
IN SCIENCE AND MATHEMATICS

QUALITATIVE THEORY OF DIFFERENTIAL EQUATIONS, V.V. Nemytskii and V.V. Stepanov. Classic graduate-level text by two prominent Soviet mathematicians covers classical differential equations as well as topological dynamics and ergodic theory. Bibliographies. 523pp. 5⅜ × 8½. 65954-2 Pa. $10.95

MATRICES AND LINEAR ALGEBRA, Hans Schneider and George Phillip Barker. Basic textbook covers theory of matrices and its applications to systems of linear equations and related topics such as determinants, eigenvalues and differential equations. Numerous exercises. 432pp. 5⅜ × 8½. 66014-1 Pa. $9.95

QUANTUM THEORY, David Bohm. This advanced undergraduate-level text presents the quantum theory in terms of qualitative and imaginative concepts, followed by specific applications worked out in mathematical detail. Preface. Index. 655pp. 5⅜ × 8½. 65969-0 Pa. $13.95

ATOMIC PHYSICS (8th edition), Max Born. Nobel laureate's lucid treatment of kinetic theory of gases, elementary particles, nuclear atom, wave-corpuscles, atomic structure and spectral lines, much more. Over 40 appendices, bibliography. 495pp. 5⅜ × 8½. 65984-4 Pa. $11.95

ELECTRONIC STRUCTURE AND THE PROPERTIES OF SOLIDS: The Physics of the Chemical Bond, Walter A. Harrison. Innovative text offers basic understanding of the electronic structure of covalent and ionic solids, simple metals, transition metals and their compounds. Problems. 1980 edition. 582pp. 6⅛ × 9¼. 66021-4 Pa. $14.95

BOUNDARY VALUE PROBLEMS OF HEAT CONDUCTION, M. Necati Özisik. Systematic, comprehensive treatment of modern mathematical methods of solving problems in heat conduction and diffusion. Numerous examples and problems. Selected references. Appendices. 505pp. 5⅜ × 8½. 65990-9 Pa. $11.95

A SHORT HISTORY OF CHEMISTRY (3rd edition), J.R. Partington. Classic exposition explores origins of chemistry, alchemy, early medical chemistry, nature of atmosphere, theory of valency, laws and structure of atomic theory, much more. 428pp. 5⅜ × 8½. (Available in U.S. only) 65977-1 Pa. $10.95

A HISTORY OF ASTRONOMY, A. Pannekoek. Well-balanced, carefully reasoned study covers such topics as Ptolemaic theory, work of Copernicus, Kepler, Newton, Eddington's work on stars, much more. Illustrated. References. 521pp. 5⅜ × 8½. 65994-1 Pa. $11.95

PRINCIPLES OF METEOROLOGICAL ANALYSIS, Walter J. Saucier. Highly respected, abundantly illustrated classic reviews atmospheric variables, hydrostatics, static stability, various analyses (scalar, cross-section, isobaric, isentropic, more). For intermediate meteorology students. 454pp. 6⅛ × 9¼. 65979-8 Pa. $12.95

CATALOG OF DOVER BOOKS

RELATIVITY, THERMODYNAMICS AND COSMOLOGY, Richard C. Tolman. Landmark study extends thermodynamics to special, general relativity; also applications of relativistic mechanics, thermodynamics to cosmological models. 501pp. 5⅜ × 8½. 65383-8 Pa. $12.95

APPLIED ANALYSIS, Cornelius Lanczos. Classic work on analysis and design of finite processes for approximating solution of analytical problems. Algebraic equations, matrices, harmonic analysis, quadrature methods, much more. 559pp. 5⅜ × 8½. 65656-X Pa. $12.95

SPECIAL RELATIVITY FOR PHYSICISTS, G. Stephenson and C.W. Kilmister. Concise elegant account for nonspecialists. Lorentz transformation, optical and dynamical applications, more. Bibliography. 108pp. 5⅜ × 8½. 65519-9 Pa. $4.95

INTRODUCTION TO ANALYSIS, Maxwell Rosenlicht. Unusually clear, accessible coverage of set theory, real number system, metric spaces, continuous functions, Riemann integration, multiple integrals, more. Wide range of problems. Undergraduate level. Bibliography. 254pp. 5⅜ × 8½. 65038-3 Pa. $7.95

INTRODUCTION TO QUANTUM MECHANICS With Applications to Chemistry, Linus Pauling & E. Bright Wilson, Jr. Classic undergraduate text by Nobel Prize winner applies quantum mechanics to chemical and physical problems. Numerous tables and figures enhance the text. Chapter bibliographies. Appendices. Index. 468pp. 5⅜ × 8½. 64871-0 Pa. $11.95

ASYMPTOTIC EXPANSIONS OF INTEGRALS, Norman Bleistein & Richard A. Handelsman. Best introduction to important field with applications in a variety of scientific disciplines. New preface. Problems. Diagrams. Tables. Bibliography. Index. 448pp. 5⅜ × 8½. 65082-0 Pa. $11.95

MATHEMATICS APPLIED TO CONTINUUM MECHANICS, Lee A. Segel. Analyzes models of fluid flow and solid deformation. For upper-level math, science and engineering students. 608pp. 5⅜ × 8½. 65369-2 Pa. $13.95

ELEMENTS OF REAL ANALYSIS, David A. Sprecher. Classic text covers fundamental concepts, real number system, point sets, functions of a real variable, Fourier series, much more. Over 500 exercises. 352pp. 5⅜ × 8½. 65385-4 Pa. $9.95

PHYSICAL PRINCIPLES OF THE QUANTUM THEORY, Werner Heisenberg. Nobel Laureate discusses quantum theory, uncertainty, wave mechanics, work of Dirac, Schroedinger, Compton, Wilson, Einstein, etc. 184pp. 5⅜ × 8½. 60113-7 Pa. $4.95

INTRODUCTORY REAL ANALYSIS, A.N. Kolmogorov, S.V. Fomin. Translated by Richard A. Silverman. Self-contained, evenly paced introduction to real and functional analysis. Some 350 problems. 403pp. 5⅜ × 8½. 61226-0 Pa. $9.95

PROBLEMS AND SOLUTIONS IN QUANTUM CHEMISTRY AND PHYSICS, Charles S. Johnson, Jr. and Lee G. Pedersen. Unusually varied problems, detailed solutions in coverage of quantum mechanics, wave mechanics, angular momentum, molecular spectroscopy, scattering theory, more. 280 problems plus 139 supplementary exercises. 430pp. 6½ × 9¼. 65236-X Pa. $11.95

CATALOG OF DOVER BOOKS

ASYMPTOTIC METHODS IN ANALYSIS, N.G. de Bruijn. An inexpensive, comprehensive guide to asymptotic methods—the pioneering work that teaches by explaining worked examples in detail. Index. 224pp. 5⅜ × 8½. 64221-6 Pa. $6.95

OPTICAL RESONANCE AND TWO-LEVEL ATOMS, L. Allen and J.H. Eberly. Clear, comprehensive introduction to basic principles behind all quantum optical resonance phenomena. 53 illustrations. Preface. Index. 256pp. 5⅜ × 8½. 65533-4 Pa. $7.95

COMPLEX VARIABLES, Francis J. Flanigan. Unusual approach, delaying complex algebra till harmonic functions have been analyzed from real variable viewpoint. Includes problems with answers. 364pp. 5⅜ × 8½. 61388-7 Pa. $7.95

ATOMIC SPECTRA AND ATOMIC STRUCTURE, Gerhard Herzberg. One of best introductions; especially for specialist in other fields. Treatment is physical rather than mathematical. 80 illustrations. 257pp. 5⅜ × 8½. 60115-3 Pa. $5.95

APPLIED COMPLEX VARIABLES, John W. Dettman. Step-by-step coverage of fundamentals of analytic function theory—plus lucid exposition of five important applications: Potential Theory; Ordinary Differential Equations; Fourier Transforms; Laplace Transforms; Asymptotic Expansions. 66 figures. Exercises at chapter ends. 512pp. 5⅜ × 8½. 64670-X Pa. $10.95

ULTRASONIC ABSORPTION: An Introduction to the Theory of Sound Absorption and Dispersion in Gases, Liquids and Solids, A.B. Bhatia. Standard reference in the field provides a clear, systematically organized introductory review of fundamental concepts for advanced graduate students, research workers. Numerous diagrams. Bibliography. 440pp. 5⅜ × 8½. 64917-2 Pa. $11.95

UNBOUNDED LINEAR OPERATORS: Theory and Applications, Seymour Goldberg. Classic presents systematic treatment of the theory of unbounded linear operators in normed linear spaces with applications to differential equations. Bibliography. 199pp. 5⅜ × 8½. 64830-3 Pa. $7.95

LIGHT SCATTERING BY SMALL PARTICLES, H.C. van de Hulst. Comprehensive treatment including full range of useful approximation methods for researchers in chemistry, meteorology and astronomy. 44 illustrations. 470pp. 5⅜ × 8½. 64228-3 Pa. $10.95

CONFORMAL MAPPING ON RIEMANN SURFACES, Harvey Cohn. Lucid, insightful book presents ideal coverage of subject. 334 exercises make book perfect for self-study. 55 figures. 352pp. 5⅜ × 8¼. 64025-6 Pa. $8.95

OPTICKS, Sir Isaac Newton. Newton's own experiments with spectroscopy, colors, lenses, reflection, refraction, etc., in language the layman can follow. Foreword by Albert Einstein. 532pp. 5⅜ × 8½. 60205-2 Pa. $9.95

GENERALIZED INTEGRAL TRANSFORMATIONS, A.H. Zemanian. Graduate-level study of recent generalizations of the Laplace, Mellin, Hankel, K. Weierstrass, convolution and other simple transformations. Bibliography. 320pp. 5⅜ × 8½. 65375-7 Pa. $7.95

CATALOG OF DOVER BOOKS

THE ELECTROMAGNETIC FIELD, Albert Shadowitz. Comprehensive undergraduate text covers basics of electric and magnetic fields, builds up to electromagnetic theory. Also related topics, including relativity. Over 900 problems, 768pp, 5⅜ x 8¼. 65660-8 Pa. $17.95

FOURIER SERIES, Georgi P. Tolstov. Translated by Richard A. Silverman. A valuable addition to the literature on the subject, moving clearly from subject to subject and theorem to theorem. 107 problems, answers. 336pp. 5⅜ × 8½. 63317-9 Pa. $7.95

THEORY OF ELECTROMAGNETIC WAVE PROPAGATION, Charles Herach Papas. Graduate-level study discusses the Maxwell field equations, radiation from wire antennas, the Doppler effect and more. xiii + 244pp. 5⅜ × 8½. 65678-0 Pa. $6.95

DISTRIBUTION THEORY AND TRANSFORM ANALYSIS: An Introduction to Generalized Functions, with Applications, A.H. Zemanian. Provides basics of distribution theory, describes generalized Fourier and Laplace transformations. Numerous problems. 384pp. 5⅜ × 8½. 65479-6 Pa. $9.95

THE PHYSICS OF WAVES, William C. Elmore and Mark A. Heald. Unique overview of classical wave theory. Acoustics, optics, electromagnetic radiation, more. Ideal as classroom text or for self-study. Problems. 477pp. 5⅜ × 8½. 64926-1 Pa. $11.95

CALCULUS OF VARIATIONS WITH APPLICATIONS, George M. Ewing. Applications-oriented introduction to variational theory develops insight and promotes understanding of specialized books, research papers. Suitable for advanced undergraduate/graduate students as primary, supplementary text. 352pp. 5⅜ × 8½. 64856-7 Pa. $8.95

A TREATISE ON ELECTRICITY AND MAGNETISM, James Clerk Maxwell. Important foundation work of modern physics. Brings to final form Maxwell's theory of electromagnetism and rigorously derives his general equations of field theory. 1,084pp. 5⅜ × 8½. 60636-8, 60637-6 Pa., Two-vol. set $19.90

AN INTRODUCTION TO THE CALCULUS OF VARIATIONS, Charles Fox. Graduate-level text covers variations of an integral, isoperimetrical problems, least action, special relativity, approximations, more. References. 279pp. 5⅜ × 8½. 65499-0 Pa. $7.95

HYDRODYNAMIC AND HYDROMAGNETIC STABILITY, S. Chandrasekhar. Lucid examination of the Rayleigh-Benard problem; clear coverage of the theory of instabilities causing convection. 704pp. 5⅜ × 8¼. 64071-X Pa. $14.95

CALCULUS OF VARIATIONS, Robert Weinstock. Basic introduction covering isoperimetric problems, theory of elasticity, quantum mechanics, electrostatics, etc. Exercises throughout. 326pp. 5⅜ × 8½. 63069-2 Pa. $7.95

DYNAMICS OF FLUIDS IN POROUS MEDIA, Jacob Bear. For advanced students of ground water hydrology, soil mechanics and physics, drainage and irrigation engineering and more. 335 illustrations. Exercises, with answers. 784pp. 6⅛ × 9¼. 65675-6 Pa. $19.95

CATALOG OF DOVER BOOKS

NUMERICAL METHODS FOR SCIENTISTS AND ENGINEERS, Richard Hamming. Classic text stresses frequency approach in coverage of algorithms, polynomial approximation, Fourier approximation, exponential approximation, other topics. Revised and enlarged 2nd edition. 721pp. 5⅜ × 8½.
65241-6 Pa. $14.95

THEORETICAL SOLID STATE PHYSICS, Vol. I: Perfect Lattices in Equilibrium; Vol. II: Non-Equilibrium and Disorder, William Jones and Norman H. March. Monumental reference work covers fundamental theory of equilibrium properties of perfect crystalline solids, non-equilibrium properties, defects and disordered systems. Appendices. Problems. Preface. Diagrams. Index. Bibliography. Total of 1,301pp. 5⅜ × 8½. Two volumes.
Vol. I 65015-4 Pa. $12.95
Vol. II 65016-2 Pa. $12.95

OPTIMIZATION THEORY WITH APPLICATIONS, Donald A. Pierre. Broad-spectrum approach to important topic. Classical theory of minima and maxima, calculus of variations, simplex technique and linear programming, more. Many problems, examples. 640pp. 5⅜ × 8½.
65205-X Pa. $13.95

THE MODERN THEORY OF SOLIDS, Frederick Seitz. First inexpensive edition of classic work on theory of ionic crystals, free-electron theory of metals and semiconductors, molecular binding, much more. 736pp. 5⅜ × 8½.
65482-6 Pa. $15.95

ESSAYS ON THE THEORY OF NUMBERS, Richard Dedekind. Two classic essays by great German mathematician: on the theory of irrational numbers; and on transfinite numbers and properties of natural numbers. 115pp. 5⅜ × 8½.
21010-3 Pa. $4.95

THE FUNCTIONS OF MATHEMATICAL PHYSICS, Harry Hochstadt. Comprehensive treatment of orthogonal polynomials, hypergeometric functions, Hill's equation, much more. Bibliography. Index. 322pp. 5⅜ × 8½. 65214-9 Pa. $9.95

NUMBER THEORY AND ITS HISTORY, Oystein Ore. Unusually clear, accessible introduction covers counting, properties of numbers, prime numbers, much more. Bibliography. 380pp. 5⅜ × 8½. 65620-9 Pa. $8.95

THE VARIATIONAL PRINCIPLES OF MECHANICS, Cornelius Lanczos. Graduate level coverage of calculus of variations, equations of motion, relativistic mechanics, more. First inexpensive paperbound edition of classic treatise. Index. Bibliography. 418pp. 5⅜ × 8½. 65067-7 Pa. $10.95

MATHEMATICAL TABLES AND FORMULAS, Robert D. Carmichael and Edwin R. Smith. Logarithms, sines, tangents, trig functions, powers, roots, reciprocals, exponential and hyperbolic functions, formulas and theorems. 269pp. 5⅜ × 8½.
60111-0 Pa. $5.95

THEORETICAL PHYSICS, Georg Joos, with Ira M. Freeman. Classic overview covers essential math, mechanics, electromagnetic theory, thermodynamics, quantum mechanics, nuclear physics, other topics. First paperback edition. xxiii + 885pp. 5⅜ × 8½.
65227-0 Pa. $18.95

CATALOG OF DOVER BOOKS

HANDBOOK OF MATHEMATICAL FUNCTIONS WITH FORMULAS, GRAPHS, AND MATHEMATICAL TABLES, edited by Milton Abramowitz and Irene A. Stegun. Vast compendium: 29 sets of tables, some to as high as 20 places. 1,046pp. 8 × 10½. 61272-4 Pa. $22.95

MATHEMATICAL METHODS IN PHYSICS AND ENGINEERING, John W. Dettman. Algebraically based approach to vectors, mapping, diffraction, other topics in applied math. Also generalized functions, analytic function theory, more. Exercises. 448pp. 5⅜ × 8¼. 65649-7 Pa. $8.95

A SURVEY OF NUMERICAL MATHEMATICS, David M. Young and Robert Todd Gregory. Broad self-contained coverage of computer-oriented numerical algorithms for solving various types of mathematical problems in linear algebra, ordinary and partial, differential equations, much more. Exercises. Total of 1,248pp. 5⅜ × 8½. Two volumes. Vol. I 65691-8 Pa. $14.95
Vol. II 65692-6 Pa. $14.95

TENSOR ANALYSIS FOR PHYSICISTS, J.A. Schouten. Concise exposition of the mathematical basis of tensor analysis, integrated with well-chosen physical examples of the theory. Exercises. Index. Bibliography. 289pp. 5⅜ × 8½.
65582-2 Pa. $7.95

INTRODUCTION TO NUMERICAL ANALYSIS (2nd Edition), F.B. Hildebrand. Classic, fundamental treatment covers computation, approximation, interpolation, numerical differentiation and integration, other topics. 150 new problems. 669pp. 5⅜ × 8½. 65363-3 Pa. $14.95

INVESTIGATIONS ON THE THEORY OF THE BROWNIAN MOVEMENT, Albert Einstein. Five papers (1905-8) investigating dynamics of Brownian motion and evolving elementary theory. Notes by R. Fürth. 122pp. 5⅜ × 8½.
60304-0 Pa. $4.95

NUMERICAL METHODS FOR SCIENTISTS AND ENGINEERS, Richard Hamming. Classic text stresses frequency approach in coverage of algorithms, polynomial approximation, Fourier approximation, exponential approximation, other topics. Revised and enlarged 2nd edition. 721pp. 5⅜ × 8½. 65241-6 Pa. $14.95

AN INTRODUCTION TO STATISTICAL THERMODYNAMICS, Terrell L. Hill. Excellent basic text offers wide-ranging coverage of quantum statistical mechanics, systems of interacting molecules, quantum statistics, more. 523pp. 5⅜ × 8½. 65242-4 Pa. $11.95

ELEMENTARY DIFFERENTIAL EQUATIONS, William Ted Martin and Eric Reissner. Exceptionally clear, comprehensive introduction at undergraduate level. Nature and origin of differential equations, differential equations of first, second and higher orders. Picard's Theorem, much more. Problems with solutions. 331pp. 5⅜ × 8½. 65024-3 Pa. $8.95

STATISTICAL PHYSICS, Gregory H. Wannier. Classic text combines thermodynamics, statistical mechanics and kinetic theory in one unified presentation of thermal physics. Problems with solutions. Bibliography. 532pp. 5⅜ × 8½.
65401-X Pa. $11.95

CATALOG OF DOVER BOOKS

ORDINARY DIFFERENTIAL EQUATIONS, Morris Tenenbaum and Harry Pollard. Exhaustive survey of ordinary differential equations for undergraduates in mathematics, engineering, science. Thorough analysis of theorems. Diagrams. Bibliography. Index. 818pp. 5⅜ × 8½. 64940-7 Pa. $16.95

STATISTICAL MECHANICS: Principles and Applications, Terrell L. Hill. Standard text covers fundamentals of statistical mechanics, applications to fluctuation theory, imperfect gases, distribution functions, more. 448pp. 5⅜ × 8½. 65390-0 Pa. $9.95

ORDINARY DIFFERENTIAL EQUATIONS AND STABILITY THEORY: An Introduction, David A. Sánchez. Brief, modern treatment. Linear equation, stability theory for autonomous and nonautonomous systems, etc. 164pp. 5⅜ × 8¼. 63828-6 Pa. $5.95

THIRTY YEARS THAT SHOOK PHYSICS: The Story of Quantum Theory, George Gamow. Lucid, accessible introduction to influential theory of energy and matter. Careful explanations of Dirac's anti-particles, Bohr's model of the atom, much more. 12 plates. Numerous drawings. 240pp. 5⅜ × 8½. 24895-X Pa. $5.95

THEORY OF MATRICES, Sam Perlis. Outstanding text covering rank, non-singularity and inverses in connection with the development of canonical matrices under the relation of equivalence, and without the intervention of determinants. Includes exercises. 237pp. 5⅜ × 8½. 66810-X Pa. $7.95

GREAT EXPERIMENTS IN PHYSICS: Firsthand Accounts from Galileo to Einstein, edited by Morris H. Shamos. 25 crucial discoveries: Newton's laws of motion, Chadwick's study of the neutron, Hertz on electromagnetic waves, more. Original accounts clearly annotated. 370pp. 5⅜ × 8½. 25346-5 Pa. $9.95

INTRODUCTION TO PARTIAL DIFFERENTIAL EQUATIONS WITH APPLICATIONS, E.C. Zachmanoglou and Dale W. Thoe. Essentials of partial differential equations applied to common problems in engineering and the physical sciences. Problems and answers. 416pp. 5⅜ × 8½. 65251-3 Pa. $10.95

BURNHAM'S CELESTIAL HANDBOOK, Robert Burnham, Jr. Thorough guide to the stars beyond our solar system. Exhaustive treatment. Alphabetical by constellation: Andromeda to Cetus in Vol. 1; Chamaeleon to Orion in Vol. 2; and Pavo to Vulpecula in Vol. 3. Hundreds of illustrations. Index in Vol. 3. 2,000pp. 6⅛ × 9¼. 23567-X, 23568-8, 23673-0 Pa., Three-vol. set $41.85

ASYMPTOTIC EXPANSIONS FOR ORDINARY DIFFERENTIAL EQUATIONS, Wolfgang Wasow. Outstanding text covers asymptotic power series, Jordan's canonical form, turning point problems, singular perturbations, much more. Problems. 384pp. 5⅜ × 8½. 65456-7 Pa. $9.95

AMATEUR ASTRONOMER'S HANDBOOK, J.B. Sidgwick. Timeless, comprehensive coverage of telescopes, mirrors, lenses, mountings, telescope drives, micrometers, spectroscopes, more. 189 illustrations. 576pp. 5⅜ × 8¼. (USO) 24034-7 Pa. $9.95

CATALOG OF DOVER BOOKS

SPECIAL FUNCTIONS, N.N. Lebedev. Translated by Richard Silverman. Famous Russian work treating more important special functions, with applications to specific problems of physics and engineering. 38 figures. 308pp. 5⅜ × 8½.
60624-4 Pa. $7.95

OBSERVATIONAL ASTRONOMY FOR AMATEURS, J.B. Sidgwick. Mine of useful data for observation of sun, moon, planets, asteroids, aurorae, meteors, comets, variables, binaries, etc. 39 illustrations. 384pp. 5⅜ × 8¼. (Available in U.S. only)
24033-9 Pa. $8.95

INTEGRAL EQUATIONS, F.G. Tricomi. Authoritative, well-written treatment of extremely useful mathematical tool with wide applications. Volterra Equations, Fredholm Equations, much more. Advanced undergraduate to graduate level. Exercises. Bibliography. 238pp. 5⅜ × 8½.
64828-1 Pa. $6.95

CELESTIAL OBJECTS FOR COMMON TELESCOPES, T.W. Webb. Inestimable aid for locating and identifying nearly 4,000 celestial objects. 77 illustrations. 645pp. 5⅜ × 8½.
20917-2, 20918-0 Pa., Two-vol. set $12.00

MODERN NONLINEAR EQUATIONS, Thomas L. Saaty. Emphasizes practical solution of problems; covers seven types of equations. ". . . a welcome contribution to the existing literature. . . ."—*Math Reviews.* 490pp. 5⅜ × 8½. 64232-1 Pa. $9.95

FUNDAMENTALS OF ASTRODYNAMICS, Roger Bate et al. Modern approach developed by U.S. Air Force Academy. Designed as a first course. Problems, exercises. Numerous illustrations. 455pp. 5⅜ × 8½.
60061-0 Pa. $8.95

INTRODUCTION TO LINEAR ALGEBRA AND DIFFERENTIAL EQUATIONS, John W. Dettman. Excellent text covers complex numbers, determinants, orthonormal bases, Laplace transforms, much more. Exercises with solutions. Undergraduate level. 416pp. 5⅜ × 8½.
65191-6 Pa. $9.95

INCOMPRESSIBLE AERODYNAMICS, edited by Bryan Thwaites. Covers theoretical and experimental treatment of the uniform flow of air and viscous fluids past two-dimensional aerofoils and three-dimensional wings; many other topics. 654pp. 5⅜ × 8½.
65465-6 Pa. $16.95

INTRODUCTION TO DIFFERENCE EQUATIONS, Samuel Goldberg. Exceptionally clear exposition of important discipline with applications to sociology, psychology, economics. Many illustrative examples; over 250 problems. 260pp. 5⅜ × 8½.
65084-7 Pa. $7.95

LAMINAR BOUNDARY LAYERS, edited by L. Rosenhead. Engineering classic covers steady boundary layers in two- and three-dimensional flow, unsteady boundary layers, stability, observational techniques, much more. 708pp. 5⅜ × 8½.
65646-2 Pa. $15.95

LECTURES ON CLASSICAL DIFFERENTIAL GEOMETRY, Second Edition, Dirk J. Struik. Excellent brief introduction covers curves, theory of surfaces, fundamental equations, geometry on a surface, conformal mapping, other topics. Problems. 240pp. 5⅜ × 8½.
65609-8 Pa. $6.95

CATALOG OF DOVER BOOKS

ROTARY-WING AERODYNAMICS, W.Z. Stepniewski. Clear, concise text covers aerodynamic phenomena of the rotor and offers guidelines for helicopter performance evaluation. Originally prepared for NASA. 537 figures. 640pp. 6⅛ × 9¼. 64647-5 Pa. $14.95

DIFFERENTIAL GEOMETRY, Heinrich W. Guggenheimer. Local differential geometry as an application of advanced calculus and linear algebra. Curvature, transformation groups, surfaces, more. Exercises. 62 figures. 378pp. 5⅜ × 8½. 63433-7 Pa. $7.95

INTRODUCTION TO SPACE DYNAMICS, William Tyrrell Thomson. Comprehensive, classic introduction to space-flight engineering for advanced undergraduate and graduate students. Includes vector algebra, kinematics, transformation of coordinates. Bibliography. Index. 352pp. 5⅜ × 8½. 65113-4 Pa. $8.95

A SURVEY OF MINIMAL SURFACES, Robert Osserman. Up-to-date, in-depth discussion of the field for advanced students. Corrected and enlarged edition covers new developments. Includes numerous problems. 192pp. 5⅜ × 8½. 64998-9 Pa. $8.95

ANALYTICAL MECHANICS OF GEARS, Earle Buckingham. Indispensable reference for modern gear manufacture covers conjugate gear-tooth action, geartooth profiles of various gears, many other topics. 263 figures. 102 tables. 546pp. 5⅜ × 8½. 65712-4 Pa. $11.95

SET THEORY AND LOGIC, Robert R. Stoll. Lucid introduction to unified theory of mathematical concepts. Set theory and logic seen as tools for conceptual understanding of real number system. 496pp. 5⅜ × 8¼. 63829-4 Pa. $10.95

A HISTORY OF MECHANICS, René Dugas. Monumental study of mechanical principles from antiquity to quantum mechanics. Contributions of ancient Greeks, Galileo, Leonardo, Kepler, Lagrange, many others. 671pp. 5⅜ × 8½. 65632-2 Pa. $14.95

FAMOUS PROBLEMS OF GEOMETRY AND HOW TO SOLVE THEM, Benjamin Bold. Squaring the circle, trisecting the angle, duplicating the cube: learn their history, why they are impossible to solve, then solve them yourself. 128pp. 5⅜ × 8½. 24297-8 Pa. $3.95

MECHANICAL VIBRATIONS, J.P. Den Hartog. Classic textbook offers lucid explanations and illustrative models, applying theories of vibrations to a variety of practical industrial engineering problems. Numerous figures. 233 problems, solutions. Appendix. Index. Preface. 436pp. 5⅜ × 8½. 64785-4 Pa. $9.95

CURVATURE AND HOMOLOGY, Samuel I. Goldberg. Thorough treatment of specialized branch of differential geometry. Covers Riemannian manifolds, topology of differentiable manifolds, compact Lie groups, other topics. Exercises. 315pp. 5⅜ × 8½. 64314-X Pa. $8.95

HISTORY OF STRENGTH OF MATERIALS, Stephen P. Timoshenko. Excellent historical survey of the strength of materials with many references to the theories of elasticity and structure. 245 figures. 452pp. 5⅜ × 8½. 61187-6 Pa. $10.95

CATALOG OF DOVER BOOKS

GEOMETRY OF COMPLEX NUMBERS, Hans Schwerdtfeger. Illuminating, widely praised book on analytic geometry of circles, the Moebius transformation, and two-dimensional non-Euclidean geometries. 200pp. 5⅜ × 8¼. 63830-8 Pa. $6.95

MECHANICS, J.P. Den Hartog. A classic introductory text or refresher. Hundreds of applications and design problems illuminate fundamentals of trusses, loaded beams and cables, etc. 334 answered problems. 462pp. 5⅜ × 8½. 60754-2 Pa. $8.95

TOPOLOGY, John G. Hocking and Gail S. Young. Superb one-year course in classical topology. Topological spaces and functions, point-set topology, much more. Examples and problems. Bibliography. Index. 384pp. 5⅜ × 8¼. 65676-4 Pa. $8.95

STRENGTH OF MATERIALS, J.P. Den Hartog. Full, clear treatment of basic material (tension, torsion, bending, etc.) plus advanced material on engineering methods, applications. 350 answered problems. 323pp. 5⅜ × 8½. 60755-0 Pa. $7.50

ELEMENTARY CONCEPTS OF TOPOLOGY, Paul Alexandroff. Elegant, intuitive approach to topology from set-theoretic topology to Betti groups; how concepts of topology are useful in math and physics. 25 figures. 57pp. 5⅜ × 8½. 60747-X Pa. $2.95

ADVANCED STRENGTH OF MATERIALS, J.P. Den Hartog. Superbly written advanced text covers torsion, rotating disks, membrane stresses in shells, much more. Many problems and answers. 388pp. 5⅜ × 8½. 65407-9 Pa. $9.95

COMPUTABILITY AND UNSOLVABILITY, Martin Davis. Classic graduate-level introduction to theory of computability, usually referred to as theory of recurrent functions. New preface and appendix. 288pp. 5⅜ × 8½. 61471-9 Pa. $6.95

GENERAL CHEMISTRY, Linus Pauling. Revised 3rd edition of classic first-year text by Nobel laureate. Atomic and molecular structure, quantum mechanics, statistical mechanics, thermodynamics correlated with descriptive chemistry. Problems. 992pp. 5⅜ × 8½. 65622-5 Pa. $19.95

AN INTRODUCTION TO MATRICES, SETS AND GROUPS FOR SCIENCE STUDENTS, G. Stephenson. Concise, readable text introduces sets, groups, and most importantly, matrices to undergraduate students of physics, chemistry, and engineering. Problems. 164pp. 5⅜ × 8½. 65077-4 Pa. $6.95

THE HISTORICAL BACKGROUND OF CHEMISTRY, Henry M. Leicester. Evolution of ideas, not individual biography. Concentrates on formulation of a coherent set of chemical laws. 260pp. 5⅜ × 8½. 61053-5 Pa. $6.95

THE PHILOSOPHY OF MATHEMATICS: An Introductory Essay, Stephan Körner. Surveys the views of Plato, Aristotle, Leibniz & Kant concerning propositions and theories of applied and pure mathematics. Introduction. Two appendices. Index. 198pp. 5⅜ × 8½. 25048-2 Pa. $6.95

THE DEVELOPMENT OF MODERN CHEMISTRY, Aaron J. Ihde. Authoritative history of chemistry from ancient Greek theory to 20th-century innovation. Covers major chemists and their discoveries. 209 illustrations. 14 tables. Bibliographies. Indices. Appendices. 851pp. 5⅜ × 8½. 64235-6 Pa. $17.95

CATALOG OF DOVER BOOKS

THE FOUR-COLOR PROBLEM: Assaults and Conquest, Thomas L. Saaty and Paul G. Kainen. Engrossing, comprehensive account of the century-old combinatorial topological problem, its history and solution. Bibliographies. Index. 110 figures. 228pp. 5⅜ × 8½. 65092-8 Pa. $6.95

CATALYSIS IN CHEMISTRY AND ENZYMOLOGY, William P. Jencks. Exceptionally clear coverage of mechanisms for catalysis, forces in aqueous solution, carbonyl- and acyl-group reactions, practical kinetics, more. 864pp. 5⅜ × 8½. 65460-5 Pa. $19.95

PROBABILITY: An Introduction, Samuel Goldberg. Excellent basic text covers set theory, probability theory for finite sample spaces, binomial theorem, much more. 360 problems. Bibliographies. 322pp. 5⅜ × 8½. 65252-1 Pa. $8.95

LIGHTNING, Martin A. Uman. Revised, updated edition of classic work on the physics of lightning. Phenomena, terminology, measurement, photography, spectroscopy, thunder, more. Reviews recent research. Bibliography. Indices. 320pp. 5⅜ × 8¼. 64575-4 Pa. $8.95

PROBABILITY THEORY: A Concise Course, Y.A. Rozanov. Highly readable, self-contained introduction covers combination of events, dependent events, Bernoulli trials, etc. Translation by Richard Silverman. 148pp. 5⅜ × 8¼.
63544-9 Pa. $5.95

THE CEASELESS WIND: An Introduction to the Theory of Atmospheric Motion, John A. Dutton. Acclaimed text integrates disciplines of mathematics and physics for full understanding of dynamics of atmospheric motion. Over 400 problems. Index. 97 illustrations. 640pp. 6 × 9. 65096-0 Pa. $17.95

STATISTICS MANUAL, Edwin L. Crow, et al. Comprehensive, practical collection of classical and modern methods prepared by U.S. Naval Ordnance Test Station. Stress on use. Basics of statistics assumed. 288pp. 5⅜ × 8½.
60599-X Pa. $6.95

DICTIONARY/OUTLINE OF BASIC STATISTICS, John E. Freund and Frank J. Williams. A clear concise dictionary of over 1,000 statistical terms and an outline of statistical formulas covering probability, nonparametric tests, much more. 208pp. 5⅜ × 8½. 66796-0 Pa. $6.95

STATISTICAL METHOD FROM THE VIEWPOINT OF QUALITY CONTROL, Walter A. Shewhart. Important text explains regulation of variables, uses of statistical control to achieve quality control in industry, agriculture, other areas. 192pp. 5⅜ × 8½. 65232-7 Pa. $6.95

THE INTERPRETATION OF GEOLOGICAL PHASE DIAGRAMS, Ernest G. Ehlers. Clear, concise text emphasizes diagrams of systems under fluid or containing pressure; also coverage of complex binary systems, hydrothermal melting, more. 288pp. 6½ × 9¼. 65389-7 Pa. $10.95

STATISTICAL ADJUSTMENT OF DATA, W. Edwards Deming. Introduction to basic concepts of statistics, curve fitting, least squares solution, conditions without parameter, conditions containing parameters. 26 exercises worked out. 271pp. 5⅜ × 8½. 64685-8 Pa. $7.95

CATALOG OF DOVER BOOKS

DE RE METALLICA, Georgius Agricola. The famous Hoover translation of greatest treatise on technological chemistry, engineering, geology, mining of early modern times (1556). All 289 original woodcuts 638pp. 6¾ × 11.
60006-8 Pa. $17.95

SOME THEORY OF SAMPLING, William Edwards Deming. Analysis of the problems, theory and design of sampling techniques for social scientists, industrial managers and others who find statistics increasingly important in their work. 61 tables. 90 figures. xvii + 602pp. 5⅜ × 8½.
64684-X Pa. $15.95

THE VARIOUS AND INGENIOUS MACHINES OF AGOSTINO RAMELLI: A Classic Sixteenth-Century Illustrated Treatise on Technology, Agostino Ramelli. One of the most widely known and copied works on machinery in the 16th century. 194 detailed plates of water pumps, grain mills, cranes, more. 608pp. 9 × 12. (EBE)
25497-6 Clothbd. $34.95

LINEAR PROGRAMMING AND ECONOMIC ANALYSIS, Robert Dorfman, Paul A. Samuelson and Robert M. Solow. First comprehensive treatment of linear programming in standard economic analysis. Game theory, modern welfare economics, Leontief input-output, more. 525pp. 5⅜ × 8½.
65491-5 Pa. $13.95

ELEMENTARY DECISION THEORY, Herman Chernoff and Lincoln E. Moses. Clear introduction to statistics and statistical theory covers data processing, probability and random variables, testing hypotheses, much more. Exercises. 364pp. 5⅜ × 8½.
65218-1 Pa. $9.95

THE COMPLEAT STRATEGYST: Being a Primer on the Theory of Games of Strategy, J.D. Williams. Highly entertaining classic describes, with many illustrated examples, how to select best strategies in conflict situations. Prefaces. Appendices. 268pp. 5⅜ × 8½.
25101-2 Pa. $6.95

MATHEMATICAL METHODS OF OPERATIONS RESEARCH, Thomas L. Saaty. Classic graduate-level text covers historical background, classical methods of forming models, optimization, game theory, probability, queueing theory, much more. Exercises. Bibliography. 448pp. 5⅜ × 8¼.
65703-5 Pa. $12.95

CONSTRUCTIONS AND COMBINATORIAL PROBLEMS IN DESIGN OF EXPERIMENTS, Damaraju Raghavarao. In-depth reference work examines orthogonal Latin squares, incomplete block designs, tactical configuration, partial geometry, much more. Abundant explanations, examples. 416pp. 5⅜ × 8¼.
65685-3 Pa. $10.95

THE ABSOLUTE DIFFERENTIAL CALCULUS (CALCULUS OF TENSORS), Tullio Levi-Civita. Great 20th-century mathematician's classic work on material necessary for mathematical grasp of theory of relativity. 452pp. 5⅜ × 8½.
63401-9 Pa. $9.95

VECTOR AND TENSOR ANALYSIS WITH APPLICATIONS, A.I. Borisenko and I.E. Tarapov. Concise introduction. Worked-out problems, solutions, exercises. 257pp. 5⅜ × 8¼.
63833-2 Pa. $6.95

CATALOG OF DOVER BOOKS

TENSOR CALCULUS, J.L. Synge and A. Schild. Widely used introductory text covers spaces and tensors, basic operations in Riemannian space, non-Riemannian spaces, etc. 324pp. 5⅜ × 8¼. 63612-7 Pa. $7.95

A CONCISE HISTORY OF MATHEMATICS, Dirk J. Struik. The best brief history of mathematics. Stresses origins and covers every major figure from ancient Near East to 19th century. 41 illustrations. 195pp. 5⅜ × 8½. 60255-9 Pa. $7.95

A SHORT ACCOUNT OF THE HISTORY OF MATHEMATICS, W.W. Rouse Ball. One of clearest, most authoritative surveys from the Egyptians and Phoenicians through 19th-century figures such as Grassman, Galois, Riemann. Fourth edition. 522pp. 5⅜ × 8½. 20630-0 Pa. $10.95

HISTORY OF MATHEMATICS, David E. Smith. Nontechnical survey from ancient Greece and Orient to late 19th century; evolution of arithmetic, geometry, trigonometry, calculating devices, algebra, the calculus. 362 illustrations. 1,355pp. 5⅜ × 8½. 20429-4, 20430-8 Pa., Two-vol. set $23.90

THE GEOMETRY OF RENÉ DESCARTES, René Descartes. The great work founded analytical geometry. Original French text, Descartes' own diagrams, together with definitive Smith-Latham translation. 244pp. 5⅜ × 8½. 60068-8 Pa. $6.95

THE ORIGINS OF THE INFINITESIMAL CALCULUS, Margaret E. Baron. Only fully detailed and documented account of crucial discipline: origins; development by Galileo, Kepler, Cavalieri; contributions of Newton, Leibniz, more. 304pp. 5⅜ × 8½. (Available in U.S. and Canada only) 65371-4 Pa. $9.95

THE HISTORY OF THE CALCULUS AND ITS CONCEPTUAL DEVELOPMENT, Carl B. Boyer. Origins in antiquity, medieval contributions, work of Newton, Leibniz, rigorous formulation. Treatment is verbal. 346pp. 5⅜ × 8½. 60509-4 Pa. $7.95

THE THIRTEEN BOOKS OF EUCLID'S ELEMENTS, translated with introduction and commentary by Sir Thomas L. Heath. Definitive edition. Textual and linguistic notes, mathematical analysis. 2,500 years of critical commentary. Not abridged. 1,414pp. 5⅜ × 8½. 60088-2, 60089-0, 60090-4 Pa., Three-vol. set $29.85

GAMES AND DECISIONS: Introduction and Critical Survey, R. Duncan Luce and Howard Raiffa. Superb nontechnical introduction to game theory, primarily applied to social sciences. Utility theory, zero-sum games, n-person games, decision-making, much more. Bibliography. 509pp. 5⅜ × 8½. 65943-7 Pa. $11.95

THE HISTORICAL ROOTS OF ELEMENTARY MATHEMATICS, Lucas N.H. Bunt, Phillip S. Jones, and Jack D. Bedient. Fundamental underpinnings of modern arithmetic, algebra, geometry and number systems derived from ancient civilizations. 320pp. 5⅜ × 8½. 25563-8 Pa. $8.95

CALCULUS REFRESHER FOR TECHNICAL PEOPLE, A. Albert Klaf. Covers important aspects of integral and differential calculus via 756 questions. 566 problems, most answered. 431pp. 5⅜ × 8½. 20370-0 Pa. $8.95

CATALOG OF DOVER BOOKS

CHALLENGING MATHEMATICAL PROBLEMS WITH ELEMENTARY SOLUTIONS, A.M. Yaglom and I.M. Yaglom. Over 170 challenging problems on probability theory, combinatorial analysis, points and lines, topology, convex polygons, many other topics. Solutions. Total of 445pp. 5⅜ × 8½. Two-vol. set.
Vol. I 65536-9 Pa. $6.95
Vol. II 65537-7 Pa. $6.95

FIFTY CHALLENGING PROBLEMS IN PROBABILITY WITH SOLUTIONS, Frederick Mosteller. Remarkable puzzlers, graded in difficulty, illustrate elementary and advanced aspects of probability. Detailed solutions. 88pp. 5⅜ × 8½.
65355-2 Pa. $3.95

EXPERIMENTS IN TOPOLOGY, Stephen Barr. Classic, lively explanation of one of the byways of mathematics. Klein bottles, Moebius strips, projective planes, map coloring, problem of the Koenigsberg bridges, much more, described with clarity and wit. 43 figures. 210pp. 5⅜ × 8½. 25933-1 Pa. $5.95

RELATIVITY IN ILLUSTRATIONS, Jacob T. Schwartz. Clear nontechnical treatment makes relativity more accessible than ever before. Over 60 drawings illustrate concepts more clearly than text alone. Only high school geometry needed. Bibliography. 128pp. 6⅛ × 9¼. 25965-X Pa. $5.95

AN INTRODUCTION TO ORDINARY DIFFERENTIAL EQUATIONS, Earl A. Coddington. A thorough and systematic first course in elementary differential equations for undergraduates in mathematics and science, with many exercises and problems (with answers). Index. 304pp. 5⅜ × 8½. 65942-9 Pa. $7.95

FOURIER SERIES AND ORTHOGONAL FUNCTIONS, Harry F. Davis. An incisive text combining theory and practical example to introduce Fourier series, orthogonal functions and applications of the Fourier method to boundary-value problems. 570 exercises. Answers and notes. 416pp. 5⅜ × 8½. 65973-9 Pa. $9.95

THE THEORY OF BRANCHING PROCESSES, Theodore E. Harris. First systematic, comprehensive treatment of branching (i.e. multiplicative) processes and their applications. Galton-Watson model, Markov branching processes, electron-photon cascade, many other topics. Rigorous proofs. Bibliography. 240pp. 5⅜ × 8½. 65952-6 Pa. $6.95

AN INTRODUCTION TO ALGEBRAIC STRUCTURES, Joseph Landin. Superb self-contained text covers "abstract algebra": sets and numbers, theory of groups, theory of rings, much more. Numerous well-chosen examples, exercises. 247pp. 5⅜ × 8½. 65940-2 Pa. $6.95

Prices subject to change without notice.
Available at your book dealer or write for free Mathematics and Science Catalog to Dept. GI, Dover Publications, Inc., 31 East 2nd St., Mineola, N.Y. 11501. Dover publishes more than 175 books each year on science, elementary and advanced mathematics, biology, music, art, literature, history, social sciences and other areas.